Annual Session of Scientific Papers (IMT Oradea 2021)

AA007276

IOP Conference Series: Materials Science and Engineering
Volume 1169

Oradea, Romania
27 - 28 May 2021

Editors:

Gavril Grebenisan
Alexandru-Viorel Pele
Florin Sandu Blaga
Horia Beles

ISBN: 978-1-7138-4629-1
ISSN: 1757-8981

Printed from e-media with permission by:

Curran Associates, Inc.
57 Morehouse Lane
Red Hook, NY 12571

CURRAN ASSOCIATES INC.
proceedings
.com

Some format issues inherent in the e-media version may also appear in this print version.

Institute of Physics
Dirac House, Temple Back
Bristol BS1 6BE UK

Phone: 44 1 17 929 7481
Fax: 44 1 17 920 0979

techtracking@iop.org

Additional copies of this publication are available from:

Curran Associates, Inc.
57 Morehouse Lane
Red Hook, NY 12571 USA
Phone: 845-758-0400
Fax: 845-758-2633
Email: curran@proceedings.com
Web: www.proceedings.com

TABLE OF CONTENTS

IOP Conf. Series: Materials Science and Engineering 1169 (2021) 011001 doi:10.1088/1757-899X/1169/1/011001

Preface

The International Conference, ANNUAL INTERNATIONAL SESSION OF SCIENTIFIC PAPERS - IMT Oradea 2021, with topics in the field of industrial engineering, took place in the last week of May 2021, for two days, Thursday 27 May, respectively Friday 28 May, in the virtual environment, as well as in 2020, given the travel restrictions decided by authorities around the world. According to the preliminary schedule, all participants presented their work as a result of their research.

The organizers, editors and members of the technical staff continuously assisted, from a logistical point of view, the entire development of the scientific event.

The Conference was opened by its Chairman, Assoc. Prof. PhD Gavril Grebenişan. He, as every year, invited Prof. PhD Alexandru Viorel Pele, as dean of the Faculty of Managerial and Technological Engineering of the University of Oradea (FIMT-UO), as the main sponsor of the scientific event, to address, to the participants and guests, a word of welcome. In his speech, the dean of FIMT-UO, wanted to express his enthusiasm generated by the possibility of holding the Conference, even if the conditions are so unfavourable. The speaker also showed that the joy of meeting on the occasion of the scientific event, known as IMT Oradea 2021, can not be overshadowed by the fact that it takes place virtually and not in such a friendly environment in Oradea-Baile Felix SPA, where we meet, with great pleasure, every year, in the last week of May. After expressing, with deep respect, thanks, both in his name and in the name of the faculty of FIMT-UO, which he leads, to those present, the dean wanted to emphasize that he wants the scientific meeting IMT Oradea 2021 to represent a great success, and at the next IMT Oradea Conference 2022, to meet in Oradea, as we were used to, before the outbreak of the Corona Virus pandemic.

After the welcome speech of the dean of FIMT-UO, the Chairman of the Conference, he invited the Coordinator of the Presentations, Prof. PhD Vesselenyi Tiberiu, to take control of the presentation program.

The first presentation was that of Prof. PhD Florin Blaga and Assist. Prof. PhD Alin Pop, as Keynote Speakers, who presented the paper "The efficiency of modelling and simulation of manufacturing systems using Petri nets". Also, as Keynote Speaker, the following day, Friday 28 May, presented works of great interest, and Prof. PhD Erdei Timotei István and Prof. PhD Husi Géza, who presented the work "Incorporation of Industry 4.0 into education at the Faculty of Engineering of the University of Debrecen", respectively Prof. PhD Vesselenyi Tiberiu who presented the paper entitled "Artificial Neural Networks Image Processing in industrial machining applications".

The conference program was strictly observed by the organizers and coordinated by the Editorial Committee, in each of the two days, Thursday 27 May, respectively Friday 28 May, and works are scheduled between 10 AM and 4 PM.

Discussions followed the presentation of each participant's paper, comments and clarifications requested by participants. In general, 15 minutes were allocated for each article, of which five minutes for discussions. It has rarely happened that these time intervals are not observed.

IOP Conf. Series: Materials Science and Engineering 1169 (2021) 011001 doi:10.1088/1757-899X/1169/1/011001

Two independent reviewers reviewed all papers, unknown to the authors, and one editor for each article. The acceptance rate of the works was 67%, which means that a third of the works were rejected.

The Editorial Board thanks, in this way, Mrs Lucy Evans, Coordinator of IOP Conference Series: Materials Science and Engineering, who was kind enough to participate in the work of the IMT Oradea 2021 Conference.

Editorial Board

IOP Conf. Series: Materials Science and Engineering 1169 (2021) 011001 doi:10.1088/1757-899X/1169/1/011001

IOP Conf. Series: Materials Science and Engineering 1169 (2021) 011002 doi:10.1088/1757-899X/1169/1/011002

Peer review declaration

All papers published in this volume of *IOP Conference Series: Materials Science and Engineering* have been peer reviewed through processes administered by the Editors. Reviews were conducted by expert referees to the professional and scientific standards expected of a proceedings journal published by IOP Publishing.

- **Type of peer review: Single-blind**
- **Conference submission management system:**

https://imt.uoradea.ro/conference/conf.detail.php?workID=0

- **Number of submissions received: 70**

- **Number of submissions sent for review: 70**

- **Number of submissions accepted:47**

- **Acceptance Rate (Number of Submissions Accepted / Number of Submissions Received X 100): 67.15%**

- **Average number of reviews per paper: 2**

- **Total number of reviewers involved:6**

- **Any additional info on review process:**

The Editorial Board of the International Conference Annual Session of Scientific Scientific manuscripts IMT Oradea appraised that the researchers who submit a manuscript to the Conference, to be published on Conference Proceedings, have the right to a careful, timely and unbiased peer-review overseen by the journal editors, who often seek the advice of referees. Editorial Board of the Conference, following the principle of transparency, accepts that the authors have to know in advance the processes by which their manuscript will be handled, and a right to be informed of the

IOP Conf. Series: Materials Science and Engineering 1169 (2021) 011002 doi:10.1088/1757-899X/1169/1/011002

grounds for the acceptance or rejection of their manuscript, including normally being given access to any referee reports that have been sought.

Every collaborator of the Conference, i.e. researchers, authors who decide to submit a manuscript, must adhere to high standards of research integrity and sound scholarship. It is the responsibility of authors to submit a well documented and well-written scientific manuscript if necessary, seeking advice if it is not written in their native language, to clearly describe any novel and non-trivial content, and to suitably acknowledge the contributions of others, including referees. Submission of a scientific manuscript to be published on to the Proceedings of the Conference implies that it is not currently under consideration by any other journal or Conference. Any substantial overlap with other published or submitted scientific manuscripts is appropriately acknowledged. In addition, authors should be responsive to correspondence with the journal. Even though a scientific manuscript has multiple authors, they should communicate fully, speak with one voice, and accept mutual responsibility in their communications with the journal. All authors are expected to have materially contributed to the scientific manuscript and be familiar with its contents.

The Editorial Board of the International Conference Annual Session of Scientific Scientific manuscripts IMT Oradea, referring to the peer-review process, acts according to fundamental principles of publishing procedure: primary among these are transparency and integrity, and last but not least the professionalism.

Consequently, by *transparency,* we mean that all the Conference's stakeholders, i.e.

readers, authors, referees, editors, publishers, etc. – are fully aware of the decision processes that affect them. Ongoing, *The Editorial Board laid down that integrity* of the publication process is overriding one. It includes maintaining an objective review process focused on scientific quality, proper acknowledgement of sources, and a respect for confidentiality. As we declared before, last but not least, p*rofessionalism* is also essential; this includes, not solely, timely handling of manuscripts at each stage of the peer-review and publishing process, and continuity of management, scope, and vision as they evolve.

The entire process of reviewing needs will be conducted confidentially. The scientific manuscript which enquired to be reviewed will not be disclosed to a third party. The Editorial Board do not share the identity of the reviewer with the Author. We encourage the reviewers, in order to help us protect their identity, to not reveal their names within the text of their review, i.e. into the "Comments to the author" particular space reserved.

The Conference Proceedings peer review process pay attention to the following steps and, as a result, the members of referees staff gives one of four possible states of the scientific manuscript:

- **Accept as is**;
- **Accept if minor revisions are made**;
- **Reconsider after major revisions are made**;
- **Meritorious but not appropriate**;
- **Reject**;

within deadline terms.

IOP Conf. Series: Materials Science and Engineering 1169 (2021) 011002 doi:10.1088/1757-899X/1169/1/011002

Article I. Step 1: (The Author)

The author who submits a manuscript shall be registered as a user of the conference platform (using "Registration- Payment information" menu, https://imt.uoradea.ro/conference/register.php). After this process is done, the author should log in on the conference platform and may submit their manuscript(s) using the "New Scientific manuscript" menu (https://imt.uoradea.ro/conference/conf.detail.php?workID=0).

Article II. Step 2: (The Editor)

Editor in chief forwards the scientific manuscript to appropriate reviewers related to that subject. Each scientific manuscript will be reviewed by three scientific personalities: two independent reviewers and one member of the Editorial Board. This is the core of the entire peer-review process. Peer review is a critical element of scholarly publication and one of the major cornerstones of the scientific process. Peer Review serves two essential functions:

-Acts as a filter: Ensures research is correctly verified before being published;

-Improves the quality of the research by a rigorous review by other experts who help hone key points and correct inadvertent errors. One of the peer-review process's primary aims consists of that to evaluate and, inasmuch as possible, certifying the correctness, importance, novelty, and clarity of a scientific manuscript.

Article III. Step 3: (The Reviewer)

The reviewer board will check the manuscript and give the status of the scientific manuscript as follows:

RECOMMENDATION

Accept as is

Accept if minor revisions are made

Reconsider after major revisions are made

Meritorious but not appropriate

Reject

The Editorial Board of the International Conference Annual Session of Scientific Papers IMT Oradea gives this conformation to the author by e-mail message generated by the system. The first status consists of accepting scientific manuscripts as is, any corrections, no needs. In this case, the editor will also give account detail for fees submission. Following the second status implies minor or major

The Annual Session Of Scientific Papers - IMT Oradea 2021 IOP Publishing
IOP Conf. Series: Materials Science and Engineering 1169 (2021) 011002 doi:10.1088/1757-899X/1169/1/011002

revision, which the author shall make as soon as possible. Last two status means the scientific manuscript is rejected. In this case, the author will be informed about the editor's decision, and the scientific manuscript rejected follow to be deleted from the conference platform.

Excepts for the first status of the scientific manuscript, the subsequent four implies that the reviewer provides a clearly identified which revision is required. This will be e-mailed to the author.

Article IV. Step 4 : (The Author)

If the scientific manuscript is accepted, then Author will deposit fees (as given in processing charges) to account detail by the editor.

The author shall send a scanned copy of the copyright form (format can download by website http://www.imt.uoradea.ro/conference/download/Copyright_statement.pdf) by mail to the editor.

- **Contact person for queries:**

Gavril Grebenisan: grebe@uoradea.ro

Călin Băban cbaban@uoradea.ro	Horia Beleş horia.beles@gmail.com	Florin Blaga fblaga@uoradea.ro
Izabela Dembinska izdem@wp.pl	Gavril Grebenişan grebe@uoradea.ro	Alexandru Viorel Pele apele@uoradea.ro

IOP Conf. Series: Materials Science and Engineering 1169 (2021) 012001 doi:10.1088/1757-899X/1169/1/012001

A study on the effect of the deformations in flexiblocks on the behavior of the vehicle axle suspension mechanisms

C Alexandru

Transilvania University of Braşov, Romania

E-mail: calex@unitbv.ro

Abstract. This work deals with a study on the effect of the deformations in flexiblocks (bushings) on the behavior of the vehicle axle suspension mechanisms. The mechanism in study is a representatiove one for the structural group/class of mono-mobile axle suspension systems, namely the guiding mechanism by five points - on five spheres (so called 5SS). In a previous work, it was established that the rigid joint model for this class of mechanisms does not provide results close to the real flexible bushing model, reason for which the research in this paper is directed towards the identification of a theoretical model of joint which, at a lower complexity than that of the real model, would still provide viable results, in terms of axle guidance accuracy in the spatial movement relative to car body. In this regard, static and kinematic models were developed and analyzed by using the MBS (Multi-Body Systems) software solution ADAMS.

1. Introduction

The suspension mechanisms of vehicles' rear (beam) axles perform two functions, namely the function of guiding the beam axle in movement relative to car body, respectively the function of transmitting the contact forces from wheels. Regarding the guiding function, this is defined by the spatial movement (position and orientation) of the axle, which is established by the positional analysis of the guiding mechanism. The guidance function is studied on models in which the car body is considered fixed (namely, kinematic and static models). In the case of the dynamic model, where the car body is a mobile part, the axle is forced, by wheels, to follow the road profile (more or less accurately, depending on the elasticity of the tires).

Therefore, the analysis of the guidance function can be performed on the kinematic model, considering the loading of the mechanism through vertical positions imposed on the wheels, respectively the static model, where the external loading is performed by forces applied to the wheels. Obviously, the determination of the guidance function on the static model is closer to reality, because the elastic elements of the suspension are also taken into account.

The (quasi)static analysis of the axle guiding mechanisms can be performed on the basis of two models, in correlation with the modelling of the connections between the guiding bars and the adjacent parts (car body and axle), which are actually made by flexiblocks/bushings (compliant joints that support linear and angular deformations), as follows: rigid models, in which the bushings are modelled by spherical couplings/joints, with 3 degrees of freedom (DOF) corresponding to the rotational movements (these models are also the basis of the structural synthesis and kinematic study of the axle guiding mechanisms [1]); flexible/compliant models, with 6 degrees of freedom restricted elastically, which considers both linear and angular movements (deformations).

IOP Conf. Series: Materials Science and Engineering 1169 (2021) 012001 doi:10.1088/1757-899X/1169/1/012001

Opting for one or the other model is, first of all, a choice dictated by the available computational tool. If the rigid models, due to the low degree of mobility they have (DOM = 1 or DOM = 2), are suitable for analysis by specialized methods and in-house programs, the elastic models, with a large number of degrees of mobility, require the use of MBS (Multi-Body Systems) commercial software solutions, which are actually used in a large variety of applications [2-6]. Approaching the compliant bushing model through classical formalisms, in addition to being extremely difficult, involves the risk of obtaining significant computation errors, a consequence of the particularly vast mathematical apparatus that should be managed. For the positional analysis of the axle guiding mechanisms, there is frequently considered the simplifying assumption that the rigid models approximate quite well the behaviour of the real models with compliant bushings. Obviously, the following problem/question can be raised: what is the deviation/error generated by the symbolic representation of bushing by spherical coupling regarding the guiding function of the mechanism? It is interesting, in fact, for which category of axle guiding mechanisms the rigid model hypothesis can be accepted, regarding the accuracy of the results provided for the axle movement, compared to the flexible model.

In the literature such a problem is addressed mainly for front wheel guiding mechanisms (four-bar, McPherson and multi-link configurations) [7-13].The conclusions of these works converge on the idea that, although there are some differences between the behaviour of rigid and flexible joint models in terms of guiding accuracy, the rigid model is a viable alternative for the positional analysis of the wheel guiding mechanisms. These approaches are especially kinematic, as it is not possible to make a quantitative assessment of the dimensional deviations in bushing (i.e. the deformations as a function of forces), much less a correlation between the deformations for all bushings in the guiding mechanism.

In a previous work by the author [14], a comparative analysis between rigid and flexible joint models for the two main categories of axle guiding mechanisms (in terms of the number of degrees of mobility, DOM = 1 and DOM = 2) was carried out. A similar study was performed in [15], by considering other structural variants of axle guiding mechanisms and respectively numerical values of the specific geometric parameters. The conclusions resulting from these papers can be summarized as follows: the hypothesis of the rigid coupling model is valid only in the case of bi-mobile (DOM = 2) axle guidance mechanisms, while in the case of mono-mobile (DOM = 1) mechanisms, it is necessary to consider models of couplings/joints closer to reality (i.e. compliant bushings), which must allow linear deformations.

Starting from the previously mentioned research, the present work deals with a more detailed study on the influence of the joints' deformations on the guidance function of the mono-mobile axle suspension mechanisms (for which the rigid joint model hypothesis is invalid), in order to identify the simplest models that ensure an appropriate behavior, close to that of the real model. The study is conducted by developing and testing MBS models made by using the virtual prototyping environment ADAMS of MSC.Software.

2. The MBS model of the suspension system

To study the influence of elasticities in bushings on the guiding function of the axle suspension mechanisms, the static model of the mechanism was initially developed, considering that the car body is fixed and the external loading of the guiding mechanism is realized by forces applied to the wheels. The suspension system in study is based on an axle guiding mechanism by five points - on five spheres (5SS), whose equivalent structural model has a single degree of mobility, corresponding to the vertical movement of one of the wheels (left or right).

The static model of the suspension system based on 5SS axle guiding mechanism, which is shown in Figure 1, contains the bodies/parts in the suspension system (axle & wheels, guiding links/bars, car body), the connections between bodies (which in the real case are made by bushings), the spring & damper assemblies (which are arranged between the axle and the car body), the compression & extension bumper stops (which are arranged inside the shock absorbers / dampers, thus limiting the relative movement between their pistons and cylinders).

IOP Conf. Series: Materials Science and Engineering 1169 (2021) 012001 doi:10.1088/1757-899X/1169/1/012001

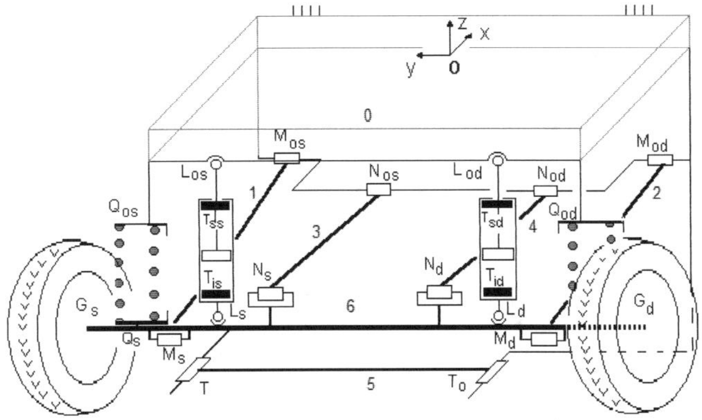

Figure 1. The static model of the 5SS suspension system.

For this type of axle suspension systems, there were modeled and analyzed rigid (where the connections of the guiding bars to axle and car body are made by spherical couplings/joints) and flexible (couplings by bushings) models. By taking into account the bushings' elasticity, all the parameters that describe the guiding function of the mechanism (i.e. the spatial movement of the axle) will obviously be influenced. The guiding function of the mechanism is reported/defined by the roll angle and the vertical coordinate of the axle's center, because these are the parameters that actually describe the global character of the spatial movement of the axle relative to the car body.

3. Results and conclusions

The comparative analysis of the rigid and flexible models was carried out for several functional situations determined by the nature and values of the wheel contact forces. In this work, the test model considered corresponds to the situation where the vehicle is suspended, with the car body locked, and it is operated on the left wheel by a variable vertical force (F_z), in the range of values $F_z \in [0, 800]$ daN. The results of these analyzes led to conclusions similar to those of previous researches [14, 15] confirming the invalidity of the rigid joint model in the case of the mono-mobile suspension system.

Further, it was started from the premise that it might be possible that for certain types of bushings, where due to high stiffnesses there are practically no linear deformations, the behavior of the flexible model is close to that of the rigid model. From the point of view of the linear (radial and axial) deformations that occur in bushings, the flexible model for the 5SS guiding mechanism led to the results shown in Figure 1, for the connections on axle of the lower longitudinal bars (1 - left, 2 - right), the upper longitudinal bars (3 - left, 4 - right), and the transversal bar (5). Similar deformations were obtained for the bushings through which the guiding bars are connected to car body. According to these diagrams, the linear deformations in bushings are very small, of the order of tenths-hundredths of a millimeter. So, there are enough small deformations in bushings for the law of motion through the flexible model to be very different from that of the rigid model. The question then arises: what should a bushing look like from a geometric point of view in which not even such small deformations appear?

According to the relations in [16], the linear rigidities in bushings are mainly influenced by the thickness of the rubber intermediate layer. High rigidities, which determine even the proximity of the condition of linear non-deformability of the elastic element (a necessary condition for the validation of the rigid model), involve an extremely small thickness of the rubber between the outer and inner metallic rings of the bushing. However, such a constructive solution would have effects on the necessary angular mobility, on the sound and anti-vibration insulation capacity of the bushing. Moreover, the symbolic representation bushing $\rightarrow$ spherical coupling, which is the base of the structural synthesis [1], would no longer be valid either, because in fact the spherical rotations could no longer occur. The analysis of a large number of constructive solutions showed that no constructive variant of bushing, at least the known ones, does not provide the necessary conditions for the validation of the rigid model for the mono-mobile axle guiding mechanisms.

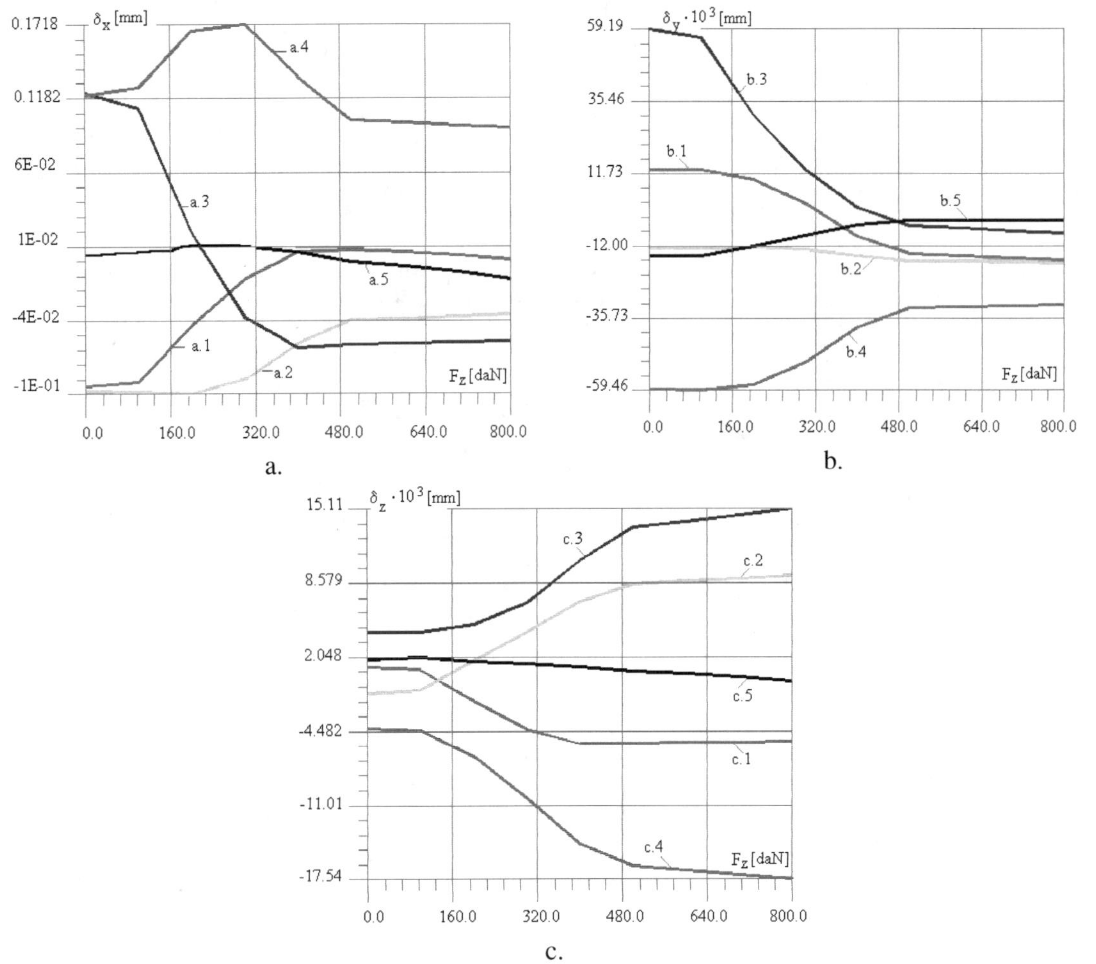

Figure 2. The radial and axial deformations in bushings for the 5SS axle guiding mechanism.

On the other hand, the analysis of the guidance function of the axle guiding mechanisms can also be performed on the kinematic model. For this, the elastic and damping elements (springs, dampers, bump stops) are removed from the suspension system, and the input (actuating) is made through vertical positions imposed on the left wheel ($\Delta Z_{Gs} \in [-80, 80]$ mm, from the static/rest position when $Z_{Gs} = -77$ mm). Because the kinematic model does not contain the elastic suspension elements, which would generate internal reaction forces, the linear deformations in bushings are much smaller than in the static model. Even under these conditions, according to the diagrams in Figure 2, the rigid model for the guiding mechanisms with DOM = 1 behaves differently from the flexible model, referring to the law of motion by the guiding mechanism, which strengthens the conclusion regarding the invalidation of the rigid joint model for this structural group of mechanisms.

Since in bushings there are, theoretically, linear deformations in all directions (axial and radial), the influence of each linear elasticity characteristic on the guiding function of the axle suspension mechanism will be further identified. In this regard, the bushings were modeled by connections with 4 degrees of freedom (DOF), considering in addition to the spherical rotations one of the linear deformations (along X, Y, Z, by case), the 4-DOF coupling models being defined by compound sphere-translation joints. Such an approach allows the identification of the simplest model for bushing, which is able to ensure a behavior close to that of the real 6-DOF bushing. The results obtained by analyzing the 4-DOF joint models for the parameters of interest that define the spatial position of the axle are shown in Figures 3 and 4 (a - static analysis, b - kinematic analysis).

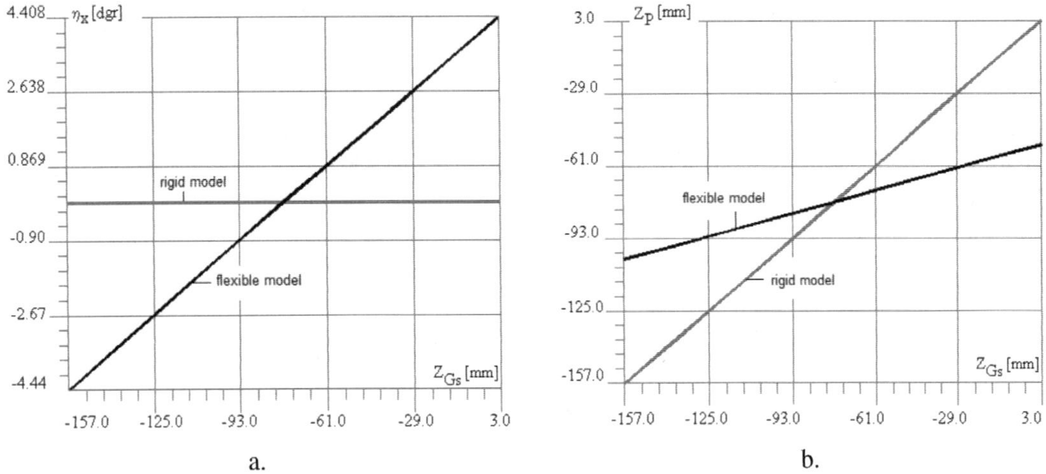

Figure 3. The results of the comparative analysis for the kinematic models.

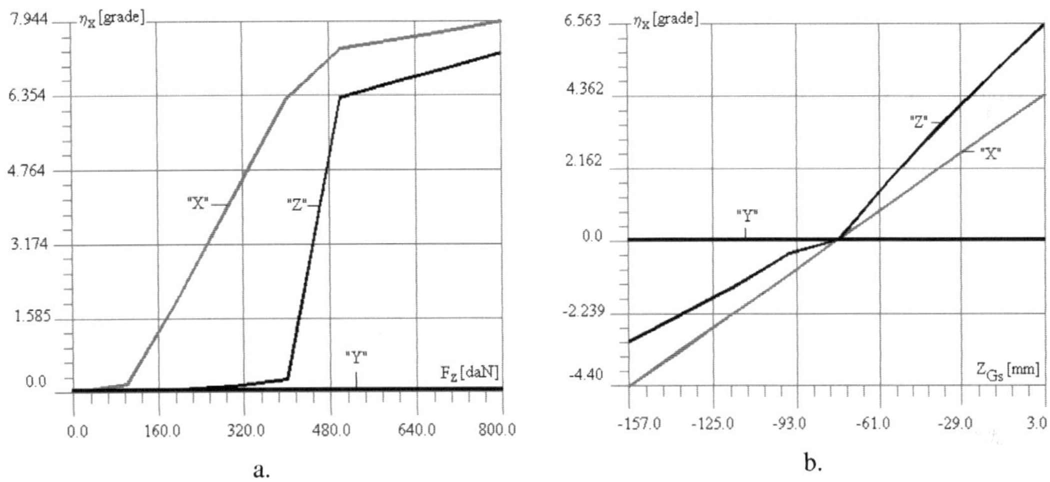

Figure 4. The roll angle of the axle for the 4-DOF joint models.

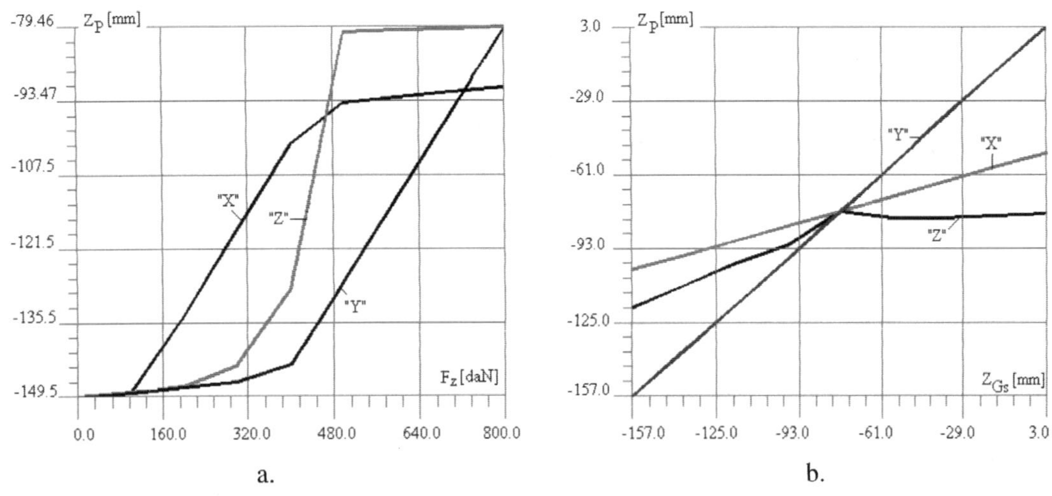

Figure 5. The vertical displacement of the axle's centre for the 4-DOF joint models.

According to these diagrams, the 4-DOF joint model "X", which allows radial linear deformation along X, leads to results close to the real compliant model (with 6-DOF bushings), while the 4-DOF joint "Y", which allows axial linear deformation along Y, is very close to the rigid model (with 3-DOF spherical joints). Therefore, the simplest viable model for rear axle bushings is the composed sphere-translation joint on the longitudinal axis, with 4 elastically restricted degrees of freedom. Consequently, the main reaction forces in bushings are the longitudinal ones, while and lateral reaction forces (along the bushing's axis) are practically insignificant. This information is also important for the proper sizing of the bushings in the 5SS axle guiding mechanism.

References

[1] Alexandru C 2009 The kinematic optimization of the multi-link suspension mechanism used for rear axle of the motor vehicle *Proceedings of the Romanian Academy - A* **10**(3) pp 244-253
[2] Alexandru C and Comşiţ M 2007 Virtual prototyping of the solar tracking systems *Renewable Energy and Power Quality Journal* **1**(5) pp 105-110
[3] Alexandru C and Pozna C 2008 Virtual prototype of a dual-axis tracking system used for photovoltaic panels *Proceedings of the IEEE International Symposium on Industrial Electronics - ISIE* pp 1598-1603
[4] Alexandru C and Pozna C 2009 Dynamic modeling and control of the windshield wiper mechanisms *WSEAS Transactions on Systems* **8**(7) pp 825–834
[5] Alexandru P, Macaveiu D and Alexandru C 2012 A gear with translational wheel for a variable transmission ratio and applications to steering box *Mechanism and Machine Theory* **52**, pp 267-276
[6] Geonea ID, Alexandru C, Margine A and Ungureanu A 2013 Design and simulation of a single DOF human-like leg mechanism *Applied Mechanics and Materials* **332** pp 491-496
[7] Attia HA 2003 Kinematic analysis of the multi-link five-point suspension system in point coordinates *Journal of Mechanical Science and Technology* **17**(8) pp 1133-1139
[8] Balike KP, Rakheja S and Stiharu I 2008 Kinematic analysis and parameter sensitivity to hard points of five-link rear suspension mechanism of passenger car *Proceedings of the Design Engineering Technical Conference* pp 755-764
[9] Hiller M and Woernle C 1985 Kinematical analysis of a five point wheel suspension *ATZ* **87** pp 59-64
[10] Knapczyk J and Maniowski M 2002 Selected effects of bushings characteristics on five-link suspension elastokinematics *Mobility and Vehicle Mechanics* **3**(2) pp 107-121
[11] Knapczyk J and Maniowski M 2006 Elastokinematic modeling and study of five-rod suspension with subframe *Mechanism and Machine Theory* **41**(9) pp 1031-1047
[12] Simionescu PA and Beale D 2002 Synthesis and analysis of the five-link rear suspension system used in automobile *Mechanism and Machine Theory* **37**(9) pp 815-832
[13] Tică M, Dobre G and Mateescu V 2014 Influence of compliance for an elastokinematic model of a proposed rear suspension *International Journal of Automotive Technology* **15**(6) pp 885-891
[14] Ţoţu V and Alexandru C (2013) Study concerning the effect of the bushings' deformability on the static behavior of the rear axle guiding linkages *Applied Mechanics and Materials* **245** pp 132-137
[15] Ţoţu V (2014) A comparative analysis between the rigid and compliant joint models for the guiding system of the cars axles *Annals of the Oradea University, Fascicle of Management and Technological Engineering* **XXIII** pp 131-134
[16] Alexandru C (2019) Method for the quasi-static analysis of beam axle suspension systems used for road vehicles *Proceedings of the Institution of Mechanical Engineers, Part D: Journal of Automobile Engineering* **233**(7) pp 1818-1833

IOP Conf. Series: Materials Science and Engineering 1169 (2021) 012002 doi:10.1088/1757-899X/1169/1/012002

Design sensitivity analysis in the kinematics of the 4SS-axle guiding mechanism with Panhard bar

C Alexandru

Transilvania University of Braşov, Romania

E-mail: calex@unitbv.ro

Abstract. This work deals with the design sensitivity analysis in the kinematics of the vehicle rear axle guided by four points - on four spheres (so called 4SS), with Panhard bar. The geometric parameters that define the kinematic scheme of the axle guiding mechanism are considered in this study, which aims to identify the influence of these parameters on the specific kinematic functions. The purpose is to achieve a separation of geometric parameters depending on their influences on the objective functions to be optimized from a kinematic point of view, so as to simplify the process of effective optimization, by taking into account (as design variables) only the main parameters.

1. Introduction

Relative to car body, the vehicle wheels can be guided independently - by means of a guiding mechanism for each wheel, or dependently - by a guiding mechanism of the rigid axle. The first solution is frequently used for the front & rear wheels of the passenger cars, while the second one is mainly used for the rear axles of larger gauge cars. For the rear axle guidance, spatial mechanisms formed by a number of binary links (bars) are interposed between axle and car body. The bars' connections to axle and car body are made by using compliant joints (i.e. bushings) [1-3]. Usually, for the kinematic study (where the car body is fixed), the bushings are modelled as spherical joints, the corresponding models having a low number of degree of mobility (DOM=1 or DOM=2) [4, 5].

The guidance of the rear axle is made by driving a number of its points on suitably chosen surfaces and curves (sphere, circle or coupler curve). By the guidance of four axle points on four spheres with centres on car body (4SS), bi-mobile (DOM=2) guiding mechanisms are obtained. According to the study carried out in [6], for this structural group/class of axle guiding mechanisms, the spherical joint model assumption can be accepted because the behaviour of these mechanisms is closer to that of the real model with compliant joints (bushings). The structural variants of 4SS axle guiding mechanism are presented in Figure 1. For the scheme shown in Figure 1.a, all the guiding bars of the mechanism are arranged in the longitudinal direction, while for the mechanism shown in Figure 1.b one of the bars (4) is arranged transversely (so called Panhard bar), which ensures a better takeover of the transversal forces from the wheels, the latest solution being addressed in this work.

The kinematic optimization of the axle guiding mechanisms is a rather complex problem, considering both the variety of kinematic parameters that need to be optimized and the multitude of geometric parameters that define the mechanism. For this reason, it would be preferable to achieve a separation of geometric parameters depending on the influence they have on the objective functions to be optimized. Such a study will allow that in the effective optimization only the main parameters, which significantly influence the behaviour of the axle guiding mechanism, to be taken into account.

IOP Conf. Series: Materials Science and Engineering 1169 (2021) 012002 doi:10.1088/1757-899X/1169/1/012002

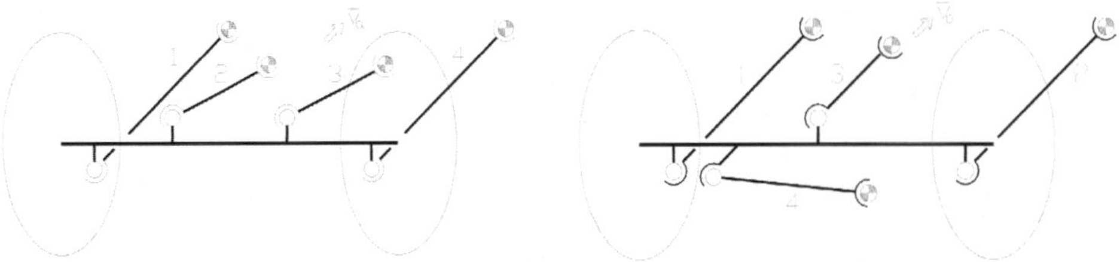

a. b.

Figure 1. Configurations of 4SS-axle guiding mechanism.

Under these terms, the present work deals with the design sensitivity analysis in the kinematics of the 4SS axle guiding mechanism with Panhard bar. The kinematic analysis is carried out through the characteristic points's method, which was depicted in [7] as part of a more detailed numerical algorithm for determining the equilibrium position of the axle suspension system. The kinematic analysis method was algorithmized using the C ++ programming environment.

2. Geometric and kinematic parameters

Its position and orientation, in relation to the global reference frame OXYZ that is fixed in car body, describe the spatial movement of the rear axle. The global coordinates X_P^0, Z_P^0 determine the initial position of the axle in OXYZ reference frame. Due to the symmetry of the guiding linkage relative to the longitudinal plane of the car, there are the following relationships between the geometrical parameters: $X_{M01}= X_{M02}$, $|Y_{M01}|= Y_{M02}$, $Z_{M01}= Z_{M02}$, $Y_{M03}= 0$, $X_{M1(P)}= X_{M2(P)}$, $|Y_{M1(P)}|=Y_{M2(P)}$, $Z_{M1(P)}= Z_{M2(P)}$, $Y_{M3(P)}= 0$. Therefore, the geometrical model of the 4SS guiding mechanisms is defined by 19 geometrical parameters, as follows (Figure 2): $X_{M01,2}$, $Y_{M01,2}$, $Z_{M01,2}$, X_{M03}, Z_{M03}, X_{M04}, Y_{M04}, Z_{M04}, $X_{M1,2(P)}$, $Y_{M1,2(P)}$, $Z_{M1,2(P)}$, $X_{M3(P)}$, $Z_{M3(P)}$, $X_{M4(P)}$, $_{M4(P)}$, $Z_{M4(P)}$, $l_{1,2}$, l_3, l_4.

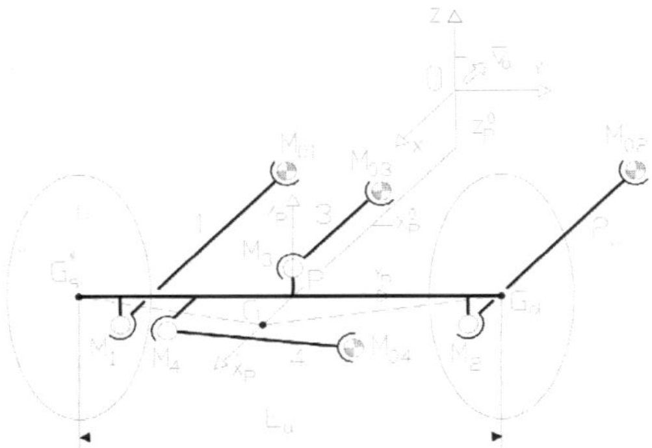

Figure 2. The geometrical model of the 4SS guiding mechanism with Panhard bar.

Relative to the car body, the rear axle must have the possibility of vertical displacement and rotation around the longitudinal axis of the vehicle (X). When the vehicle is moving, besides the above mentioned necessary motions, secondary undesirable motions can occur, as follows: displacements of the axle centre P along the longitudinal and transversal axes, axle pivoting movement, and axle rotation around its own axis). The minimization of the undesirable motions represents the goal of the kinematic optimization [4, 5].

The spatial position & orientation of the axle are defined by the following kinematic parameters:

- the linear displacements of the axle's centre:

$$\Delta X_P = X_P - X_P^0,\ \Delta Y_P = Y_P - Y_P^0,\ \Delta Z_P = Z_P - Z_P^0; \tag{1}$$

- the rotations of the rear axle in the global coordinate system's plans:

IOP Conf. Series: Materials Science and Engineering 1169 (2021) 012002 doi:10.1088/1757-899X/1169/1/012002

$$\eta_{xy} = arctg\frac{X_{Gd} - X_{Gs}}{Y_{Gd} - Y_{Gs}},\ \eta_{yz} = arctg\frac{Z_{Gd} - Z_{Gs}}{Y_{Gd} - Y_{Gs}},\ \eta_{xz} = arcsin\frac{\left[Z_G - Z_G^0\right] - \left[Z_P - Z_P^0\right]}{\left|X_{G(P)}\right|}. \quad (2)$$

The global coordinates of the characteristic points (G_s, G_d, G) and of the axle's center (P) are determined in accordance with the numerical method depicted in [**]. The vertical coordinates of the wheels' centers / axle's ends (Z_{Gs}, Z_{Gd}) are independent kinematic parameters.

Usually, the coordinates of the guiding points on axle $(X, Y, Z)_{Mi(P)}$ are established by constructive criteria. Afterwards, it will be analyzed the influence of the other parameters on the kinematic behavior of the guiding mechanisms (in fact, on the undesirable motions). The spatial configuration of the mechanism is defined by taking into account the disposing of the guiding bars, in accordance with the schemes shown in Figure 3 (the longitudinal bars - 1, 2, 3) and Figure 4 (the transversal bar - 4).

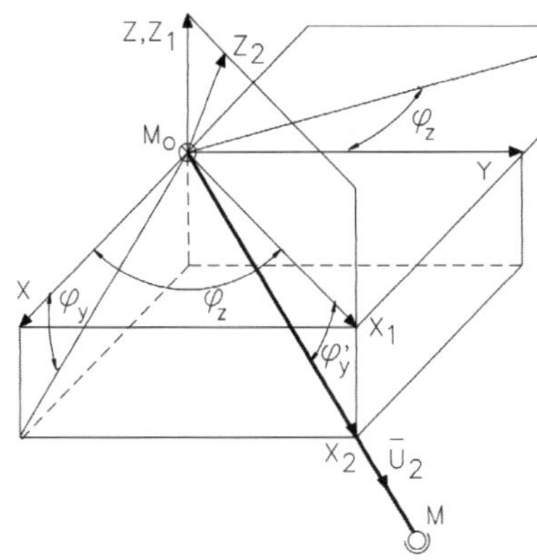

Figure 3. The positioning of the longitudinal bars.

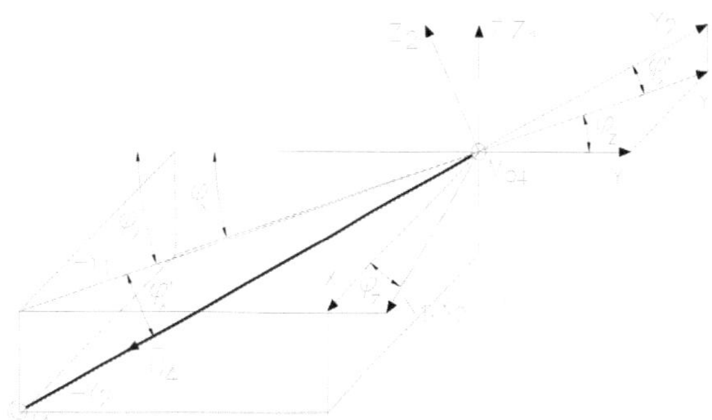

Figure 4. The positioning of the transversal bar.

The unitary position vectors $u_{1\text{-}3}$ (for the longitudinal bars) and u_4 (for the transversal bar) are defined by the following equations:

$$u_i = \begin{bmatrix} u_i^x \\ u_i^y \\ u_i^z \end{bmatrix} = \begin{bmatrix} cos\varphi_{iz} \cdot cos\varphi'_{iy} \\ sin\varphi_{iz} \cdot cos\varphi'_{iy} \\ -sin\varphi'_{iy} \end{bmatrix},\ i = 1...3,\ u_4 = \begin{bmatrix} u_4^x \\ u_4^y \\ u_4^z \end{bmatrix} = \begin{bmatrix} sin\varphi_{4z} \cdot cos\varphi'_{4x} \\ -cos\varphi_{4z} \cdot cos\varphi'_{4x} \\ -sin\varphi'_{4x} \end{bmatrix}, \quad (3)$$

where:

$$\varphi'_{iy} = arctg(tg\varphi_{iy} \cdot cos\varphi_{iz}),\ i = 1...3,\ \varphi'_{4x} = arctg(tg\varphi_{4x} \cdot cos\varphi_{4z}). \quad (4)$$

The global coordinates of points/joints on car body (X_{M0}, Y_{M0}, Z_{M0}) will be:

$$X_{M0i} = X_{Mi} - l_i\, cos\varphi_{iz} \cdot cos\varphi'_{iy},\ Y_{M0i} = Y_{Mi} - l_i\, sin\varphi_{iz} \cdot cos\varphi'_{iy},\ Z_{M0i} = Z_{Mi} + l_i \cdot sin\varphi'_{iy},\ i = 1...3, \quad (5)$$
$$X_{M04} = X_{M4} - l_4 \cdot sin\varphi_{4z} \cdot cos\varphi'_{4x},\ Y_{M04} = Y_{M4} + l_4 \cdot cos\varphi_{4z} \cdot cos\varphi'_{4x},\ Z_{M04} = Z_{M4} + l_4 \cdot sin\varphi'_{4x},$$

where the global coordinates of the guiding points on axle (X_M, Y_M, Z_M) correspond to the known initial position of the mechanism.

By noting $k = l_3/l_{1(2)}$ (the ratio between the lengths of the upper and lower longitudinal bars), there are obtained the following geometrical parameters (whose influence on the kinematic behaviour of the axle guiding mechanism will be analysed): k, l_1, l_4, φ_{1y}, φ_{1z}, φ_{3y}, φ_{4x}, φ_{4z}.

3. Results and conclusions

The results of the design sensitivity study are presented in the diagrams shown in Figures 5-9, corresponding to the functional case $Z_{Gs}=Z_{Gd}$, considering the vertical travel of the wheels in the interval $\Delta Z\in[-80, 80]$mm relative to the initial position (vehicle in rest). The influence of parameters l_4, φ_{1z} and φ_{4z} on the undesirable motions ΔX_P, ΔY_P, η_{xy} and η_{xz} is insignificant; on the other hand, all geometrical parameters have a small influence on the axle pivoting motion η_{xy} (for this reason, the corresponding variation diagrams are no longer presented). The following values of the geometrical parameters were considered: $\varphi_{1y} = \{-6, 0, +6\}°$, $\varphi_{3y} = \{-6, 0, +6\}°$, $l_1 = \{400, 500, 600\}$ mm, $k = \{0.4, 0.6, 0.8\}$, $\varphi_{4x} = \{-6, 0, +6\}°$.

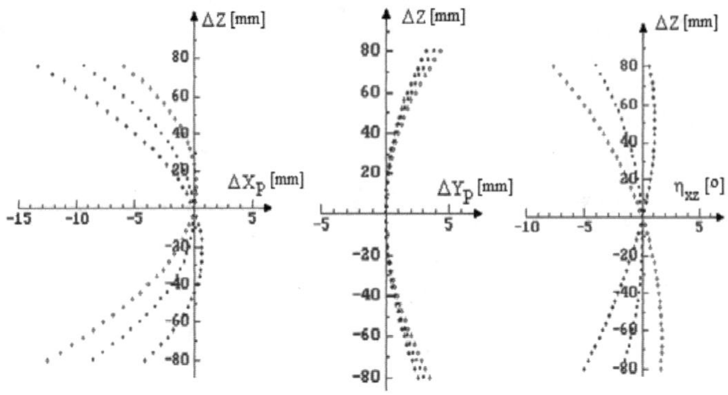

Figure 5. The influence of parameter φ_{1y}.

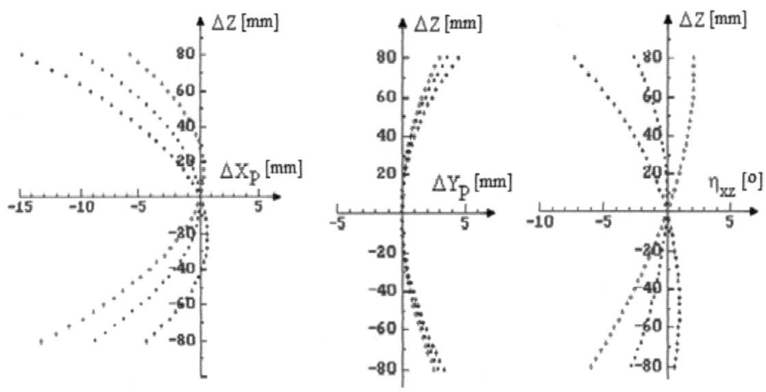

Figure 6. The influence of parameter φ_{3y}.

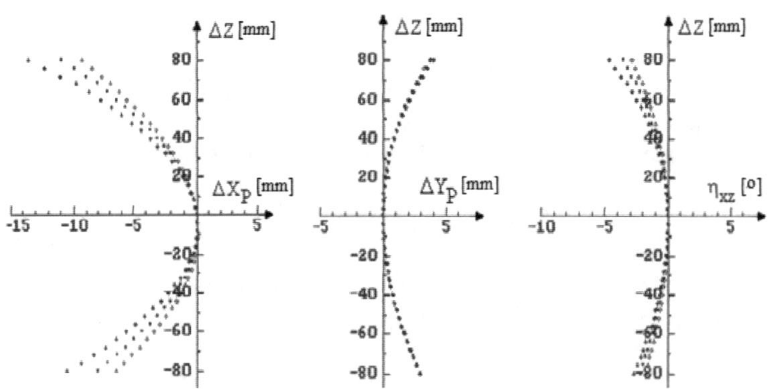

Figure 7. The influence of parameter l_1.

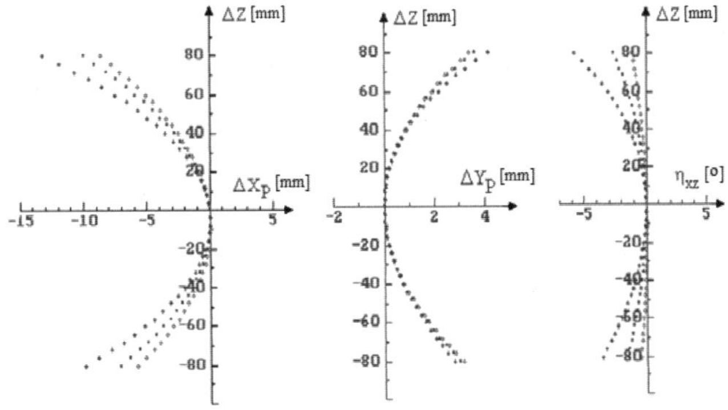

Figure 8. The influence of parameter k.

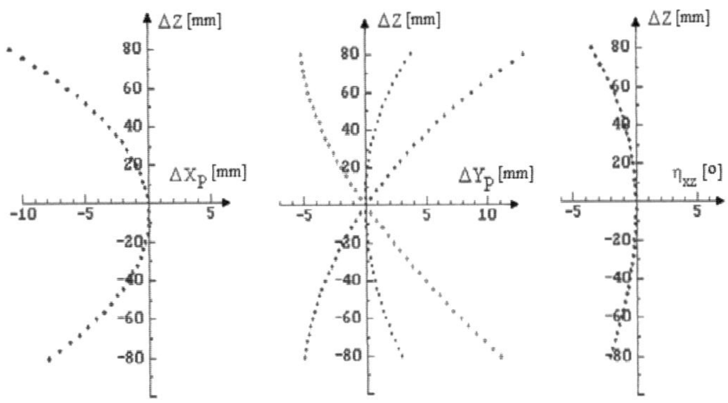

Figure 9. The influence of parameter φ_{4x}.

According to these results, a separation of the geometrical parameters was obtained, as follows: main parameters, with great influence on kinematic behaviour of the axle guiding mechanism: k, φ_{1y}, φ_{3y}; secondary parameters, with small influence: $l_{1(2)}$, l_4, $\varphi_{1(2)z}$, φ_{4x}, φ_{4z}. There can be considered that the undesirable motions are given by the following functions, in pairs [kinematic parameters; (geometric parameters)]: $\Delta X_P \cong F_1[Z_{Gs,d}; (k, \varphi_{1y}, \varphi_{3y})]$, $\Delta Y_P \cong F_2[Z_{Gs,d}; (\varphi_{4x})]$, $\eta_{xy} \cong F_3[Z_{Gs,d}]$, $\eta_{xz} \cong F_4[Z_{Gs,d}; (k, \varphi_{1y}, \varphi_{3y})]$. Afterwards, the kinematic synthesis of the axle guiding mechanism can be carried out on the basis of the main geometrical parameters, by neglecting the secondary parameters, which simplifies the optimal design process, with beneficial effects on the allotted time.

References

[1] Knapczyk J and Maniowski M 2002 Selected effects of bushings characteristics on five-link suspension elastokinematics *Mobility and Vehicle Mechanics* **3**(2) pp 107-121

[2] Knapczyk J and Maniowski M 2006 Elastokinematic modeling and study of five-rod suspension with subframe *Mechanism and Machine Theory* **41**(9) pp 1031-1047

[3] Tică M, Dobre G and Mateescu V 2014 Influence of compliance for an elastokinematic model of a proposed rear suspension *International Journal of Automotive Technology* **15**(6) pp 885-891

[4] Alexandru C 2009 The kinematic optimization of the multi-link suspension mechanism used for rear axle of the motor vehicle *Proceedings of the Romanian Academy - A* **10**(3) pp 244-253

[5] Simionescu PA and Beale D 2002 Synthesis and analysis of the five-link rear suspension system used in automobile *Mechanism and Machine Theory* **37**(9) pp 815-832

[6] Țoțu V and Alexandru C (2013) Study concerning the effect of the bushings' deformability on the static behavior of the rear axle guiding linkages *Applied Mechanics and Materials* **245** pp 132-137

[7] Alexandru C 2019 Method for the quasi-static analysis of beam axle suspension systems used for road vehicles Proceedings of the Institution of Mechanical Engineers, Part D: Journal of Automobile Engineering 233(7) pp 1818-1833

IOP Conf. Series: Materials Science and Engineering 1169 (2021) 012003 doi:10.1088/1757-899X/1169/1/012003

The effect of structural aspect for planar systems with 2DOF upon the stability of motion

S Alaci[1], F-C Ciornei[1], C-C Suciu[1] and I-C Romanu[1]

[1]Mechanics and Technologies Department, „Stefan cel Mare" University of Suceava, Suceava, Romania

E-mail: stelian.alaci@usm.ro

Abstract. The paper analyses all planar chains with 2DOF from the point of view of stability of motion. For the rotation-prismatic structural solution first there are obtained the equations of motion and then, the numerical integration procedure is applied. A strong instability of the system can be noticed. The same dynamical system is modelled using dynamical analysis software and the instability is confirmed.

1. Introduction

There are uncommon the cases when a part of a mechanical structure moves unrestrained, without interacting with other parts from its proximity. The majority of the mobile mechanical structures are made of solid elements that come into contact to each other. When two of the components of a system make contact the number of degrees of freedom of the two elements reduces. One of the main classification criteria of the constraints which take place between two parts, named kinematical joints, is the class of the pair, defined as the number of degrees of freedom taken for one of the element of the pairs, when the other element is considered fixed, [1]. Considering the fact that a rigid body has 6 degrees of freedom (DOF) it results that the class of the pair can have values from 1 to 5, the limit case of 0 representing the lack of linkage and the pair of class 6 represents the relative rigidity of the two parts that behave as a single part.

The second especially important notion when analyzing the mobility of a structure is the family, concept that illustrates the number of common restraints imposed to each element of the structure. The monographs of mechanisms theory consider as main categories of mechanisms: spatial mechanisms, in the general significance, spherical mechanisms that have as characteristic the existence of the same fixed point for all the elements of the mechanism and planar mechanisms that have as characteristic the fact that all the elements move parallel to the same plane. Both spherical and planar mechanisms are mechanisms of family 3, [2]. Since a kinematical chain of a certain family cannot contain pairs of a class inferior or equal to the family of the chain, for a planar kinematical chain - that will be the referred next, there is possible the existence of pairs of class 5 - rotation R and prismatic T and also of superior pairs of class 4. The dynamical systems with two degrees of freedom are the simplest systems where the chaos phenomenon can appear.

2. Possible structural solutions for planar kinematical chains with 2 DOF

The effect of chaos for a double physical pendulum, figure 1, is presented in [3] for the situation when the launch position differs substantially from the position of static equilibrium, that corresponds to the vertical orientation of both rods and with the center of mass in the lowest position. To illustrate the effect of chaos that occurs in the system, two identical pendula were considered, positioned symmetrically with respect to a vertical plane. The motion of the pendula was modeled using dynamical simulation software. It was observed that for a relatively short time the motions of the pendula were identical, but after that, the oscillations of the pendula differed. From experiments performed with the double pendulum, figure 1, it was concluded that the attempt to obtain identical motions from two similar launches, with the same angular amplitudes, $\theta_{10} \geq 90^o$, $\theta_{20} \geq 90^o$, is unsuccessful. The explanation of the phenomenon resides in the strong dependency of the evolution of the system on the initial conditions. But, in spite of the efforts, the precision of obtaining the initial position cannot be superior to the precision of the instruments of measure used in the estimation of the positional parameters of the system.

In figure 2 is presented a system having in structure a prismatic pair and a rotation pair for which Voinea [4] presents the differential equations of motion. In order to appreciate how stable the system is, the manner used for the double pendulum is applied. For that reason, using dynamical simulation software, two identical systems were modeled, positioned in mirror, with the same initial position, figure 4. The angle characterizing the initial position of the rod is $\theta_0 = 135^o$. Unlike the double pendulum from figure 1, where after a few seconds the motions of the pendula are completely different, in this case, the motions of the two systems are rigorously identical, as shown in figure 5 where the variations of velocities of the prismatic bodies were plotted. To validate this affirmation, in figure 6 is presented the dependency of the velocity of prism 2 as a function of the velocity of the prism 1. The straight line plot from figure 6 shows the equality of the two velocities.

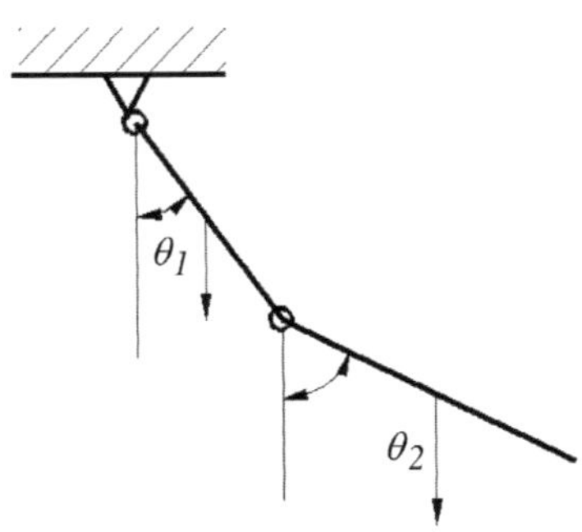

Figure 1. Kinematical chain 2 DOF with two rotation pairs

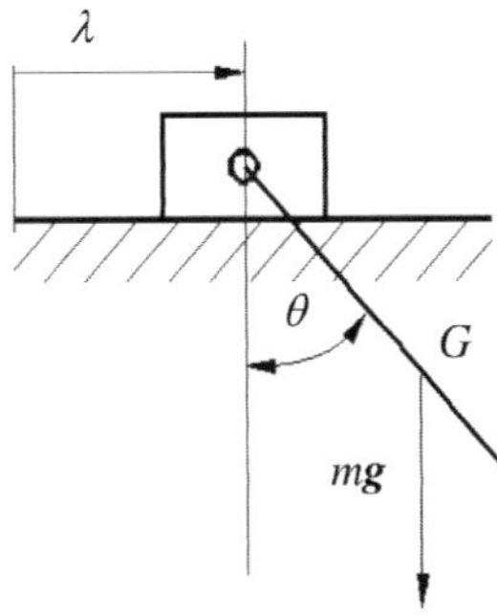

Figure 2. Kinematical chain 2 DOF with one rotation pair and one prismatic pair

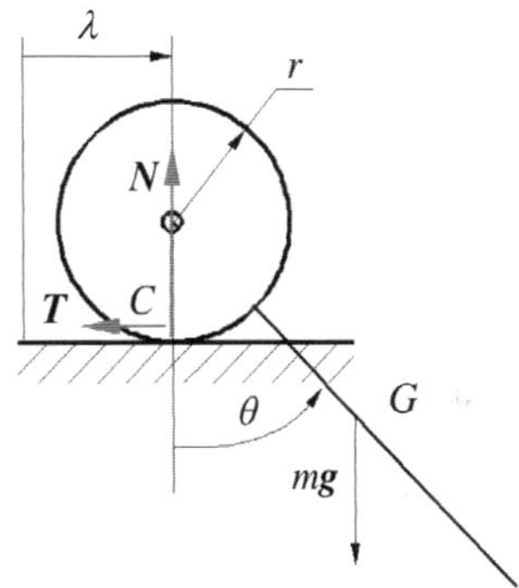

Figure 3. Kinematical chain 2 DOF with higher pair

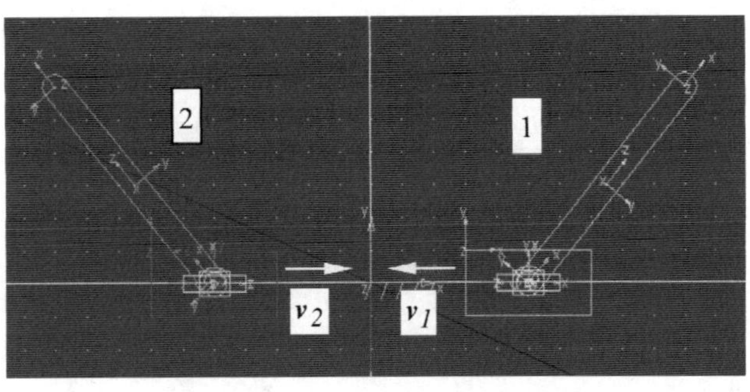

Figure 4. Mirrored identical systems

IOP Conf. Series: Materials Science and Engineering 1169 (2021) 012003 doi:10.1088/1757-899X/1169/1/012003

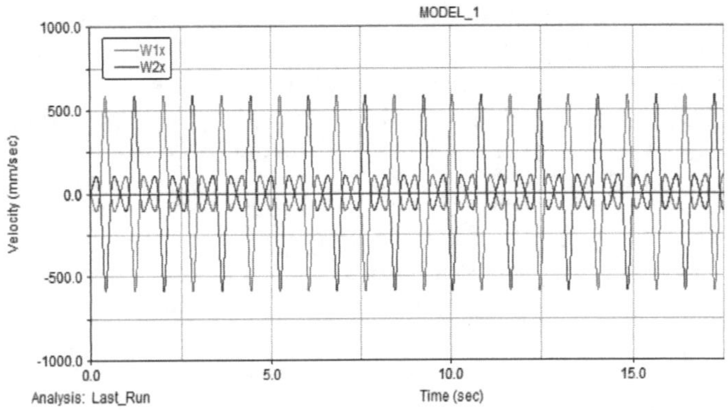

Figure 5. Variation of the velocities of the prisms

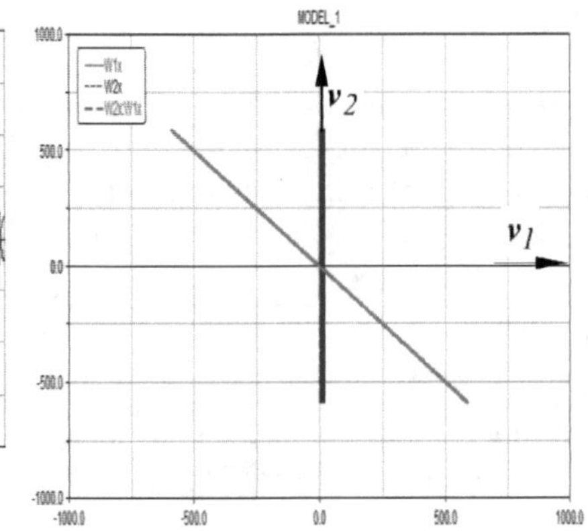

Figure 6. The dependency v_2 *vs* v_1

A kinematical chain that contains a higher kinematic pair is presented in figure 3. The system has two degrees of freedom λ, θ because when the friction is absent, the fundamental theorems of dynamics provide three scalar equations (the momentum theorem gives two scalar equations and the moment of momentum theorem gives a single equation) which allow for finding three unknowns, here the two positional parameters λ and θ and the magnitude of the normal reaction $\boldsymbol{N}$. But particularly interesting is the case when friction occurs in the higher pair. The magnitude of the friction force $\boldsymbol{T}$ is added besides the three unknowns λ, θ and N. To solve the problem, a supplementary equation is required. Two cases can be distinguished, depending on the friction type:

a) when rolling friction is present, the supplementary equation is $\dot{\lambda} = r\dot{\theta}$ and the system has 1 DOF, and in fact it is a cycloidal pendulum with the dynamics presented in [5] for which there is no risk of chaos phenomenon occurrence.

b) when dry friction is present, the fourth equation is given by the relation between the magnitude of friction force and the magnitude of normal reaction, $T = \mu N$ according to Coulomb's law. In this case the system has again 2 DOF.

3. The 2DOF linkage with rotation and prismatic pairs

Next it is presented the effect produced by reciprocate changing the pairs from the kinematical chain from figure 2, as in figure 7, problem proposed by Spiegel [6]. It is considered the particular case when the homogenous rod is jointed to the ground in the center of mass. A sleeve with negligible dimensions slides without friction on the rod.

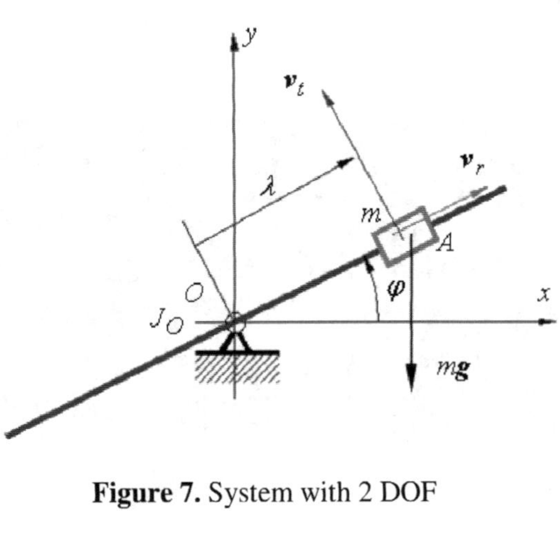

Figure 7. System with 2 DOF

IOP Conf. Series: Materials Science and Engineering 1169 (2021) 012003 doi:10.1088/1757-899X/1169/1/012003

The rod of length L and diameter d has the joint in the center of mass O, and can rotate about an horizontal axis. The moment of inertia of the bar with respect to the axis of rotation is J_O. The position of the rod is specified using the angle φ made with the vertical direction. A sleeve of negligible dimensions, but of mas m is placed the rod, having the position précised by the distance λ from the joint of the bar to the center of the sleeve. The hypothesis of ideal linkages is accepted and therefore the motion of the system can be found using the Lagrange equations of the second kind [7]:

$$\frac{d}{dt}\frac{\partial E_c}{\partial \dot{q}_k} - \frac{\partial E_c}{\partial q_k} = Q_k \tag{1}$$

where E_c is the kinetic energy of the system, q_k are the generalized coordinates and Q_k are the generalized forces. In this case the generalized coordinates are the angle φ and the distance λ. The kinetic energy of the system consists in the kinetic energy of rotation of the rod:

$$E_{C_1} = \frac{1}{2} J_O \dot{\varphi}^2 \tag{2}$$

and the translation energy of the sleeve:

$$E_{c_2} = \frac{1}{2} m v^2 \tag{3}$$

As noticed form figure 7, the absolute velocity of the sleeve is the vector sum between the transport velocity $\boldsymbol{v}_t$ and the relative velocity $\boldsymbol{v}_r$, the two components being perpendicular: the first is normal to the rod and the second is oriented along the rod.

$$\boldsymbol{v} = \boldsymbol{v}_t + \boldsymbol{v}_r \tag{4}$$

The magnitude of the transport velocity is:

$$v_t = \lambda \dot{\varphi} \tag{5}$$

and of the relative velocity is:

$$v_r = \dot{\lambda} \tag{6}$$

The final from of the kinetic energy of the system is:

$$E_c = \frac{1}{2} J_O \dot{\varphi}^2 + \frac{1}{2} m(\lambda^2 \dot{\varphi}^2 + \dot{\lambda}^2) \tag{7}$$

The single force that produces mechanical work in the system is the force of gravity of the sleeve.

$$\boldsymbol{G} = -m\boldsymbol{g} = -mg\boldsymbol{j} \tag{8}$$

where g is the gravitational acceleration. The position vector of the center of mass of the sleeve is:

$$\boldsymbol{r} = \lambda(\boldsymbol{i} \cos\varphi + \boldsymbol{j} \sin\varphi) \tag{9}$$

The elementary variation of the position of the center of mass is:

$$\delta\boldsymbol{r} = (\boldsymbol{i} \cos\varphi + \boldsymbol{j} \sin\varphi)\delta\lambda + \lambda(-\sin\varphi \boldsymbol{i} + \cos\varphi \boldsymbol{j})\lambda\, \delta\varphi \tag{10}$$

The virtual work of the mass is:

$$\delta L = \boldsymbol{G} \cdot \delta\boldsymbol{r} = -mg \sin\varphi \delta\lambda - \lambda mg \cos\varphi \delta\varphi j \tag{11}$$

For the generalized coordinate φ it is obtained:

IOP Conf. Series: Materials Science and Engineering 1169 (2021) 012003 doi:10.1088/1757-899X/1169/1/012003

$$\frac{\partial E_c}{\partial \dot{\varphi}} = J_O \dot{\varphi} + m\lambda^2 \dot{\varphi} \tag{12}$$

$$\frac{d}{dt}\frac{\partial E_c}{\partial \dot{\varphi}} = J_O \ddot{\varphi} + m\lambda^2 \ddot{\varphi} + 2m\lambda\dot{\varphi}\dot{\lambda} \tag{13}$$

$$\frac{\partial E_c}{\partial \varphi} = 0 \tag{14}$$

The generalized force Q_φ is:

$$Q_\varphi = \frac{\delta L}{\delta \varphi} = -mg\lambda \sin\varphi \tag{15}$$

For the generalized coordinate λ it is obtained:

$$\frac{\partial E_c}{\partial \dot{\lambda}} = m\dot{\lambda} \tag{16}$$

$$\frac{d}{dt}\frac{\partial E_c}{\partial \dot{\lambda}} = m\ddot{\lambda} \tag{17}$$

$$\frac{\partial E_c}{\partial \lambda} = m\lambda\dot{\varphi}^2 \tag{18}$$

The generalized force Q_λ is:

$$Q_\lambda = \frac{\delta L}{\delta \lambda} = -mg \sin\varphi \tag{19}$$

The motion of the system is described by the system of differential equations:

$$\begin{cases} (J_O + m\lambda^2)\ddot{\varphi} + 2m\lambda\dot{\varphi}\dot{\lambda} = -mg\lambda \cos\varphi \\ m\ddot{\lambda} - m\lambda\dot{\varphi}^2 = -mg \sin\varphi \end{cases} \tag{20}$$

The system (20) is written under the final form:

$$\begin{cases} \ddot{\varphi} = -\lambda \dfrac{2\dot{\varphi}\dot{\lambda} + g\cos\varphi}{\lambda^2 + J_O / m} \\ \ddot{\lambda} = \lambda\dot{\varphi}^2 - g \sin\varphi \end{cases} \tag{21}$$

In order to integrate the system (20), two new variables are introduced:

$$u = \dot{\varphi} \tag{22}$$

$$v = \dot{\lambda} \tag{23}$$

Now, using the above notations, the system (20) can be written in matrix form:

$$\begin{bmatrix} \dot{\varphi} \\ \dot{\lambda} \\ \dot{u} \\ \dot{v} \end{bmatrix} = \begin{bmatrix} u \\ v \\ -\dfrac{2uv + g\cos\varphi}{J_O / m + \lambda^2}\lambda \\ \lambda u^2 - g \sin\varphi \end{bmatrix} \tag{24}$$

The system (24) is a system of differential equations of first order with four unknowns and it was written under this form because it allows for applying the Runge-Kutta [8] integration methodology. In order to integrate it, it is required the stipulation of the initial conditions. Explicitly, the values of the functions φ, λ, u and v must be specified for the initial moment. Aiming to complete a comparison with an experimental test, it is considered that the system is initially at rest:

$$u = 0, \quad v = 0 \tag{25}$$

and only the position parameters for the initial moment, must be specified:

$$\varphi = \varphi_0, \quad \lambda = \lambda_0 \tag{26}$$

It was avoided the case of a system in motion for the initial conditions because in practical tests it is difficult to ensure an imposed initial motion to a system.
For arbitrary values of the inertial characteristics of the rod and sleeve, the motion of the sleeve is easy to guess, as due to own weight it slides along the rod and at a certain time it leaves the rod. In the meantime, the rod performs a rotation motion of amplitude 2-3 rad. The trajectory of the center of mass of the sleeve is presented in figure 8a-b, obtained after integration of the equations (22), where the characteristics considered are: $L = 1m$, $d = 0.01m$, $J_O = 0.051050880620834 kg \cdot m^2$, $m = 0.548703825640 81kg$, $\lambda_0 = L/2$, $\theta_0 = \pi/3$, $\dot{\lambda}_0 = 0$, $\dot{\theta}_0 = 0$.
In order to highlight the instability of the solution, in figure 8b it is represented the trajectory of the center of mass obtained for the same conditions, but for $\theta_0 = 0.99999\pi/3$.

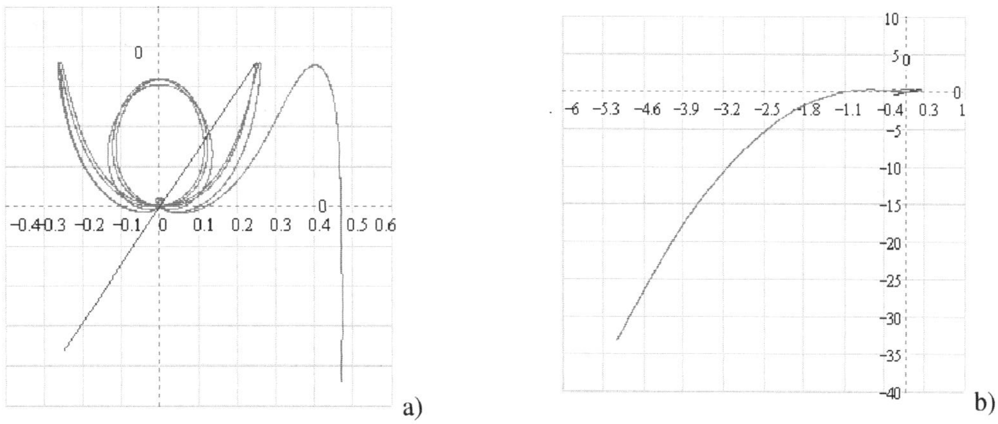

Figure 8. The trajectory of the center of mass of the sleeve

The same conclusion was reached after modelling the behaviour of the system using the dynamic simulation software, as presented in figure 9.

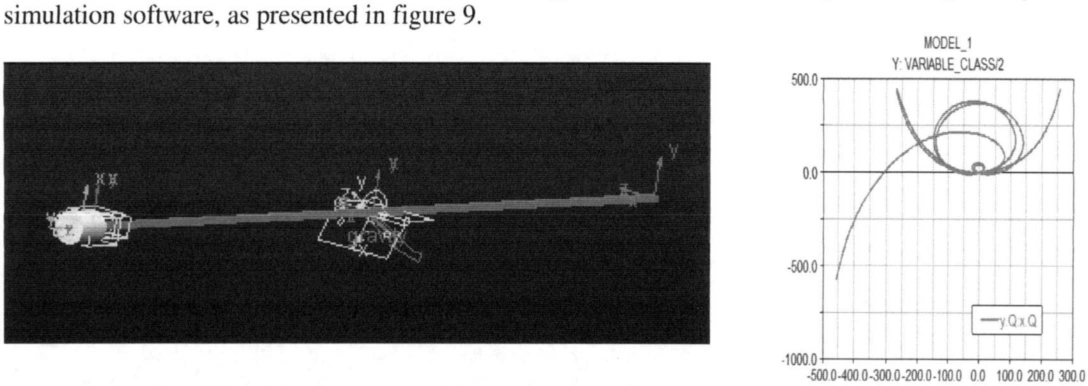

Figure 9. Simulation of the motion of the system using specialized software

IOP Conf. Series: Materials Science and Engineering 1169 (2021) 012003 doi:10.1088/1757-899X/1169/1/012003

4. Conclusions

The paper examines all kinematical planar chains with 2 DOF from the point of view of motion stability. There are four possible structural solutions: rotation-rotation, prismatic-rotation, rotation-prismatic and higher pair linkage, from which the first two are classical examples in dynamical systems monographs and the last two solutions are seldom met – most probable due to the deficiency of practical usage.

Though for the structural possible solutions that have lower pairs the friction from the pairs was neglected, for the system with higher pair, the effects of friction were considered because the type of friction (rolling or sliding) defines the DOF of the system (one or two, respectively) which is essential upon the stability of the motion of the system. For the rotation-prismatic 2DOF system with the rotation pair placed on the ground, the equations of motion were obtained applying the Lagrange's equations of second kind. The system of differential equations was numerically integrated and a strong instability of the obtained solution was highlighted. Actually, when altering any of the initial conditions with a fraction of $10^{-4}\%$, a complete different law of motion of the system is obtained.

The same conclusion was reached after studying the system with specialised software for dynamical analyses.

The paper underlines the necessity of correct structural design, of large interest for practical devices and it also is suitable for didactical activity in exemplification of simple yet very instable dynamical systems.

References

[1] Uicker J, Pennock G and Shigley J 2010 Theory of Machines and Mechanisms (Oxford UP)
[2] McCarthy J M 1990 *Introduction in Theoretical Kinematics* (MIT Press) p 130
[3] Alaci S, Alexandru C, Ciornei F-C, Doroftei I and Irimescu L 2020 Chaos illustrations in dynamics of mechanisms *Mechanisms and Machine Science* **89** pp 297-304
[4] Voinea R P, Voiculescu D and Ceaușu V 1983 *Mechanics* 2nd Ed (EDP București) p 615
[5] Ciornei M C, Alaci S, Ciornei F C and Romanu I C 2017 A method for the determination of the coefficient of rolling friction using cycloidal pendulum *IOP Conference Series: MSE* **227** (1) 012027
[6] Speigel M, 1980 *Schaum's outline of theory and problems of theoretical mechanics: With an introduction to Lagrange's equations and Hamiltonian theory* (McGaw Hill) p 368
[7] Ardema M 2005 *Analytical Dynamics* (Springer US) p 340
[8] Hermann M and Saravi M 2014 *A First Course in Ordinary Differential Equations: Analytical and Numerical Methods* (Springer India) p 288

IOP Conf. Series: Materials Science and Engineering 1169 (2021) 012004 doi:10.1088/1757-899X/1169/1/012004

The model of a permanent magnet DC motor in time domain and frequency domain based on Bond Graph modelling and design of position control using PID controller

Gy Korsoveczki[1], G Husi[1]

[1]Mechatronics Department, Faculty of Engineering, University of Debrecen, Debrecen, Hungary

korsoveczki.gyula@gmail.com

Abstract. During the research work, the modelling of a permanent magnetic DC (PMDC) motor was realised in time domain and frequency domain via 4 different types of model construction. To achieve this, the Bond Graph model of the motor was determined, and based on them the system equations were defined in time domain firstly and then in frequency domain by using the Laplace transform. According to the system equations in time domain, the state-space model of the motor was also realised and implemented. Based on the system equations in frequency domain, the transfer function of the position control loop was realised. As the further part of the research work, PID control-based position control was designed in both time domain and frequency domain. The controller was tuned according to the Ziegler-Nichols method. Finally, the models were compared from the point of view of behaviour during the tuning process. The models were implemented in LabVIEW environment produced by the National Instruments.

1. Introduction

Nowadays the technical life is producing high-rate development. As the available technological background evolves, the increase in complexity of systems becomes indispensable. Systems can operate only in one domain, but it is often necessary to establish a connection between systems of different domains. The field of mechatronics seeks to bridge the differences between systems. It is important to have accurately detailed planning and system modelling that represents the behaviour of the real physical system. The purpose of this research is to show the process and the possibilities of the modelling of a mechatronic system by using different modelling and implementation methods [1], [2].

2. The Bond Graph modelling

System modelling in engineering should distinguish between Single Domain systems and Multi-Domain systems. Single Domain systems can be purely mechanical, electrical, hydraulic or thermal etc. systems. They have their own description languages that can be used for modelling, understanding and analysing. If a system (such as a mechatronics system) contains more than one Single Domain system which is different types, it is called Multi-Domain system. It is necessary to realise physical or informational connection between them, as it happens in real physical work. It is difficult to establish mathematical relations between domains because the only way is to find a variable that is present in all systems. This physical variable is energy, and the language describing its change is Bond Graph modelling, developed by Henry Paynter in 1959. Bond Graph is an energy-based description language that tracks the change of energy across the domains. It is an effective solution for modelling Multi-Domain systems such as mechatronic systems. The Bond Graph is based on the fact that the transmitted power is the multiplication of the power variables, which are effort and flow. The role of them is unchangeable and it is defined for all systems. The Bond Graph has its own graphical description language that contains basic one-port passive elements (R, C and I element) that do not add new energy to the system but dissipate (R elements) or store it as potential energy (C elements) or kinetic (I elements) energy of the system. There are active one-port elements (Se, Sf), which can produce new effort or flow for the system. Junctions are also parts of Bond Graph, which can be two-port (TF, GY) or three-port (0, 1) junctions. The most important element of the Bond Graph modelling is the causality that defines the directions of effort and flow and allows us to determine the system equations [2], [3].

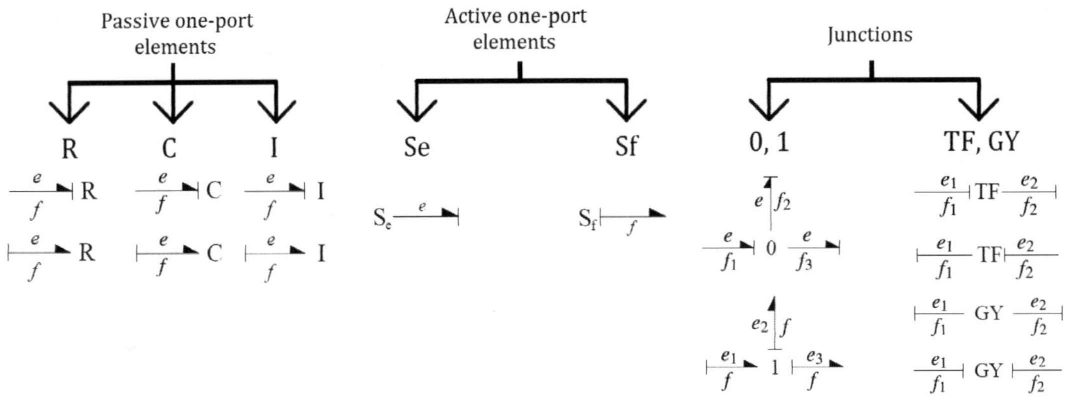

Figure 1. The basic elements of the Bond Graph [3].

3. The Bond Graph model of an ideal PMDC motor and its modelling in time domain

3.1. The Bond Graph model of the PMDC motor

A permanent magnet DC motor (PMDC motor) converts electrical power into mechanical rotating motion. It is a typical Multi-Domain system as it uses both electrical and mechanical power. The electrical part can be modelled as a simple serial circuit that is the equivalent circuit of the motor. The circuit consists of a voltage source (*Ua*), an armature resistance (*Ra*) and an armature coil (*La*) inductance. Since the circuit itself is a serial connection, the current is constantly flowing in it, but each electrical unit has the own voltage value. In an electrical system, the voltage is effort, the current is flow, so the electrical part has constant flow but variable effort, which can be modelled with 1 junction. The voltage source is a source of effort (*Se*), the armature resistance is an R element, and the armature coil is an I element. The strong Bond of the system is the I element, and in terms of causality, the voltage source has flow-in-effort-out, the armature resistance also flow-in-effort-out, while the armature coil has an effort-in-flow-out causality [3].

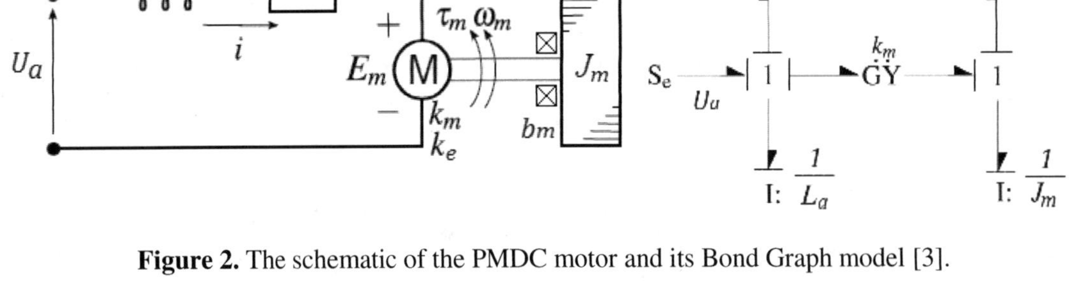

Figure 2. The schematic of the PMDC motor and its Bond Graph model [3].

The electromechanical conversion is represented by a Gyrator (GY) element that has flow-in-effort-out causality and connects the flow of the electrical domain to the effort of the mechanical domain and the flow of the mechanical domain to the effort of the electrical domain. The ratio is the motor constant *km* [Vs/rad] or [Nm/A]). The mechanic domain includes a bearing, rotational inertia and the electromechanical conversion of the motor. Bearing (*bm*) is represented by an R element with flow-in-effort-out causality, the rotational inertia (*Jm*) is in I element with effort-in-flow-out causality [3].

3.2. *The space-state model of the motor*

As a result of the Bond graph modelling the system equations could be realised. These differential equations can be written in state-space structure, which is a matrix structure for a better process of a larger number of system equations. The derivate of state variables ($\dot{\mathbf{y}}$), the state matrix (A), the state variables (y), the input matrix (B) and the input variables (u) were realised [4].

$$\frac{di}{dt} = \frac{U_a}{L_a} - \frac{R_a}{L_a} \cdot i - \frac{ke}{L_a} \cdot \omega_m \tag{1}$$

$$\frac{d\omega_m}{dt} = \frac{km}{Jm} \cdot i - \frac{b_m}{Jm} \cdot \omega_m \tag{2}$$

$$\frac{d}{dt}\begin{bmatrix} i \\ \omega_m \end{bmatrix} = \begin{bmatrix} -\frac{R_a}{L_a} & -\frac{ke}{L_a} \\ \frac{ke}{Jm} & -\frac{b_m}{Jm} \end{bmatrix} \cdot \begin{bmatrix} i \\ \omega_m \end{bmatrix} + \begin{bmatrix} \frac{1}{L_a} \\ 0 \end{bmatrix} \cdot U_a \tag{3}$$

4. Modelling the motor in frequency domain

The difficulty of modelling in time domain is to manage a large number of differential equations. In the case of the frequency domain, the Laplace transform simplifies the differential equations into simple algebraic equations, which can be easier processed and simulated. Using the Laplace transform, time domain system equations were transformed into frequency domain. In the frequency domain, the transfer function allows us to model a system. The transfer function defines the relation between the system output in frequency domain and the system input in frequency domain [4].

$$L_a \cdot i(s) \cdot s = U_a(s) - R_a \cdot i(s) - ke \cdot \omega_m(s) \tag{4}$$

$$J_m \cdot \omega_s(s) = km \cdot i(s) - b_m \cdot \omega_m(s) \tag{5}$$

$$G(s) = \frac{i(s)}{U_a(s)} = \frac{1}{L_a \cdot s + R_a} \tag{6}$$

$$G(s) = \frac{\omega_m(s)}{i(s)} = \frac{1}{J_m \cdot s + b_m} \cdot km \tag{7}$$

$$G(s) = \frac{\theta(s)}{U_a(s)} = \frac{k_m}{J_m \cdot L_a \cdot s^3 + (R_a \cdot J_m + b_m \cdot L_a) \cdot s^2 + (k_m{}^2 + R_a \cdot B_m) \cdot s} \tag{8}$$

5. The design of position control using PID controller in different domains

5.1. *Generally about PID controllers*

The control of linear systems is important in control theory. In this research, a position control loop for PMDC motor was designed in time domain and frequency domain using PID controller. It operates based on error signal, which is the difference between the reference value and the output value. The controller uses parallel compensation, thus the execution signal is proportional to the error signal (P), to the integral of the error signal (I) and to the derivate of the error signal (D). The system equations of PID controller in time domain and frequency domain are the following [4], [5]:

$$e(t) = K_p \cdot \left(e(t) + \frac{1}{T_i} \cdot \int_0^t e(t)dt + T_d \cdot \frac{de(t)}{dt}\right) \tag{9}$$

$$e(s) = K_p \cdot \left(1 + \frac{1}{T_i \cdot s} + T_d \cdot s\right) \tag{10}$$

IOP Conf. Series: Materials Science and Engineering 1169 (2021) 012004 doi:10.1088/1757-899X/1169/1/012004

5.2. The design of the position control loop

The system of the PMDC motor was modelled with 4 different system models in 2 different domains (time and frequency). All four system models were based on the description language of the Bond Graph. The software environment of the implementation was LabVIEW, which is provided by National Instruments. In the first case, the system model was built based on Block diagram structure in time domain. The second model was the state-space model, which is a numerical solution and it is equivalent to the Block diagram. Both system models represent the system in the time domain. In frequency domain, the best modelling method is to determine the transfer functions. The system was modelled based on its transfer functions using block structure and numerical implementation in MathScript Node. Each system model contains position control loop using PID controller that ensures the stable behaviour of the system. In the case of the model, the θ angle of the rotation is the output variable, And the input variable is the *Ua* input armature voltage [4], [5].

6. The tuning of the PID controller in different domains using the Ziegler-Nichols method

In cases of controllers which are based on the parallel compensation of linear systems such as PID, several methods are known for tuning such as the testing-based Trial and Error Method, the Cohen-Coon method that requires to open the loop of the control loop, and the Ziegler-Nichols method that is suitable and effective for tuning closed control loops. The Ziegler-Nichols method was developed by John G. Ziegler and Nathaniel B. Nichols. The method is when the closed control loop needs to be tuned up to the limit of instability, while the elements of the PID are determined one by one based on the critical amplification and the critical period time. The method is excellent for simulation modelling, where system unstableness does not cause physical damage and the single running of the control loop takes a short time. Changing the value of the proportional (P) part, the system must be controlled to the limit of instability, where critical amplification and its critical period time can be determined. In the meantime, the effects of the integrating (I) and derivative (D) part should be eliminated from the system by increasing the *Ti* integration time to infinite and reducing the *Td* derivative time to zero. At the limit of stability, the system will resonate with constant amplitude around the reference signal. After defining the critical amplification (K_{Pkrit}) and the critical period time (T_{krit}), the values of the parts can be determined as the following relations show [4], [5]:

$$K_P = 0{,}6 \cdot K_{Pkrit} \quad (11)$$

$$T_i = 0{,}5 \cdot T_{krit} \quad (12)$$

$$T_D = 0{,}12 \cdot T_{krit} \quad (13)$$

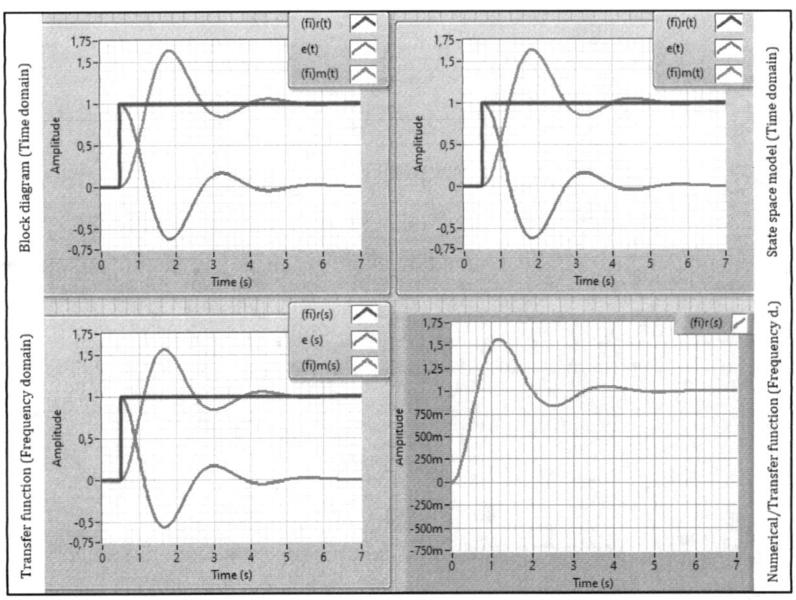

Figure 3. The recognition of the pentagon object.

7. The simulation results

The simulation data were exported to Microsoft Excel environment, where it was evaluated. In the simulation, all 4 models produced the same behaviour, the simulation results were equivalent. The Ziegler-Nichols method was used in every case, the critical amplification ($\boldsymbol{K_{Pkrit}}$) and critical period time ($\boldsymbol{T_{krit}}$) at the limit of instability were the same.

Table 1. The evaluated values of the controlling using PID controller.

	ODE Solver	Step Size (s)	Rise Time (s)	Overshoot (%)	Settling Time (s)	Steady-State Error
Block Diagram model (time)	Runge-Kutta 4	0,001	0,48	40	3,31	0,005
State-Space model (time)	Runge-Kutta 4	0,001	0,47	40	3,29	0,005
Transfer function (Block) (frequency)	Runge-Kutta 4	0,001	0,49	40	3,09	0,005
Transfer function (Numerical) (frequency)	Runge-Kutta 4	0,001	0,53	40	3,22	0,01

Conclusion

In this study, an ideal PMDC motor was modelled using 4 different system models based on the graphical description language of the Bond Graph. In the case of every model, position control loop was implemented using PID controller. The controller was tuned using the Ziegler-Nichols method, in which the system had to be amplified to the limit of instability. In the tested domains, the systems had the same behaviour, provided equivalent data and their computing resource requirements were low.

References

[1] B. Rokseth, E. Pedersen, A Bond Graph Approach for Modelling of Robotic Manipulators, July 2014., [Online]. Available: https://www.researchgate.net/publication/309410097_A_Bond_Graph_Approach_for_Modelling_of_Robotic_Manipulators. [Accessed: 10.05.2021.].

[2] C. Apostoaia, Multi-domain System Models Integration for Fault Detection in Induction Motor Drives, June 2014, [Online]. Available: https://www.researchgate.net/publication/262974154_Multi-domain_System_Models_Integration_for_Faults_Detection_in_Induction_Motor_Drives. [Accessed: 10.05.2021.].

[3] J. A. Kypuros, System Dynamics and Control with Bond Graph Modeling, CRC Press Taylor & Francies Group LLC, U.S., 2013.

[4] M. Sulaiman, M. S. M. Aras, Tuning Methods of PID Controller for DC Motor Speed Control, August 2016, [Online]. Available: https://www.researchgate.net/publication/311588542_Tuning_Methods_of_PID_Controller_for_DC_Motor_Speed_Control. [Accessed: 10.05.2021.].

[5] M. M. Sabir, J. A. Khan, Optimal Design of PID Controller for the Speed Control of DC Motor by Using Metaheuristic Techniques, 2014, [Online]. Available: https://www.researchgate.net/publication/277951362_Optimal_Design_of_PID_Controller_for_the_Speed_Control_of_DC_Motor_by_Using_Metaheuristic_Techniques. [Accessed: 10.05.2021.].

Acknowledgement

The publication was supported by the project EFOP-3.6.1-16-2016-00022. This project was co-funded by the European Union, from the European Social Fund.

IOP Conf. Series: Materials Science and Engineering 1169 (2021) 012005 doi:10.1088/1757-899X/1169/1/012005

The efficiency of modeling and simulation of manufacturing systems using Petri nets

F S Blaga[1], A Pop[1], V Hule[1] and C I Indre[1]

[1]University of Oradea, Faculty of Management and Technological Engineering, str.Universității nr.1, Oradea 410087, Romania

E-mail: fblaga@uoradea.ro

Abstract. The paper is a partial synthesis of the work that a team within the Faculty of Managerial and Technological Engineering has in the field of modeling and simulation of systems with Petri nets. The paper deals with three issues. The first part is Modeling and simulation of manufacturing systems with timed Petri nets. The first part presents the way in which the performances of some manufacturing systems are evaluated using timed Petri nets. The first example is a manufacturing system specific to the textile industry. The system has three production lines. In this case, the modeling with timed Petri nets allows the analysis of the operation of the flow lines and highlights the bottleneck. Another modeled manufacturing system is specific to a series production, more precisely it refers to the manufacture of gas cookers within a company from Oradea. The model with Petri nets also describes the supply sequences of the stocks of parts and subassemblies. The next modeled system is an automated manufacturing system for the packaging of detergents. The complexity of the Detergent Packing System and the high level of automation, require the use of efficient modeling and simulation methods to verify the validity of the solutions adopted. In this chapter is present the model of a robotic welding cell. The second part of the paper, the transition from timed Petri nets to hierarchical colored Petri nets, shows how Petri nets were developed for a robotic cell for metallographic analysis. The third part is dedicated to modeling with colored Petri nets timed with complex colors of the flexible manufacturing cell A007 within the Department of Industrial Engineering.

1. Introduction

The concept of Petri nets was introduced by Carl Adam Petri in his work Kommunikation mit Automaten (dissertation) [1]. Further the method of modeling and simulation of Petri nets was developed and found more and more fields of application. [2], [3], [4], [5] and [6].

At the University of Oradea, within the department of industrial engineering there are significant concerns in the field of using Petri nets in modeling and simulation of systems. In particular, research has focused on modeling and simulating manufacturing systems.

The paper [7] proposes the use of modeling and simulation with timed Petri nets in evaluating the performance of manufacturing lines in the textile industry. The paper [8] presents the results of research in the field of manufacturing process management through modeling and simulation. The research object is the manufacturing process that is specific to the mass production, more specifically refers to the gas cooker manufacturing within a company in Oradea.

An automated manufacturing system is present in [9] for the packaging of detergents can be modeled using Petri nets. The complexity of the Detergent Packing System and the high level of automation, require the use of efficient modeling and simulation methods to verify the validity of the solutions adopted. For this purpose, a modern modeling and simulation method based on Petri nets is used. The models associated with the system are developed into a hierarchical structure: the model of some modules in a line; the model of a line; system model as a whole.

The paper [10] presents the model with timed Petri nets of a robotized manufacturing cell specific to the automotive industry.

There is a concern to model the same system using several types of Petri nets. Thus, the robotic cell destined for the preparation and analysis of metallographic samples was modeled with: timed Petri nets [11], colored Petri nets [12] and hierarchical colored Petri nets [13]. Also in this context are the works [14] and [15]. They propose two models with colored Petri nets with complex colors of the flexible manufacturing cell A007 from the University of Oradea.

The paper [16] proposes an approach of modeling and simulation of a manufacturing process based on timed colored Petri nets, by developing an application based on XML (eXtensible Markup Language) technology.

There is a concern to develop models with Petri nets in which to find the techniques specific to fuzzy sets [17]. Also, models of hierarchical colored Petri nets are developed for the analysis of the performances of flexible manufacturing systems [18], [19], [20], [21]. An analysis of the modules in which it can be used state space is presented in [22].

2. Modeling and simulation of manufacturing systems with timed Petri nets

2.1. Evaluation of the textile industry fabrication lines performances using Petri networks models

The paper [7] presents an approach to the matters of the flux lines which are specific to the industry of textile clothes, from the point of view of their balance. Frequently, it happens that the time periods of the operations are much different, and this leads to jams and extended waiting times, with negative influence on the efficiency. This paper proposes the use of a modeling and simulation method based on timed Petri networks, which will allow the analysis of the flux lines operation, the highlighting of the bottlenecks and to yield information to the personnel that take managerial decisions. Three lines, which manufacture the following products: jacket, skirt and trousers, are analyzed. For each case, the modeling and simulation procedure yields solutions for jam solving.

The specific products of the company SC TRICOUL SRL belong to the category of (ready-made) clothes for children, women and men.

Figure 1 shows specific products of this company of type “jacket”, “skirt” and “trousers”.

Figure 1. Products of company SC TRICOUL SRL [7]

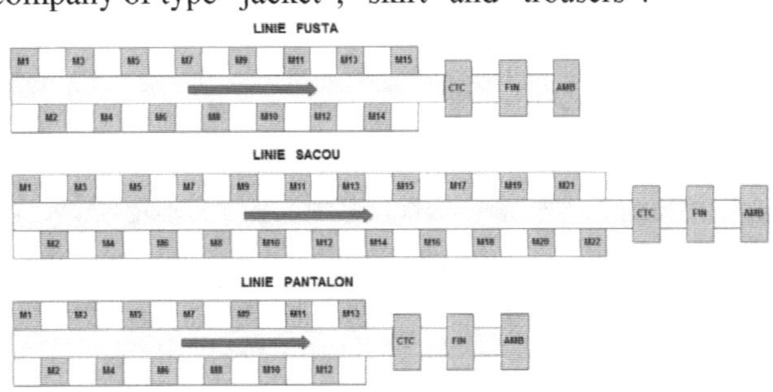

Figure. 2. The fabrication lines in the company [7]

Figure 3 shows the Petri networks model for the skirt fabrication lines. Each operation is modelled by a transition. The temporisation of a transition corresponds to the duration of the modelled operation.

Figure 4 shows the Petri networks model for the jacket fabrication and figure 5 shows the model for trousers fabrication lines

IOP Conf. Series: Materials Science and Engineering 1169 (2021) 012005 doi:10.1088/1757-899X/1169/1/012005

In order to evaluate the production on the three fabrication lines, the time period is set to 8 hours (one shift), that is 28800 seconds.

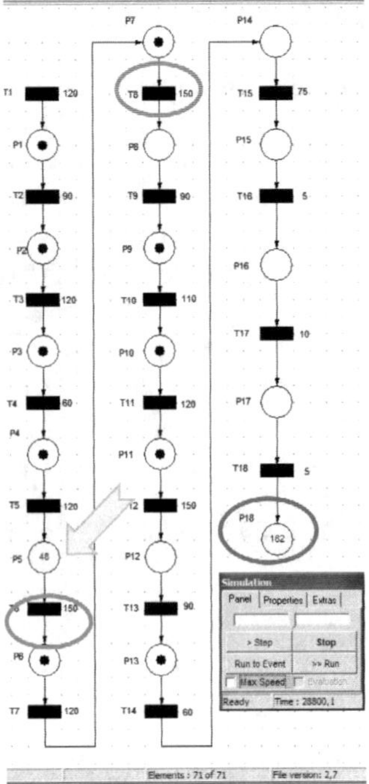

Figure. 3. Petri networks model for the skirts fabrication lines [7]

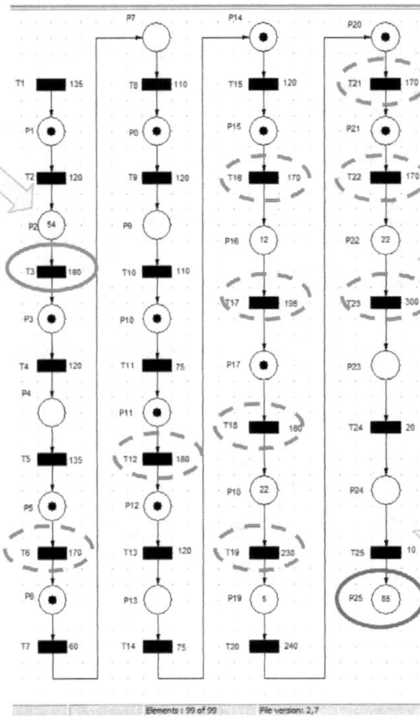

Figure. 4. Petri networks model for jacket fabrication line [7]

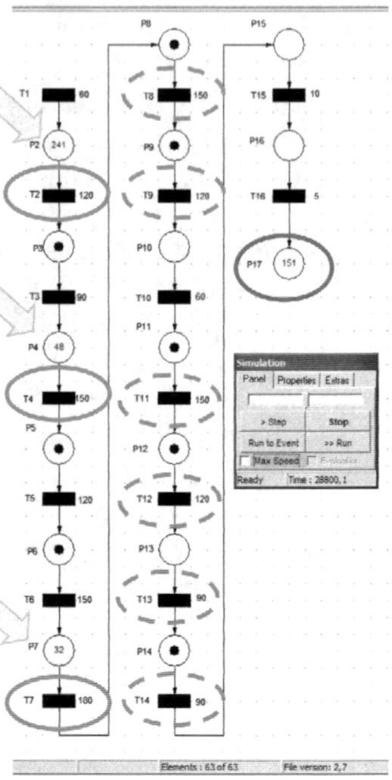

Figure. 5. Petri networks model for trousers fabrication lines [7]

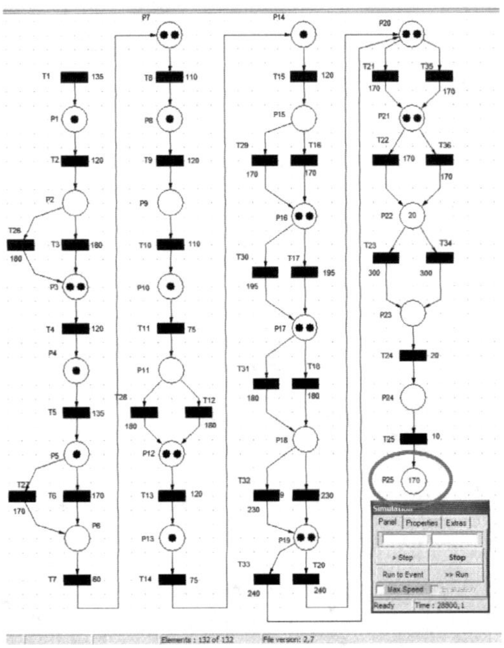

Figure 6. Petri networks model for jacket fabrication line after the multiplication of the work places [7]

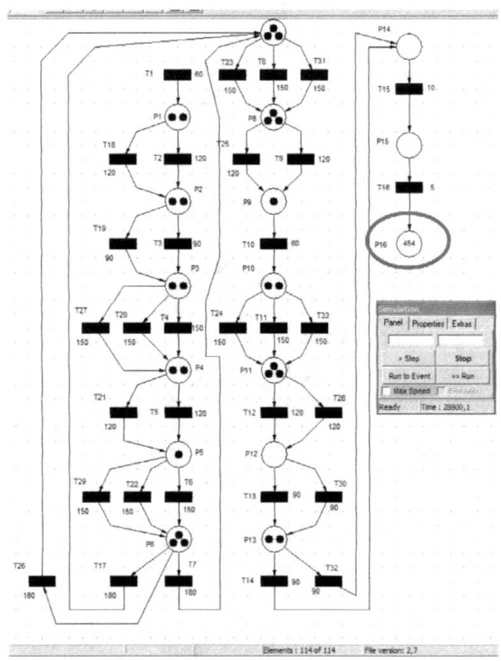

Figure 7. Petri networks model for trousers fabrication lines after the multiplication of the work places [7]

2.2. Description of product and assembly line [8]

The *Metalica 1685 F4-S1* gas cooker is a household appliance intended for the preparation of food. Through construction and execution, it is a superior finish product with easy use and maintenance, with economic functioning.

The gas cooker is equipped with 4 burners on the hob and a burner in the oven, gray glass cover of the hob. The oven is equipped with a thermocouple fuse. The gas cooker has 60 types of parts and subassemblies.

There are 26 operations on the assembly line. Each operation, in turn, has several phases. Moving the work object from one workstation to another, is made by a conveyor belt with imposed speed and therefore the speed of the transport means requires a certain rhythm of work. Figure 8 shows the panoramic view of the assembly line.

The model was built in four versions using Visual Object Net ++ software. For each version, the line operation was simulated during a work shift (8 hours).The results of the simulation have been analyzed and can be used as information in the decision-making process specific to production management.

Figure 8. Panoramic view of the assembly line [8]

We will present the fourth version, the most complex. The assembly operations performed on the manufacturing line involve the existence of storage spaces for the parts, subassemblies and materials that are part of the finished product component. Some components have a larger gauge and cannot be stored in the quantities needed for a shift. It is therefore necessary to supply the stocks during the manufacturing process. The supply procedure was modeled using generalized Petri nets. The arc capacities define the amount of supply components. In the detail of figure 9 there is an example of modeling the supplying for operation 6 – *mounted sidewalls* (transition 6). Position P19 models the stock of sidewalls.

This has the initial value of 80. For each of the finished product required two pieces of side wall component. This is modeled by the fact that the P19 → T6 arc has a load of 2. After each execution of the T6 transition from position P19, two tokens are withdrawn and two tokens are also deposited in the P19 'position. When the position P19' is equal to 80, the transition T6' is executed, which results in the deposition of 80 jets in position P19. In the real system this means supplying the stock of sidewalls.

In the case of version 4, after the 8-hour simulation of the line, the estimated output production is 238 gas cookers.

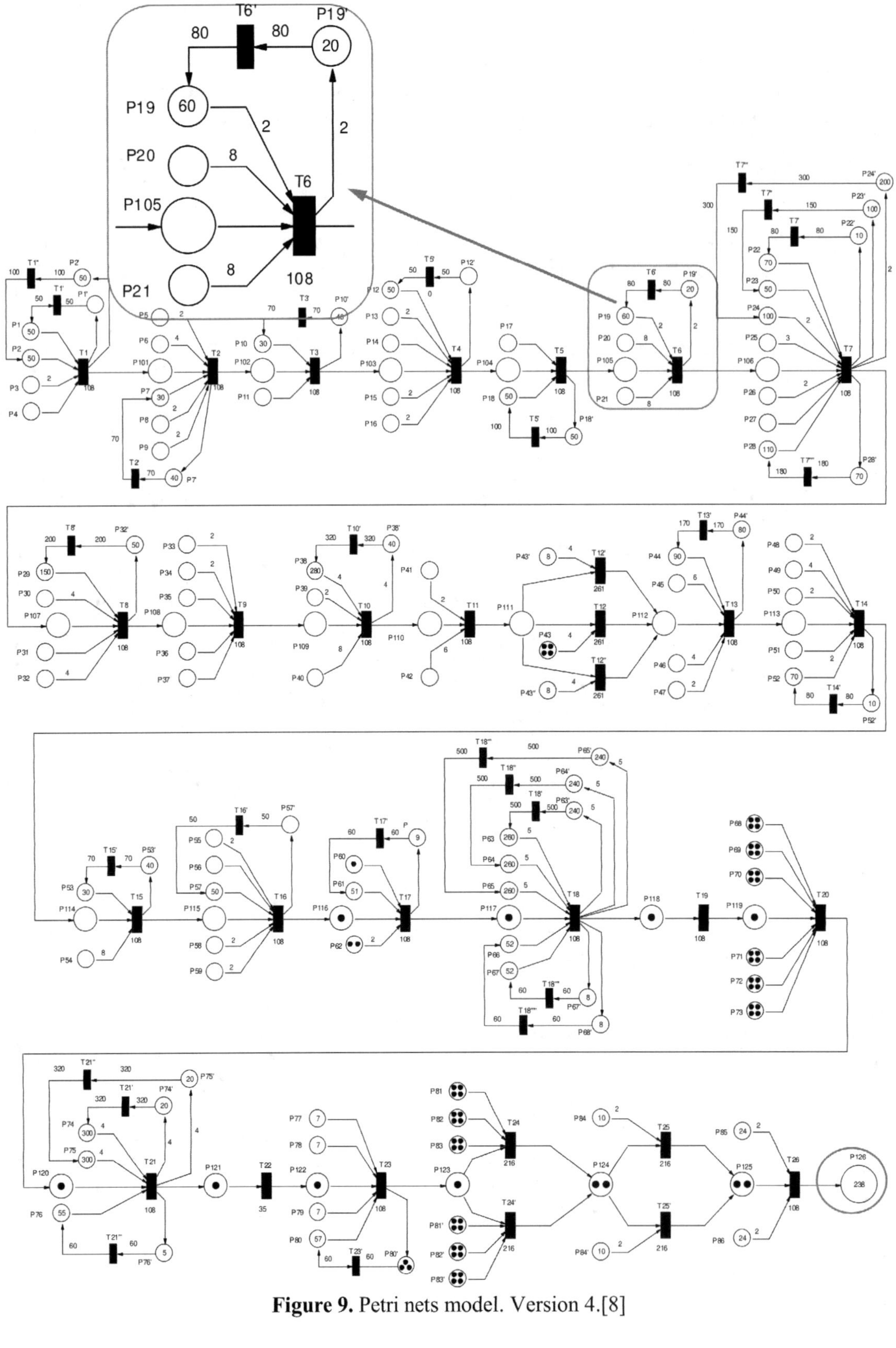

Figure 9. Petri nets model. Version 4.[8]

2.3. Modeling and simulation of a flexible packaging system for detergents [9]

2.3.1 Description of the manufacturing system

The automatic detergent packaging system consists of six manufacturing lines. Figure 10 shows a schematic diagram of a manufacturing line structure. For each line, the following subsystems can be highlighted: I – the bag filling subsystem; II- the subsystem for grouping bags in boxes; III – the box grouping in a row subsystem; IV – palletizing subsystem; V – full pallet evacuation subsystem; VI – supply subsystem with empty euro pallets

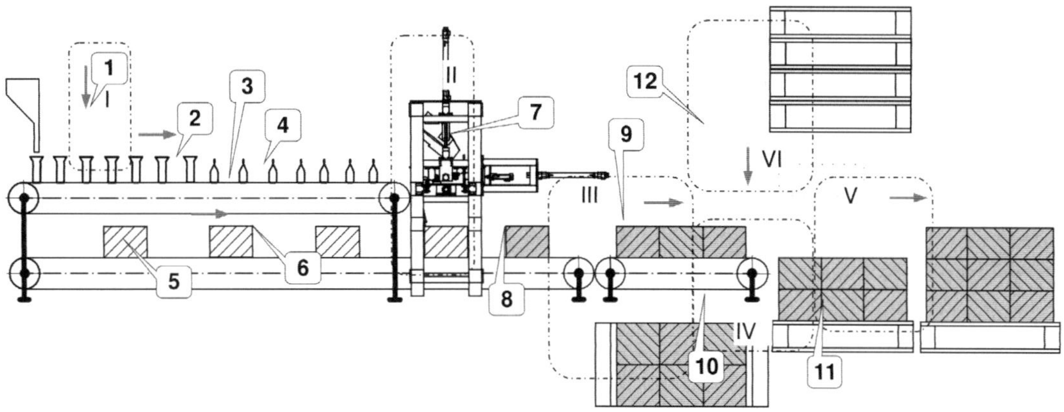

Fig. 10. Layout of a packing line and its subsystems. [9]

Figure 10 also describes the operation of the detergent packaging line. The detergent is dosed from the hopper 1 into the open bags 2. They are transferred through the transfer system 3 (conveyor belt). During the transfer, the closing of the bags is also carried out (4-closed bag). Under the bag conveyor belt there is a transfer system of roller conveyor type (6) for transferring boxes (6). Through the case packer equipment 7, the filled bags are grouped in boxes. The filled boxes (8) are grouped in rows (9). When a row is complete, it is transferred to pallet 10. The pallet is evacuated from the system. After evacuation of the full pallet, an empty pallet is inserted in the pallet supply subsystem (12).

Figure 11 shows the case packer system made at [9]. Also, a part of a packing line is shown in figure 12 [9].

Figure 11. Case packer system [9]

Figure 12. Packaging line [9]

2.3.2. Modeling and simulation of the automatic detergent packaging system

The models associated with the system are developed in a hierarchical structure: the model of the system as a whole, the model of a line, the model of some modules in the composition of a line.

IOP Conf. Series: Materials Science and Engineering 1169 (2021) 012005 doi:10.1088/1757-899X/1169/1/012005

In these conditions, the operation mode is checked step by step and the necessary corrections can be made until the optimal solution is found.

The model with Petri nets consists, in fact, of six submodels (figure 12), these correspond to the modules in the line structure: I- the submodel of filling the bags; II- the submodel for grouping bags into boxes; III- the submodel for grouping boxes in a row; IV - the palletizing submodel; V - the submodel for the evacuation of full euro pallets; VI - the sub-model of supply with empty euro pallets.

Two model versions were built with Petri nets:

Version I. Each line packs detergents in bags of different sizes:

- Line I: 1 kg/bag; 45 bags/min; 12 bags/box;
- Line 2: 2 kg/bag; 45 bags/min; 6 bags/box;
- Line 3: 2.2 kg/bag; 45 bags/min; 6 bags/box;
- Line 4: 3 kg/bag; 40 bags/min; 4 bags/box;
- Line 5: 3.3 kg/bag; 40 bags/min; 4 bags/box;
- Line 6: 4 kg/bag; 30 bags/min; 3 bags/box.

Version II. Detergents are packed on all lines in this way: 3.3 kg/bag; 40 bags/min; 4 bags/box.
Figure 13 shows the version I- on each line detergents are packed in bags of different sizes. The model is a Hybrid Petri Nets: it has a continuous component and a discrete component.

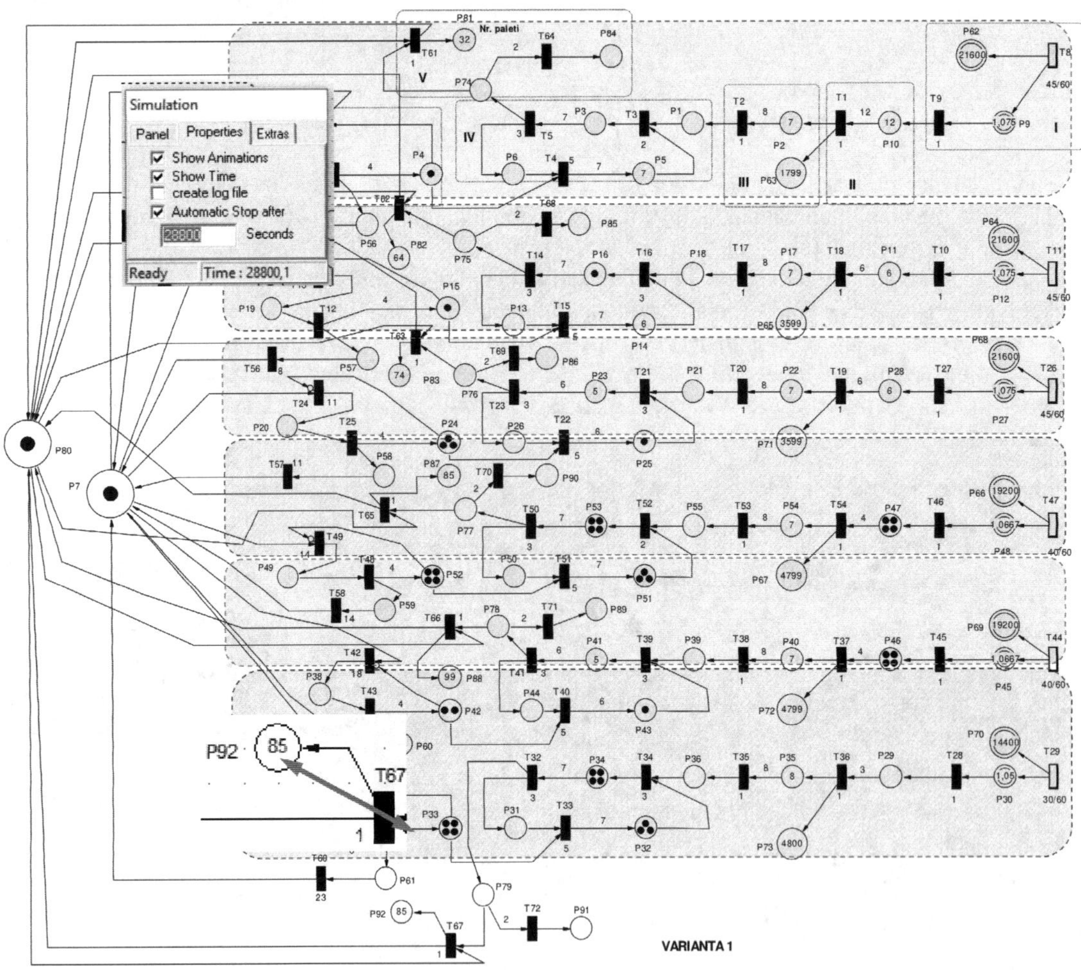

Figure 13 Detergents packing system modeled with Petri nets [9]

The model is designed so that potential conflicts are avoided. By simulation, it is possible to determine the number of euro pallets that can be made during an 8-hour work shift (figure 13). The simulation result of the system operation during the 8 hours is presented in table 1.

Table 1. Summary of information regarding the automatic detergent packaging system. Estimated productivity. Simulation results. [9]

LINE	**SIZE [kg/bag]**	**Frequency of bags [bags/min]**	**Nbr. of bags box [bags/box]**	**The frequency of the boxes [boxes/min]**	**Nbr. of boxes in a row [boxes/row]**	**Nbr. of rows on europallets [rows/europallets]**	**Simulation results 8 hours**	
							boxess	**europallets**
1	1	45	12	3.75	8	7	1799	32
2	2	45	6	7.5	8	7	3599	64
3	2.2	45	6	7.5	8	6	4799	74
4	3	40	4	10	8	7	4799	85
5	3.3	40	4	10	8	6	4799	99
6	4	30	3	10	8	7	4800	85

2.4.Evaluation of the performance of robotic welding cells by modeling with timed Petri nets [10]

Welding is a very common technological process used in the automotive industry, especially in assembling car bodies. Modern assembly lines are characterized by the fact that they have complex devices in their component and in most cases are robotic.

The welding cell is intended for assembling the structure Central lower floor subassembly. This subassembly has three components: sheet 1, welded subassembly and sheet 2 (Figure 14)

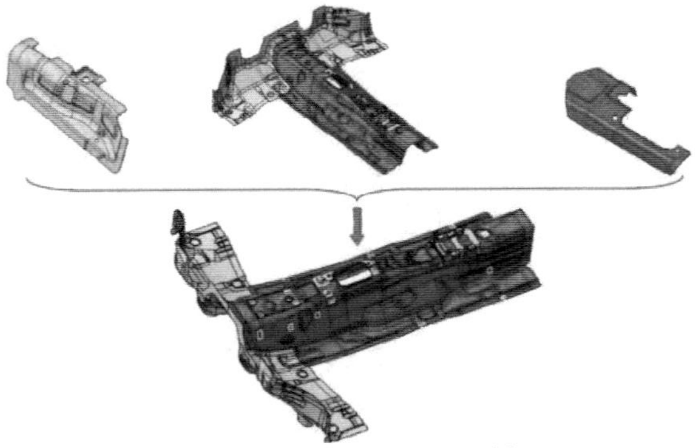

Figure 14. *Central lower floor subassembly* components [10]

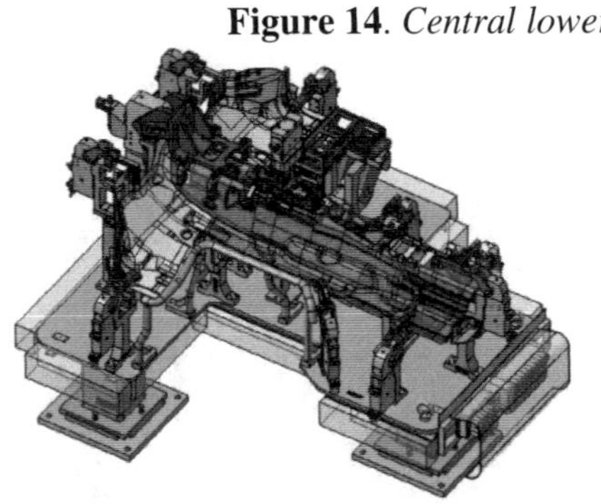

Figure 15. The components to be welded are positioned and fixed [10]

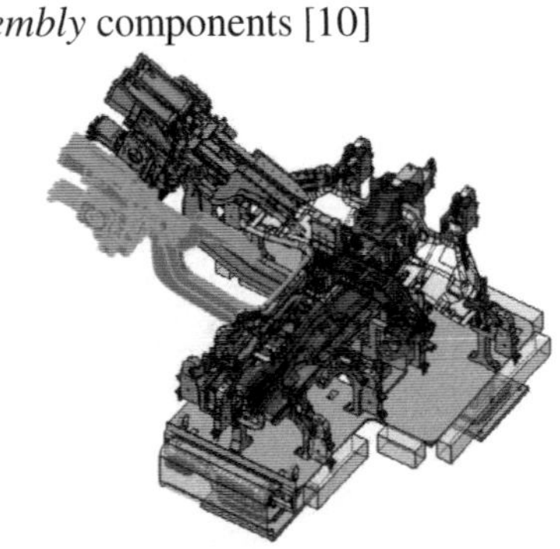

Figure 16. Type C gun welding [10]

The automated welding cell is of human-robot type and works sequentially, ie first, the operator loads the sheet elements to be welded after which the robot performs the welding operation. The body elements to be assembled are oriented by means of the pins and fixed with the help of pneumatic clamps in the geometric welding station. Figure 15 shows the components to be welded positioned and fixed in the welding station. Welding is performed with the help of welding pliers. All welding points defined on the road elements can be accessed by the welding pliers, so there will be no need for respot points. Welding a point takes 4 sec. Welding is done successively, only one welding point at each step until all welding points are executed. The robot starts spot welding with type X guns, and then with type C guns (figure 16). Figure 17 shows the station and the numbering of the component units. A complete cell operating cycle is composed of a sequence of sequences.

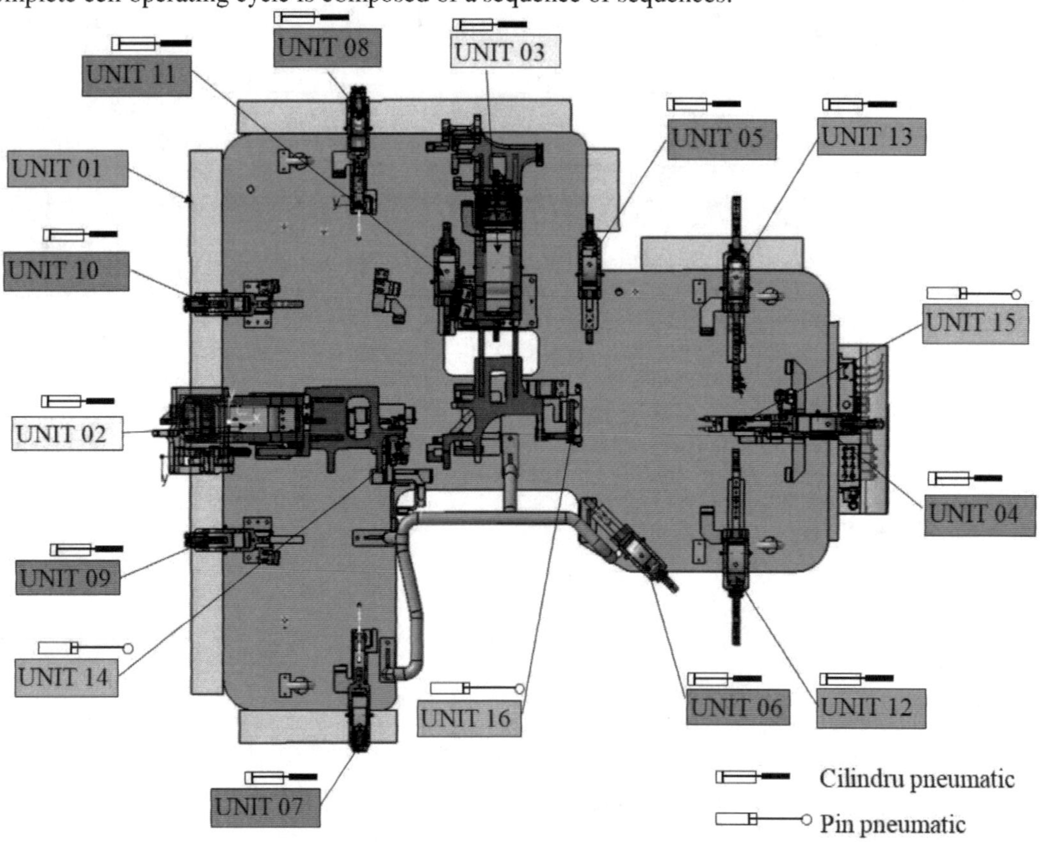

Figure 17. The units within the station [10]

To evaluate the performance of the welding cell and to identify possible blockages, a model with Petri nets was developed. Three models with Petri nets were made. Table 2 shows the sequences performed by the cell components. These correspond to the transitions from the model with Petri nets. The states of the cell components are modeled by positions (places).

Table 2. Petri net positions and transitions [10]

No.	Symbol	Type	Description	Characteristics		
				Version 1	Version 2	Version 3
1	P1	Position	The operator is free	m_0(P1)=1	m_0(P1)=1	m_0(P1)=1
2	P2	Position	Stock of type 1 sheets	m_0(P2)=150	m_0(P2)=160	m_0(P2)=20
3	T1	Transition	Loading the sheet 1	d_1=10,5 sec	d_1=10,5 sec	d_1=10,5 sec

No.	Symbol	Type	Description	Characteristics		
				Version 1	Version 2	Version 3
4	P3	Position	Table 1 is on the station	m_0(P4)=0	m_0(P4)=0	m_0(P4)=0
...	...	...	...	...	...	...
19	P11	Position	The robot is in position	m_0(P11)=0	m_0(P11)=0	m_0(P11)=0
20	T7	Transition	Welding the 12 (8) points with X-type welding pliers	d_7=48 sec	d_7=32 sec	d_7=32 sec
21	P12	Position	All 8 points are welded	m_0(P12)=0	m_0(P12)=0	m_0(P12)=0
22	T8	Transition	The robot rotates 180 °	d_8=2,7 sec	d_8=2,7 sec	d_8=2,7 sec
23	P13	Position	The robot is in position	m_0(P13)=0	m_0(P13)=0	m_0(P13)=0
24	T9	Transition	Replacing the welding pliers X with the tee type C	d_9=15 sec	d_9=15 sec	d_9=15 sec
25	P14	Position	Type C welding pliers are on the robot	m_0(P14)=0	m_0(P14)=0	m_0(P14)=0
26	T10	Transition	The robot rotates 180 °	d_{10}=2,7 sec	d_{10}=2,7 sec	d_{10}=2,7 sec
27	P15	Position	The robot is in position	m_0(P15)=0	m_0(P15)=0	m_0(P15)=0
28	T11	Transition	Welding the 10 (6) points with type C welding pliers	d_{11}40 sec	d_{11}=24 sec	d_{11}=24 sec
...	...	...	...	...	...	...
52	P30	Position	Consumption counter table 2	-	-	m_0(P30)=0
53	T22	Transition	Supply of sheet metal stock 2	-	-	-
54	P31	Position	Sheet stock reload number 2	-	-	m_0(P31)=0
55	P32	Position	Confirmation of completion of sheet metal assembly	-	-	m_0(P32)=0
56	T23	Transition	The waiting time of the station until the operator is free	-	-	-

The models were gradually developed. Thus, in the first variant the number of welding points was 22, having a cell cycle duration of 215 sec, and after simulating the operation of the station for 8 hours, it was found that 133 pieces were welded to the lower central floor subassemblies.

In the second version the number of welding points was reduced from 22 to 14, and the cycle time decreased to 183 sec, is by 32 sec. After simulating the operation of the 8h station, it was found that 157 pieces of Central lower floor subassemblies were welded.

Version 3 (figure 18) included in the model with timed Petri nets and stock refueling sequences with components that are welded. Thus, it is considered that the storage space of the sheets, respectively of the subassemblies, which are welded, is not enough to store the necessary for 8 hours. Consequently, the stocks, modeled with positions P2 (Table 1), P4 (welded semi-finished product) and P6 (Table 2), will be replenished periodically during the 8 hours. Items P26, P28 and P30 count the number of Tables 1, Welded Subassemblies, Tables 2 consumed from stock. When the number of tokens in these positions (the number of components used) is equal to the capacity of the stocks, their replenishment is done. Stock replenishment is modeled, through transitions T20, T21 and T22The model becoming a generalized timed Petri nets.

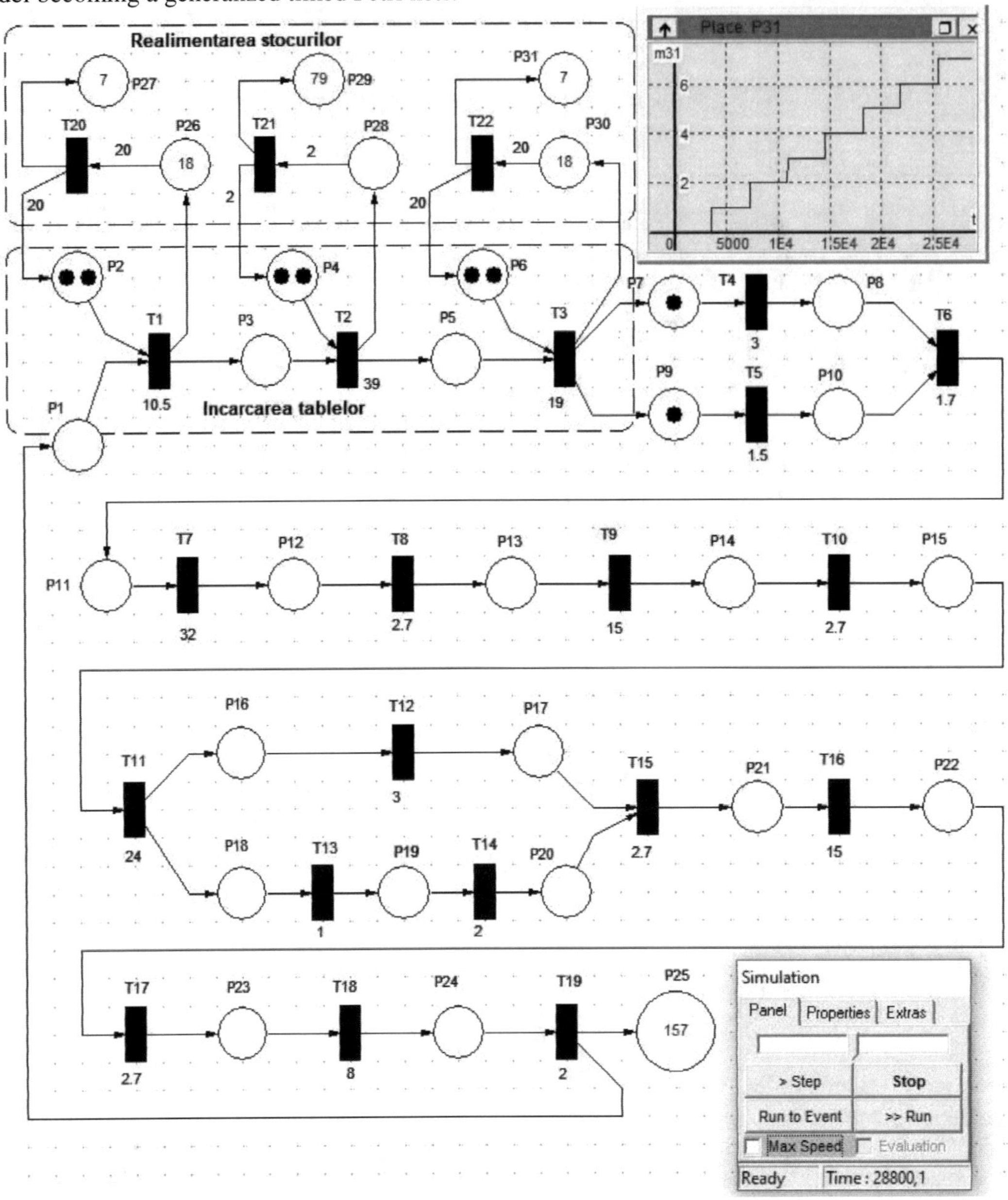

Figure 18. Model with timed Petri nets. Variant 3 [10]

3. The transition from timed Petri nets to hierarchical colored Petri nets

3.1. Description of the automated system for preparation and analysis of metallographic samples

The automated system (flexible cell) for the preparation and analysis of metallographic samples has been designed taking into account the technical and material resources of the Mechatronics and Science of Materials laboratories (University of Oradea) [11]. The layout of the designed cell is shown in figure 19. Two subsystems can be seen:

- subsystem $S1$, which prepares the metallographic samples;
- subsystem $S2$, which performs the metallographic analysis.

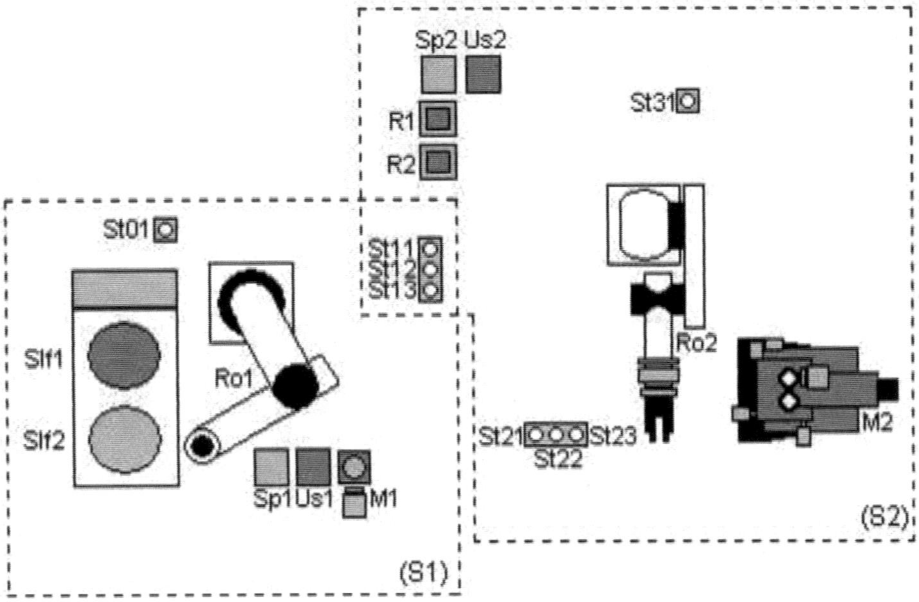

Figure 19. The layout of the automated system (flexible cell) for the preparation and analysis of metallographic samples. [11]

The following components of the automated system can be seen in figure 19:

- samples polishing posts: *Slf*1, *Slf*2;
- storage devices:
    - input storage device: *St*01;
    - buffer storage devices between subsystem *S*1 and subsystem *S*2: *St*11, *St*12, *St*13;
    - storage devices for the samples that have been prepared for analysis with microscope *M*1: *St*21, *St*22, *St*23;
    - output storage device: *St*31.
- industrial robots: *Ro*1, *Ro*2;
- samples washing posts: *Sp*1, *Sp*2;
- samples drying posts: *Us*1, *Us*2;
- reagent attack posts: *R*1, *R*2;
- surface assessment post (microscope): *M*1;
- metallographic structure assessment post (microscope): M2.

The two subsystems are served by robots in the following manner:

- posts *Slf*1, *Slf*2, *Sp*1, *Us*1, *M*1, *St*01 of subsystem *S*1 are served by robot *Ro*1;
- posts *R*1, *R*2, *Sp*2, *Us*2, *M*2, *St*21, *St*22, *St*23, *St*31 of subsystem *S*2 are served by robot *Ro*2.

3.2 The realization of the generalized timed Petri nets model [11]

A generalized timed Petri nets model was made in order to assess the performances of the automated system and to identify the probable bottlenecks. The model takes into account the analysis of two types of metallographic samples, A and B. The sequences performed by the robots are modelled by transitions, to which timings were attached, equal to the time intervals needed by the robots to perform these transitions. Other activities were modelled, such as coarse grinding, polishing, reagent attack, washing, microscope examination. The execution of one transition, equivalent to the execution of a real sequence, is possible if certain conditions are fulfilled. These conditions are modelled in the Petri net by means of positions. Some of the transitions and positions of the model are shown in table 3.

Table 3. Positions and transitions of Petri nets model. [11]

Nr.	Symbol	Type	Meaning	Characteristics
1.	P1	Pos.	*Ro*1 is free	m_0(P1)= 1
2.	P2	Pos.	*St*01 holds sample A	m_0(P2)= 1
3.	T1	Tr.	*Ro*1 grips sample A	d_1= 5 sec.
4.	P3	Pos.	*Ro*1 holds sample A in its grip	m_0(P3)= 0
5.	T2	Tr.	*Ro*1 transfers sample A to *Slf*1	d_2= 2 sec.
6.	P4	Pos.	*Ro*1 placed sample A to *Slf*1	m_0(P4)= 0
7.	T3	Tr.	Coarse grinding sequence (*Slf*1)	d_3= 120 sec.
8.	P5	Pos.	*Ro*1 takes sample A polished on *Slf*1	m_0(P5)= 0
9.	T4	Tr.	*Ro*1 transfers sample A to *Slf*2	d_4= 2 sec.
10.	P6	Pos.	*Ro*1 placed sample A to *Slf*2	m_0(P6)= 0
...	...	...	...	...
75.	P40	Pos.	Confirmation – *Ro*2 has sample B from *St*12	m_0(P40)= 0
76.	P41	Pos.	Confirmation – Ro2 has sample B from *St*13	m_0(P41)= 0
77.	T36	Tr.	*Ro*2 has one sample B (logical transition)	d_{36}= 0 sec.
78.	P42	Pos.	Confirmation – *Ro*2 has one sample B	m_0(P42)= 0
79.	T37	Tr.	*Ro*2 transfers sample A to *R*1 and processes it	d_{37}= 12 sec
80.	P43	Pos.	*Ro*2 has sample A processed by *R*1	m_0(P43)= 0
81.	T38	Tr.	*Ro*2 transfers sample A to *R*2 and processes it	d_{38}= 12 sec
82.	P44	Pos.	*Ro*2 has sample A processed by *R*2	m_0(P44)= 0
83.	T39	Tr.	*Ro*2 transfers sample A to *Sp*2 and washes it	d_{39}= 12 sec
84.	P45	Pos.	*Ro*2 has sample A washed	m_0(P45)= 0
85.	T40	Tr.	*Ro*2 transfers sample A to *Us*2 and dries it	d_{40}= 12 sec
...	...	...	...	...
153.	P79	Pos.	Sample B (*St*23) was examined at *M*2	m_0(P79)= 0
154.	T75	Tr.	*Ro*2 grips sample B (*St*21) from *M*2	d_{80}= 5 sec
155.	T76	Tr.	*Ro*2 grips sample B (*St*22) from *M*2	d_{81}= 5 sec
156.	T77	Tr.	*Ro*2 grips sample B (*St*23) from *M*2	d_{82}= 5 sec
157.	P80	Pos.	*Ro*2 has sample B (*St*21) examined	m_0(P80)= 0
158.	P81	Pos.	*Ro*2 has sample B (*St*22) examined	m_0(P81)= 0
159.	P82	Pos.	*Ro*2 has sample B (*St*23) examined	m_0(P82)= 0
160.	T78	Tr.	*Ro*2 evacuates sample B (*St*21)	d_{83}= 5 sec
161.	T79	Tr.	*Ro*2 evacuates sample B (*St*22)	d_{84}= 5 sec
162.	T80	Tr.	*Ro*2 evacuates sample B (*St*23)	d_{85}= 5 sec
163.	P83	Pos.	Storage device *St*31 – samples B	m_0(P83)= 0

In fact, the Petri nets model consists of two sub-models (figure 20): sub-model for samples A (colored blue) and sub-model for samples B (colored yellow). The resources of the system consist of robots, storage devices, polishing devices and reagent attack sections, which must be shared between the two sample types.

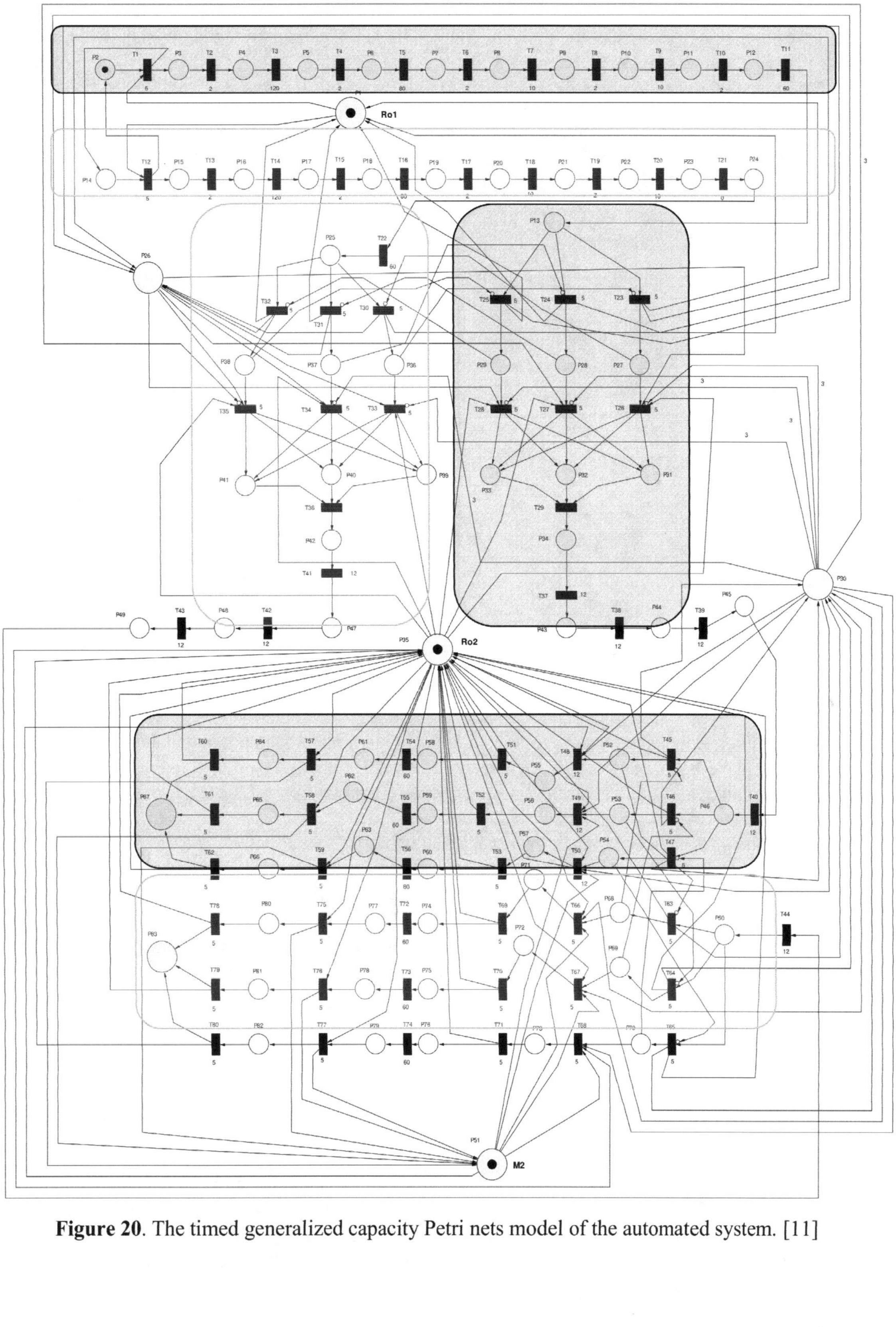

Figure 20. The timed generalized capacity Petri nets model of the automated system. [11]

In the first version of the automated system simulation it has been assumed that the two types of samples enter the system alternatively as soon as the initial storage device is free. A time interval of 8 hours (28800 seconds) is considered for simulation. The simulation shows that 47 samples A and 47 samples B have been processed and analysed.

In the next model version, we will assume that the samples are input at certain time intervals. Thus, samples A are input at 20 minutes (1200 seconds) each, and samples B are input at 15 minutes (900 seconds) each. The following modifications have been performed in order to simulate this:

- for sample A, the discrete transition T_A has been added, with 1200 second timing and position *Sample A* (figure 21);
- for sample B, the discrete transition T_B has been added, with 900 second timing and position *Sample B* (figure 22);

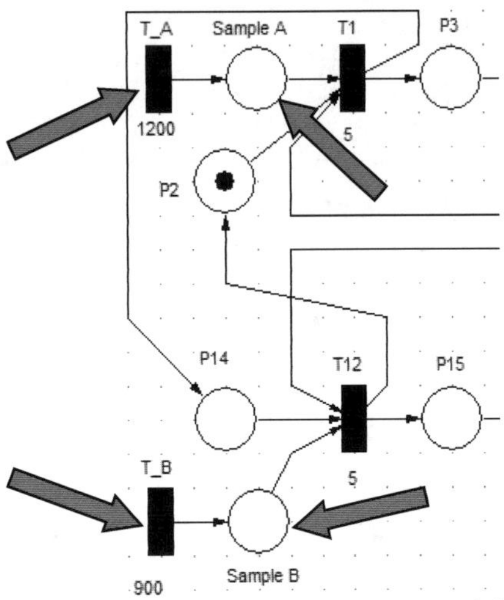

Figure 21. Modelling of the inputs of the system – version 2. [11]

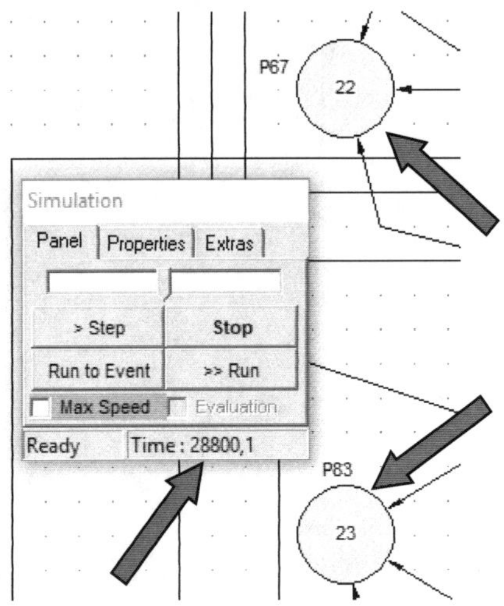

Figure 22. Simulation results – version 2. [11]

The simulation shows that 22 samples A and 23 samples B have been processed and analysed. Both versions have been designed in such manner that the two types of samples are processed and analysed alternatively.

3.3. Elaboration of Simple Colored Petri Net Model [12]

A simple colored Petri net model was made in order to assess the performance of the automated system and to identify any possible bottlenecks. Thus, the colours that can be attached to several system entities are defined.

We define the set of the colours that will describe the fact that, at a given time, the buffer storage devices St11, St12 and St13 are able to store one sample (*colset antiplace=with e*):

$$antiplace=\{e\} \tag{1}$$

Also, we define the set of colours that will be associated to the samples to be analyzed in the cell. Such colour will be named *sample* and will have value *A* for sample A and value *B* for sample B (*colset sample=with A|B*):

$$sample=\{A,B\} \tag{2}$$

The colour *robot* has been defined for the two robots, with value *r*1 for robot Ro1 and value *r*2 for robot Ro2 (*colset robot=with r1|r2*):

$$robot=\{r1, r2\} \tag{3}$$

The colour that is associated to the metallographic microscope is *m* (*colset microscope=with m*):

$$microscope=\{m\} \tag{4}$$

Within the program CPN Tools a variable is assigned to every arch. Thus, variable *r* may have value *r*1 or *r*2:

$$r\ \{r1, r2\} \tag{5}$$

Variable *i* may have value *A* or *B*:

$$i\ \{A, B\} \tag{6}$$

Some of the transitions and positions of the model are presented in table 4.

Table 4. Positions and transitions of the simple colored Petri net model. [12]

Nr. crt.	Symbol	Type	Meaning	Characteristics
1.	P1	Pos.	St01 stored one sample A or B	1\`A++1\`B
2.	T1	Tr.	Ro1 grips one sample A or B	
3.	P2	Pos.	Ro1 has one sample A or B in the grip	
4.	T2	Tr.	Ro1 transfers sample A or B to Sf1	
5.	P3	Pos.	Ro1 placed the sample A or B onto Sf1	
6.	T3	Tr.	Coarse grinding sequence (Slf1)	
7.	P4	Pos.	Ro1 has one sample A or B polished onto Slf1	
8.	T4	Tr.	Ro1 transfers samples A or B to Sf2	
9.	P5	Pos.	Ro1 positioned sample A or B to Sf2	
...	...	...	...	...
35.	T17	Tr.	Ro2 take sample A or B to St13	
36.	P20	Pos.	Confirmation- Ro2 has a sample A	
37.	T18	Tr.	Ro2 transfers sample A or B to R1 and treats it	
38.	P21	Pos.	Ro2 has sample A or B treated by R1	
39.	T19	Tr.	Ro2 transfers sample A or B la R2 and treats it	
40.	P22	Pos.	Ro2 has sample A or B treated by R2	
41.	T20	Tr.	Ro2 transfers sample A or B to Sp2 and washes it	
42.	P23	Pos.	Ro2 has sample A or B washed	
43.	T21	Tr.	Ro2 transfers sample A or B to Us2 and dries it	
44.	P24	Pos.	Ro2 has a dry sample A or B	
45.	P25	Pos.	The M2 microscope is free	1\`m
...	...	...	...	...
75.	T36	Tr.	Ro2 take the sample A or B (St21) to M2	
76.	P41	Pos.	Ro2 has one sample A sau B (St21) examined	
77.	P42	Pos.	Ro2 has one sample A sau B (St22) examined	
78.	P43	Pos.	Ro2 has one sample A sau B (St23) examined	
79.	T37	Tr.	Ro2 evacuates sample A or B (St21)	
80.	T38	Tr.	Ro2 evacuates sample A or B (St22)	
81.	T39	Tr.	Ro2 evacuates sample A or B (St23)	
82.	P44	Pos.	Storage device St31- samples A	
83.	P45	Pos.	Ro1, Ro2 are available	1\`r1++1\`r2

Figure 23 shows the simple colored Petri nets model and the results of the simulation after 200 steps.

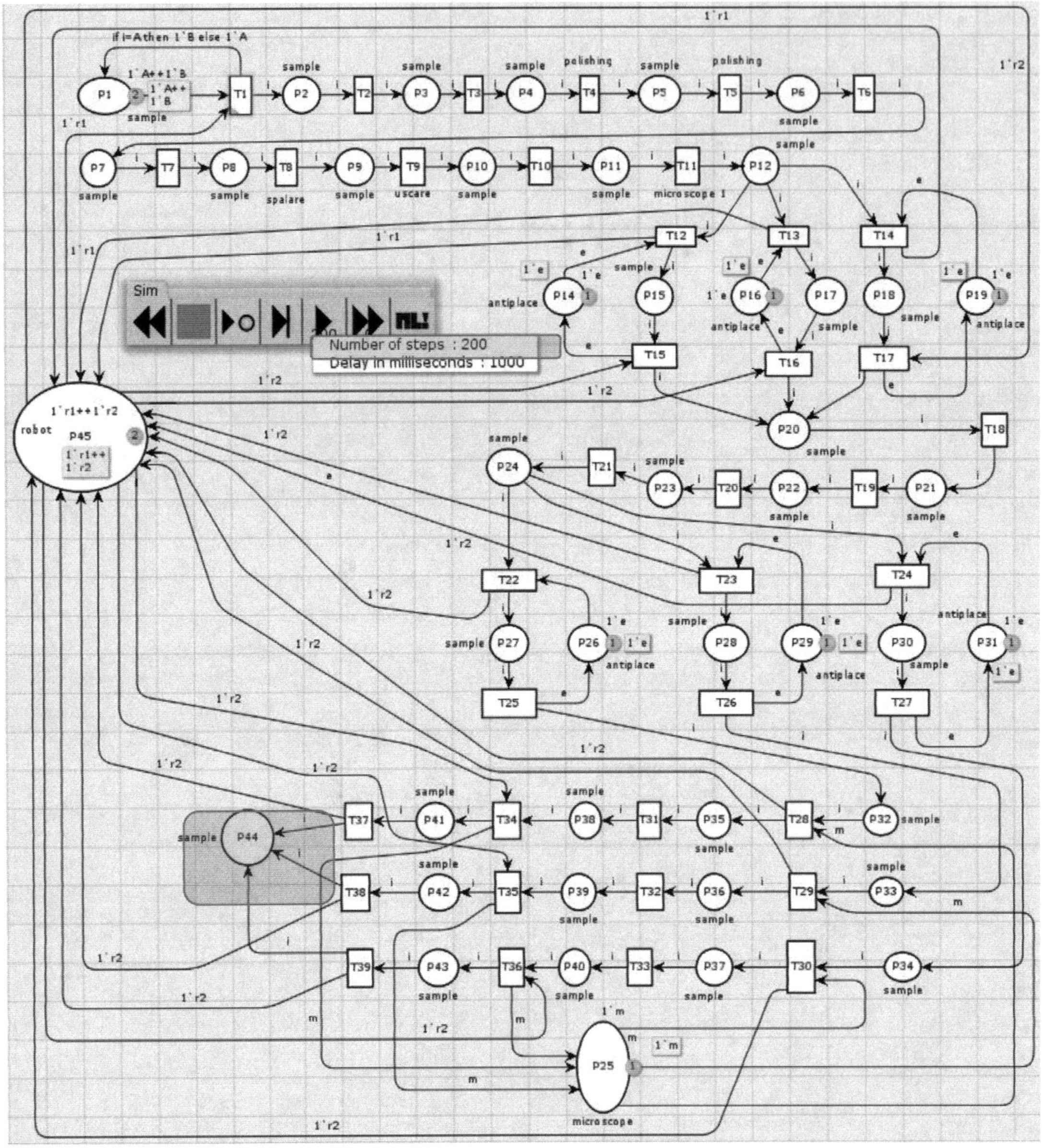

Figure 23. The simple colored Petri nets model. [12]

One can see that the model is viable and there are no bottlenecks. Also, the two types of samples are alternatively analyzed. The simulation showed that four A-type samples and four B-type samples were analyzes, as four colours A and four colours B are at position P44 (figure 23).

3.4. Realization of timed complex colored Petri nets model

The positions and transitions of the timed complex colored Petri nets are partially shown in table 5. Compared to the simple colored Petri nets model, a smaller number of elements (positions and transitions) can be noticed. The number of positions and transitions decreased from 83 to 51.

Table 5. The positions and transitions of the timed complex colored Petri net model. [12]

Nr. crt.	Symbol	Type	Meaning	Characteristics
1.	P1	Pos.	St01 stored a sample A or B	1\`A++1\`B
2.	T1	Tr.	Ro1 grips a sample A or B	5 sec
3.	P2	Pos.	Ro1 has the sample A sau B in the gripper	
4.	T2	Tr.	Ro1 transfers sample A or B to Sf1	
...	...	...	...	...
21.	P11	Pos.	Ro1 positioned with specimen A or B at M1	
22.	T11	Tr.	Structural analysis of specimen A or B on M1	
23.	T12	Tr.	Ro1 places a sample A or B to St11	5 sec
24.	P13	Pos.	A sample A or B is in St12	1\`1++1\`2++1\`3
25.	P14	Pos.	Available places left in St11	
26.	P13	Pos.	Sample A or B is in St12	
...	...	...	...	...
47.	T22	Tr.	Ro2 grips sample A or B (St21) from M2	5 sec
48.	P26	Pos.	Ro2 are sample A or B (St21) examined	
49.	T23	Tr.	Ro2 evacuates sample A or B (St21)	5 sec
50.	P27	Pos.	Storage St1- samples A	
51.	P28	Pos.	Ro1, Ro2 is available	1\`r1++1\`r2

In this version, we defined the set of colours that will describe the fact that the buffer storage devices St11, St12 and St13 are able to store a single sample at a given time (*colset antiplace=int 1..3*):

$$antiplace=\{1,2,3\} \quad (7)$$

In order to highlight that at a link between the sample type and the buffer storage on which it is stored before entering the analysis subsystem, the set of complex colors is defined *storage* (*colset storage = product sample*antiplace*). This is the cartesian product of the colors *sample* (relation (1)) and *antiplace* (relation (7)), having the following values:

$$storage=\{(A, 1), (A, 2), (A, 3), (B, 1), (B, 2), (B, 3)\} \quad (8)$$

The variables r and i are defined by relations (5) and (6). In the version with complex colours one more variable e is defined which can have the values 1, 2 sau 3:

$$e \ \{1,2,3\} \quad (9)$$

Also, there is a complex variable defined directly in the model (e, i), which can have the following values:

$$(e, i) \ \{(A, 1), (A, 2), (A, 3), (B, 1),(B, 2), (B, 3)\} \quad (10)$$

This variable is associated to arches T13→P12 and P12→T14 (figure 24).

The introduction of time factor in the model has been made in two ways. Timings have been associated to the transitions that model sequences (operations) of the preparing and analysis processes for the metallographic samples. For example, a 120 seconds timing was associated to transition T3 which models the coarse grinding operation. This is enclosed in the model by expression *@+120* (figure 24).

IOP Conf. Series: Materials Science and Engineering 1169 (2021) 012005 doi:10.1088/1757-899X/1169/1/012005

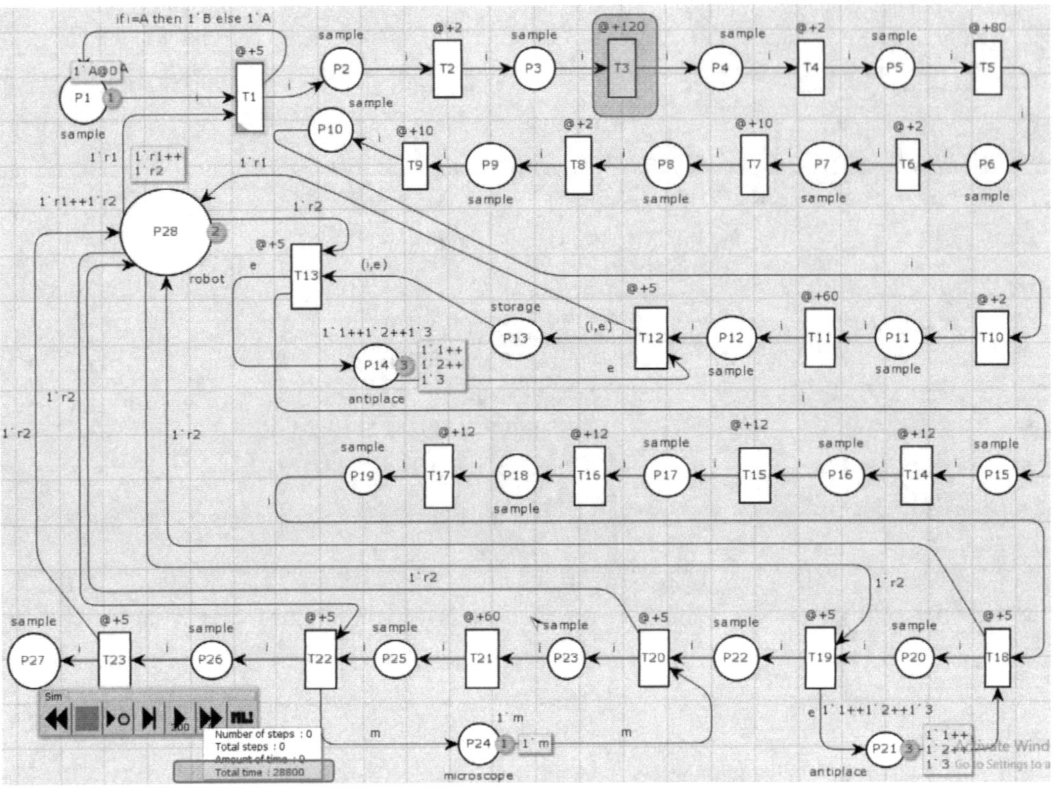

Figure 24. The timed complex colored Petri model. [12]

3.5. Defining hierarchical levels [13]

Two hierarchical levels of the model with colored Petri nets are defined:

- Level 1: is the model of the entire flexible system for preparation and analysis of metallographic samples;
- Level 2 contains three sub-models:
  - the sub-model of the system for the preparation of metallographic samples;
  - the sub-model of the system for reagent attack of metallographic samples;
  - the sub-model of the system for the analysis of metallographic samples.

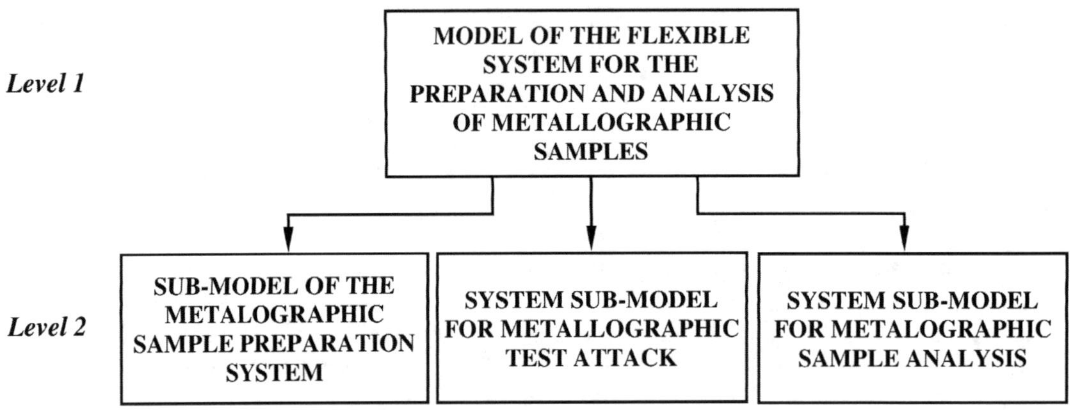

Figure 25. Hierarchical levels of the model with colored Petri nets. [13]

IOP Conf. Series: Materials Science and Engineering 1169 (2021) 012005 doi:10.1088/1757-899X/1169/1/012005

This hierarchical structure corresponds to both the model with colored Petri nets with simple colors and the model with complex colors.
Figure 26 shows the main model (*Level 1*) called *System*. Level 2 sub-models are highlighted: Sample *Preparation, Sample Attack and Sample Analysis*. The model and sub-models are displayed on independent pages (in windows), as follows:

- the main model is displayed on the *System* page;
- the sub-model of the system for the preparation of metallographic samples is displayed on the *Preparation* page;
- the sub-model of the system for reagent attack of metallographic samples is displayed on the *Attack* page;
- the sub-model system for the analysis of metallographic samples is displayed on the *Analysis* page.

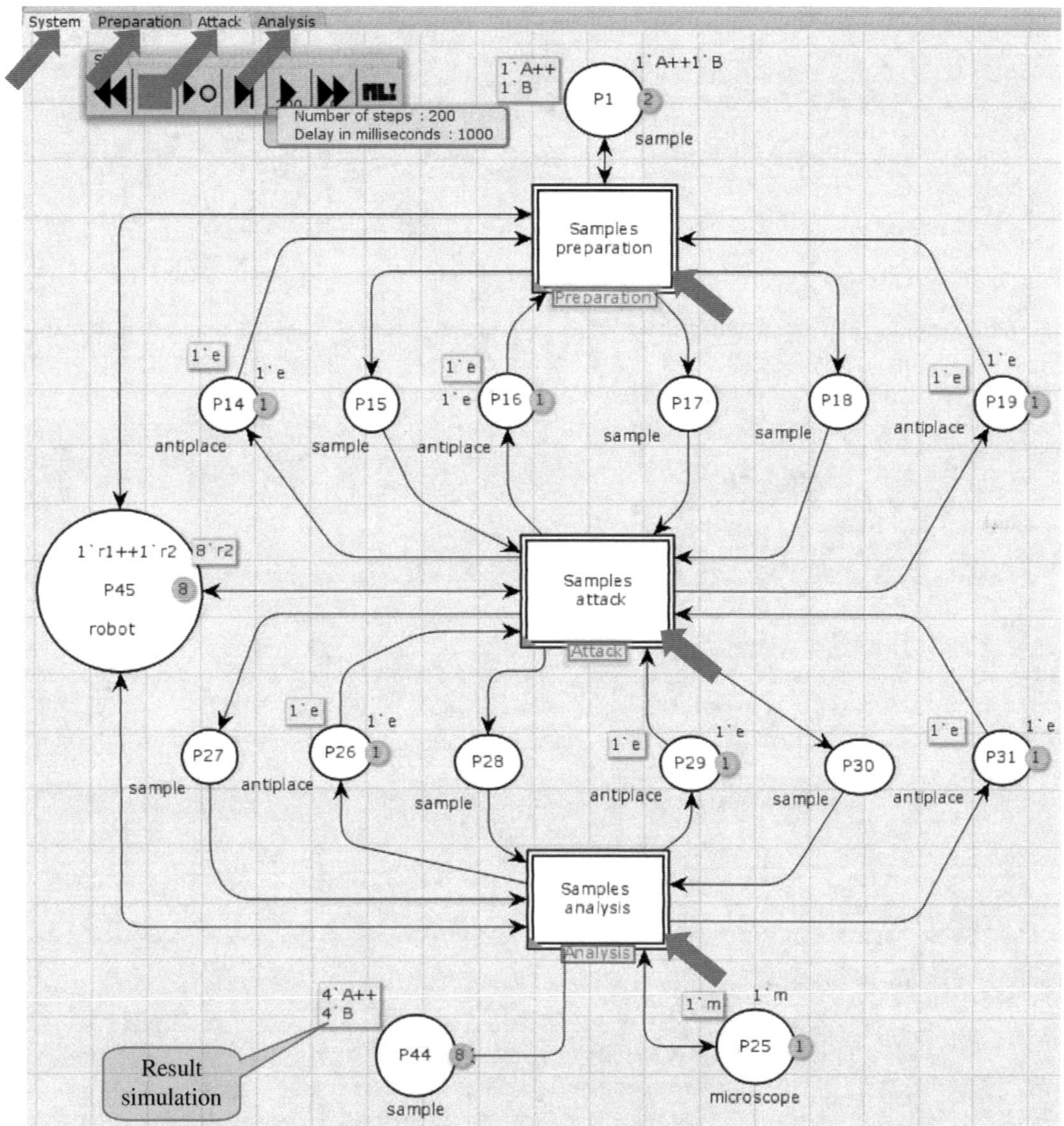

Figure 26. The main model with colored Petri nets with simple colors. [13]

Figure 27 shows the sub-model of the system for preparing metallographic samples. The positions through which the sub-model connects to the main model are highlighted. These positions can be of three types: input positions (P14, P16, P19), output positions (P15, P17, P18) and input / output positions (P1, P45).

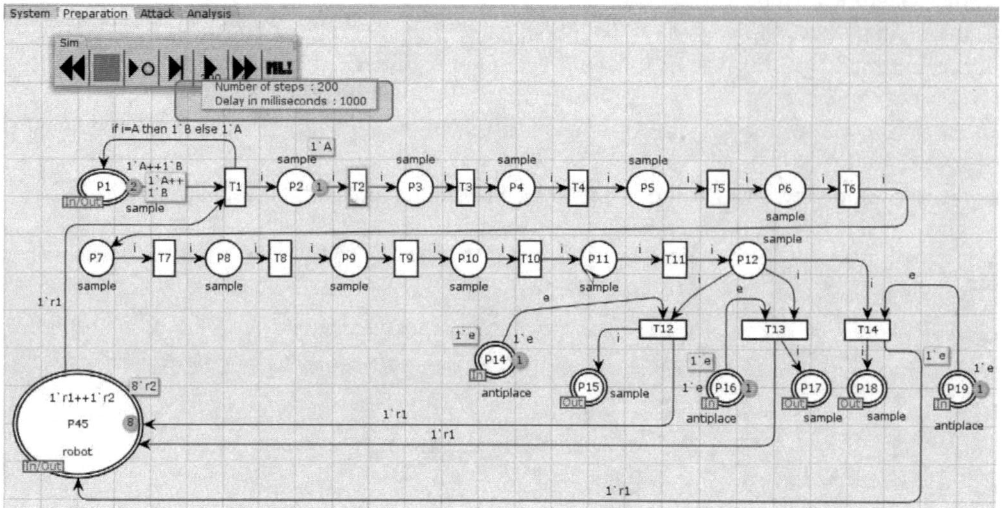

Figure 27. Sub-model of the system for the preparation of metallographic samples. [13]

Figure 28 shows the sub-model of the system for reagent attack of metallographic samples. In this sub-model, the input positions (P15, P17, P18, P26, P29, P31), the output positions (P14, P16, P19, P27, P29, P30) and the input / output position (P45) also appear.

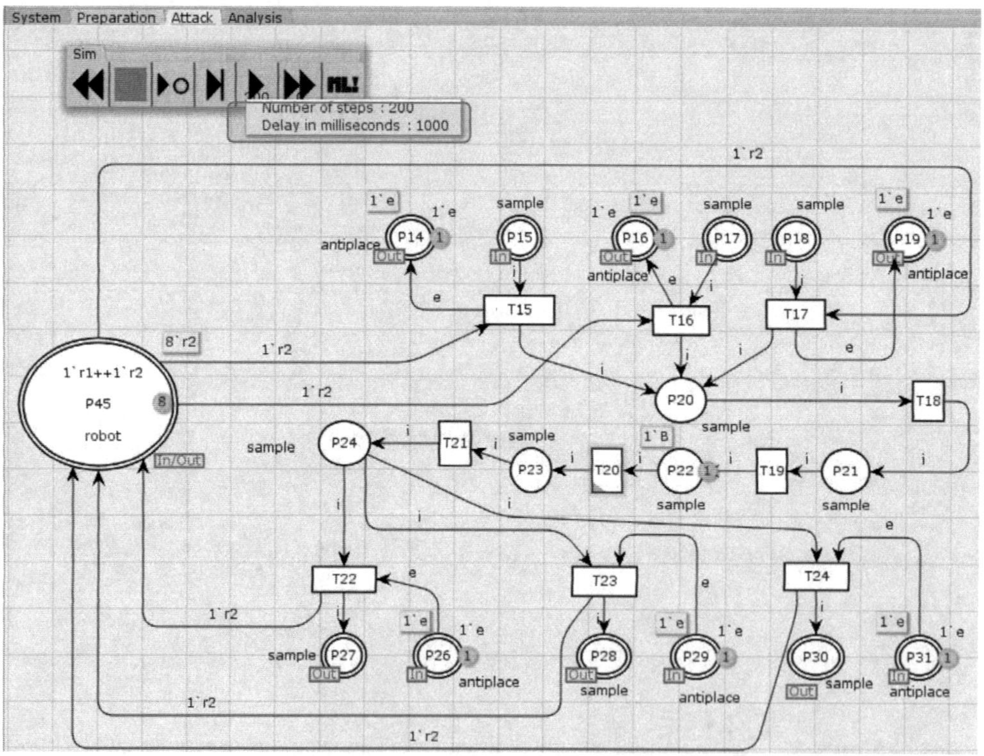

Figure 28. Sub-model of the system for the attack with reagents of metallographic samples. [13]

The hierarchical model with timed colored Petri nets with complex colors has the structure presented in section 3. The main model is presented in figure 29.

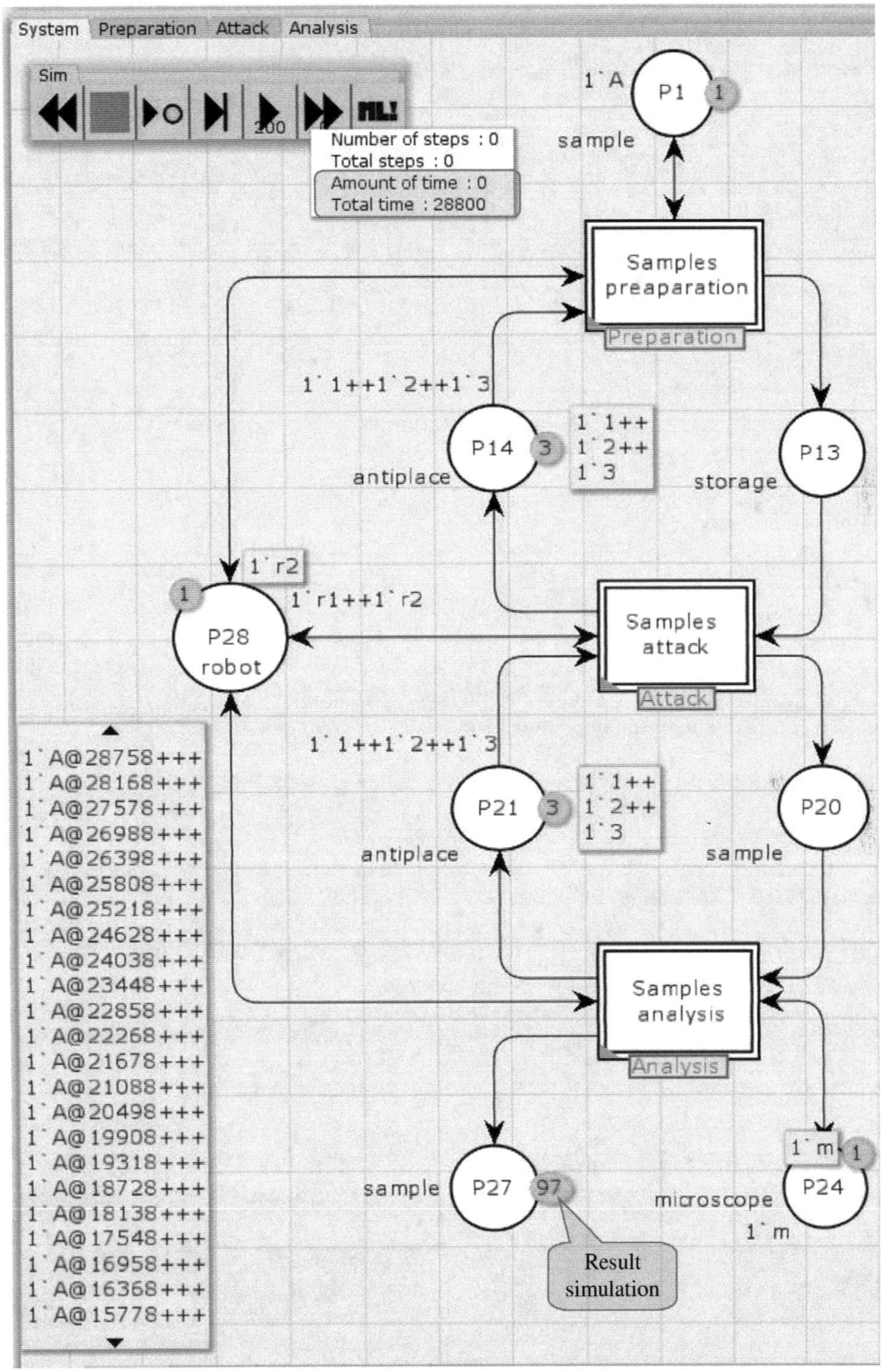

Figure 29. The main model with timed colored Petri nets with complex colors. [13]

In the case of the hierarchical model with colored Petri nets timed with complex colors, as in the case of the model with simple colors, the interconnection between the main model and the sub-models is made through input positions, output positions and input / output positions.
Figure 29 shows the sub-model for the analysis of metallographic sample.

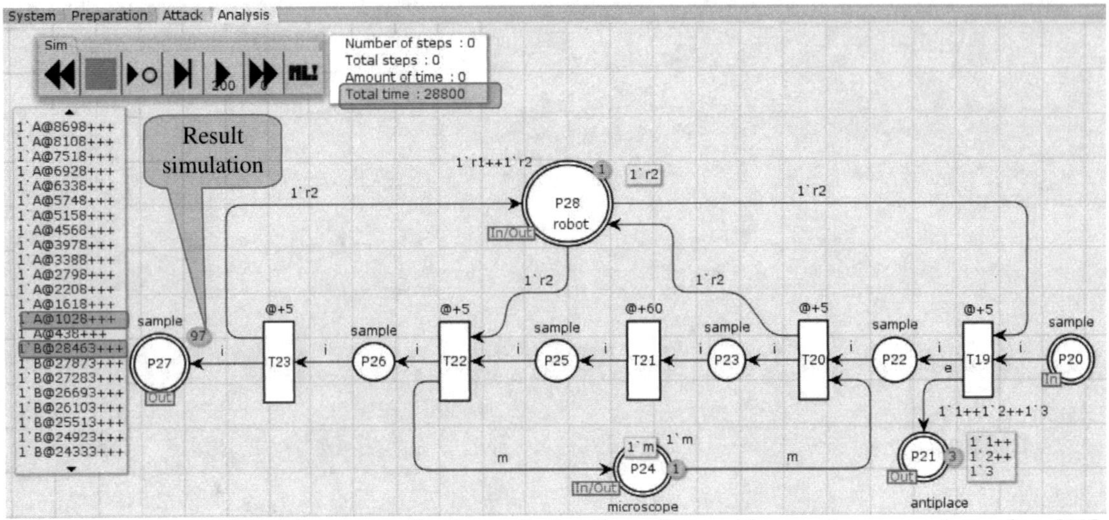

Figure 29. Sub-model of the system for the analysis of metallographic samples - complex colors [13]

For simulation, a time interval of 8 hours (28800 sec) is considered. The simulation duration is set in the *Simulation* window. During the simulation, 2236 steps were performed. The program also provides information on when the analysis of each sample is completed. For example, the last type B test was completed at time 28463, and the second type A test at the time.

It can also be noted that during the 8 hours a number of 97 samples were analyzed.

4. The Flexible Manufacturing System A007 model colored Petri Nets

4.1 The Flexible Manufacturing System A007. Components [14]

The flexible manufacturing system A007 of Industrial Engineering Department of Oradea University is shown in Figure 31. The cell is composed of: *Power station (1); Evacuation station (2); Lathe CNC CONCEPT TURN 55 (3); Control equipment (FANUC or SIMESS) of milling machine (4); Milling machine CNC CONCEPT MILL 55 (May); Control equipment (or FANUC SIMESS) the lathe (6); Mitsubishi RV-2AJ robot (7); Handle the moving robot (8*

Flexible manufacturing system of Industrial Engineering Department Laboratory (SFF A007) is intended for processing cylindrical parts witch have surfaces that are obtained by milling operations (Figure 30).

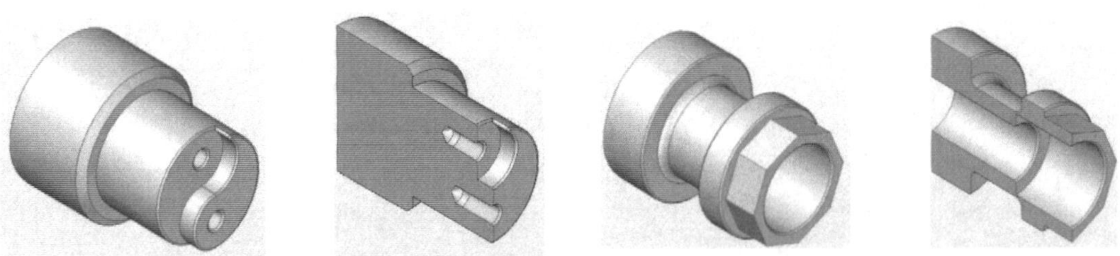

a) b)

Figure 30. Types of parts that can be processed in the flexible manufacturing system [14]

Figure 31 *The Flexible Manufacturing System A007*

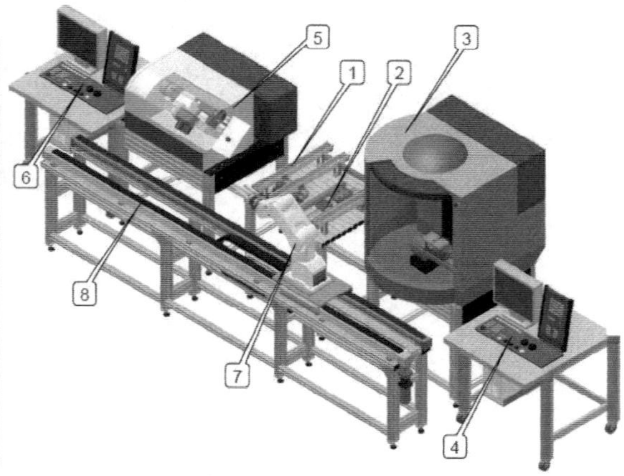

Figure 32. *The Flexible Manufacturing System A007. Components*

4.2. Model 1

A version of route that the piece can traverse the system is: the piece is positioned by the installation whip in place where it will be processed by the robot, this brings the piece in input position if the CNC program is load.

The piece is clamped in the lathe to be processed by the program. After processing, the robot take the piece from the lathe (CNC CONCEPT TURN 55) and brings it to the milling machine (*CNC* CONCEPT MILL 55), with the condition that the milling machine is not 0. The piece is attached to the device and processed by the program. After processing, the robot take a piece moved it to EI (exhaust installation finished parts). The Conveyor transfer finished parts in the storage.

Model 1 is built with Petri nets colored with complex colors. The complex color used is called complex and has three components: *piece*- P1 (means the generic part), *history* (means the place where the piece is at a given time) and *position* (position of the part in the manufacturing load, in batch);

$$Complex= product\ piece*history*position \qquad (11)$$

Table 6 shows the values that the history component can take and their significance. These values are also important from the perspective that the robot is the component that connects the component subsystems of the manufacturing system and the part. Thus, it is necessary to know the previous state of the part in order to know where the part should be transferred by the robot.

Table 6. History component values [14]

History value	Signification
0	The part is in the Filing Station
1	The part was taken from the filling station, to be loaded on the CNC lathe CONCEPT TURN 55
2	The part was unloaded from the CNC lathe CONCEPT TURN 55, and will be loaded on the CNC milling machine CONCEPT MILL 55
3	The part was unloaded from the CNC milling machine CONCEPT MILL 55 and is deposited in the evacuation station
4	The piece is in the batch of the 12 pieces that must be evacuated from the system

The position component takes values from 1 to 12 and defines the position of the part in the batch launched in production.

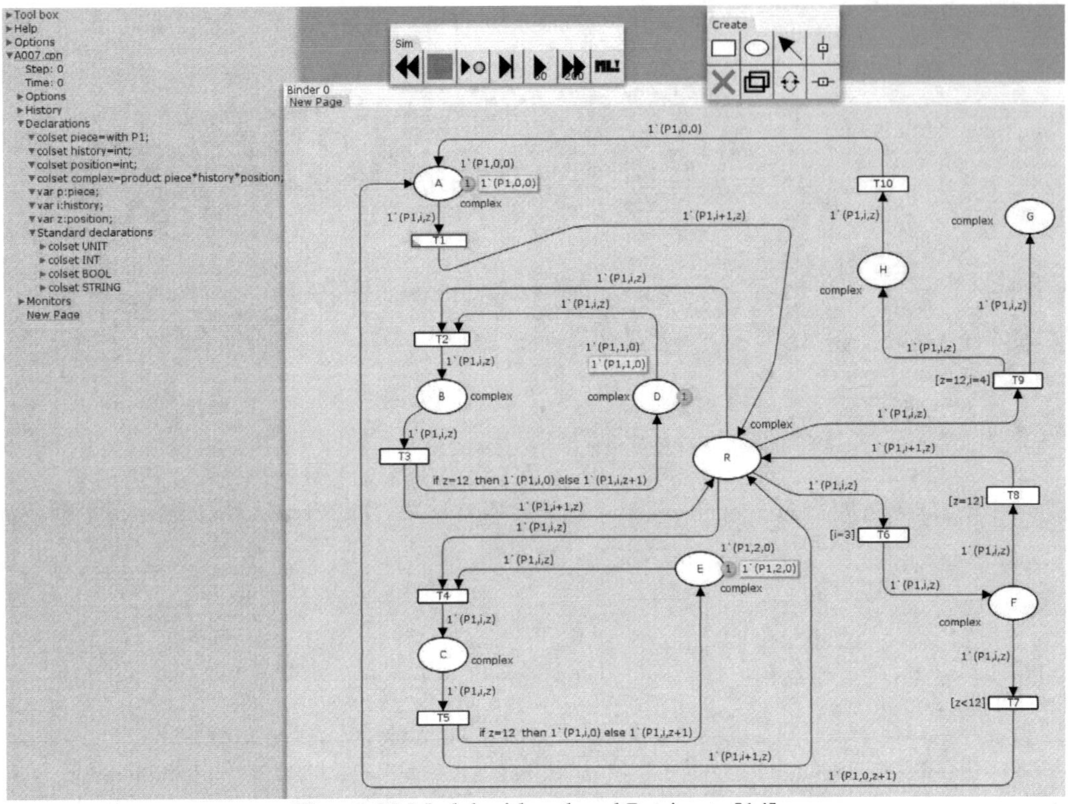

Figure 33 Model with colored Petri nets [14]

4.3. Model 2 [15]

This model was designed for the case in which the systems process the parts of figure 34:

1) two cylinder parts (figure 34a and figure 34b), these will be processed on CONCEPT TURN 55 CNC lathe;
2) two prismatic parts (figure 34c and figure 34d), these will be processed on CNC milling machine CONCEPT MILL 55.

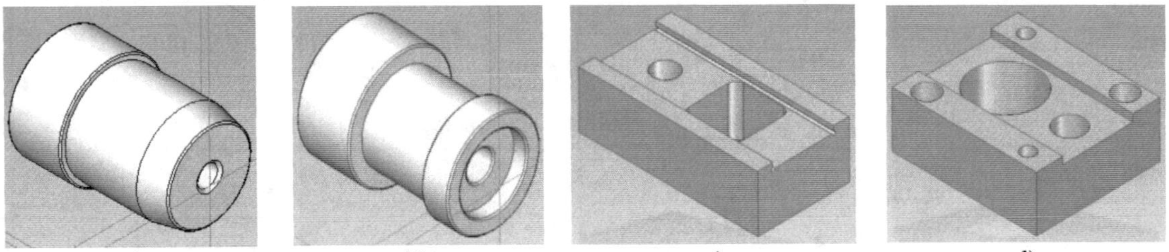
a) b) c) d)

Figure 34 The parts that will be processed on the fabrication cell [15]

Each of the four parts, that will be included in the manufacturing process has a timing, if we consider a referral moment t0= 0. Therefore: Part 1 (first cylinder part – figure 34a) is the part that comes first in the cell; Part 2 (second cylinder part- figure 34 b) is the fourth part that comes into the cell; part 3 (figure 34 c) is the second and part 4 (figure 34 d) is the third part that comes into the cell.

Another important parameter in the evaluation of a manufacturing system is time processing. The processing time of part 1 on CNC lathe is 7 minutes, the processing time of part 2 is 9 minutes, the processing time is 18 minutes for part 3 and part 4 is processed in 21 minutes.

Timing will be developed from three perspectives:

1) Color timing (marks)
2) Arcs timing.

The simulation time associated to colors are used to highlights the arrival of parts in the system. Thus, we define the simulation time (P1, M1) @ 0; (P3, M2) @ 400; (P4, M2) @ 1450 and (P2, M1) @ 2000 (figure 35).

The timings associated to transitions are used to define the processing time. The transition which shapes the processing of the part is transition T2. To this transition was associated in fact four timings, one for each part. Actual value of the timing at a time, is the result of the instruction if: @ + (if (p, m) = (P1, M1) then 420 else if (p, m) = (P2, M1) then 540 else if (p, m) = (P3, M2) then 1080 else 1260. It has been considering the second to be unit measurement.

The timings associated to arcs was defined to show the manipulation times corresponding to Mitsubishi RV-2AJ robot. To the variable (function) which load the arcs: P7→T1; T2→P7; P7→T3 si T5→P7 it was associated the simulation time r @+8.

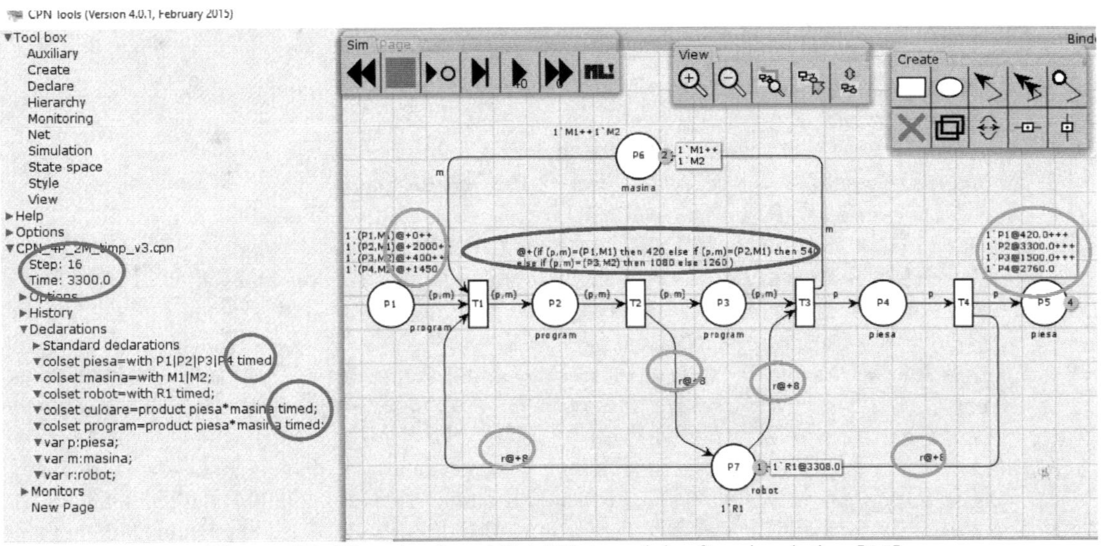

Figure 35. Timed colored Petri Net model, after simulation [15]

Considering that the information associated to model is built with colored Petri nets it is simulate the cell processing of the four parts. It can be see the end of processing times of each part: P1, 420 seconds; P2-3300 seconds; P3-1500 seconds; P4-3300 seconds. It also can evaluate the total functioning time of the robot: 3308 seconds (figure 35).

Simulation involves the execution of 16 steps; the number of executions transitions is 16.

5. Conclusions

Modeling and simulation with Petri nets are powerful tools for evaluating system performance. The paper presented how Petri net models have been developed for a wide variety of manufacturing systems: manufacturing lines in the textile industry, assembly lines in the consumer goods industry, automated lines in the detergent industry and robotic welding cells in the industry auto. Generalized hybrid timed Petri nets were used for these models.

The paper also presented how the same system can be used with several types of Petri nets. The modeled system was the robotic cell for the preparation and analysis of metallographic samples. The first model variant was a model with timed Petri nets. This is a very complex model with many elements (positions, transitions and arcs). The simplification of the model was achieved by switching to colored Petri nets with simple colors and then colored Petri nets with complex colors. The use of hierarchical colored Petri nets (the third modeling variant) allowed a structural organization of the model by considering the component subsystems of the cell. This type of model facilitates the monitoring of the cell's functioning through the submodes that make up the main model.

Colorful Petri nets with complex colors offer various model solutions for the same system. This was demonstrated by the development of the two models associated with the flexible manufacturing cell A007.

Future research will follow the development of models with colored fuzzy Petri nets and stochastic Petri nets.

6. References

[1] Petri C A 1962 *Kommunikation mit Automaten*, Dissertation, Schriften des IIM 2, Rheinisch-Westfälisches Institut für Instrumentelle Mathematik an der Universität Bonn (Bonn)
[2] Diaz M 2007 Petri Nets: Fundamental Models, Verification and Applications (London- ISTE Ltd)
[3] David, R and Alla, H 1992 *Du Grafcet aux réseaux de Petri,* (Paris- Ed. Hermes)
[4] Kurt J. 1992, *Colored Petri Nets:Basic Concepts Analysis Methods and Practical Use* (Springer Verlag)
[5] Peter J. Haas 2002, *Stochastic Petri Nets Modelling, Stability, Simulation* (New York –Springer Verlag)
[6] Păstrăvanu O Matcovschi M and Mahulea C 2002 *Aplicaţii ale reţelelor Petri in studierea sistemelor cu evenimente discrete* (Iasi-Editura Gh. Asachi)
[7] Blaga F Hule V Bodea R and Stanasel I 2013 *U. O. Fasc. Man. Tech. En.* **22** 5
[8] Blaga F Stanasel I Pop A Hule V and Karczis 2018 *IOP Publishing, IOP Conf. Series: Materials Science and Engineering* Oanta E and at Constanta Romania IOP Publishing 10
[9] Blaga F Pop A Stănăşel I Hule V and Indre C 2019 *MATEC Web Conf.Volume 299* N. Balc and D. Leordean Cluj Napoca Romania 6
[10] Blaga F Pop A Hule V Mocan C Popa C 2020 *Proceedings of the 7th Review of Management and Economic Engineering International Management Conference „Management Challenges With in Globalization"* Lungu F Cluj-Napoca România 173
[11] Blaga F S, Vesselenyi T, Ursu M P, Hule V and Stanasel I 2019 IOP *Conference Series: Materials Science and Engineering* Raval K R and at. Iasi Romania IOP Publishing 10
[12] Pop A Blaga F S Ursu M P Vesselenyi and Bungău C 2020 *IOP Conf. Ser.: Mater. Sci. Eng. 916 012088* Jun Ni at al Iasi Romania 166
[13] Blaga F S Pop A Vesselenyi T Hule V and Indre C I 2020 *IOP Conf. Ser.: Mater. Sci. Eng. 898 012017* Grebenisan G at al Oradea Romania **898** 12
[14] Pop A Blaga F and Stansel I 2010 *U. O. Fasc. Man. Tech. En.* **19** 3.137
[15] Blaga F Stanasel I Pop A Hule V and Buidos T 2014 *A. U. O. Fasc. Man. Tech. En.* **23** 299
[16] Pop A Blaga F S Ursu M P Bungau C and Hule V 2020 *IOP Conf. Ser.: Mater. Sci. Eng. 916 012089* Jun Ni at al Iasi Romania 8
[17] Saren S K, Blaga F and Vesselenyi T 2018 IOP *Conference Series: Materials Science and Engineering* Oanta E and at Constanta Romania IOP Publishing 17
[18] Saren S K, Blaga F, Dzitac S and Vesselenyi T 2017 *Procedia Computer Science* Ahuja V Shi Y Khazanchi D Abidi N Tian Y Berg D and Tien M J New Delhi Elsevier B.V. 122 pp 253-260
[19] Saren S K, Blaga F and Vesselenyi T 2017 *MATEC 2017 Web of Conferences 137, 01011* N. Balc Cluj Napoca Romania 6
[20] Saren S K, Blaga F and Vesselenyi T 2017 *MATEC 2017 Web of Conferences 137, 01012* N. Balc Cluj Napoca Romania 6
[21] Saren S K, Blaga F, Vesselenyi T 2017 *ATN Series: App. Math., Mech*, and Eng. **60** 293
[22] Saren S K and Blaga F *MATEC Web of Conferences 94, 02015 G.* Oancea and M.V. Drăgoi Brasov Romania 14

IOP Conf. Series: Materials Science and Engineering 1169 (2021) 012006 doi:10.1088/1757-899X/1169/1/012006

3D Modelling and FEM Analysis on Metal Coin Edge Punching Error

C C Gavrila[1] and M T Lates[1]

[1]Product Design, Mechatronics and Environment Department, "Transilvania" University of Brasov, Brasov, Romania

E-mail: cgavrila@unitbv.ro

Abstract. Minting metal coin, mean to impress a given relief model on both coin faces and also on the edge. Due to the high striking pressure, the round blank material is forced to fill the inner space between dies. In the past, for less coin value, the coin edge remained cylindrical, without any inscription. Following the increased coin value, it was introduced a security feature against forgery; on the coin edge model could be toothed or inscribed with relief or incuse text logo. Besides the coin edge basic design, a various amount of coin edge error types are released. In the paper there are studied some errors on the metal coin edge impressed with incuse text by punching. First, there are presented some particular examples and general aspects about the coin edge punching are also introduced. Then, the simplified 3D model of the parts is presented. Also, the finite element analysis on this model is realized and is achieved for different initial conditions. In the final part, the analysis results and conclusions are presented.

1. Introduction

As manufacturing process, the metal coin production is followed in most of the situations by a various amount of error types. The each error type depends on the coin manufacturing step where it could appear; there are known errors on the coin metal alloy, coin blank manufacture, at the striking tools manufacturing and, respectively, on the properly coin striking [1, 2]. Since the coins striking include also the edge manufacturing, as toothed or impressed with various text devices and symbols, some error on the edge inscription could appear. The pieces wearing some mint errors are most valuable on the collectors market: for increased spectacular error on coin, the increased value for the subjected piece [1, 3, 4].

Each metal coin is manufactured by pressing at high loads the coin round blank with hardened steel negative dies, which forms a closed space to be filled by coin material [2]. In most cases, the model on the negative die is incused, so after the coin striking, result the relief model on coin.

Introduced as security feature, the coin edge lettering was as relief positive lettering or incuse negative lettering. The relief edge lettering was usually obtained on the properly coin striking, when the coin material had to fill also the negative edge die relief.

To increase productivity and reduce costs, the incuse coin edge lettering was introduced; usually was manufactured before properly coin striking, after forming the coin's blank rim. A straight steel die was used, against which the blanks were pressed with great force and rotated (figure 1). This straight steel die contains the hardened letter punches to be impressed on coin edge [3, 4]. Together with toothed edge, the incuse edge lettering was the most commonly used on circulation coins.

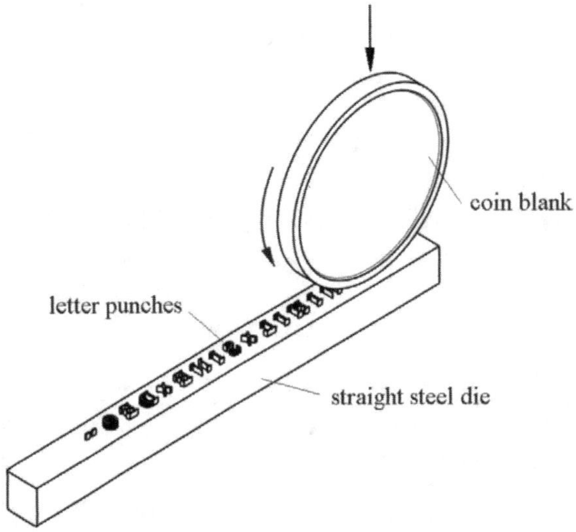

Figure 1. The incuse coin edge lettering

In figure 2, there are presented two examples of coins which bears incuse device on edge: an old Romanian 500 lei silver coin, minted in year 1944, and a former German coin 5 mark, minted in year 1975. In Romania, the incuse device on coin edge was introduced starting 1939. Many devices were used; one of them was "NIHIL SINE DEO", massively manufactured on different coins edge, from 1941 until 1947 [6, 7]. The mentioned period covers the war and the after war years, in which, due to major inflation, a large amount of metallic coins were minted. Depending by coin value many alloys were used, as nickel plated iron, silver, brass.

a) b)

Figure 2. The incuse letters on metal coin edges

If some failures or appeared on the straight steel die, the pressed coins resulted with error on edge inscription. The most frequently error is the punch letter crack, witch led to an incomplete incuse letter (named cracked or broken letters) impressed on the coin edge. In the most cases, the literature doesn't reveal exactly the causes of this failure in the coins manufacturing process; moreover, the pieces wearing this error are often unknown or, if known, some of them are considered variants from the basic design [5]. In the following figures there are presented various examples of coins with edge inscription "NIHIL SINE DEO" having broken letters. The studied inscription is composed by twelve letters, included in three words, each separated by ornaments. The letter shape is: open and sharp (for N, I, H, L, E), closed and sharp (D), open and round (S), and, respectively, closed and round (O). The separating ornament shape is usually less complicated.

In figure 3 are presented two Romanian nickel plated iron coins, 100 lei, minted in years 1943 and 1944, with broken letters on the edge inscription. In figure 4, a, the damaged letters are N, I, I from NIHIL, S, N, E, from SINE and D, E, O from DEO. In figure 4, b the damaged letters are I, H, I from NIHIL and I, from SINE.

a) b)

Figure 3. Broken letters impressed on nickel plated iron coin edge

In figure 4 are presented two Romanian silver coins, 25000 lei and 100000 lei, minted in year 1946, with broken letters on the mentioned edge inscription. In figure 4, a, all the inscription damaged letters are I. In figure 4, b the damaged letters are N, I, I from NIHIL.

a) b)

Figure 4. Broken letters impressed on silver coin edge

a) b)

Figure 5. Broken letters impressed on brass coin edge

In figure 5 are presented two Romanian brass coins, 2000 lei and 10000 lei, minted in years 1946 and 1947 with broken letters on the edge inscription. In figure 5, a, all the inscription letters and ornaments are damaged, except S from Sine and the final separation ornament. In figure 5, b all letters and ornaments are damaged, except N from NIHIL, E from SINE, D from DEO and the final separation ornament.

Analyzing previous figures, it can be observed that, the most frequently damaged letter or ornament portion on the coin edge are the thin sections (letter serif for sharp letter shape), firstly intersected by the blank during its motion. If the coin blank is rolling without misalignments, the damaged letter portion is on the top and bottom serif, as is presented in figure 6, a. If some misalignments appear, as is presented in figure 6, b, the damages appear on bottom serif on letter. Due to its shape, the letter O has approximately the same intersected portion with coin blank.

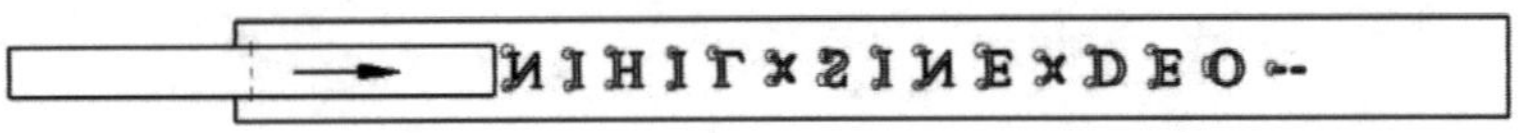

a)

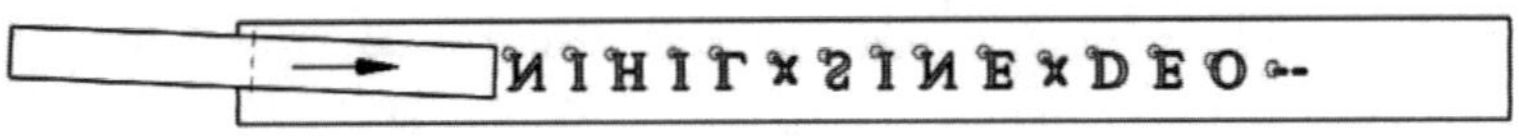

b)

Figure 6. The most damaged letter portion

2. Theoretical aspects

According to the mechanisms theory, the path of a point belonging on a straight line *Δ* which rolls without slip along the circumference of a circle with radius ***r*** is a planar curve named involutes, as presented in figure 7. If the point belongs to a parallel straight line *Δ''*, which intersects the circle, it describes elongated involutes path. If the parallel straight line *Δ'* is outside the circle, the path involutes are shorted [8, 9].

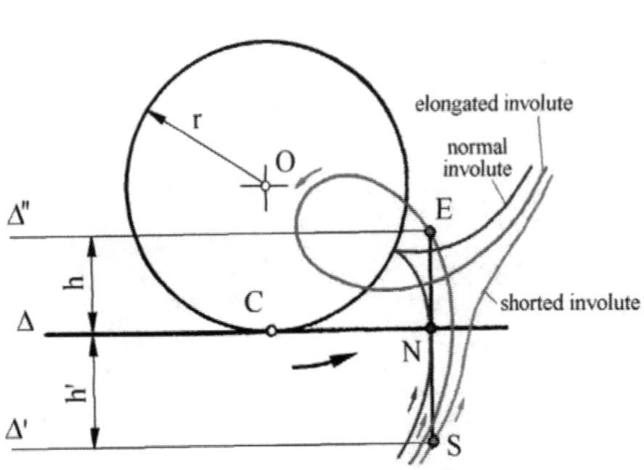

Figure 7. The involute path types between circle and straight line

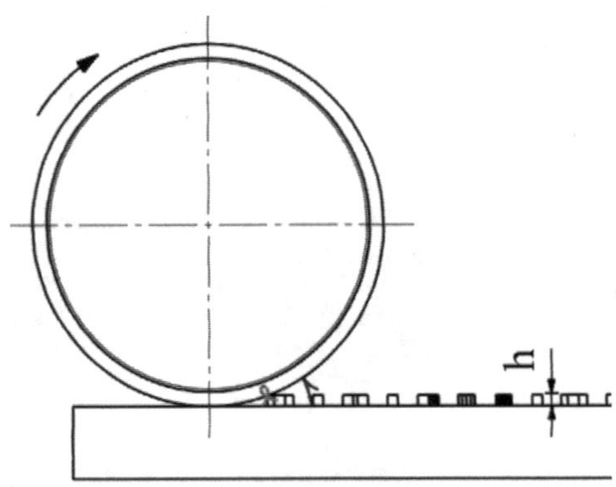

Figure 8. The involute path between coin blank and straight die with punches

When the coin blank is rolling without slip along the straight die, each point on straight die describes involutes path; the letter lower base point on die must be included. Taking account by the punches letter height ***h***, the letter top points will describe elongated involutes path, as presented in figure 8. Since the letter height is reduced, the elongated involutes amplitude is also reduced.

Usually, the distance between letters on inscription ***e*** is constant, but the each letter width ***s*** is not (figure 9). This generates a different angular pitch on the blank coin's circumference, corresponding to different letter shape. Due to the reduced letter dimension, it will be considered just the equidistant angular pitch between letters, given by relation

$$\tau = 2\pi/z\,, \tag{1}$$

where z is total number of the inscription letter and symbols.

For the studied case, the inscription is composed by twelve letters, included in three words, each separated by one ornament. The result is fifteen characters, equiangular disposed on coin edge; taking account by previous relation the angular pitch is $\tau = 24^\circ$.

Also, the relation (1) led to the circular pitch between letters, given by

$$p = 2\pi r/z\,. \tag{2}$$

From relation (2) result that, for the inscription with same number characters used, the letter punches distance on the corresponding straight die depends by different coins diameter.

On straight die arrangement, the relief negative inscription has a beginning letter or symbol punch and also, an ending one. The studied inscription has, as beginning letter the N from NIHIL and, as ending symbol, the separating ornament which precedes the mentioned word. Before and after the relief negative inscription punches on die, only the straight die surface can be found. So, at coin blank rolling on this surface, its circumference will intersect the first inscription punch letter, as in figure 9.

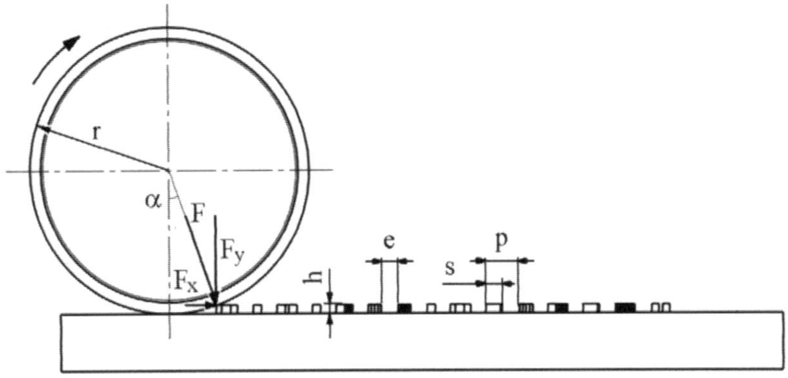

Figure 9. The contact between the rolling blank coin and letter punch

The center angle corresponding to this coin blank position depends by the blank radius and letter punch height and is given by relation

$$\alpha = arccos\left(1 - {}^{h}/_{r}\right). \tag{3}$$

Considering the relation (1) and (3) the number of letter punches simultaneously intersected by coin blank can be determined

$$n = 2\alpha/\tau, \tag{4}$$

except for first letter, until second letter enters in contact and, also for the last letter after the precedent is out of contact. Taking account by the relation (3) the number of punches simultaneously in contact depends by letter height $\boldsymbol{h}$, coin blank radius $\boldsymbol{r}$ and also by the inscription total number of letters and symbols z.

The force $\boldsymbol{F}$ which appears on contact between the blank circumference and the letter punch edge has, as its components

$$F_x = Fsin\alpha \tag{5}$$

and

$$F_y = Fcos\alpha. \tag{6}$$

From relation (3), (5) and (6) result the influence of the center angle, the blank radius and also the letter punch height on the force components.

In the coin edge punching real case, the blank circumference has to maintain contact with the straight die surface, which means that the punch letter needs to dislocate the intersected blank material, in order to obtain the incuse letter on it. In figure 10, the punching steps are presented. Firstly, the coin blank circumference is intersecting the corner point $\boldsymbol{P}$ on the punching letter profile. From this moment, the forces given by relation (5) and (6) start their load to dislocate the coin blank material. The second step can be considered when the letter symmetry axis is coincident with the blank vertical axis. In this position, the most of the coin blank material is dislocated, except some material on top and lateral letter side. The third step is given by the point $\boldsymbol{P}$ last contact position with the coin blank; when all material is already dislocated and the loading force ends their action. The both, first and last contact, between the coin blank and the punching letter profile are given by the same point $\boldsymbol{P}$, located on the top corner on the letter profile. So, the most loaded letter punch side is given by this point position.

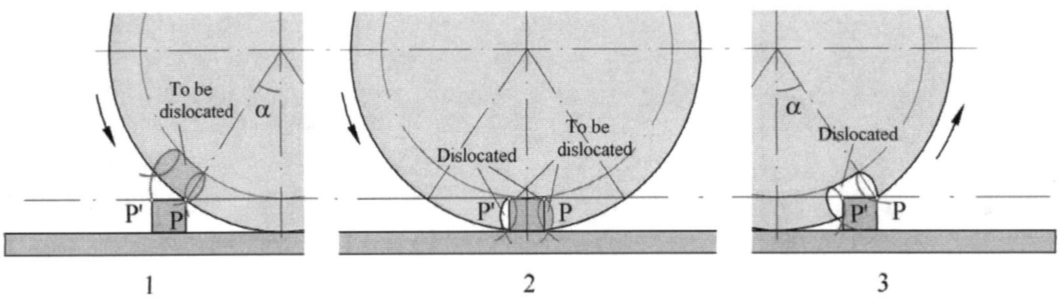

Figure 10. The punching steps between the rolling blank coin and letter punch

3. The virtual model

As was presented, the studied inscription is composed by twelve letters, included in three words, each separated by ornaments. Since on the entire straight die there are multiple complex details, the simplified punches models for each of the mentioned letters will be computed for study. The each punch simplified model consist in a cylinder with relief extruded letter on top, as is presented in figure 11. To elaborate the 3D models, the CATIA software is used [10, 11].

Taking account by punching steps, previously introduced, the most loaded position is the first contact between the coin blank circumference and the punching letter highest profile. From figure 6 results that the letter portion firstly intersected by coin blank is the same for letters N, I, H, L, E, D and consist in the thin portion defined by each letter serifs on left side. From these group of letters, were chosen to be computed the letter N, as the first inscription letter and, also, the letter I, the narrowest in the inscription. Another computed letters are S, where the contact may be on the round lateral portion or on the bottom left serif, and, also the letter O, where any contact with the coin blank is on the round left portion.

Letter N Letter I Letter S Letter O

Figure 11. The letter punches 3D model

Each punch model (exception on letter O) is loaded in two cases, for aligned rolling coin blank on die, when the force value is equal divided between the top and bottom letter profile and respectively for misaligned rolling coin blank on die, when the force value is entirely applied on the bottom letter profile. For the letter O punch, the entire force is applied on a corresponding portion of the outside round edge. The loading cases are exemplified in figure 12, for letters N and O.

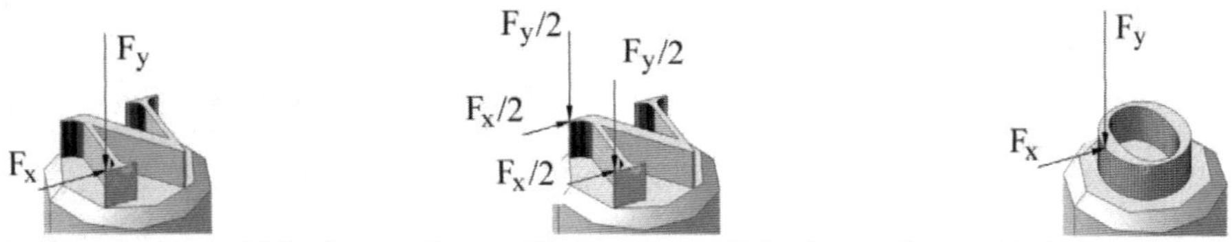

Letter N, for misaligned blank Letter N, for aligned blank Letter O, for any blank position

Figure 12. The loading cases on letter punches

4. Finite element analysis, simulation and results.

For the analysis, the ANSYS software is used [12]. The analysis objective is to determine the letter punches behaviour under the load. For the analysis, the previous punches virtual model is used. The letter height is 1.5 mm and the considered blank radius is 25 mm, which led to angle value $\alpha = 20^{o}$. For all the punches models, the considered material is steel; the material properties, as Young modulus, Poisson coefficient, Tensile Yield Strength, Ultimate Strength should be defined [6, 9, 10]. In the contact area it is chosen a smooth mesh with the minimum edge length equal with 0.001 mm.

To dislocate the blank material, the punching force value must exceed the coin blank material allowable stress [2]. The applied normal force ***F*** on each punch is equal with 500N, in order to obtain high contact pressures, over the coin blank material allowable stress – 1600 MPa [1, 2, 5]. The results, presented in figures 13 to 19 and also in table 1, consist in the contact pressure maximum values and also the maximum values of the punch material deformation.

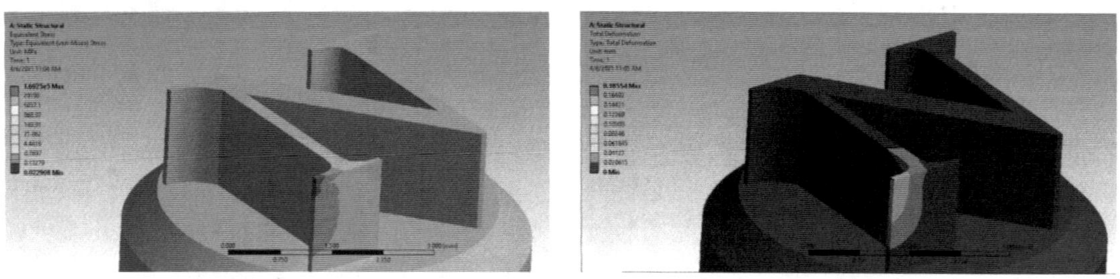

a) b)

Figure 13. The letter punch N loaded for misaligned coin blank on die

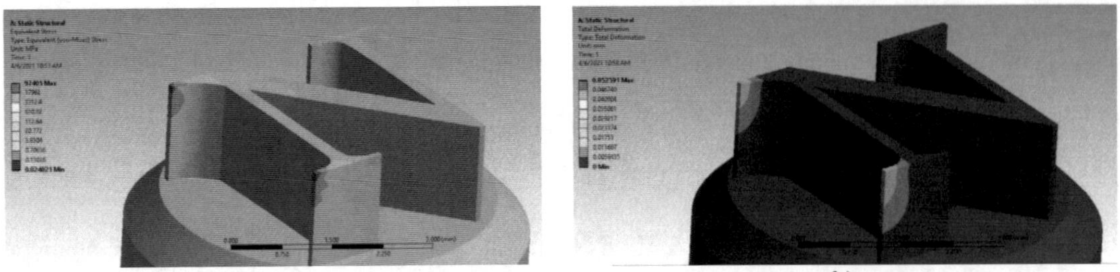

a) b)

Figure 14. The letter punch N loaded for aligned coin blank on die

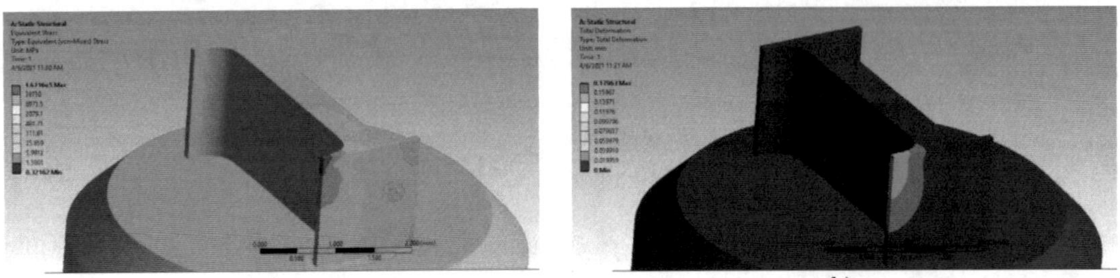

a) b)

Figure 15. The letter punch I for misaligned coin blank on die

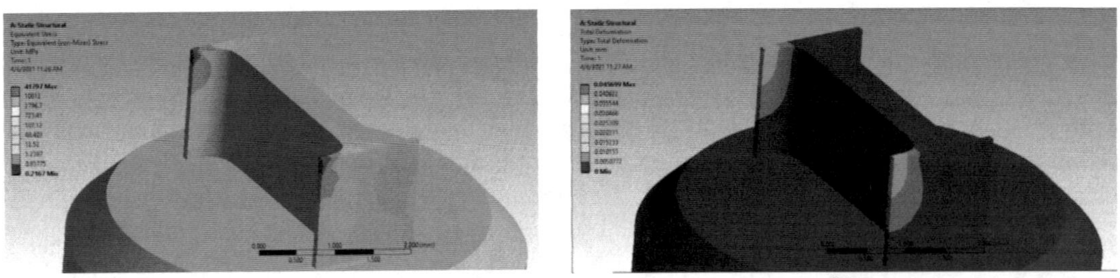

a) b)

Figure 16. The letter punch I loaded for aligned coin blank on die

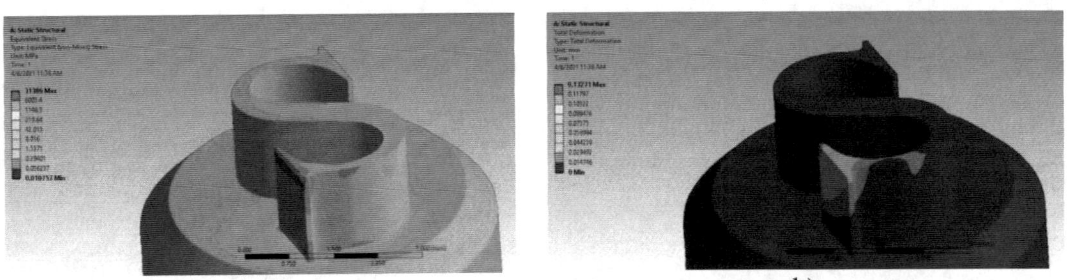

a) b)

Figure 17. The letter punch S loaded for misaligned coin blank on die

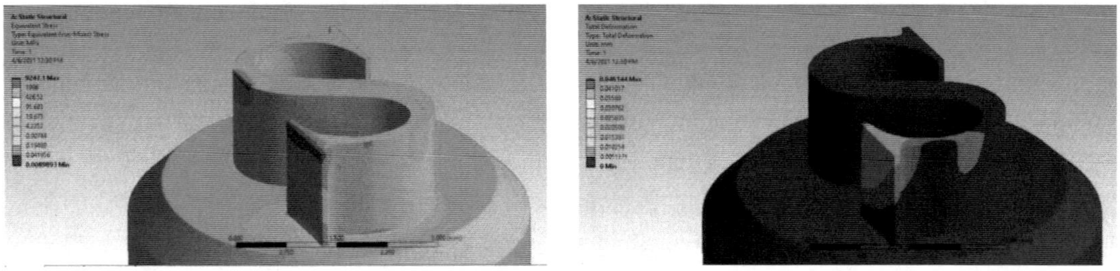

a) b)

Figure 18. The letter punch S loaded for aligned coin blank on die

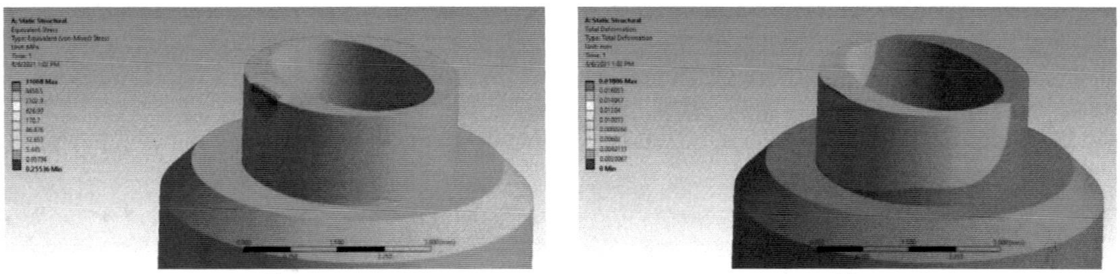

a) b)

Figure 19. The letter punch O loaded for both misaligned and aligned coin blank on die

5. Conclusions

The most frequently damaged letter or ornament portion on the coin edge are the thin sections of the punch letter (letter serif for sharp letter shape), firstly intersected by the coin blank during its motion. If the coin blank is rolling without misalignments, the damaged letter portion is on the top and bottom serif. If some misalignments appear, the damages appear on bottom serif on letter. Due to its shape, the letter O has approximately the same intersected portion with coin blank in both situations.

When the coin blank is rolling without slip along the straight die, each point on straight die describes involutes path; the letter lower base point on die describes an involutes path and, respectively, the letter top point describes an elongated involutes path. The elongated involutes trace gives the incuse letter cavity shape inside the coin blank edge.

Depending by the letter angular pitch and center angle, the presented punching steps led to different combinations on simultaneously letter punches in action, to obtain the corresponding loading cases for various coin edge inscriptions. Some characteristics, as the loading case, blank radius, letter punch height and shape and also the misalignments inside the pressing machine, have a direct influence on the force components and on the contact area. To impress the incuse letter on coin blank, the punching force value must exceed the blank material allowable stress but also, to be less than the punches material allowable stress. Instead of this, the increased loads led to worse the punch letter profile behaviour, so the coin edge result impressed with broken letter having a missing part.

The obtained values presented in table 1 should be considered as relative values, to be used to compare the different studied cases. Due to the dramatically reduced contact area between the studied letter punch and coin blank, the contact pressure and the material deformation are highly increased; the thin details of the punch letter are exposed to be damaged. For the same loading force, the round letters S and O have better behaviour (less contact pressure and less deformations) than the sharp letters N and I. For all studied letters, the loading case without rolling misalignments led also to a better behaviour on each punch letter.

Table 1. The contact pressure and the deformation, maximum values

The letter punch	Loading case	Contact pressure, MPa	Deformation, mm
N	For misaligned coin blank on die	169250	0.18554
N	For aligned coin blank on die	97405	0.052591
I	For misaligned coin blank on die	167160	0.17963
I	For aligned coin blank on die	41797	0.045699
S	For misaligned coin blank on die	31386	0.13271
S	For aligned coin blank on die	9247.1	0.046144
O	For misaligned coin blank on die	31607	0.01806

References

[1] Gavrilă C C and Lateş M T 2020 3D Modelling and FEM Analysis on Die Clash Mint Error *IOP Conf. Ser.: Mater. Sci. Eng.* **898** 012040 pp 1 – 7.
[2] Iliescu C and Tureac O 1987 Cold Pressing Technology (Tehnologia presării la rece) *Editura Universităţii din Braşov* (in romanian) pp 266-267.
[3] *** 2007 Coin Punch Presses *Mint Errors News **17*** Mike Byers U.S.A. pp 37 – 38.
[4] *** 2007 Euro 4, Coins and Banknotes (Euro4, Monnaies et Billets) *Edition Les Chevau-Lègers* Paris France (in french) pp 203 – 204.
[5] Gavrilă C C and Lateş M T Metal Coin Error Study Using 3D Modelling and FEM Analysis, 2019 *IOP Conf. Ser.: Mater. Sci. Eng.* **568** 012002 pp 1 – 6.
[6] *** 1945 Finance Ministry, National Mint, Ten Years of Activity (Ministerul Finanţelor, Monetăria naţională. Zece ani de activitate) *Monitorul oficial şi Imprimeriile statului* Bucureşti (in romanian) p. 75.
[7] Buzdugan G Luchian O and Oprescu C C 1977 Romanian Coins and Banknotes (Monede şi bancnote româneşti) *Editura Sport Turism* Bucureşti (in romanian) pp 256 - 277.
[8] Jula A et al 1989 Involute Gearing Design (Proiectarea angrenajelor evolventice) *Editura Scrisul Românesc* Craiova (in romanian) pp 108 – 120.
[9] Vişa I et al 2004 Mechanism Functional Design (Proiectarea funcţională a mecanismelor) *Editura Lux Libris* Braşov (in romanian) pp 118 - 126.
[10] Ghionea I G 2009 CATIA v5. Application in Mechanical Engineering (CATIA v5. Aplicaţii în ingineria mecanică) *Editura Bren* Bucureşti (in romanian).
[11] Lateş M T 2008 Finite Elements Method. Applications (Metoda elementelor finite. Aplicaţii) *Editura Universităţii Transilvania Braşov* (in romanian).
[12] Lee H H, 2012 Finite Element Simulations with ANSYS Workbench 14. Theory, Applications, Case Studies. *Schroff Development Corporation* Kansas USA.

IOP Conf. Series: Materials Science and Engineering 1169 (2021) 012007 doi:10.1088/1757-899X/1169/1/012007

Oil condition monitoring and predicting actions using an Artificial Intelligence technique: Principal Components Analysis algorithm

N. Salem

Zarqa University, Jordan

nsalem@zu.edu.jo

Abstract. Like other lubricants, oils play a crucial role in providing the space needed to separate moving parts. In addition, the lubricants ensure the suspension and transport of contaminants, neutralize corrosive acids, protect surfaces likely to wear, ensure heat dissipation, and provide the performance and increasing performance characteristics of industrial equipment and beyond. During use, lubricating oils undergo several chemical transformations due to oxidation at high temperatures (regime temperatures) due to degradation and contamination, water, ethylene glycol, coolants, waste residues. Finally, oils reach a lifetime due to a wide variety of degradation mechanisms, which lead to increased oxidation and nitration, base depletion, acid build-up, water contamination, cooling fluids, and viscosity changes. The rather complex nature of lubricants, along with the distinct variety of industrial equipment, especially the latest generation, equipped with high-performance techniques and artificial intelligence, make it very difficult, if not impossible, to predict all possibilities of generating defects. This study was intended to show how to expect action-time series data using Artificial Intelligence techniques on a set of data collected using direct/ indirect sensors and computational determinations based on empirical relationships. The algorithm Principal Components Analysis (PCA) has been approached to predict the values of the next steps of any sequence by support vector machines (SVM) models. The PCA approach was considered a favourable experiment. The answers obtained characterize and equate the training sequences with values changed by a step of the time. This means that the data structure learns to predict the next step's output value at each stage of the input sequence.

1. Introduction

Typically, an oil change is based on criteria, often based on empirical determinations, calculations that refer to hours of operation, etc. Based on sound analysis, there are fewer criteria based on chemical composition and existing lubricants resulting from tests or online records. Generally, oils are tested in specialized laboratories, the tests mainly being time-consuming and energy-intensive. Due to the tests often performed by laboratories, and operations have, for this reason, disadvantages related to the date of acquisition instantaneity, only basic degradation features coverage. Therefore, using laboratory tests to monitor the condition of oils is not a practical solution, nor is it generally applicable. It is the time factor that requires that oil state analyses be almost simultaneous, if not concurrent with sampling time. Only in this way do oil analyses reflect the equipment's actual condition.

Electrochemical analytical techniques (electrochemical impedance spectroscopy (EIS), micro acoustic viscosity), chemical (pH measurement, thin-film contaminant monitor), physical (the kinematic viscosity, ultrasonic techniques, thermo-conductivity, ferrography, optical (optical absorption), and other types of analysis, especially those combining different kinds of scenarios, are best suited for data collection using high-performance sensors, [1].

2. Experimental setup

The experiment was conducted in several phases, being part of a Research Program supported by private companies from Romania and Italy, in collaboration with researchers and academics staff, [1], [2], [3]. The data set was collected for six months, continuously validating the data in several 258646 instances for 19 operating parameters. A first test phase of an experimental model took place at a private Italian company on a simulated demo system in the company's lab.

The tested elements (i.e., physicochemical parameters) were evaluated in the first stage on a single level. As data was collected, based on predefined scenarios, "event/ error modules" were interleaved at a higher level to test the system's response speed [3]. All the data was then processed and interpolated. At that time, the "Artificial Intelligence" contribution became essential to evaluate the infinitesimal transformations (those that can escape the vigilance of a human operator) of multiparametric elements. The system has been tested to detect between the "n" elements that need to be considered simultaneously and to evaluate in real-time the evolution and trend of the monitored parameters, being "taught" to detect singularities, outliers, essential anomalies, or possible failures, [19].

Data processing and computational work have yielded positive results to obtain the necessary data points for advancement in effectively monitoring a correctly identified mechanical system. The overcoming of this early phase has allowed the decision to install the hydraulic system containing sensor sets and transducers designed and physically established in industrial installations (called pilot systems) at a private company in Romania, as well as in four other locations in Italy, to partners of the Research Program, named above, [1], [2], [3].

The data set contains 19 parameters, not all independent, whose values have been collected over six months or determined empirically. From this set, the data collected for three status parameters (which determines the degree of degradation of the lubricating oil) and a bunch of values corresponding to the collection of target values were selected, the latter being used as a comparison element, training, learning, testing, and validation, respectively. The number of time steps is estimated at 258646 items. Outliers detection techniques have replaced missing or erroneous values by comparison with the closest mean value set. The values with which these abnormalities have been replaced complies with the condition that they are interpolated linearly over a strictly determined range in the valid set of values.

This experiment uses the data set collected within the Romanian National Program PN II project, ERA MANUNET: NR 13081221. These experiment models and analyzes at the same time the behaviour of a mechanical system, studying its behaviour based on data collected from applied sensors the hydraulic system of automatic machine lubrication. Computational determinations have been made for a small set of parameters to estimate the degradation state of the lubricating oil (A, B, C), respectively, for the complete set of data (are used all nineteen independent parameters, or not, collected during the experiment). The latter analysis required an excessive computational effort in the necessary resources, involving parallel computing techniques on an i7 computer with eight kernels. For this approach, it was also necessary to process the data provided by the last data column, which contains calculated estimates of the qualitative degradation of the oil (empirical determination, [4-11]) so that the data column reflects the correct classification of the qualitative estimates as well as the values collected from the system sensors for the other parameters of the lubricating oil. In this sense, the last data column in the data set matrix contains binary values (0 and 1) for the two possible states of the oil in the plant: the logical value "1" corresponds to a degraded oil and needs to be changed, respectively the logical value "0", represents a suitable quality oil, and can be used in the lubrication system. The data set is partitioned as follows: 70% of the data volume will be used for artificial network training, and 15% of the collected data set is used for network testing, and 15% of the dataset is used to validate the results. The working procedure states that the response values of the system are the network training data sequence. Carefully prevent situations of divergence of the forecast; it will be necessary to standardize the training data are so

IOP Conf. Series: Materials Science and Engineering 1169 (2021) 012007 doi:10.1088/1757-899X/1169/1/012007

organized that it has zero mean and unit variance. Also, the data set used for the test will be standardized in the same way as the training data. The long-term memory (LSTM) regression network was used, which has been specified to have an architecture of 10 hidden units and train for 100 epochs. This will prevent the gradients from exploding the gradient threshold. The initial learning rate was defined to 0.005 and provided to drop the learning rate after 25 epochs multiplied by a factor of 0.2. The network is initialized when the predictive training data is made. Next should be the first prediction using the last time step of the training response. The network function will be helpful when the forecast and outputs of the network are updated sequentially using predicted data. The training phase is monitored permanently by the calculated root-mean-square error (RMSE), which measures errors induced by the network. This was the preparatory stage of the work

3. Principal Components Analysis algorithm and application

The Principal Components Analysis (PCA) is a very commonly used method of extracting statistical features [4], [12-16]. Fortunately, in variable data sets, varying groups evolve concurrently. An explanation for this would be that several variables could follow the same evolutionary principle that governs the behaviour of a mechanical system. There are only a few such driving forces in technical systems, but plenty of instrumentation allows one to measure dozens of system variables. When this happens, the immediate danger would be to create a redundancy of information. In this instance, the subject appears hard to be solved because of the surplus of information. Still, it can be simplified by replacing the variable set with a new set of variables, generally smaller in number and size. Thus, considering a collection of measured data (a matrix with m lines and n columns), $\boldsymbol{X_{mxn}}$, where each column is a single sample (or an instance) of the data set (ie $\vec{\boldsymbol{X}}$), and either $\boldsymbol{Y_{kxn}}$ another matrix, which is a linear transformation ***P*** of the first matrix. In conclusion, ***X*** is the original set of recorded data, and ***Y*** is a re-representation of this dataset. Then, expressing the change of the base ***X***, with the transformation ***P***, in ***Y***, results [4], [13-18]:

$$P*X=Y \tag{1}$$

where ***P*** is a matrix with ***k***-rows and ***m*** columns, $\boldsymbol{P_{kxm}}$And the elements of this matrix (the vectors $\boldsymbol{p_1, p_2, \dots, p_m}$) are the set of vectors in the new base. Thus, by applying the transformation matrix ***P*** to the matrix ***X***, then the matrix ***Y*** is formally obtained:

$$\begin{bmatrix} p_{11} & p_{12} & \cdots & p_{1m} \\ \vdots & & \ddots & \vdots \\ p_{k1} & p_{k2} & \cdots & p_{km} \end{bmatrix} * \begin{bmatrix} x_{11} & x_{12} & \cdots & x_{1n} \\ \vdots & & \ddots & \vdots \\ x_{m1} & x_{m2} & \cdots & x_{mn} \end{bmatrix} = \begin{bmatrix} y_{11} & y_{12} & \cdots & y_{1n} \\ \vdots & & \ddots & \vdots \\ y_{k1} & y_{k2} & \cdots & y_{kn} \end{bmatrix} \tag{2}$$

alternatively, in a more simplified form:

$$\begin{bmatrix} p_1 \\ p_2 \\ \vdots \\ p_m \end{bmatrix} * \begin{bmatrix} x_1 & x_2 & \cdots & x_n \end{bmatrix} = \begin{bmatrix} p_1*x_1 & p_1*x_2 & \cdots & p_1*x_n \\ p_2*x_1 & p_2*x_2 & \cdots & p_2*x_n \\ \vdots & \ddots & \cdots & \vdots \\ p_m*x_1 & p_m*x_2 & \cdots & p_m*x_n \end{bmatrix} \tag{3}$$

It should remark that each coefficient of $\boldsymbol{y_i}$ is a dot product of $\boldsymbol{x_i}$ with the corresponding row in ***P***:

$$y_i = \begin{bmatrix} p_1*x_i \\ p_2*x_i \\ \vdots \\ p_m*x_i \end{bmatrix} \tag{4}$$

In other words, the coefficient of jth of y_i is a projection on the jth line of the matrix P. The matrix P rows are a new vector base indeed for representing the columns of X. Hence, the elements of Y are linear combinations of those vectors. This is why the PCA is a method that projects a set of data into a new coordinate system by determining its eigenvectors and eigen-matrix values. This involves calculating a covariance matrix of a data set to minimize redundancy and maximize the variance. Mathematically, the PCA is defined as a linear orthogonal transformation and assumes that all the base vectors are an orthogonal matrix [4], [13], [16], [17]. Therefore, the PCA is targeted towards finding variations and coefficients of a data set by computing and determining its eigenvalues and eigenvectors of the covariance matrix. The PCA is calculated by determining the eigenvectors and the eigenvalues of the covariance matrix. The covariance matrix evaluates how much each dimension differs from the relative median reported to the differences between them. The covariance of two random variables (dimensions) represents their tendency to vary together:

$$\mathrm{cov}(X, Y) = E[E[X] - X] \cdot E[E[Y] - Y] \tag{5}$$

where $E[X]$ moreover, $E[Y]$ are (expected) values of X and Y, respectively.
For a sample of values

$$\mathrm{cov}(X, Y) = \sum_{i=1}^{N} \frac{(x_i - \mu_x)(y_i - \bar{y})}{N}$$

where:

$$\bar{x} = \frac{1}{N}\sum_{1}^{N} X; \quad \bar{y} = \frac{1}{N}\sum_{1}^{N} Y \tag{6}$$

$\bar{x}$ and $\bar{y}$ is the means of X and Y, respectively.

Equation (1), from a geometric point of view, represents rotation and stretch, which transforms X into Y.

The first principal component is a single axis in space. When the projection is constructed, each observation on that axis, then the resulting values form a new variable. The variation of this variable is the maximum of all possible options of the first axis.

The second principal component is another axis in space, perpendicular to the first one. The projection of the observations on this axis generates a new variable. The variance of this variable is the maximum of all possible options of this second axis. The complete set of principal components is as large as the original set of variables. However, the sum of variations in the first few principal components is usually expected to exceed 80% of the total variance of the original data.

In MATLAB®, owned by The MathWorks Inc., one may use the *pca* MATLAB function to find the principal components. To use the *pca*, the measured data one wants to analyze should be matrix organized. For example, suppose you do not have the actual data, but you have a sample for which you know the correlation matrix or the data covariance matrix. In that case, you can use the *pcacov* MATLAB function to perform a principal component analysis.

A very intuitive method for calculations and PCA visualization is presented in detail [16], [17]. In addition, the authors have implemented an interactive PCA view, and a text viewer called Text Variation Explorer is developed. It allows the user to interactively study the outcome of the PCA and provides a better understanding of the process.

An exciting and practically experienced perspective in exploring medical data is presented in [18], [19]. Since the primary purpose of the PCA is to extract new uncorrelated characteristics, it is logical to introduce a correlation-based criterion with the possibility of defining a threshold value. Such a criterion is Kaiser-Meyer-Olkin (KMO), a benchmark that explains both the total correlation and the partial correlation:

$$KMO = \frac{\sum_i \sum_j r_{ij}^2}{\sum_i \sum_j r_{ij}^2 + \sum_i \sum_j a_{ij}^2} \qquad (7)$$

in which: $r_{ij} = r(x^{(i)}, x^{(j)})$ it is an element of the correlation matrix $\boldsymbol{R}$, and $\boldsymbol{a_{ij}}$ is an element of partial matrix correlation A: $a_{ij, X^{(i,j)}} = \frac{-R_{ij}}{\sqrt{R_{ii} R_{jj}}}$ is a partial correlation coefficient for $x^{(i)}$ and $x^{(j)}$ without considering the effect of all of the other variables, and the characteristic $X^{(i,j)}$ is fixed and $\boldsymbol{R_{kl}}$ is the algebraic complement for $\boldsymbol{r_{kl}}$ from the determinant of the correlation matrix $\boldsymbol{R}$. It is easy to note that if two characteristics have a common correlation factor with other characteristics, it results that partial correlation $\boldsymbol{a_{ij}}$, is small, indicating a corresponding variation. Thus, if $\boldsymbol{a_{ij}}$ approaches zero, then KMO approaches the unit value, while if $\boldsymbol{a_{ij}}$ approaches the unit value (indicating that the variables do not have a common correlation factor) then KMO approaches zero value. The author recommends using PCA if KMO> 0.5.

Since the experiment results contain data on many variables (19 parameters), it was chosen to evaluate these data using the Principal Components Analysis model, being retained for the study of three principal components in a 3D graphic representation. The names and meaning of the symbols of the collected parameters are further presented for this analysis in the Classification Learning approach and the Artificial Intelligence techniques of this paper Table 1:

Table 1- Parameters' definitions and symbols' significance

Crt. nr.	Parameter Symbol	• Significance of parameters
1	icm_rh	• Relative humidity of the oil (%)
2	icm_flow	• Oil flow (ml/min)
3	icm_temp	• Temperature of the sensor (°C)
4	icm_iso4	• Disaggregated ISO Cleanliness Code 4406:1999
5	icm_iso6	• Disaggregated ISO Cleanliness Code 4406:1999
6	icm_iso14	• Disaggregated ISO Cleanliness Code 4406:1999
7	icm_pc4	• Particle counter 4µ
8	icm_pc6	• Particle counter 6µ
9	icm_pc14	• Particle counter 14µ
10	fps_vcst	• Kinematic viscosity (cSt)
11	fps_v	• Dynamic viscosity (cPoise)
12	fps_density	• Density (g/cm^3)
13	fps_dielectric	• Dielectric
14	fps_temp	• Oil temperature (°C)
15	oh_temp	• The temperature of the sensor (°C)
16	oh_parama	• Param A used for oil degradation calculation
17	oh_paramb	• Param B used for oil degradation calculation
18	oh_paramc	• Param C used for oil degradation calculation
19	oh_od	• Oil degradation

Figure 1 shows the relative distribution of sensor data collected or analytically determined:

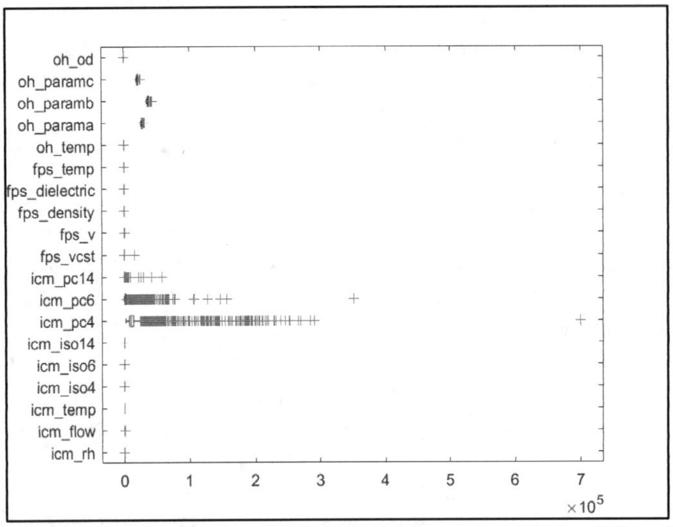

Figure 1- The distribution of the" AllOilParametersMonitor" data file

The Principal Components are represented bi-dimensionally in figure 2, also in this figure are indexed, several outliers, using the interactive features of digital graphical interfaces. Thus, you can easily see the contour of the data volume on two main components and the "scattered" set of independent data (outliers) in the isolated, but concentrated, area of the principal components.

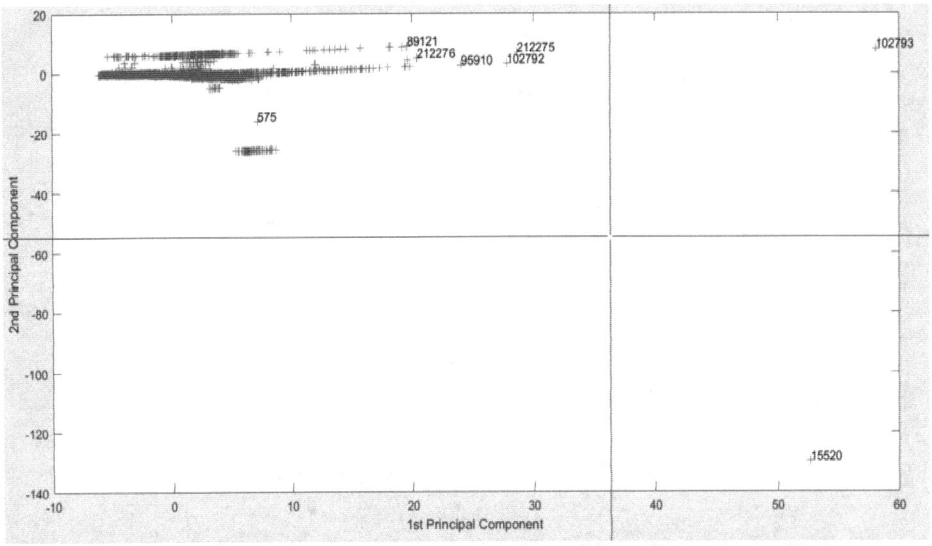

Figure 2- The Principal Components plot, with outliers outstanding

In figure 2, it can be easy to observe the volume of data falling within the "abnormalities" or "outliers" category, data to be eliminated, or for which special conditions should be imposed when processing the results.

The two-dimensional approach of the Principal Components, for all variables, offers the possibility of processing the data with the appropriate tools for the study, figure 3. It can be seen from figure 3 that the collected data is grouped quite uniformly over a relatively narrow area and that some parameters have distributed values in well-defined ranges. Anomalies being concentrated in isolated areas allow conclusions about the occurrence of anomalies and confirm the existence of "outliers" values.

IOP Conf. Series: Materials Science and Engineering 1169 (2021) 012007 doi:10.1088/1757-899X/1169/1/012007

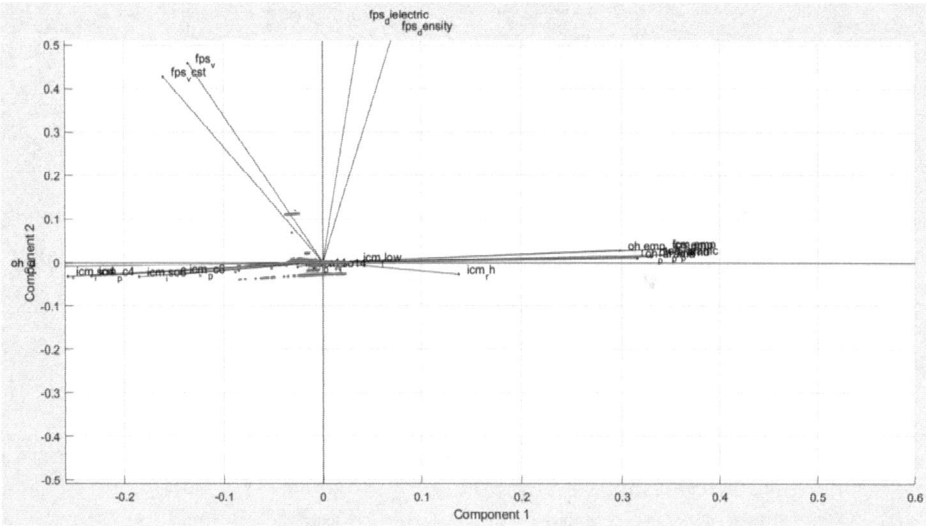

Figure 3- The two Principal Components plot, with all 19 parameters outstanding

Highlighting outliers is particularly well worth the 3D Graphic representation of Principal Components, using as bases the three main components, analytically determined, figure 4.

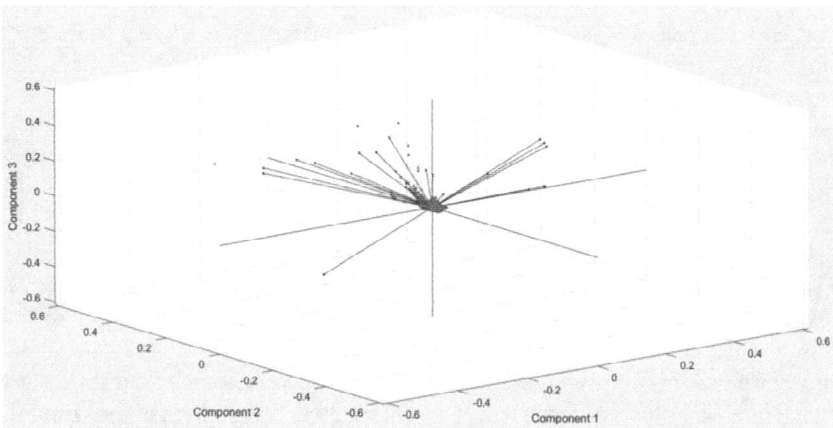

Figure 4-Three Principal Components, 3D plot representation

The variance of the three Principal Components is graphically displayed in figure 5. Again, observation is required: the figure puts into perspective the variance of the first ten components, which fall within the 95% threshold of the total variation. It is to be noticed that the first principal component, the most important one, represents a variance of almost 40%, the other components having much more comparable weights, their variance being almost insignificant, the first three components covering two-thirds of the total variance. This is why we can make the decision dimensionally reduced components, that is, consider only three of them, most obviously critical.

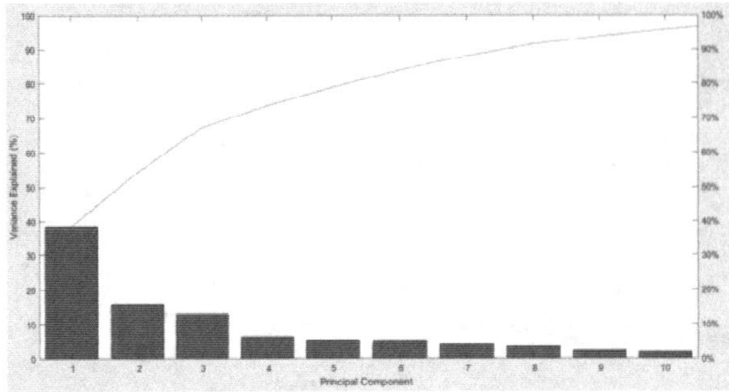

Figure 5- The amount of variance accounted for each component

4. Conclusions

Condition monitoring and diagnosis and predicting the estimation of some output parameters for a technical system, respectively, can be framed in the "extrapolation" chapter as a mathematical evaluation, testing, and validation operation [13], [15]. Doubtless, the time series for which mathematical operations of this kind can be applied are subject to errors due to the statistical nature of these mathematical classes. Therefore, a prediction is understood as an estimate of the values of a time function, based on values of a time series, values that can be, or can not be, affected by random errors. Thus, for example, a prediction problem could be expressed as follows: Given a series of time, ***S (t)***, which consists of a set of values, and a random set of disturbing signals assimilated to a set of noise, ***Z(t)***, it is proposed to estimate a future value, a prediction, therefore, ***P(t+τ)***, where ***τ*** is a positive constant, the forecast is also a continuous function of time.

The success of a classification system depends significantly on the effectiveness of the extracted observation sequence to represent a particular condition or condition of the machine. Therefore, significant efforts have been made in research works, developing various techniques for extracting features and monitoring the machinery and installations. Characteristics extraction algorithms can be grouped, according to the basic methods used, into three main categories: the Principal Components Analysis, which is based on the identification of the axes on which data is most variable, the learning approach (Neural Networks, Pattern Classification, which have the advantages of superior learning, noise suppression, and parallel computational skills) and signal processing (Fourier transform, Hidden Markov Models or wavelet, which allow the classification based on modulus maxima distribution).

The Principal Components Analysis is a rigorous quantitative mathematical procedure for making such a simplification. The PCA is a stable and efficient method for finding an algorithm for structuring a multidimensional data set. The PCA is based on orthogonal transformations that convert a group of multidimensional values into linear-uncorrelated variables, called core components. The main disadvantage of the PCA approach is that the procedure and the outcome are often challenging to understand. The link between input and output can be confusing, a slight change of inputs can generate an utterly different result, and the user can often ask if the PCA does the right thing. The PCA generates a new set of variables, called main components, and each principal component is a linear combination of the initial variables. The fundamental property of these main components is that they are orthogonal, so there is no redundant information. The quantifiable results of this algorithm are that all the main components as a whole form an orthogonal basis for the data space.

References

[1] G. Adriani, M. Campatelli, M. Paoli, -Monitoil® - *Online sensors for efficient and cost-saving oil conditions' monitoring*, Mecoil Diagnosi Meccaniche s.r.l., Florence, Italy, http://www.mecoil.net/wp-content/uploads/2013/04/20120914-Articolo-colonne_Monitoil_eng.pdf

[2] Grebenişan G., Salem N., and Bogdan S., *The lubricants parameters monitoring and data collecting,* MATEC Web of Conferences, **Volume** 126, 2017, https://doi.org/10.1051/matecconf/201818403008.

[3] Grebenişan G., Salem N., and Bogdan S. -*An approach of classification and parameters estimation, using the neural network, for lubricant degradation diagnosis*, MATEC Web of Conferences, **Volume** 184, 2018, https://doi.org/10.1051/matecconf/201818403009.

[4] Statistics and Machine Learning Toolbox™ User's Guide, https://www.mathworks.com/help/releases/R2018a/pdf_doc/stats/stats.pdf, Accessed in February 2019.

[5] Lihui Wang and Robert X. Gao (Eds.) Condition Monitoring and Control for Intelligent Manufacturing, 2006, Springer-Verlag.

[6] I.T. Jolliffe, Principal Component Analysis, Second Edition, Springer Series Statistics, 2010

[7] Chandola, V., Banerjee, A., and Kumar, V. 2009. Anomaly detection: A survey. ACM Comput. Surv., 2009, 58 pages. http://doi.acm.org/10.1145/1541880.1541882

[8] Victoria J. Hodge and Jim Austin, A Survey of Outlier Detection Methodologies, Artificial Intelligence Review 22: 85–126, 2004, Kluwer Academic Publishers, DOI: 10.1023/B:AIRE.0000045502.10941.a9

[9] B.C. Sharma and O.P. Gandhi, Performance evaluation and analysis of lubricating oil using parameter profile approach, Industrial Lubrication and Tribology 60(3):131-137, DOI: 10.1108/00368790810871057

[10] Junda Zhu, David He, Eric Bechhoefer, Survey of Lubrication Oil Condition Monitoring, Diagnostics, and Prognostics Techniques and Systems, Journal of Chemical Science and Technology July 2013, Vol. 2 Iss. 3, PP. 100-115

[11] N.K. Myshkin and L.V. Markova, Online Condition Monitoring in Industrial Lubrication and Tribology, Applied Condition Monitoring 8, Springer International Publishing AG, 2018, DOI 10.1007/978-3-319-61134-1_1

[12] George E. P. Box, Gwilym M. Jenkins, Gregory C. Reinsel, Greta M. Ljung, Time Series Analysis: Forecasting and Control, Fifth Edition, 2015, John Wiley & Sons.

[13] Jon Shlens, A Tutorial on Principal Component Analysis, 2014, International Journal of Remote Sensing 51(2).

[14] Parinya Sanguansat ed., Principal Component Analysis – Engineering Applications, Rijeka, 2012, InTechOpen, DOI: 10.5772/2693.

[15] Dong Hyun Jeong, Caroline Ziemkiewicz, William Ribarsky and Remco Chang, Understanding Principal Component Analysis Using a Visual Analytics Tool, *Charlotte visualization center UNC Charlotte*, 2009.

[16] Harri Siirtola, Tanja Säily, Terttu Nevalainen, Interactive Principal Component Analysis, 2017 21st International Conference Information Visualisation (IV), IEEE Xplore, **DOI:** 10.1109/iV.2017.39.

[17] Mykola Pechenizkiy, Alexey Tsymbal, Seppo Puuronen, PCA-based Feature Transformation for Classification: Issues in Medical Diagnostics, 2004, Proceedings of the IEEE Symposium on Computer-Based Medical Systems 17:535- 540, IEEE Xplore, DOI: 10.1109/CBMS.2004.1311770

[18] J. Edward Jackson, A user's guide to principal components, 1991, John Wiley & Sons Inc., DOI:10.1002/0471725331, New York.

[19] Bernhard Schölkopf, Alex J. Smola, Klaus-Robert Müller, Nonlinear Component Analysis as a Kernel Eigenvalue Problem, Neural Computation 10(5):1299-1319, DOI: 10.1162/089976698300017467, 1998

IOP Conf. Series: Materials Science and Engineering 1169 (2021) 012008 doi:10.1088/1757-899X/1169/1/012008

Compression testing of samples printed on Delta and Cartesian 3D printer

M Ratiu[1], M A Prichici[1], D M Anton[2] and D C Negrau[2]

[1] Department of Mechanical Engineering and Automotive, University of Oradea, 410087 Oradea, Romania
[2] Doctoral School of Engineering Sciences, University of Oradea, 410087 Oradea, Romania

E-mail: mratiu@uoradea.ro

Abstract. This paper presents a short part of a larger research and consists in the investigation of the compression behavior of samples printed on two 3D printers, one Delta and one Cartesian, from three types of the commonly used materials, Polylactic acid (PLA), Carbon fiber PLA (CF-PLA), and Polyethylene terephthalate glycol (PETG). After a specific introduction regarding the subject, the printing parameters used for printing the samples are exposed, in a table, followed by a paragraph on the compression testing, both with general and particular information, on this case. The research being under development, the first set of results obtained is presented, further tests and analysis following in order to realize a comprehensive characterization of the 3D printed materials' behavior.

1. Introduction

In the context of using, at a larger and larger scale, of 3D printing in manufacturing processes, mainly, due to its advantages and accessibility, different kind of researches are developed in this area in the last period, in order to find the best solutions for certain situations. From a large range of 3D printing processes, or additive manufacturing technologies (AMT) existing today, the Fused Filament Fabrication (FFF), or Fused Deposition Modeling (FDM) is the most widely used. Many types of materials, increasingly newer and better, are used for 3D printing, the most commonly used today being Polylactic Acid (PLA), Acrylonitrile Butadiene Styrene (ABS), Polycarbonate (PC), Polyethylene Terephthalate (PET), Polyethylene Terephthalate Glycol (PET-G), Polyether Ether Ketone (PEEK) or High-Impact Polystyrene (HIPS), base materials or with various inserts, such as carbon fiber, to improve the properties of the filament. A state-of-the-art guide to modern plastics is represented by [1], which contains a comprehensive approach, an A to Z guide, to the plastics' processes, forms, formulations, design, equipment, testing, and recycling. Many properties and testing methods of the plastic materials are, generally, common to these materials, and the standard testing methods are globally accepted, as presented in [2]. But, despite the increasing use of 3D printed parts, mainly through FDM printing technology, there are no defined standards, specially dedicated to this category of plastic parts. It is generally accepted in this case and is used by the majority of authors, the standard ISO 527 for tensile test [3] and ISO 604 for compression test [4].

A large and consistent analysis of the mechanical properties of parts produced through FDM additive manufacturing technology and the influence of the factors such as material composition, extruding temperature, printing parameters, and environment temperature, on their behavior to tensile and compressive tests, are presented in [5]. A lot of experimental research has been done, and numerous scientific articles have been published, each addressing, in a different manner, the 3D printing and testing of certain standardized test specimens, mainly at tensile and compression. Thus, the experimental characterization of the mechanical properties of the 3D-printed test specimens, by using different printing parameters, are presented, depending on the material used, in: [6], [7] and [8] with PLA; [9] with PLA+ and CF-PLA; [10] and [11] with ABS, [12] with ABS and PC; [13] with PLA, ABS, PETG, and HIPS; [14] with multi-material interfaces. In [15] is presented an extensive state-of-the-art review regarding the optimization of the FDM additive manufacturing process in order to achieve the maximum tensile strength, through a comparative study with PLA, PC, PEEK and ABS.

After studying the few recent and relevant researches in this field, mentioned above, and taking into consideration the available material capacities, we resorted to our own research, by using two 3D printers, and three types of materials, for the printing two models of samples with different printing parameters and working conditions. The experimental procedures and first results obtained at compression testing for one type of the printed samples are further presented in this paper.

2. Printing parameters of the samples

The 3D printers used for printing the samples, available in our laboratory, being part of the medium printing category, are Delta model FLSun QQ-S PRO and Cartesian model Tevo Tornado. Technical specifications of these printers and the software used, as well as an extensive explanation of the printing settings for printing the chosen samples, can be found in [16].

The samples printed for compressive testing are a cube-type model, 20x20x20 mm, with one single piece on the bed, and with the printing settings according to the printing capacity of the printers and filaments used. Three types of the material, PLA, Carbon fiber PLA (CF-PLA), with 20% carbon fiber insertion and PETG were used for printing the samples, with the parameters presented in Table 1.

Table 1. Printing parameters of the samples

3D printer / Material / Printing parameters	Delta model FLSun QQ-S PRO			Cartesian model Tevo Tornado		
	PLA	PLA-CF	PET-G	PLA	PLA-CF	PET-G
Infill density	100 %	100 %	100 %	100 %	100 %	100 %
Layer thickness	0.2 mm	0.2 mm	0.2 mm	0.2 mm	0.2 mm	0.2 mm
Number of shells	4	4	4	4	4	4
Print speed	85 mm/s	85 mm/s	85 mm/s	65 mm/s	65 mm/s	65 mm/s
Nozzle temperature	225 °C	225 °C	225 °C	225 °C	225 °C	225 °C
Bed tempera-ture	60 °C	60 °C	80 °C	70 °C	70 °C	80 °C

3. Compression testing of samples

The testing of compression is made to determine the material behavior under the compression loading. The test specimens are compressed until a specified value of compression loading, until the cracking, or until the breaking. How there are no defined standards, specially dedicated to the parts printed by using AMT, the compression tests of this type of parts are carried out by using the standard ISO 604:2002 Plastics - determination of compressive properties.

There are a wide variety of machines and equipment available for compression testing, standard or that can be adapted to some specific applications if it is necessary. In this case was used a SCHMIDT Servo Press, with PressControl 5000 HMI, a compact system control for intelligent process control. A sequence during the testing of a PET-G sample, as well as the visualization of the loading diagram, on the display of the control system, are shown in Figure 1.

IOP Conf. Series: Materials Science and Engineering 1169 (2021) 012008 doi:10.1088/1757-899X/1169/1/012008

The types and dimensions of the test specimens from plastic materials, as well as their preparation for testing, are specified in the standard ISO 20753:2018 Plastics-Test specimens. How the enumeration in the standard does not exclude other kinds of test specimens, and there is no special standard for test specimens produced through FDM, other models are sometimes used, cylindrical or prismatic in shape. In Figure 2 are presented three sets of samples, from PLA, PLA-CF and PET-G, after the compression.

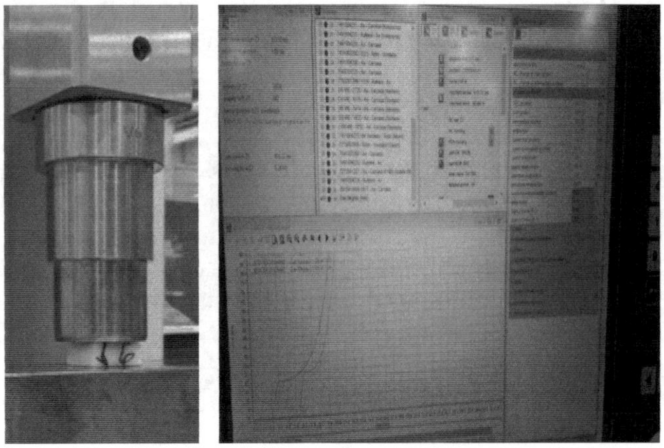

Figure 1. Compression testing of a PET-G sample.

Figure 2. Compressed samples.

4. Results of the compressive testing of samples

By using the control system and software installed on the press, a few parameters can be displayed and analyzed. The first set of them, namely the minimum and maximum forces reached at the ends of the working stroke, previously set, can be seen in Table 2. A detailed analysis and an extensive characterization of the materials' behavior will be conducted in the near future.

Table 2. Compression loading: Reached Force Minim [N] and Reached Force Maxim [N]

Nr. sample	Delta model FLSun QQ-S PRO			Nr. sample	Cartesian model Tevo Tornado		
	PLA	PLA-CF	PET-G		PLA	PLA-CF	PET-G
1	11152	28540	18800	8	35250	23873	22500
	82512	90240	90480		75072	85632	90576
2	12350	22584		9	35810	21848	22600
	84432	61392			74592	83280	90363
3	8706	22872		10	35760	21657	21200
	83232	62832			73584	79104	90384
4	8905	22426	23800	11	31824	22176	23400
	78240	71904	90576		73344	83664	90768
5	35654	22511	21980	12	29587	21908	16800
	84816	72720	90336		82080	83750	90384
6	8950	22850	18900	13	37820	22018	
	69744	72864	90624		71184	82224	
7	21507	23528	20110	14	30257	20154	
	75600	83856	90384		75456	83040	

5. Conclusion

A short part from a larger research under development was presented in this paper. It consists of the compression testing of a type of sample, in a cubic shape, printed on two 3D printers, one parallel and one Cartesian, from three types of the commonly used materials, Polylactic acid (PLA), Carbon fiber PLA (CF-PLA), with 20% carbon fiber insertion, and Polyethylene terephthalate glycol (PETG). In order to realize a comprehensive characterization of the materials' behavior, further tests and analysis will be developed and presented in future work.

References

[1] Harper C A 2000 Modern Plastics Handbook McGraw-Hill Companies, Inc
[2] Shrivastava A 2018 Plastic Properties and Testing Introduction to Plastics Engineering Plastics Design Library 2018, pp. 49-110
[3] ISO 527-1:2019 Plastics — Determination of tensile properties - Part 1: General principles
[4] ISO 604:2002 Plastics — Determination of compressive properties
[5] Banjanin B, Vladic G, Pál M, Balos S, Dramicanin M, Rackov M and Knezevic I 2018 Consistency analysis of mechanical properties of elements produced by FDM additive manufacturing technology Matéria (Rio J.) 23(4)
[6] Letcher T and Waytashek M 2014 Material property testing of 3D printed specimen in PLA on an entry level 3D printer *Proceedings of the ASME 2014 International Mechanical Engineering Congress & Exposition IMECE2014*-**39379**
[7] Brischetto S and Torre R 2020 Tensile and compressive behavior in the experimental tests for PLA specimens produced via Fused Deposition Modelling technique *Journal of Composites Science* **4**, 140
[8] Guessasma S, Belhabib S and Altin A 2020 On the tensile behaviour of bio-sourced 3D-printed structures from a microstructural perspective *Polymers* **12**, 1060
[9] Devsingh D, Reddy A R and Arjula S 2018 Characterization of fused deposition modeling of PLA+ and Carbon Fiber -PLA components *International Journal of Latest Trends in Engineering and Technology* **10** (2), pp. 010-016
[10] Cwikła G, Grabowik C, Kalinowski K, Paprocka I and Ociepka P 2017 The influence of printing parameters on selected mechanical properties of FDM/FFF 3D-printed parts *IOP Conf. Series: Materials Science and Engineering* **227** (2017) 012033
[11] Kanu R C, Hale C and Piper P 2016 The use of 3D printing to introduce students to ASTM Standards for testing tensile properties of Acrylonitrile-Butadiene-Styrene (ABS) plastic material *American Society for Engineering Education*
[12] Cantrell J et all 2017 Experimental characterization of the mechanical properties of 3D-Printed ABS and polycarbonate parts *Rapid Prototyping Journal* **23**(4)
[13] Hasan M S, Ivanov T, Tanovic D, Simonovic A and Vorkapic M 2020 Dimensional accuracy and experimental investigation on tensile behavior of various 3D printed materials *9th International Scientific Conference on Defensive Technologies* (OTEH 2020, Belgrade, Serbia)
[14] Lumpe T S, Mueller J and Shea K 2019 Tensile properties of multi-material interfaces in 3D printed parts *Materials and Design* **162** (Elsevier)
[15] Gordelier T, Thies P R, Johanning L and Turner L 2019 Optimising the FDM additive manufacturing process to achieve maximum tensile strength *Rapid Prototyping Journal* 2019
[16] Ratiu M, Anton D M and Negrau D C 2021 Experimental study on the settings of Delta and Cartesian 3D printers for samples printing *International Engineering Conference Annual session of scientific papers "IMT Oradea 2021"* in press

IOP Conf. Series: Materials Science and Engineering 1169 (2021) 012009 doi:10.1088/1757-899X/1169/1/012009

Choice of Offers by means of Multi-criteria Decision Systems, using Fuzzy Logic. Case Study

A M Prada[1] , F S Blaga[1] and M Ursu[2]

[1]Universitatea din Oradea, Facultatea de Inginerie Manageriala si Tehnologica, str.Universității nr.1, Oradea 410087, Romania
[2]Universitatea din Oradea, Facultatea de Inginerie Electrica si Tehnologia Informatiei, str.Universității nr.1, Oradea 410087, Romania

E-mail: andrei.prada.97@gmail.com

Abstract. This paper shows in detail the steps for the creation of an application in Matlab, using multi-criteria decision systems based on fuzzy logic. The aim of the application is to find the optimal offer for a client, based on the scores attained by simulation. In order to make the optimal offer in terms of quality and price for thermoplastic material parts, the main elements that are taken into account for the final price are cost of raw material, its quality and the price of the mould. The price of the raw material is mainly influenced by the annual necessary volume and by the project duration. The dimensions, the tolerances of dimensions, shapes and positions, the surface roughness, the coordinates of the critical points and the warranty of the mould have all a great influence on the price of the mould. This paper has two main chapters, of which the first one presents the selection procedure of an offer, and the definition of the multi-criteria decision systems that are used to evaluate the final offer, based on the accuracy of the choice for raw material and mould. The second chapter presents a case study which demonstrates the possibility to make such an application. The input quantities of the decision system must be carefully defined in order to attain results as coherent as possible.

1. Introduction

When there are multiple possible alternatives, the choice of an offer is a very complex process which depends on many criteria.

Papers [1] and [2] are the basic theoretical support for the conceiving of the application. Paper [3] states "Multi-Attribute Decision Making is the most well-known branch of decision making. It is a branch of a general class of Operations Research (or OR) models which deal with decision problems under the presence of a number of decision criteria. This super class of models is very often called multi-criteria decision making (or MCDM). In cases in which the number of attributes is large (e.g., more than a few dozens), attributes may be arranged in a hierarchical manner. That is, some attributes may be major attributes. Each major attribute may be associated with several sub-attributes. Similarly, each sub-attribute may be associated with several sub-sub-attributes and so on. Although some MADM methods may explicitly consider a hierarchical structure in the attributes of a problem, most of them assume a single level of attributes (e.g., no hierarchical structure)".

Paper [4] states "there have been a variety of decision support systems, with definitional differences across disciplines and countries. These include a system focus, a model focus, an application focus, and decision aids (the latter a European term). Each of these views is described with examples. The focus of the paper is on model-focused decision support systems using multicriteria models, and decision aids, of which differences between multi-attribute utility theory, analytic hierarchy process, French and Belgian outranking methods, and other approaches are discussed. The underlying preference function assumptions of each approach are compared. The many very rich applications throughout the world are reviewed".

Paper [5] describes the fuzzy logic concept. It belongs to the much larger field of the fuzzy mathematics, a non-binary kind of logic where the truth values of the variables can be anywhere between 0 and 1 inclusively. The fuzzy logic proposes the concept of partial truth, which can vary between 100% true or 100% false. This approach contrasts with the Boolean logic, where the truth values can be only 0 for "false" and 1 for "true".

The fuzzy logic is based on the fact that, in everyday life, humans make decisions by means of imprecise non-numerical information. Thus, the fuzzy sets are mathematical means to represent the imprecision of this information, so that these models will be able to recognize, represent, manage, use and interpret vague and uncertain information.

The fuzzy logic can be applied in many fields, of which the most well-known are the control theory and the artificial intelligence.

Paper [6] shows a multi-attribute decision system based on fuzzy sets which selects a project of many more available ones, for a construction company

2. The Procedure for the Selection of an Offer

2.1. General Considerations

Starting from the next premises, the elaboration of an application in Matlab with Fuzzy Logic toolbox will begin.

The most important elements that are taken into account for the evaluation of an offer are the price of the material, the prices for material tests (when necessary, the compatibility of the material with the standards of the client is checked), the price of the mould and its dimensions.

The process begins from the design of the part, of which the following information is selected in order to make the offer:

- ✓ dimensions, tolerances for dimensions, position and shape, roughness, coordinates of critical points etc. – the mould specifications will be formulated using this information, which will influence the price and dimensions of the mould;
- ✓ specifications about the material and about its validation tests, several standards imposed or agreed by the client – the costs of the raw material investment for the fabrication of the part will be assessed.

An application will be made for this purpose, based on three decision structures which are useful in order to evaluate the costs of the offer. The output quantities of the decision systems "*Choice of Material*" ("*Alegere material*") and "*Choice of Mould*" ("*Alegere matriță*") are the input quantities for the third decision structure, of which the output quantity will be a score that allows the evaluation of the offer as objectively as possible. All these are shown in Figure 1. Next, the elaboration steps of the three decision structures will be shown.

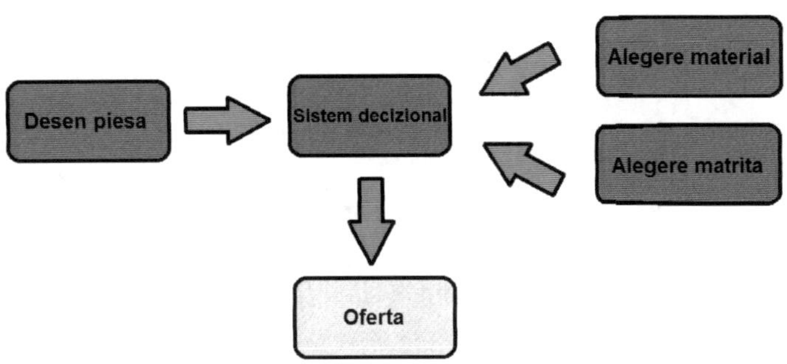

Figure 1. Basic diagram.

IOP Conf. Series: Materials Science and Engineering 1169 (2021) 012009 doi:10.1088/1757-899X/1169/1/012009

2.2. Definition of the decision systems based on fuzzy sets

2.2.1. Decision system for the choice of material

Figure 2 shows the decision system for the choice of the material, based on fuzzy sets. In order to implement the multi-criteria decision system for the evaluation of the material type the following steps must be taken:

A. Definition of the criteria (quantities for the decision process) which will be used for the evaluation of the material type:

1. *Price of material* ("*Pret material*");
2. *Compatibility to standard* ("*Compatibilitate cu standard*");
3. *Price of material tests* ("*Pret teste de material*").

B. Definition of the variation ranges for the input quantities.

The range for *Price of material* is shown by equation (1)

$$D_{Price\ of\ material} = [0;10] \quad (1)$$

The range for *Compatibility to standard* is shown by equation (2)

$$D_{Compatibility\ to\ standard} = [0;\ 100] \quad (2)$$

The range for *Price of material tests* is shown by equation (3)

$$D_{Price\ of\ material\ tests} = [2600;\ 2800] \quad (3)$$

C. Definition of the linguistic variables associated to each input quantity

The linguistic variable *price_of_material* is associated to the input quantity *Price of material* (Figure 2.a).

The linguistic variable *compatibility_to_standard* is associated to the input quantity *Compatibility to standard* (Figure 2.b).

The linguistic variable *price_of_material_tests* is associated to the input quantity *Price of material tests* (Figure 2.c).

D. Definition of the linguistic terms associated to each input quantity

The linguistic terms associated to the input linguistic variable *price_of_material* are shown by relation (4):

$$TL_{price\_of\_material} = \{very\_low,\ low,\ average,\ high,\ very\_high\} \quad (4)$$

The linguistic terms associated to the input linguistic variable *compatibility_to_standard* are shown by relation (5):

$$TL_{compatibility\_to\_standard} = \{very\_low,\ low,\ average,\ high,\ very\_high\} \quad (5)$$

The linguistic terms associated to the input linguistic variable *price_of_material_tests* are shown by relation (6):

$$TL_{price\_of\_material\_tests} = \{very\_low,\ low,\ average,\ high,\ very\_high\} \quad (6)$$

E. Definition of the ranges for the output quantity

The range for *Choice of material* is shown by relation (7)

$$D_{Choice\ of\ material} = [0,100] \quad (7)$$

F. Definition of the linguistic variables associated to the output quantity

The linguistic variable *Choice_of_material* is associated to the output quantity *Choice of material* (Figure 3).

G. Definition of linguistic terms associated to the output quantity

The linguistic terms associated to the output linguistic variable *Choice_of_material* are shown in relation (8):

$$TL_{Choice\_of\_material} = \{very\_unsuitable,\ unsuitable,\ satisfactory,\ suitable,\ very\_suitable\} \quad (8)$$

IOP Conf. Series: Materials Science and Engineering 1169 (2021) 012009 doi:10.1088/1757-899X/1169/1/012009

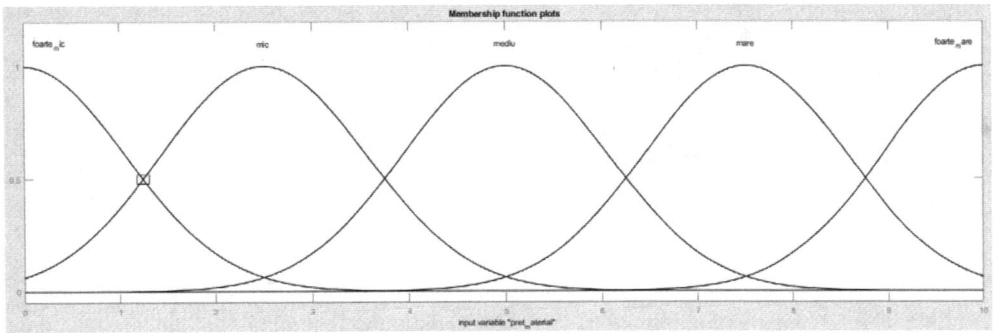

(a)

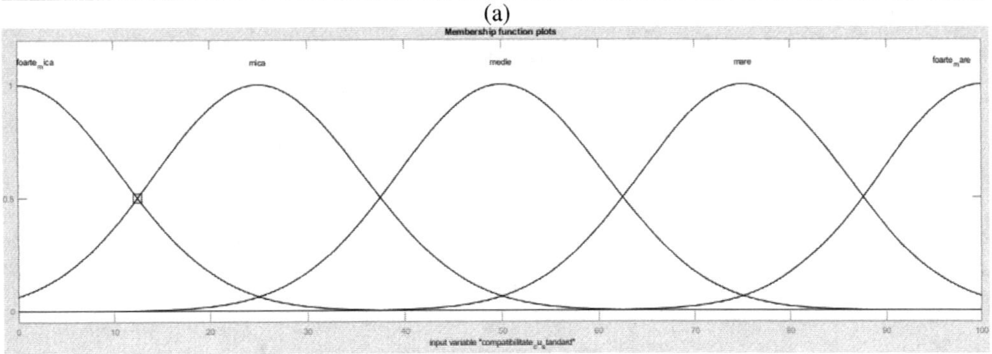

(b)

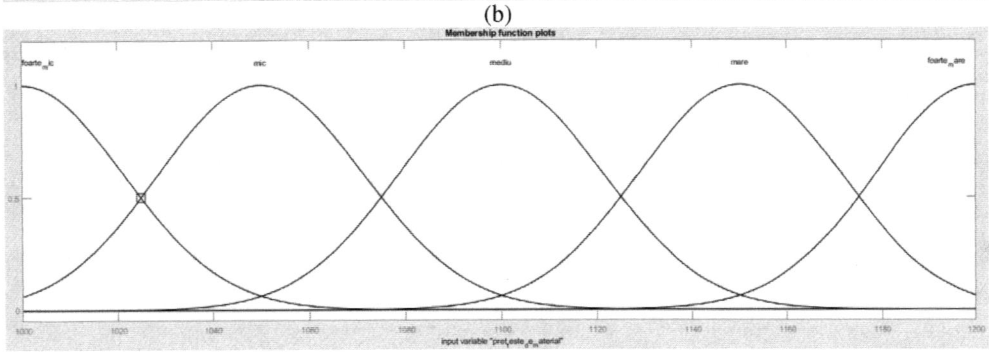

(c)

Figure 2. Input quantities of the decision system *Choice of material*: a) *price_of_material*; b) *compatibility_to_standard*; c) *price_of_material_tests*

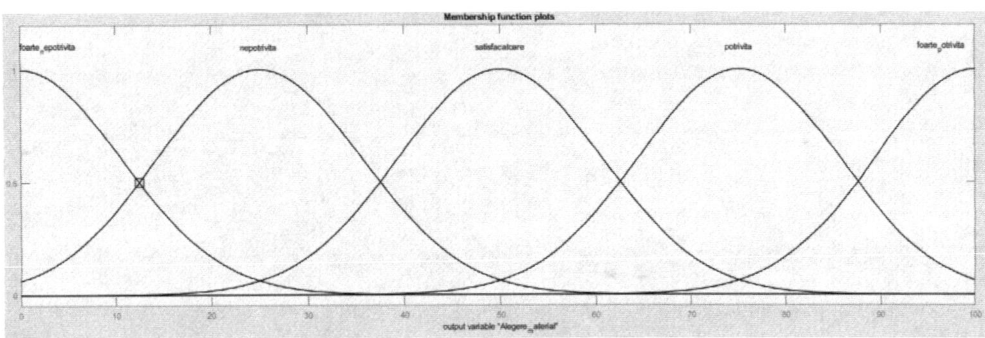

Figure 3. Score of *Choice_of_material.*

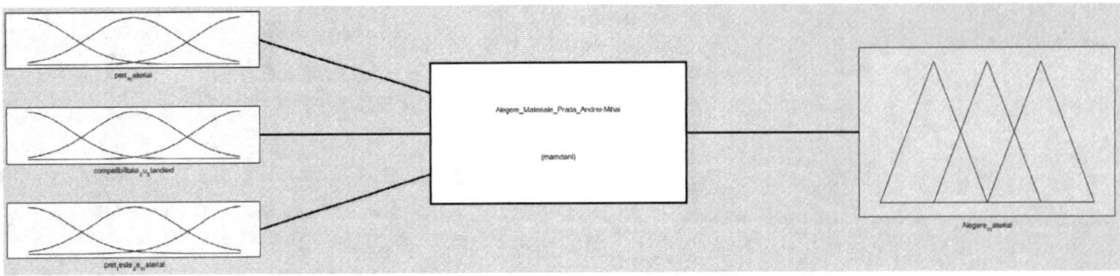

Figure 4. Decision system for the choice of material.

The dependance between the input quantities and the output quantity was defined by means of 125 inference rules:

1. *If (price_of_material is very_low) and (compatibility_to_standard is very_low) and (price_of_material_tests is very_low) then (Choice_of_material is satisfactory)*
2. *If (price_of_material is very_low) and (compatibility_to_standard is very_low) and (price_of_material_tests is low) then (Choice_of_material is satisfactory)*
3. *If (price_of_material is very_low) and (compatibility_to_standard is very_low) and (price_of_material_tests is average) then (Choice_of_material is unsuitable)*

....

54. *If (price_of_material is average) and (compatibility_to_standard is very_low) and (price_of_material_tests is high) then (Choice_of_material is very_unsuitable)*
55. *If (price_of_material is average) and (compatibility_to_standard is very_low) and (price_of_material_tests very_high) then (Choice_of_material is very_unsuitable)*
56. *If (price_of_material is average) and (compatibility_to_standard is low) and (price_of_material_tests is very_low) then (Choice_of_material is satisfactory)*

....

123. *If (price_of_material is very_high) and (compatibility_to_standard is very_high) and (price_of_material_tests is average) then (Choice_of_material is satisfactory)*
124. *If (price_of_material is very_high) and (compatibility_to_standard is very_high) and (price_of_material_tests is high) then (Choice_of_material is unsuitable)*
125. *If (price_of_material is very_high) and (compatibility_to_standard is very_high) and (price_of_material_tests is very_high) then (Choice_of_material is very_unsuitable)*

Figure 5 shows the surface that was generated by the 125 rules of the decision system *Choice of material*.

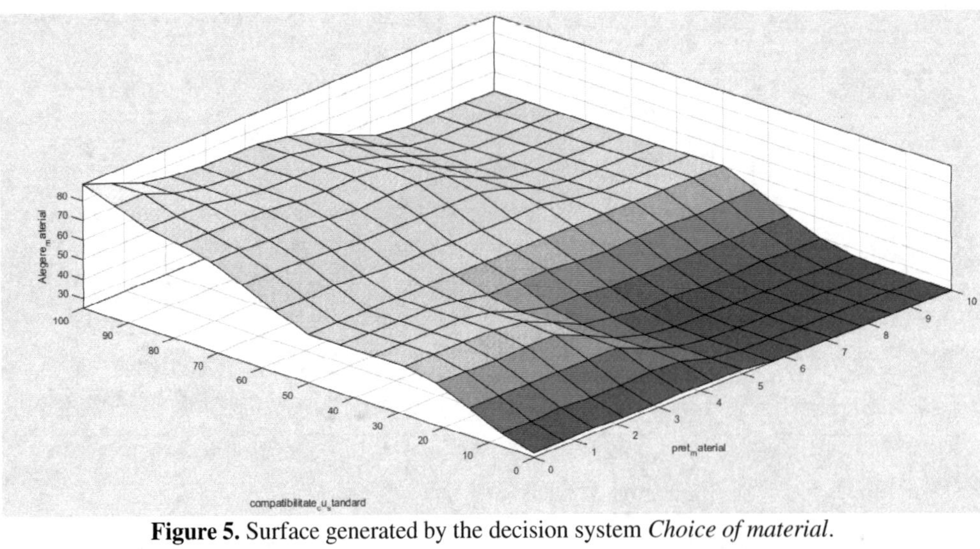

Figure 5. Surface generated by the decision system *Choice of material*.

2.2.2. Decision system for the choice of mould

Figure 6 shows the decision system for the selection of the mould, based on fuzzy sets. In order to implement the multi-criteria decision system for the evaluation of the mould the following steps must be taken:

A. Definition of the criteria (input quantities of the decision process) which will be used to evaluate the mould:
1. *Dimensions of mould*;
2. *Price of mould.*

B. Definition of the ranges for the input quantities

The range of *Dimensions of mould* is shown in relation (9):

$$D_{Dimensions\ of\ mould} = [1100,1300] \tag{9}$$

The range of *Price of mould* is shown in relation (10):

$$D_{Price\ of\ mould} = [50000,70000] \tag{10}$$

C. Definition of the linguistic variables associated to each input quantity

The linguistic variable *dimensions_of_mould* is associated to the input quantity *Dimensions of mould* (Figure 6.a).

The linguistic variable *price_of_mould* is associated to the input quantity *Price of mould* (Figure 6.b).

D. Definition of the linguistic terms associated to each input quantity

The linguistic terms associated to the input linguistic variable *dimensions_of_mould* are shown in relation (11):

$$TL_{dimensions\_of\_mould} = \{very\_low, low, average, high, very\_high\} \tag{11}$$

The linguistic terms associated to the input linguistic variable *price_of_mould* are shown in relation (12):

$$TL_{price\_of\_mould} = \{very\_low, low, average, high, very\_high\} \tag{12}$$

E. Definition of the ranges for the output quantity

The range for *Choice of mould* is shown in relation (13):

$$D_{Choice\ of\ mould} = [0,100] \tag{13}$$

F. Definition of the linguistic variables associated to the output quantity

The linguistic variable *Choice_of_mould* is associated to the output quantity *Choice of mould* (Figure 7).

G. Definition of the linguistic terms associated to the output quantity

The linguistic terms associated to the output linguistic variable *Choice_of_mould* are shown in relation (17):

$$TL_{Choice\_of\_material} = \{very\_unsuitable, unsuitable, satisfactory, suitable, very\_suitable\} \tag{14}$$

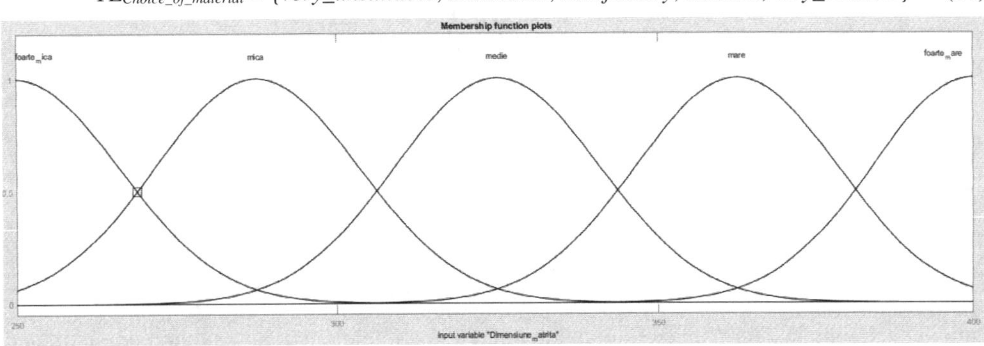

(a)

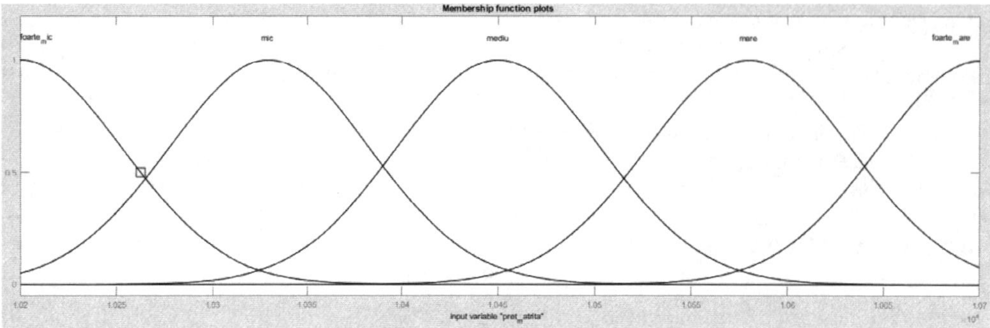

(b)

Figure 6. Input quantities for the decision system *Choice of mould*:
a) *dimensions_of_mould*; b) *price_of_mould*

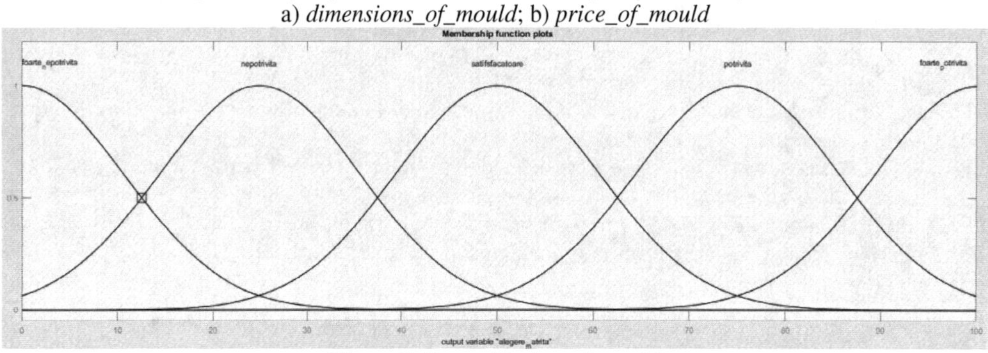

Figure 7. Score *Choice_of_mould*

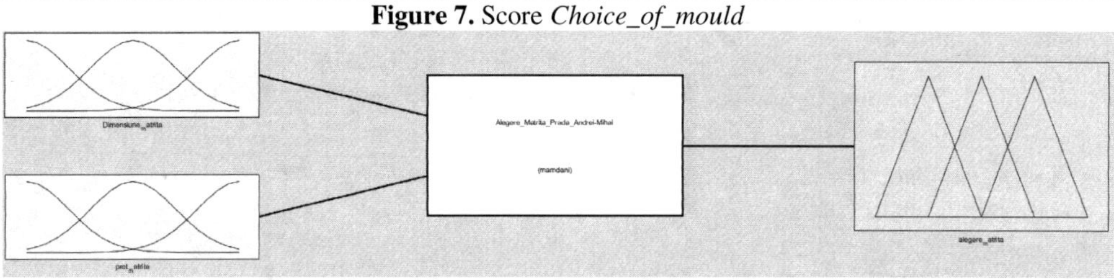

Figure 8. Decision system for the choice of the mould.

The dependance between the input quantities and the output quantity was defined by means of 25 inference rules:

1. If (dimensions_of_mould is very_low) and (price_of_mould is very_low) then (Choice_of_mould is very_suitable)

2. If (dimensions_of_mould is very_low) and (price of mould is low) then (Choice_of_mould is suitable)

3. If (dimensions_of_mould is very_low) and (price_of_mould is average) then (Choice_of_mould is satisfactory)

....

11. If (dimensions_of_mould is average) and (price_of_mould is very_low) then (Choice_of_mould is suitable)

12. If (dimensions_of_mould is average) and (price_of_mould is low) then (Choice_of_mould is suitable)

13. If (dimensions_of_mould is average) and (price_of_mould is average) then (Choice_of_mould is satisfactory)

....

23. If (dimensions_of_mould is very_high) and (price_of_mould is average) then (Choice_of_mould is unsuitable)

IOP Conf. Series: Materials Science and Engineering 1169 (2021) 012009 doi:10.1088/1757-899X/1169/1/012009

24. If (dimensions_of_mould is very_high) and (price_of_mould is high) then (Choice_of_mould is very_unsuitable)
25. If (dimensions_of_mould is very_high) and (price_of_mould is very_high) then (Choice_of_mould is very_unsuitable)

Figure 9 shows the surface generated by the 25 rules of the decision system *Choice of mould.*

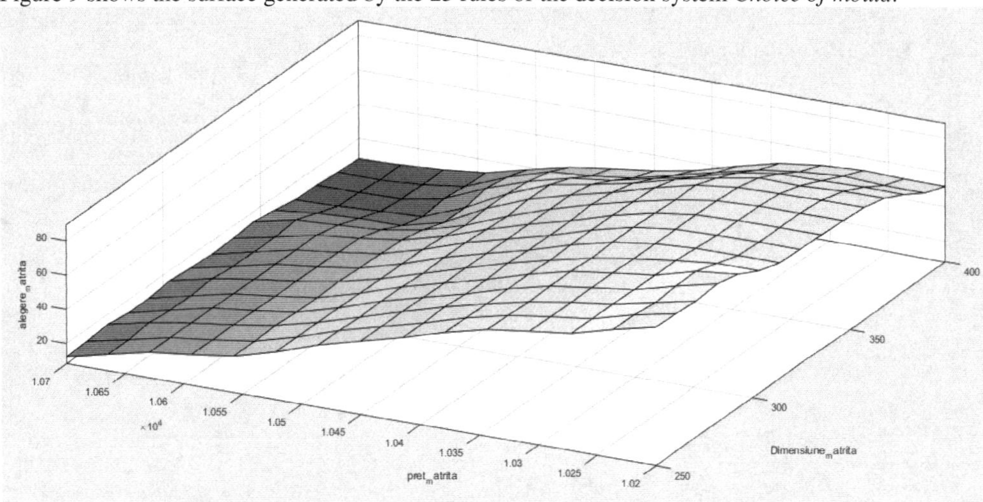

Figure 9. Surface generated by the decision system *Choice of mould.*

2.2.3. Decision system for the choice of offer

Figure 12 shows the decision system for the evaluation of the offer, based on fuzzy sets. In order to implement the multi-criteria decision system for the evaluation of the offer the following steps must be taken:

A. Definition of the criteria (input quantities of the decision process) which will be used to evaluate the offer:
1. *Choice of material*
2. *Choice of mould*

B. Definition of the ranges for the input quantities

The range for *Choice of material* is shown in relation (15)

$$D_{Choice\ of\ material} = [0,100] \quad (15)$$

The range for *Choice of mould* is shown in relation (16)

$$D_{Choice\ of\ mould} = [0,100] \quad (16)$$

C. Definition of the linguistic variables associated to each input quantity

The linguistic variable *Choice_of_material* is associated to the input quantity *Choice of material* (Figure 10.a).

The linguistic variable *Choice_of_mould* is associated to the input quantity *Choice of mould* (Figure 10.b).

D. Definition of the linguistic terms associated to each input quantity

The linguistic terms associated to the input linguistic variable *Choice_of_material* are shown in relation (17):

$$TL_{Choice\_of\_material} = \{very\_unsuitable, unsuitable, satisfactory, suitable, very\_suitable\} \quad (17)$$

The linguistic terms associated to the input linguistic variable *Choice_of_mould* are shown in relation (14).

IOP Conf. Series: Materials Science and Engineering 1169 (2021) 012009 doi:10.1088/1757-899X/1169/1/012009

E. Definition of the ranges for the output quantity
The range for *Offer* is shown in relation (18)

$$D_{Offer} = [0,100] \tag{18}$$

F. Definition of the linguistic variables associated to the output quantity
The linguistic variable *Offer* is associated to the output quantity *Offer* (Figure 11).

G. Definition of the linguistic terms associated to the output quantity
The linguistic terms associated to the output linguistic variable *Offer* are shown in relation (19):

$$TL_{Offer} = \{very\_unsuitable, unsuitable, satisfactory, suitable, very\_suitable\} \tag{19}$$

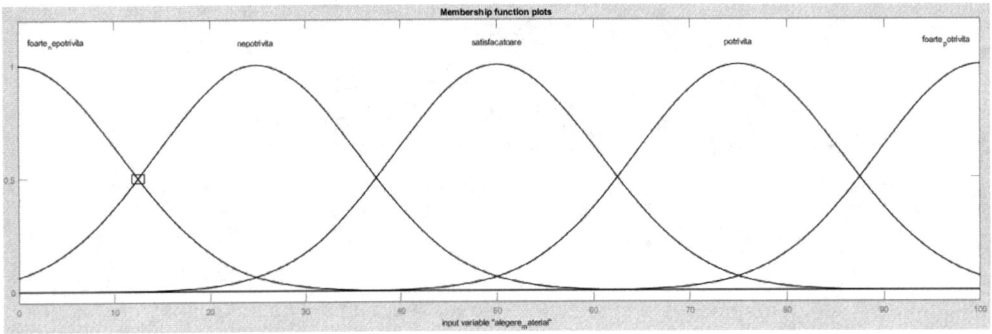

(a)

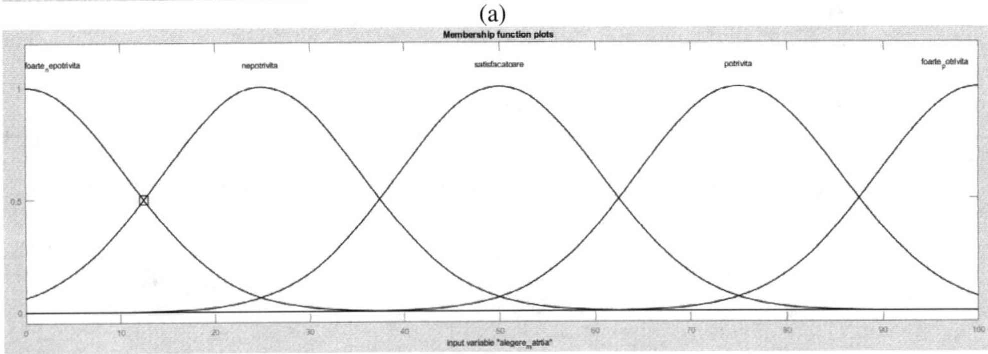

(b)

Figure 10. Input quantities for the decision system *Offer*: a) *Choice_of_material*; b) *Choice_of_mould*

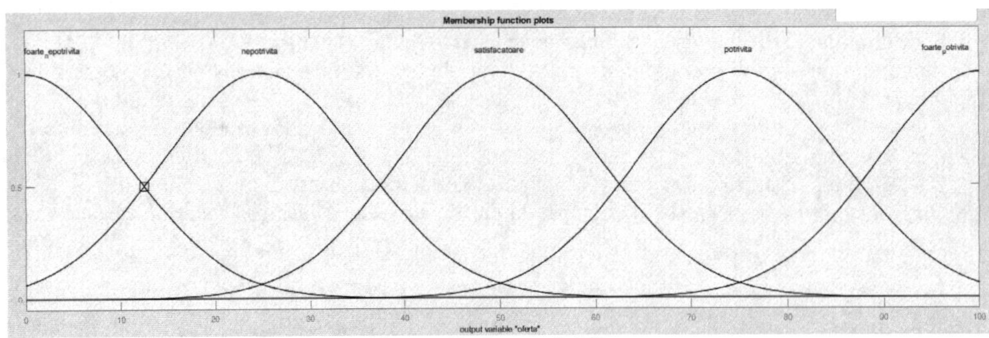

Figure 11. Score of *Offer*.

IOP Conf. Series: Materials Science and Engineering 1169 (2021) 012009 doi:10.1088/1757-899X/1169/1/012009

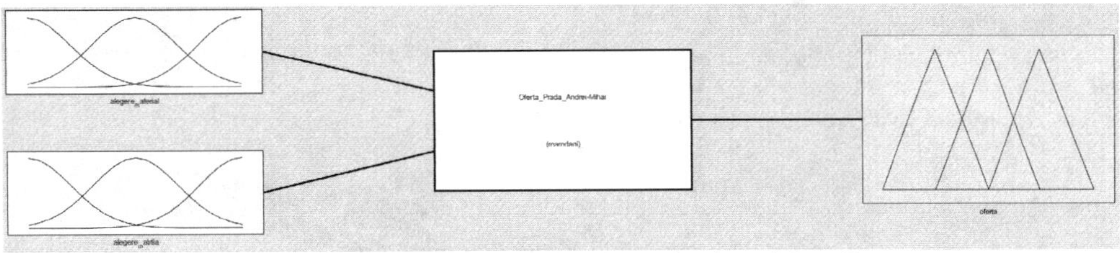

Figure 12. Decision system for the evaluation of the offer.

The dependance between the input quantities and the output quantity was defined by means of 25 inference rules:

1. If (Choice_of_material is very_unsuitable) and (Choice_of_mould is very_unsuitable) then (Offer is very_unsuitable)

2. If (Choice_of_material is very_unsuitable) and (Choice_of_mould is unsuitable) then (Offer is unsuitable)

3. If (Choice_of_material is very_unsuitable) and (Choice_of_mould is satisfactory) then (Offer is unsuitable)

....

11. If (Choice_of_material is satisfactory) and (Choice_of_mould is very_unsuitable) then (Offer is unsuitable)

12. If (Choice_of_material is satisfactory) and (Choice_of_mould is unsuitable) then (Offer is satisfactory)

13. If (Choice_of_material is satisfactory) and (Choice_of_mould is satisfactory) then (Offer is satisfactory)

....

23. If (Choice_of_material is very_suitable) and (Choice_of_mould is satisfactory) then (Offer is suitable)

24. If (Choice_of_material is very_suitable) and (Choice_of_mould is suitable) then (Offer is very_suitable)

25. If (Choice_of_material is very_suitable) and (Choice_of_mould is very_suitable) then (Offer is very_suitable)

Figure 13 shows the surface generated by the 25 rules of the decision system *Offer*.

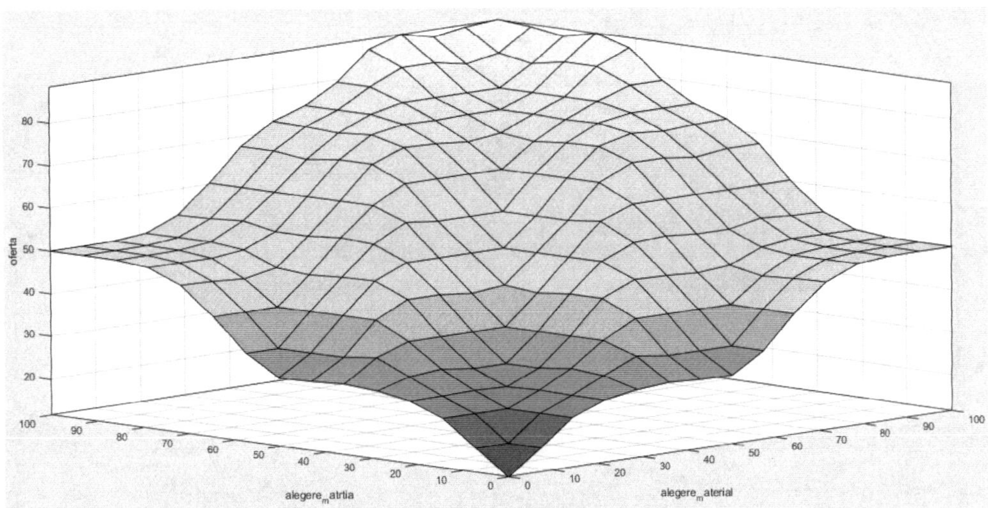

Figure 13. Surface generated by the decision system *Offer*.

3. Case Study

The part from which the application conceiving started is **Bezel LH FN**, an ornamental component of the trunk assembly for the client BOS GmbH, final client Volkswagen. Figure 14 shows the three-dimensional CAD model of this part.

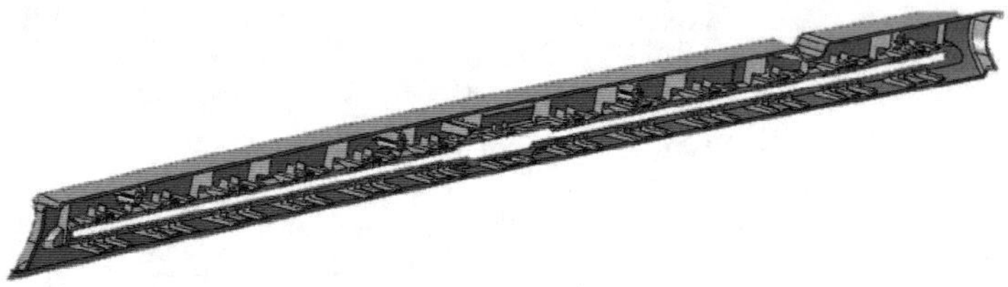

Figure 14. Part Bezel LH FN

The initial premises about the evaluation of the raw material proposed for offering are presented next. The price and the compatibility with the standard imposed by the client will also be considered.
Normally, only the materials 100% compatible to the standard are accepted, but, in crisis or force-majeure situations, alternative materials are also accepted, if they have similar properties to the standard and, of course, if they are agreed by the final customer, in our case Volkswagen.
According to Table 1, it is preferrable to choose the first version of material because it is 100% compatible to the standard, although it is more expensive than the other two options. The table shows the prices of the materials and their compatibility-to-standard coefficient.

Table 1. Prices of materials and their compatibility to standard.

Name	Price (EUR/kg)	Compatibility to standard (%)
Luran S KR 2863C Q463 UV soul	6,04	100
PC/ASA Gebablend A soul black 4PK	3,40	60
BADALAC ASA/PC 120 H UV soul	3,95	60

As for the testing of the materials, only the testing price will be considered, and the testing laboratory with the lowest prices will be selected. The testing prices are shown in Table 2.

Table 2. Prices for material testing.

Test	Laboratory	Price (EUR)
Flammability	SGS	444,00
	ASO	419,80
	IMAT-UVE	478,00
Light exposure test	SGS	1560,00
	ASO	1602,50
	IMAT-UVE	1657,50
Colorimetric evaluation	SGS	138,00
	ASO	134,80
	IMAT-UVE	140 ,00
Scratch resistance	SGS	197,00
	ASO	230,00
	IMAT-UVE	205,00
Mar resistance	SGS	311,00
	ASO	309,80
	IMAT-UVE	305,00

Total SGS = 2.650,00 EUR; Total ASO = 2.696,90; Total IMAT-UVE = 2.785,50 EUR;

Table 3 shows the ranges of the input quantities.

Table 3. Ranges of the input quantities.

Input quantity	Range
Price of material	[0;10]
Compatibility to standard	[0;100]
Price for material testing	[2.600;2.800]

The scores of Table 4 were attained by means of simulation.

Table 4. Scores attained by simulation.

Material	Testing laboratory	Price of material (EUR)	Compatibility to standard [%]	Price of material testing [EUR]	Score	Interpretation
Luran S KR 2863C Q463 UV soul	SGS	6,04	100	2.650,00	76,5	Suitable choice
Luran S KR 2863C Q463 UV soul	ASO	6,04	100	2.696,90	63	Suitable choice
Luran S KR 2863C Q463 UV soul	IMAT-UVE	6,04	100	2.785,50	47,3	Satisfactory choice
PC/ASA Gebablend A soul black 4PK	SGS	3,40	60	2.650,00	67,5	Suitable choice
PC/ASA Gebablend A soul black 4PK	ASO	3,40	60	2.696,90	58,1	Satisfactory choice
PC/ASA Gebablend A soul black 4PK	IMAT-UVE	3,40	60	2.785,50	43	Satisfactory choice
BADALAC ASA/PC 120 H UV soul	SGS	3,95	60	2.650,00	62,1	Suitable choice
BADALAC ASA/PC 120 H UV soul	ASO	3,95	60	2.696,90	58,3	Satisfactory choice
BADALAC ASA/PC 120 H UV soul	IMAT-UVE	3,95	60	2.785,50	43,3	Satisfactory choice

Figure 15 shows the best score attained by simulation.

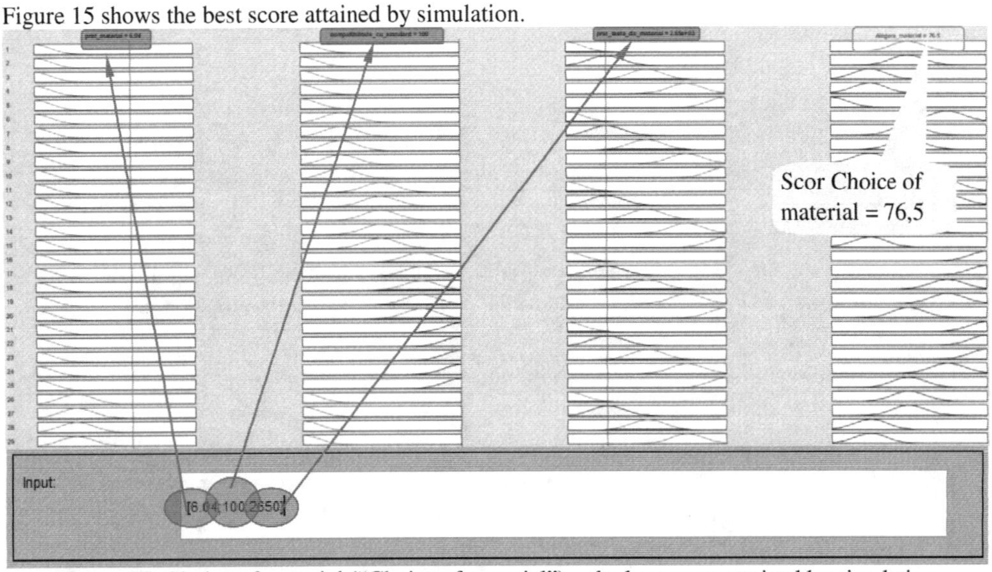

Figure 15. Choice of material ("Choice of material") – the best score attained by simulation.

As for the evaluation of the mould, the price will be considered and the average or under average prices will be selected, but also the dimensions of the mould will be taken into account. In our case, it is preferrable that the dimensions (length and width) of the mould do not exceed 1200 mm, so that it can be mounted on the table of the injection machine with 400 tons closing force. Otherwise, it will be necessary to mount the mould on a bigger injection machine, which will increase the price of the part because of the standard hourly rate of each machine. Considering the number of cavities of the mould is specified by the Reinert company according to the dimensions, geometry, mass of the part and annual quantity requested by the client, and the fact that the average value of the injection cycle is only estimated in the offering phase, these two aspects will be neglected. Thus, in our case it is preferable to take the offer based on the price specified by ZMT company (60.400 EUR), and the application was made using this consideration. Table 5 shows the specifications for the moulds.

Table 5. Specifications for the moulds.

Mould maker	Dimensions of mould [mm]	Number of cavities	Injection cycle duration (s)	Price (EUR)
DZK	1150*350*630	1	50	67.500,00
ZMT	1100*500*680	1	50	60.400,00
KK	1200*700*770	1	55	63.500,00
RYD	1300*500*670	1	50	52.980,00

Table 6 shows the ranges of the input quantities.

Table 6. Ranges for the input quantities.

Input quantity	Range
Price of mould	[50.000; 70.000]
Dimensions of mould	[1.100; 1.300]

The scores shown in Table 7 were attained by means of simulation.

Table 7. Scores attained by simulation.

Mould maker	Dimensions of mould [mm]	Price of mould [EUR]	Score	Interpretation
DZK	1.150	67.500	22,1	Unsuitable choice
ZMT	1.100	60.400	49,1	Satisfactory choice
KK	1.200	63.500	32,6	Unsuitable choice
RYD	1.300	52.980	48,3	Satisfactory choice

Figure 16 shows the best score attained by simulation.

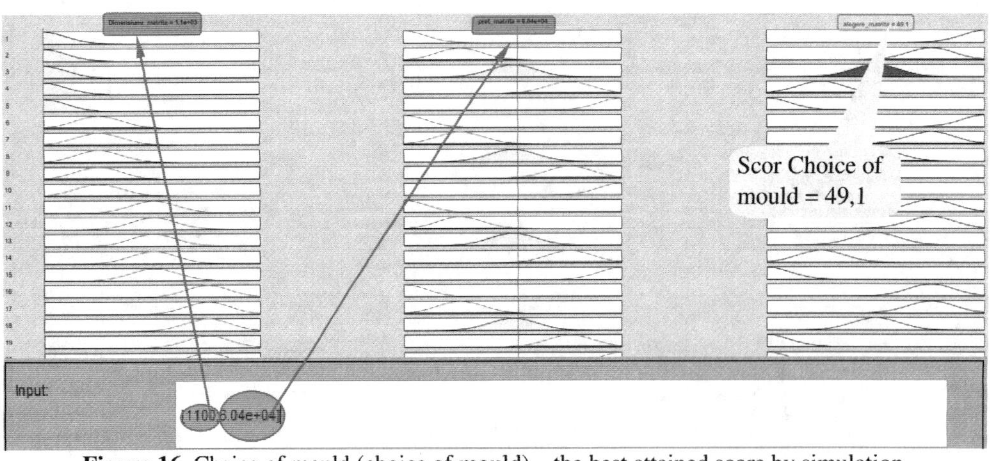

Figure 16. Choice of mould (choice of mould) – the best attained score by simulation.

IOP Conf. Series: Materials Science and Engineering 1169 (2021) 012009 doi:10.1088/1757-899X/1169/1/012009

Table 8 shows the ranges of the input quantities.

Table 8. Ranges for input quantities.

Input quantity	Range
Choice of material	[0;100]
Choice of mould	[0;100]

The score shown in Table 9 was attained by simulation.

Table 9. Score attained by simulation.

Score Choice of material	Score Choice of mould	Score Offer	Interpretation
76,5	49,1	72	Suitable choice

Figure 17 shows the score attained by simulation.

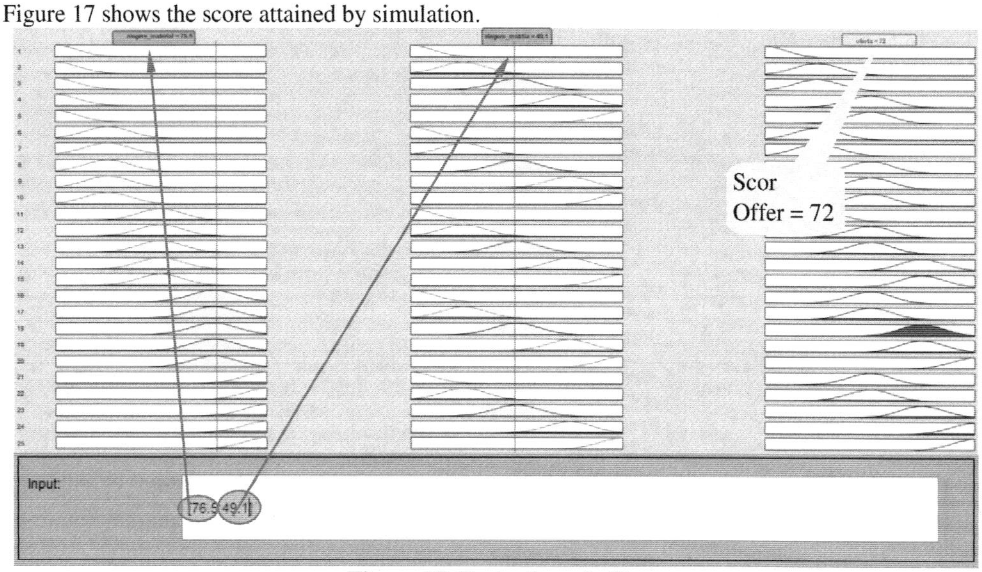

Figure 17. Evaluation of offer.

4. Conclusions

Considering the case study presented in chapter 3, it can be concluded that the designed application gave consistent results.

According to the attained score, the choice of the offer proved itself to be a suitable one. The relevance of the results depends greatly on the accurate definition of the ranges for the input quantities and on the right number of the proposed linguistic variables.

Concerning the chosen raw material, the quality of material (represented by the standard compatibility coefficient) proved itself to be more important than the price, which was almost twice the price of the other two alternative materials.

About the choice of the mould, its dimensions were decisive at the expense of the price. Considering that this is a project with low annual requirement, the choice of a too big mould would have increased the final cost of the part because it would have been necessary to mount the mould on a bigger and stronger injection machine.

The choice of the offer was evaluated as suitable according to the two attained scores: suitable choice of material and satisfactory choice of the mould.

The future research will follow the increase of the decision quality using fuzzy sets by refining the inference rules. Also, a user interface will be designed.

References

[1]http://imt.uoradea.ro/cursuri/pluginfile.php/2069/mod_resource/content/1/PPUPAC_Note%20de%20Curs.pdf, accessed 20.12.2020

[2]http://imt.uoradea.ro/cursuri/pluginfile.php/2070/mod_resource/content/1/PPUPAC_Laborator.pdf accessed 20.12.2020

[3] E Triantaphyllou, B Shu, S Nieto Sanchez and T Ray 1998 *Multi-criteria decision making: An operations research approach (Encyclopedia of Electrical and Electronics Engineering Vol 15)*, New York p 175

[4] D L Olson, F Burstein and C W Holsapple 2008 *Multi-Criteria Decision Support (Handbook on Decision Support Systems Vol 1)* p 299

[5] https://en.wikipedia.org/wiki/Fuzzy_logic accessed 1.05.2021

[6] Blaga F, Prada M, Bungau C and Stanasel I 2011 *The Selection Procedure of the Projects Construction, Based on Fuzzy Logic (Proceedings of the 2nd Review of Management and Economic Engineering International Management Conference:Management of Crisis or Crisis of Management)*, Technical University of Cluj-Napoca, Romania p 27

IOP Conf. Series: Materials Science and Engineering 1169 (2021) 012010 doi:10.1088/1757-899X/1169/1/012010

Analysis of defects in the braking systems of commercial vehicles

M F Popa[1], D M S Capătă[1] and N Burnete[1]

[1]Technical University of Cluj-Napoca, Department of Automotive Engineering and Transports, 103-105 B-dul Muncii Street, Cluj Napoca 400641, Romania

madalin.popa@live.com

Abstract. The field of commercial vehicles has consolidated an indispensable place in society through aspects such as: the continuous development of delivered performance, increased reliability, raising comfort standards and increasing traffic safety to unprecedented levels. However, increasing performance and improving the braking process are aspects that require increased attention and can be achieved by timely identification of system defects and analysis of determinants. The complexity of current systems, both the complexity of braking systems and the overall complexity of commercial vehicles, sometimes make it difficult to diagnose and repair, but they come with significant advantages. The current study analyses data on the wear of the elements of the braking system at several trucks and identifies the main trends. In addition to data on wear and maintenance, this study also examines errors in electronic control units, looking to identify dangerous situations that can lead to issues such as: reduced braking capacity; premature decommissioning of the elements in the braking system; increasing downtime in the workshop or increasing maintenance costs. The identified solutions are based on the analysis of the frequency of errors and their distribution.

1. Introduction

At present, motor vehicles have an important position in society and this position will not only be maintained but will be further strengthened as motor vehicles evolve. New technologies will significantly increase performance and further increase vehicle dependence. However, it seems that vehicles will also undergo radical changes in terms of mainly propulsion and driving systems. If the problems related to pollutant emissions, increasing energy efficiency, or eliminating dependence on fossil fuels need to be carefully considered; on the other hand, related to driving assistance (or even autonomous driving) things are much more complicated because the interaction between the electronic elements and the mechanical elements of vehicles is only at an early stage.

The development of current braking systems has gone from the simple analysis of the braking of a rotor in a friction torque, to complex mechanisms that combine mechanical braking and electronic control in a complex system, whose performance depends on the exact operation of each component. There is a high potential for the development of the elements in the braking system so as to eliminate almost all situations of premature brake failure and further to significantly reduce maintenance costs. These issues would also lead to a significant increase in traffic safety. Increasing braking performance by combining mechanics and electronics also increases the complexity of the systems. This aspect leads to the need to analyse defects on both the mechanical and electronic side.

IOP Conf. Series: Materials Science and Engineering 1169 (2021) 012010 doi:10.1088/1757-899X/1169/1/012010

2. Elements of analysis and methodology

The current paper analyzes the defects encountered in the operation of trucks, during certain periods of time. Mechanical defects (wear, cracks, blockages) and errors generated in the brake command and control units were analyzed. The data obtained were grouped and analyzed following aspects such as: identification and analysis of the main causes that lead to the failure of braking systems; identification and analysis of the main failure areas; analysis of the frequency of errors in the electronic command and control units (ECCUs) to determine the extent to which mechanical defects are detected by the ECCU and how new methods of fault monitoring can be identified.

Current braking systems deliver extremely high performance based on the interaction between mechanical and electronic systems. At present, the following can be involved in the braking process: wheel grip control systems, stability control systems, friction torque wear monitoring systems, actuation pressure control elements depending on various external factors. The sketch in figure 1 shows the complexity of the current braking systems and marks the main areas prone to failure that were analyzed in the paper (Z1-Z7 are zones of defects, a-air supply zone, b-wiring, c-unmonitored zone).

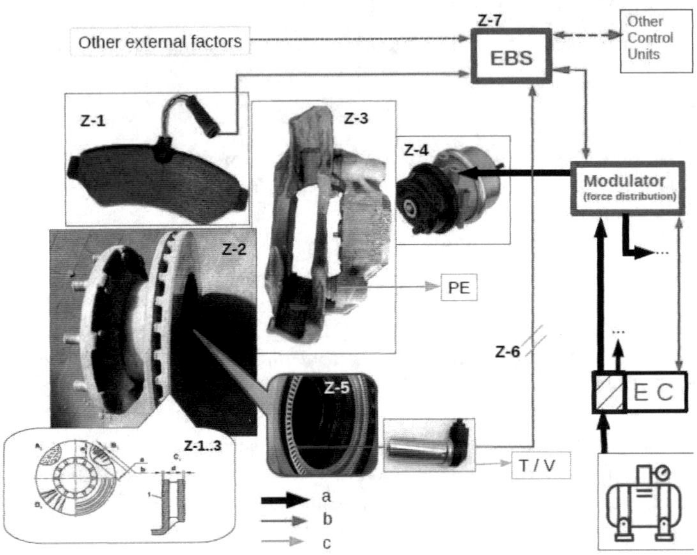

Figure 1. Schematic diagram and main fault areas of the braking systems: Z1, Z2–the area of the friction torque –characterized by different types and forms of wear. Z3–suppression of the brake caliper freedom –causes massive wear in areas Z1 and Z2, as there is no monitoring system for these defects. Z4–defects in this area lead to brake inefficiency. Z5–mechanical defects related to the wheel grip control system. The effects of these defects decrease performance and traffic safety. Z6–wiring defects. Z7–software errors.

Failure of parts of the braking system is a topic discussed in various articles [1-8]. Popa M, et all in [1], review the main defects in the braking system and in other articles [2,3] promptly analyze a series of wears in brake elements for heavy trucks and some performance of braking systems are evaluated. Defects in the brake disc area are also analyzed in other articles [4-7]. Reyes A M et all, in [4] identifies and analyzes the cracks that appeared on the surfaces of the brake discs of a bus running in areas with frequent ascents and descents. Da Silva S A M et all [5] identifies 4 types of brake discs used in vehicles and creates a mathematical model for their analysis, resulting from the simulation that the most prone to wear zone is in the area of attachment of the disc to the hub. Other studies [6,7] identify and analyze the self-loosening and breaking of the disc fixing screws on the hub on high-speed trains. However, similar situations were identified among the dump trucks used in heavy construction conditions between the trucks analyzed in the current study.

Figure 2 shows some critical situations encountered in the trucks analyzed. Starting from these situations, the main objective of the study is to analyze the defects and identify the main causes that trigger the defects and at the same time establish a link between mechanical defects and electrical defects. It is desired to identify new solutions for monitoring the operation in areas identified as high risk. For example, if Jegadeeshwaran R, in [8], identifies vibrations as the main cause of breaking the disc fastening screws, a sensor could be used to identify braking vibrations. In another study [6], for the same screws, a weakening of the screw structure due to local heating is identified as the main cause of breakage of bolts and damage of the brake disk.

Elimination of such defects could be possible if the ABS sensor in Figure 1 (Z5-T/V) would also provide the temperature in the reading area and if it could also identify some of the vibrations (by positioning the sensor relative to the polar wheel, which is attached to the brake disc). Also, in this order of ideas, monitoring the degrees of release from the brake caliper (figure 1, Z-3, PE) would help a more accurate identification of defects that lead to extreme wear situations such as those shown in figure 2.

Figure 2. The main defects identified in the braking systems: **A-** cracks in the brake disc in the area of the fixing screws; **B-** breaking the disc due to cracks; **C-** deposits that block the movement of caliper; **D, F-** thermal overload on disc surface; **E-** overloads effects on the brake pads.

In the current study, in order to prevent situations of advanced wear, the defects reported by the control units of the braking systems are analyzed. It was found that although both the diagnostic equipment and sensors present on the trucks provide accurate data, there are still unidentified defects which cause major damage in time. Based on the data obtained, the aim is to significantly reduce the incidence of serious defects while reducing maintenance costs and increasing traffic safety.

3. Results and discussions

The data analysis was performed on a number of 873 reports at the entrance to the workshop, for heavy trucks, during 14 months, starting with January 2020. During this interval, in a first stage were analysed at each entrance in workshop, the number of errors in the control units responsible for the operation and maintenance of the braking system. The data were divided into the following categories: errors related to the operation of the braking system; errors related to the maintenance part (wear); electrical errors (reports electrical faults such as: missing signal, signal interruption, erroneous signal, short circuit, etc.); mechanical errors (differences in wear, blockages, or defective parts).

Figure 3 shows the initial distribution of the defects encountered and the following main data can be outlined: the monthly distribution of the number of errors, the variation of the number of errors according to the date, the ratio between the number of errors in electronic units for operation and maintenance, the ratio between the errors that signal electrical defects and those that signal mechanical defects. It was found that both mechanical and electrical errors can be generated in both the maintenance unit and the EBS unit. Electrical errors in the maintenance unit refer to defects of the position sensors in the brake calipers or defects of the pad wear sensors (if applicable). The other errors relate to other defects in the operation of the braking systems.

Figure 4 allows the identification of periods when the frequency of defects is much higher and even the peaks are captured as the number of errors encountered in the winter months. The evolution of the number of tests performed may suggest to a certain extent the economic situation from that date, especially if we consider the fact that the study refers to commercial vehicles used for the transport of goods. Figure 5 shows the ratio between the errors reported by the central control unit of the braking system and the errors reported by the maintenance unit. There is a much higher share of errors in the central unit due to the complexity of the activities it manages. However, at a closer analysis of trends, it can be seen that, in the winter months, there are peaks in both the number of tests and the number of defects identified, which highlights the existence of influencing factors.

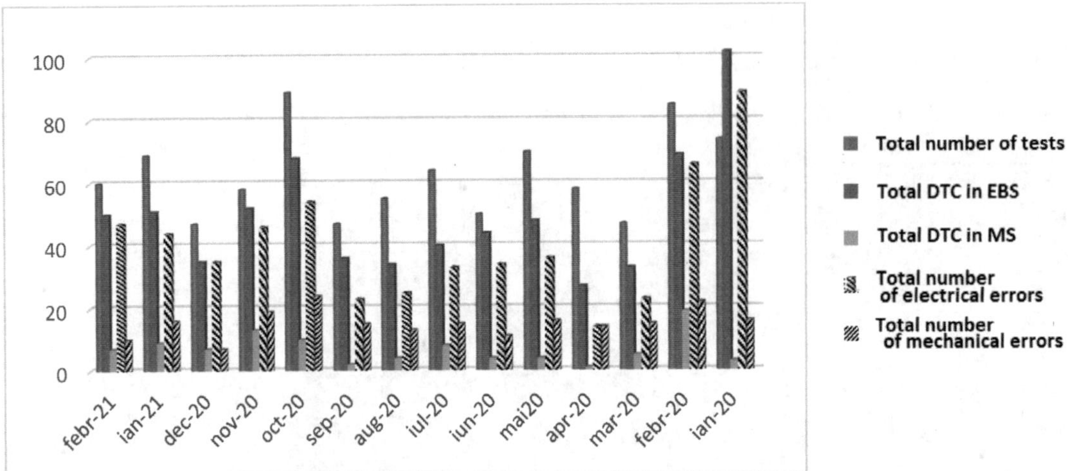

Figure 3. Distribution of defects in the first sorting (DTC- Diagnostic Trouble Codes; EBS-brake command and control unit; MS-control unit for monitoring wear on brake pads).

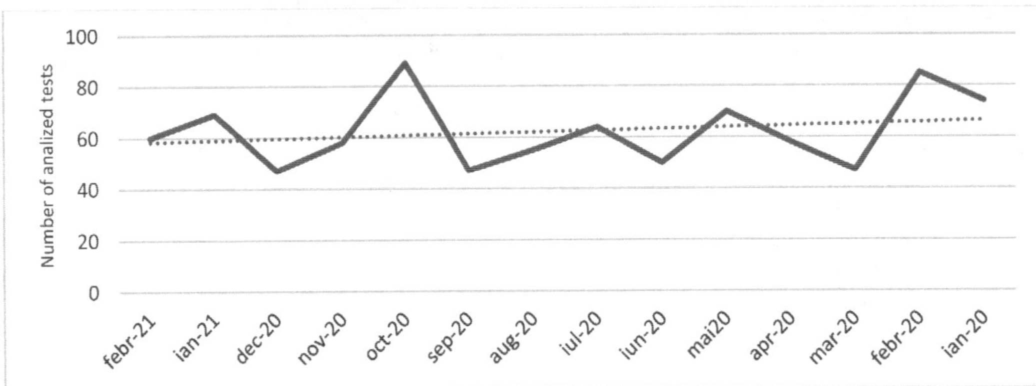

Figure 4. Number of tests analysed each month and linear trend.

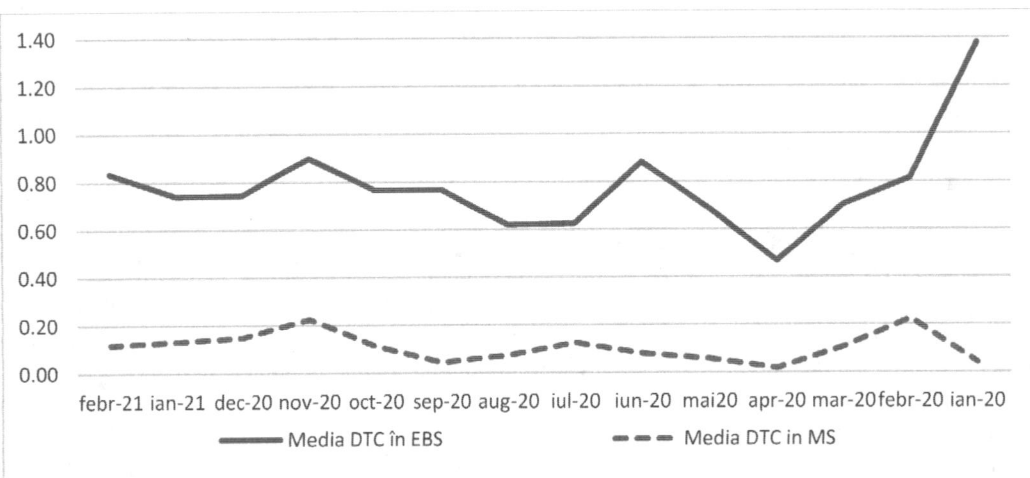

Figure 5. Average error distribution between the main brake control unit and the maintenance unit (DTC- Error Codes encountered; EBS- brake control unit; MS-control unit for wear monitoring).

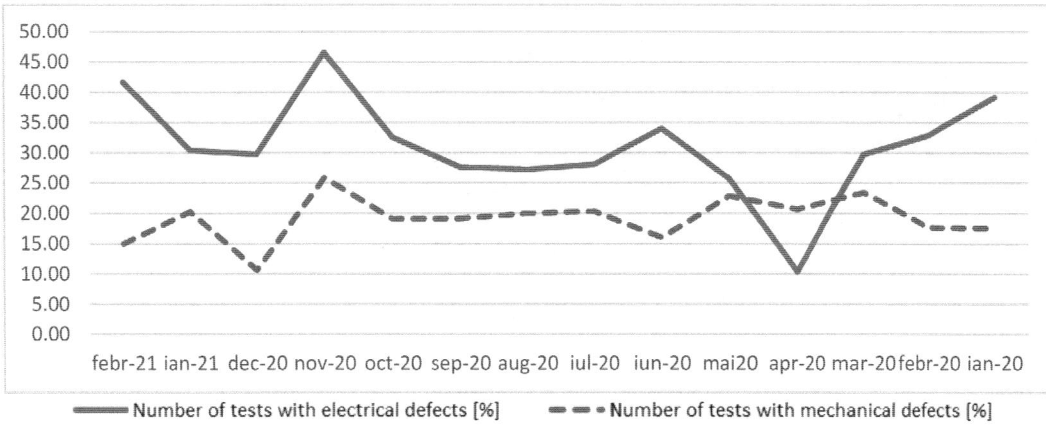

Figure 6. Defects distribution based on number of tests [%].

The analysis of the distribution of errors by type of error (mechanical or electrical) brings to the fore a significant increase in errors in the winter months. This can be explained based on the analysis of defects encountered in service and leads to the idea that humidity factors and low temperatures in the cold season favor the occurrence of electrical defects. The winter conditions on the road (humidity, mud, anti-skid materials) favor the increase of the failure rate on the exposed parts. Figure 6, in addition to the increasing trends of tests containing electrical and mechanical defects of the braking system in the winter months, also captures the effects of the corona-virus pandemic. In Romania, March and April represent the beginning of the pandemic and were characterized by a sharp decline in activity.

Another part of the study focused on the in-depth analysis of truck tests in four of the months analyzed (Figure 7). During these months, the tests were carefully analyzed following aspects such as: identification of all errors encountered and grouping by categories, identification of parts that have a high incidence of failure, identification of areas where errors occur (sensors, actuators, wheels, left / right, truck or trailer), identification of the type of error and the causes leading to the error, etc. The months of February (2020 and 2021), June (2020) and November (2020) were analyzed. A total of 240 tests were analyzed in which errors were identified in 46% of cases, which represents an average of 1.43 errors/test. From the analyzed tests, the errors encountered could be grouped into 15 categories according to the area of influence of the errors (figure 8).

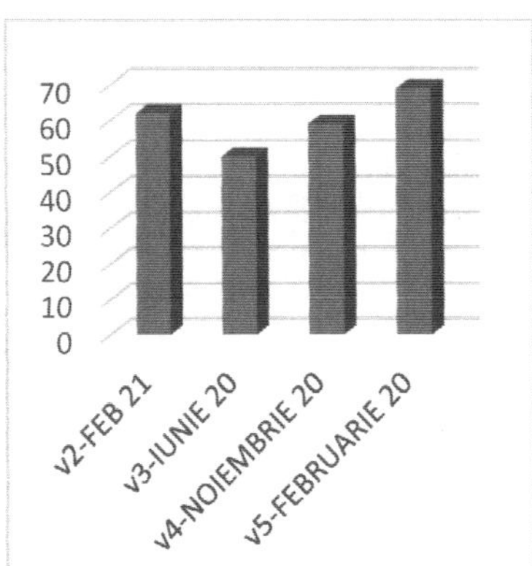

Figure 7. Total number of tests analyzed

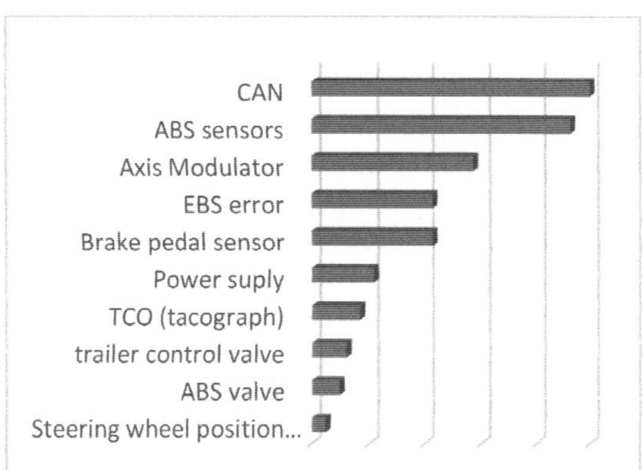

Figure 8. Distribution of defects in the 4 months analyzed [%]

IOP Conf. Series: Materials Science and Engineering 1169 (2021) 012010 doi:10.1088/1757-899X/1169/1/012010

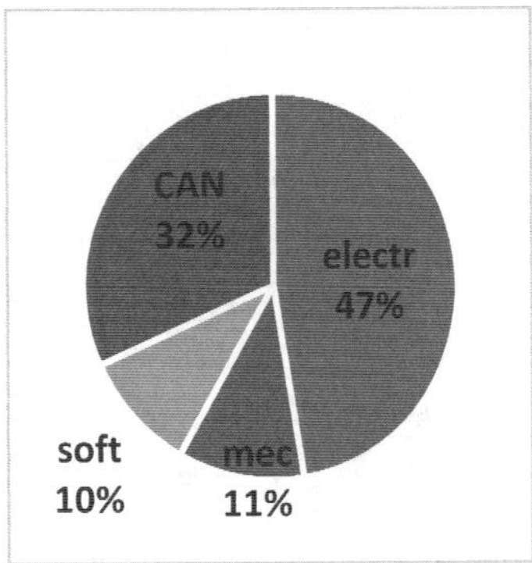

Figure 9. Error distribution

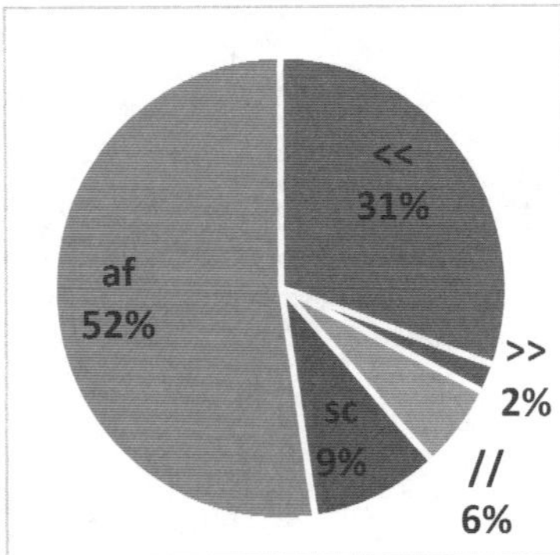

Figure 10. Causes of electrical errors

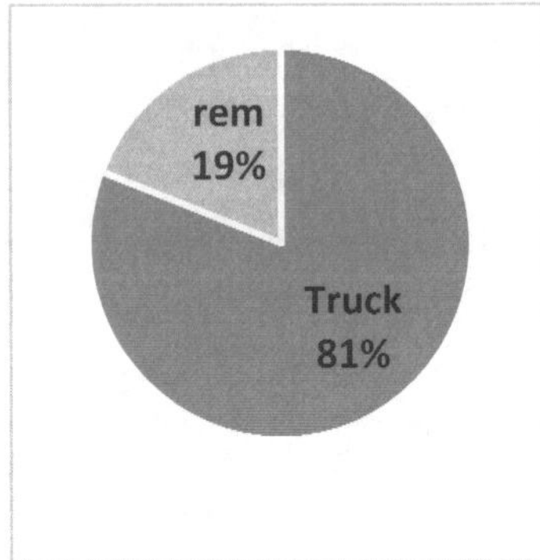

Figure 11. Frequency of errors

It was observed that the largest share is the communication errors between different command and control elements followed by defects in the ABS (anti-lock brake system) sensors. The study also divided the errors by type of error, cause of occurrence and area of occurrence. These aspects are illustrated in Figures 9-11. In figure 9 the distribution is the following: 47% -electrical errors; 32% -errors on the CAN communication bus; 11% - errors related to mechanical defects; 10% -software errors. In figure 10 the main identified causes that led to the errors were grouped as follows: 52% -external causes; 31% -insufficient voltage; 9% -short circuit; 6% -interrupted wiring; 2% -voltage too high. In figure 11 the errors were grouped by the area in which they appeared: 81% were identified on the towing vehicle and 19% on the trailer or semi-trailer.

In Figure 9, the defects are grouped according to the type of error in: electrical, mechanical, software or communication. A much higher share of electrical and communication errors was again identified to the detriment of mechanical errors. It should be noted that: electrical errors refer to faults of the electrical circuits of the type (short circuit, interrupted wire or voltage value); mechanical errors refer to mechanical defects of the parts (wear, blockages, ruptures), communication errors refer to the lack of information on a CAN bus; software errors can be fixed by updating the programming language. Analysing electrical defects, it was found that the largest share is external errors, which come from improper operation or use of non-compliant appliances. It has not yet been identified whether these errors affect the operation of the system in any way. Figure 10 shows the distribution of electrical faults caused by: external factors, too low voltage, short circuit, broken wiring, and too high voltage. In 81% of cases the defects were identified in the truck and only in 19% of the cases errors related to the operation of the trailer were reported.

The high weight of errors caused by too low a voltage once again underlines the sensitivity of current systems to voltage variations. The months analysed in the second part of the study include both winter periods and a summer month. However, a low voltage in the batteries does not necessarily have to be related to a fault. It was found that even in the hot season there can be frequent voltage drops in trucks. These issues were identified mainly against the background of declining transport demand during the covid-19 pandemic. This is explained by the fact that the driver stays in the cab and uses electricity from the batteries for a long time. The solution to eliminate such errors is to continuously monitor the truck's batteries and, in the event of a voltage drop, to display a message indicating the engine to start. This solution was found in the trucks analysed. However, during the winter, the voltage varies in much higher limits and in much shorter time intervals, an aspect that emphasizes once again the high number of errors during the winter and the sensitivity of electrical systems to certain external factors.

4. Conclusions

This study analyses the errors that occur in the electronic units of heavy trucks and identifies based on them the main trends of defecation. The reports from the entry into workshop of some trucks were analysed, regardless of the reason for entering, and in the case of intervention in the braking system, the condition of the elements in the system was analysed. Otherwise the errors related to the braking system stored in the electronic units were analyzed. The high frequency of electrical errors emphasizes the sensitivity of electrical parts and highlights the need to develop new systems to monitor the braking process in the context of vehicles becoming increasingly dependent on electronics. Based on this analysis, risk factors can be identified that, once eliminated, would significantly contribute to increasing the performance of braking systems, increasing traffic safety and significantly reducing maintenance costs for commercial vehicles.

The main determining factors in increasing the number of errors in the braking system can be listed: high humidity, anti-skid materials used in winter, low ambient temperatures, how to use, mud and other impurities on the road. The sensitivity of electrical conductors and sensors to the identified factors, significantly favors the increase of the number of errors encountered in the braking system.

Reducing the number of errors caused by winter conditions and unforeseen wear and tear can be achieved by improving electrical wiring, by monitoring the circuit voltage more closely, or by introducing new sensors to monitor areas at high risk of failure (such as suppression of degrees of release to the caliper or pads, monitoring of temperatures or vibrations in the friction torque).

Although the share of electrical errors is much more significant compared to mechanical defects, the latter should not be neglected because they have much more serious effects and are most often detected too late, after having already caused significant wear to other parts. These issues lead to a significant increase in maintenance costs and a decrease in traffic safety and an improvement in the situation can be achieved by introducing additional monitoring systems for areas identified as high risk.

References

[1] Popa M and Burnete N V 2018 *Study of brake system parameters for commercial vehicles*, AMMA, Cluj-Napoca

[2] Popa M Capata D Burnete N 2019 *Aspects of wear braking process on brake discs and brake pads in the construction of commercial vehicles*, IMT, Oradea

[3] Popa M Capata D Burnete N 2020 *Experimental research on brake behavior for different types of commercial vehicles*, SMAT, Craiova

[4] Reyes A M, et all, *Microstructure evaluation of the damage and wear characteristics of a failed disc brake of a provincial bus*, ScienceDirect, Procedings 16(2019) 1789-1795

[5] Da Silva S A M, Kallon D V V *FEA on different disc brake rotors*, ScienceDirect Procedia Manufacturing 35(2019) 181-186

[6] Xunliang S, et all, *Fatigue failure analysis of high strength bolts used for high-speed railway vehicle braking discs*, ScienceDirect, Engineering Failure Analysis 115(2020)104661

[7] Zhou Suxia, Guo Z and Xiaoyu B, *Fatigue fracture analysis of brake disc bolts under continuous braking condition*, ScienceDirect, S1350-6307(2019)31992-2

[8] Jegadeeshwaran R, Sugumaran V, *Brake fault diagnosis using Clonal Selection Classification Algorithm (CSCA)-A statistical learning approach*, Engineering Science and Technology, an International Journal 18 (2015) 14e23

An analysis of European Union's circular economy indicators with focus on materials: implications for the manufacturing industry

R L Păcurariu[1,2,3], E S Lakatos[1,2], L M Nan[1,2], L Bacali [1,2] and D Seitoar[2]

[1] Faculty of Machine Building, Technical University of Cluj-Napoca, Cluj-Napoca, Romania
[2] Institute for Research in Circular Economy and Environment "Ernest Lupan", Cluj-Napoca, Romania

[3] Corresponding author: roxana.pacurariu@ircem.ro

Abstract. The paper aims to emphasize the contribution that manufacturing industry can have to meet the Circular Economy (CE) scope related to the use of resources. First, we presented the CE concept and the framework in Europe, the monitoring framework and also, we noted some aspects of manufacturing industry related to CE. Then, we performed an analysis of the secondary raw material data, which is one of the representative indicators of CE, in order to see the state of circular economy related to the resources in European Union Member States. The results showed that the resource circularity is lowest in Europe, the states which can be remarked are Germany, Netherlands, and Belgium. Highlighting these aspects, some conclusions were formulated, the most evident is that the manufacturing industry has a very important role on the CE transition, consisting in the resources fact implicated. The manufacturing industry can choose to use secondary raw materials instead of virgin raw materials, can improve resource efficiency through the technologies used in production processes and can address the circularity of a product, being an important CE actor.

1. Introduction

The predominantly linear economic model is no longer sustainable because the resources are fast depleting and this model does not take into account the social and environmental damages [1]. The Circular Economy (CE) is a widespread concept, that is seen as a solution to the classic model of production and consumption called linear economy, and it draws attention both at European and global level as it is an alternative that meets together the economic, environmental and social needs [2]. CE can be seen as a regenerative economic system that replaces the concept of end of life with concepts such as reduction, reuse, recycling and recovery materials in the production / distribution and consumption processes, in order to achieve sustainable development, while creating environmental quality, economic prosperity and social equity, for the benefit of current and future generations [3-5]. From the European Commission perspective, a circular economy maintains the value of products, materials and resources in the economy for as long as possible and minimize the waste generation [6-7]. A key element of the CE is natural resources, for which the aims is to maximize their efficiency, transform waste into resources, recirculating the resources from products at the end-of-life through CE practices as recycling thus conserving natural capital [8].

The first European Commission initiative which aimed the circular economy (resource efficiency) materialized in July 2014 through a communiqué ("Towards a circular economy: a zero waste program for Europe") which established the framework for implementing circularity in the economy [9]; in December 2015, the European strategy for the transition from the linear to the circular economy has taken shape through the launching of an ambitious European Union Plan, named "Closing the loop - an action plan for circular economy"[10]. In order to facilitate the transition to a circular economy, the proposed measures by the European Commission cover the entire product life cycle so that resources remain in the economy as long as possible by closing the loop [11-12]. The action plan was improved in 2018, when was launched a new set of measures for CE monitoring framework [13]. On 11 March 2020, in the New Circular Economy Action Plan is outlined the immediate need for the transition to circularity, which will reduce resource consumption within the planet, accelerate towards a circular rate of material use in the next decade and decreasing dependence on virgin raw materials [14].

The system of indicators developed allows the evaluation of transition process and it is meant to complete the existing Dashboard on resource efficiency and the Dashboard of raw material resources from the European level, which have been recent implemented by the European Commission. The EU monitoring framework for the circular economy from 2018 includes 10 key indicators, which cover each phase of the product life cycle but also the major aspects of competitiveness [13,15,16]. All indicators are updated regularly, and they can be seen online, on Eurostat site. The taxonomy and the description of the indicators are presented in the Commission' official document, as an annex to the Communication from the Commission to the European Parliament on the monitoring framework for the circular economy [15].

Manufacturing industry faces the current environmental concerns which are addressed through regulation regarding the resources consumption and other issues, that are meant to drive toward the CE [17]. The manufacturing industry can affect the entire lifecycle of a product, it can introduce in the market green or circular products, made on sustainable raw materials, with high green technologies [18]. The circularity of the product can be addressed from the design phase, where the product can be designed for circularity integrating disassembly, repairability, modularity, adaptability and durability [19-20], thus they can close the loop, obtaining new resources from the transformation of waste in secondary raw materials.

Most studies that address the potential of waste use as secondary raw materials in manufacturing industry refer to construction sector [21-24]. These show the environmental benefits if the company take into account the value of waste as secondary raw material [25]. In addition to environmental benefits, the shift to CE can bring to companies' competitive advantage on the market and other economic gains, environmental benefits and social opportunities as jobs [26]. Another study in the field highlights the benefit of reducing dependence on imported raw materials and the importance in achieving the goals of the circular economy [27]. Another study underlines the importance of information system in the sense of collaboration between the parties in order to exchange secondary raw materials (wastes, by-products etc.), proposing a virtual marketplace for efficient exchange between parties, with associated CE benefits [28], and another propose for a better management of secondary raw materials is the industrial symbiosis [17]. Exploiting the potential of secondary raw materials is seen as an opportunity to close the loop, leading to environmental, social and economic gains, and ensure a secure supply to the industry [29]. In this regard are recommended measures to improve the European market for secondary raw materials, which is not very developed [30]. Another useful fact for improving the European market for secondary raw materials would be to intensify research on materials availability [31].

The manufacturing companies have the most important role in the CE transition, which can be successfully achieved if it acts in two directions: the first concerns internal evaluation and adaptation to the CE requirements, and the second is to be a driver factor towards the CE and to direct the collaborators to the CE [32]. Studies show that CE implementation in manufacturing companies cannot be done at the entity level, it implies the contribution and commitment of all stakeholders;

they are an important part of a larger system [18]. As global pressures increase, cooperation between manufacturing companies and their upstream and downstream supply chain partners is needed in order to increase product quality in terms of eco-design and eco-efficiency and at the same time substantially reduce energy, resources, emissions and waste per unit of production [33].

2. Research methodology

The analysis aims to assess the progress towards a circular economy with a focus on secondary raw materials. The analysis was performed using the available statistics from Eurostat, European Union' statistical office. There are available a lot of statistical categories, we chose the category of indicators specific to the circular economy. The circular economy indicators are grouped in four sections as follows (Figure 1): Production and consumption, Waste management, Secondary raw materials and Competitiveness and innovation.

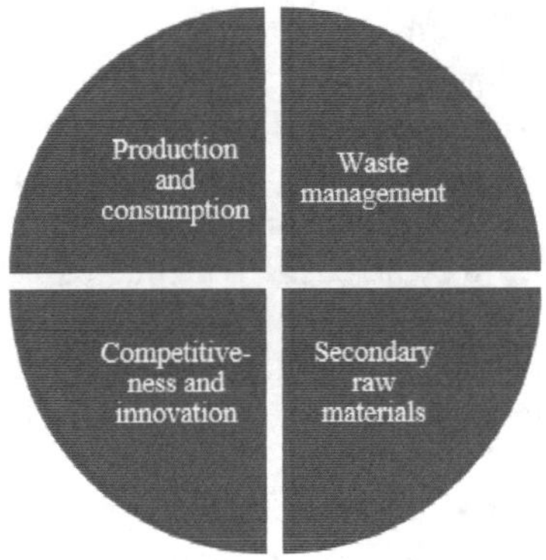

Figure 1. Circular economy indicators available on Eurostat

We analysed the third category of indicators, those related to the secondary raw materials. The databases were downloaded and the information were grouped to generate comparative graphs. Secondary raw materials indicators are organised also in two subsections: Subsection 1 - Contribution of recycled materials to raw materials demand, with two indicators End-of-life recycling input rates (EOL-RIR) (percentage) and Circular material use rate (percentage); the second subsection - Trade in recyclable raw materials, that consists in three indicators: Imports from non-EU countries, Exports to non-EU countries and Intra EU trade. The steps of the analysis can be seen in the Figure 2.

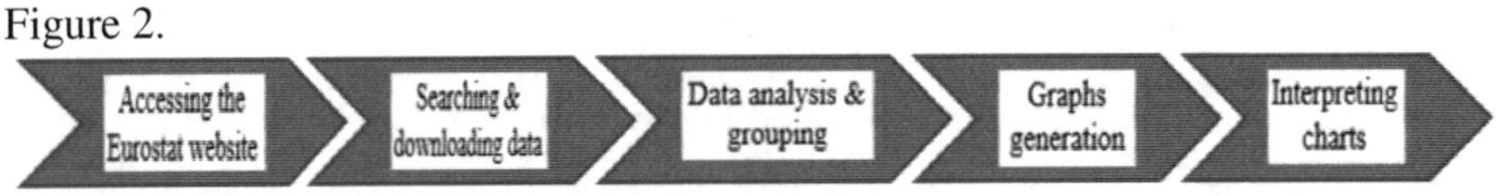

Figure 2. Steps of secondary raw materials indicators analysis

3. Results & discussion

The use of sections to divide the text of the paper is optional and left as a decision for the author. Where the author wishes to divide the paper into sections the formatting shown in table 2 should be used.

3.1. Contribution of recycled materials to raw materials demand

3.1.1. End-of-life recycling input rates (EOL-RIR) (percentage). This indicator refers to the amount of a type of raw material introduced into the production process and obtained from recycling of a material that has reached the end of its life, and his evolution in time can be seen in the following graph.

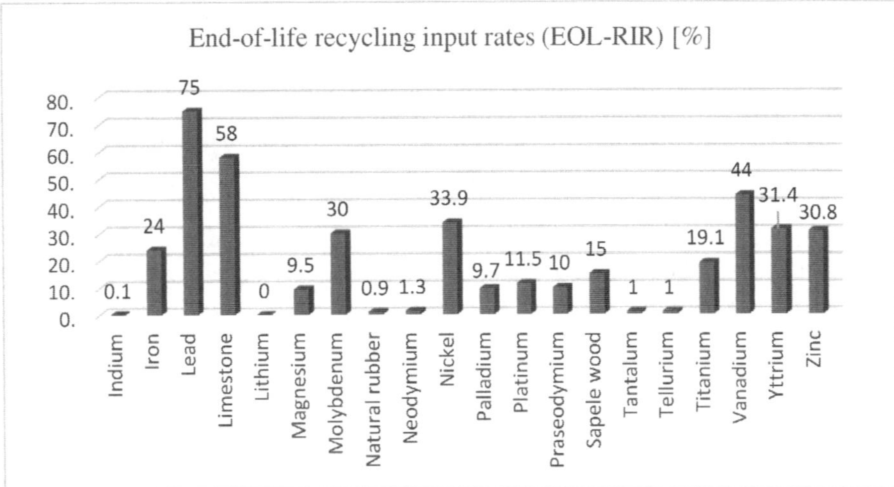

Figure 3. End-of-life recycling input rates [%] in 2016

The above figure shows that the material with the high rate of reintroducing in production process at the end of its lifespan, at the European level is Lead, with 75%. In other words, 75% of the lead's input into the production comes from recycling. The next material with a high rate of a provenience from recycling process is limestone with a rate of 58%, followed by Vanadium (44%). The lowest rate we can see for the Natural rubber (0.9%), Indium (0.1%) and Lithium (0). Unfortunately, the most recent data available are for the year 2016.

Relevance of the indicator for CE: due to the ascending demand on raw materials, the secondary raw materials are a solution for avoiding new virgin materials dislocation. The development of a secondary raw materials can help the EU in raw materials supply and contribute to a sustainable and circular economy in national and European terms. This indicator also shows the recycling's contribution of the type of material to materials demand.

3.1.2. Circular material use rate (percentage). This indicator shows the percentage of material recovered and reintroduced into the economy and the contribution to overall materials demand with the recycled materials - thus closing the loop in the product life cycle and avoiding the dislocation of new virgin raw materials.

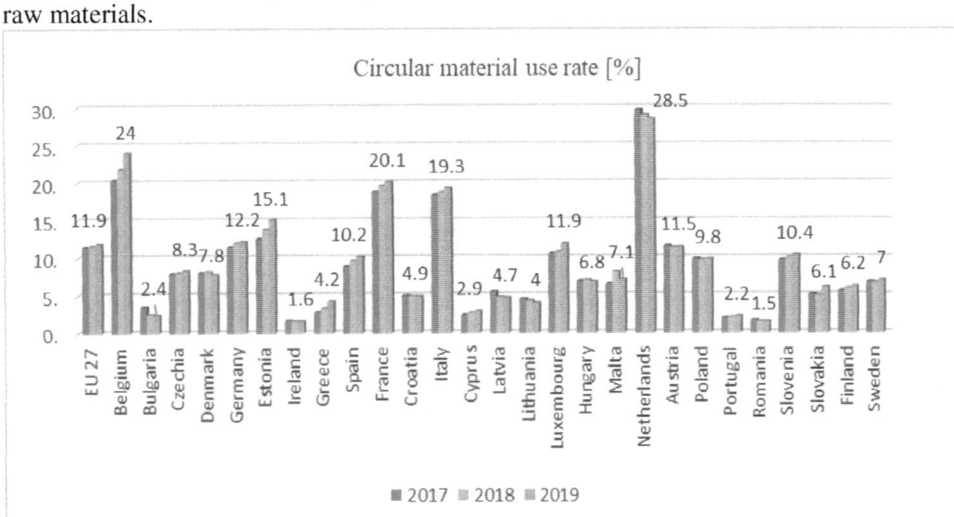

Figure 4. Circular material use rate [%]

In the figure 3 we can see the trend of the circular material use rate at the Europe level, for the three consecutive years – 2017, 2018 and 2019. We put the labels with rates' value correspond to the year 2019, to make it easier to compare. The graph shows the EU 27 mean of 11.9% in 2019 having an a slightly upward trend, the highest values for 2019 are recorded in Netherlands that is on a descendent trend, followed by Belgium but with an ascendent trend and France, also with an ascendent trend. The lowest values of the indicator we can see in Portugal, Ireland and Romania, that are well below the European average value.

Relevance of the indicator for CE: in a circular economy, waste is considered as a resource, so that can be traduced in a big material recovering and feed back into the economy from the end-of-life products, avoiding virgin resources extraction. A high value of the indicator shows a decrease in the need for new virgin material and a low level of waste.

3.2. Trade in recyclable raw materials

The indicators from this category taken into account the following five groups of materials: plastic; paper and cardboard; precious metal; iron and steel; copper, aluminium and nickel and they refer to the imported / exported quantities.

Relevance of the indicator for CE: the transformation of waste into resources and reintroducing it into European economy is an important aspect at the EU level as well as the imports / exports of certain categories of waste and by-products from non-EU countries, that can improve the supply with raw material, reduce the waste and preserve the natural capital.

3.2.1. Imports from non-EU countries. As the name suggests, the indicator is about the quantities of certain categories of waste and by-products imported by EU Member States from non-EU countries.

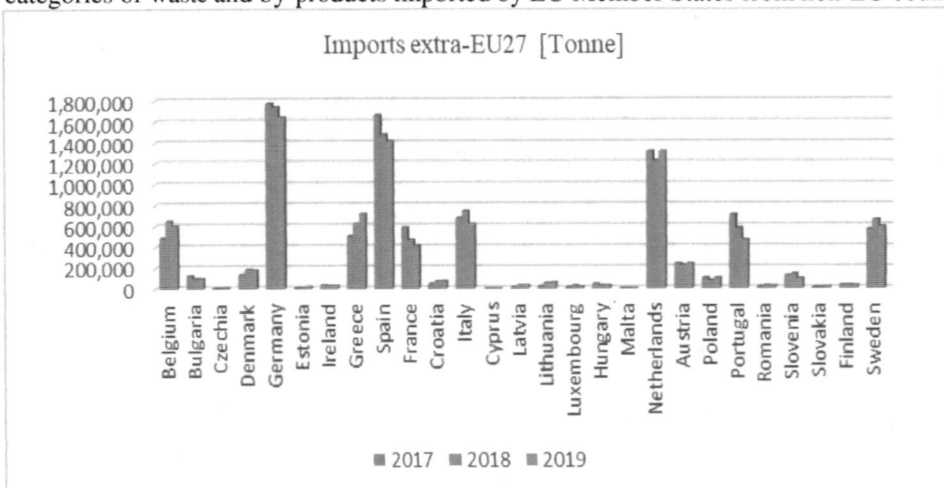

Figure 5. Imports extra-EU27 [Tonne]

The figure shows that the imports of the recyclable materials from the non-EU countries are almost non-existent or very low in many states, including Romania. It can be remarked in some countries like Germany, in a descendent trend in last three years available data, Spain with a descendent trend, also Netherlands, Italy, Sweden, Portugal, Greece, France and Belgium.

3.2.2. Exports to non-EU countries. Like the above indicator, this is about the quantities of certain categories of waste and by-products but exported by EU Member States to non-EU countries.

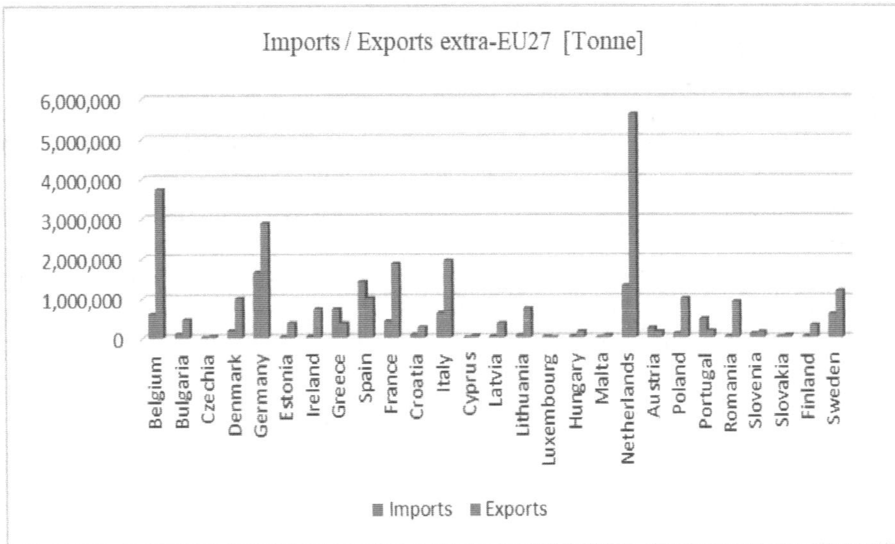

Figure 6. Imports / Exports extra-EU27 in 2019 [Tonne]

In the above graph (Figure 5) we can observe a comparison between the imports and exports of secondary raw materials in 2019. The level of the exports is bigger than the imports for most of the states, excepting Greece, Spain, Portugal and Austria. The biggest levels of exports are recorded by Netherlands, Belgium, Germany, Italy and France. In the other states, exports are almost 1000 tons or less, as in the case of Romania.

3.2.3. Intra EU trade. The indicator is about the quantities of certain categories of waste and by-products imported between Member States of European Union.

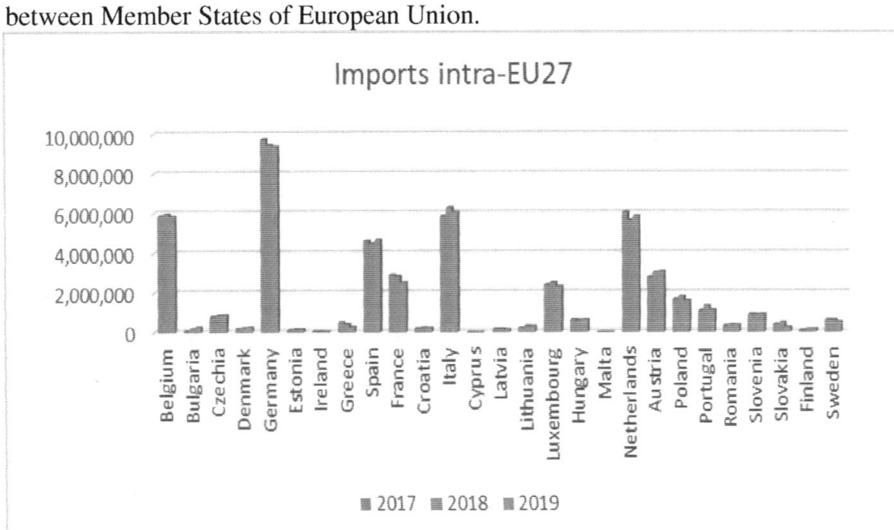

Figure 7. Imports intra-EU27 between 2017 - 2019 [Tonne]

The figure 7 reveals the statistics on intra-EU imports on last 3 years available on Eurostat – 2017, 2018 and 2019. We can see that Germany is the country more implicated in intra-EU imports with waste and by-products, registering a decreased trend, followed by Italy, Belgium and Netherlands. Romania has a lower level of this type of import, but the lower imports were in Ireland, Malta and Cyprus.

IOP Conf. Series: Materials Science and Engineering 1169 (2021) 012011 doi:10.1088/1757-899X/1169/1/012011

4. Further directions for the development of the raw materials indicator system

The EU monitoring framework is based on indicators that refer to the preserving of raw materials and materials, through recycling and waste production. Thus, the main focus of European transition policies is represented by material resources and waste [34]. Therefore, we consider that the monitoring framework indicators should address the three pillars of sustainability, environment and society harmonized with economic processes.

In order to boost the transition, we can see a need of improving the data collection system for indicators. Between the collection / processing of information and the elaboration of decisions for their change / application there must be an optimal correlation, to address the specific measures to improve the indicators within a reasonable time, or we can see that the recent data for End-of-life recycling input rates indicator are from 2016.

For an accurate image of the state of secondary raw material of each country, would be good if the statistics will be available for a certain type of material and by-products.

We must underline the fact that the circular economy is not just about the preservation of materials, we say that because most of the indicators are related to materials and recycling, the fact that is concluded also in many studies of CE issues [16]. Monitoring framework must cover and address the entire potential of the CE through specific indicators, and we suggest the expanding of indicators system is needed in order to have comprehensive measures and strategies for the circularization.

5. Conclusions

The highest EOL-RIR of materials in Europe are corresponding to materials like Lead, Limestone and Vanadium. Regarding the circular material use rate at European level, the average of EU 27 is on ascendent trend but to a low level (11.9%). The highest rate of this indicator is recorded by Netherlands, while the lowest is in Romania.

In terms of trade of recyclable materials in 2019, the exports and imports are a low level in Europe. The biggest quantities of recyclable materials from non-EU states are imported by Germany, Spain, Netherlands. Just a few countries practice this type of import, and in Romania it is almost non-existent. The level of the recyclable materials exported is bigger than the imports for most of the states, and the highest level can be seen in Netherlands. The imports between Member States are remarkable at a high level in Germany, Italy, Netherlands and Belgium.

The analysed indicators highlight a need of improving the data collection system, in order to boost the transition in a reasonable time horizon, and the implementation of decisions should not be made too late.

Due to the presented facts above, we consider that the manufacturing industry can have a significant apport to the transition to the circular economy through the phases of design and production in which they can impregnate sustainability of the product, and the second raw materials can be seen as one of the necessary instruments needed to meet the CE scope. Regarding the fostering the transition to a circular economy, an important condition for success is the collaboration with stakeholders.

6. References

[1] Lakatos E S, Cioca L I, Bacali L and Păcurariu R L 2020 *Introducere în economia circulară,* 1, Cluj-Napoca, Casa de Editură Colorama

[2] Goyal S, Chauhan S and Mishra P 2021 Circular economy research: A bibliometric analysis (2000–2019) and future research insights. *J Clean Prod,* **287**, 125011

[3] EMF 2013 Towards the Circular Economy. EMF, **1**, 1–96

[4] Geissdoerfer M, Savaget P, Bocken N M P and Hultink E J 2017 The Circular Economy – A new sustainability paradigm? *J Clean Prod*, **143**, 757–68

[5] Kirchherr J. et al. 2017 Conceptualizing the circular economy: An analysis of 114 definitions. *Resour Conserv and Recycl*, **127**, 221–32

[6] European Commission 2015 Press release - Closing the loop: Commission adopts ambitious new Circular Economy Package to boost competitiveness, create jobs and generate sustainable growth, 13–15
[7] Hughes R 2017 The EU Circular Economy Package - Life Cycle Thinking to Life Cycle Law? Procedia CIRP, **61**, 10–16
[8] Loon P.bvan, Diener D and Harris S 2021 Circular products and business models and environmental impact reductions: Current knowledge and knowledge gaps. *J Clean Prod*, **288**, 125627
[9] European Commission 2014 Communication from the Commission - Towards a circular economy: A zero waste programme for Europe, European Commission, **398**, 1–14
[10] Lazarevic D and Valve H 2016 Narrating expectations for the circular economy: Towards a common and contested European transition, Energy Res Soc Sci, **31**(May), 60–69
[11] Lakatos E S, Dan V, Cioca L I, Bacali L and Ciobanu A M 2016 How supportive are Romanian consumers of the circular economy concept: A survey. Sustain (Switzerland), **8**(8)
[12] Lakatos E S, Cioca L I, Dan V, Ciomos A O, Crisan O A and Barsan G 2018 Studies and investigation about the attitude towards sustainable production, consumption and waste generation in line with circular economy in Romania, Sustain (Switzerland), **10**(3), 1–25
[13] European Commission 2018 Report from the Commission to the European Parliament, the Council, the European Economic and Social Committee and the Committee of the Regions
[14] European Commission 2020 Communication from the Commission to the European Parliament, the Council, the European Economic and Social Committee and the Committee of the Regions, *A new Circular Economy Action Plan For a cleaner and more competitive Europe*
[15] European Commission 2018 Measuring progress towards circular economy in the European Union – Key indicators for a monitoring framework
[16] Moraga G, Huysveld S, Mathieux F, Blengini G A, Alaerts L, Van Acker K, Dewulf J 2019 Circular economy indicators: What do they measure? Res. Cons. and Recycling, **146**(January), 452–61
[17] Ferreira I A, Godina R and Carvalho H 2021 Waste valorization through additive manufacturing in an industrial symbiosis setting, *Sustain* (Switzerland), **13**(1), 1–21
[18] Bjørnbet M M, Skaar C, Fet A M and Schulte K Ø 2021 Circular economy in manufacturing companies: A review of case study literature. *J Clean Prod*, **294**, 126268
[19] Mestre A and Cooper T 2017 Circular Product Design. A Multiple Loops Life Cycle Design Approach for the Circular Economy. D. J., **20**(sup1), S1620–35
[20] EMF 2021 *Universal circular economy policy goals*
[21] Ayati B, Ferrándiz-Mas V, Newport D and Cheeseman C 2018 Use of clay in the manufacture of lightweight aggregate. *Constr and Build Mater*, **162,** 124–31
[22] Cantero B, Sáez del Bosque I F, Matías A, Sánchez de Rojas M I and Medina C 2019 Inclusion of construction and demolition waste as a coarse aggregate and a cement addition in structural concrete design. *Arch Civ Mec Eng*, **19**(4), 1338–52
[23] Jim A and Garc J 2017 *Characterization of quality recycled gypsum and plasterboard with maximized recycled content*, **67**(328)
[24] Quina M J, Bontempi E, Bogush A, Schlumberger S, Weibel G, Braga R and Lederer J 2018 Technologies for the management of MSW incineration ashes from gas cleaning: New perspectives on recovery of secondary raw materials and circular economy, *Sci Total Environ*, **635**, 526–42
[25] Migliore M, Carpinella M, Paganin G, Paolieri F and Talamo C 2018 Innovative use of scrap and waste deriving from the stone and the construction sector for the manufacturing of bricks. Review of the international scenario and analysis of an Italian case study *Environ Eng Manag J*, **17**(10), 2507–14

[26] Barbudo A, Lozano-Lunar A, López-Uceda A, Galvín A P & Ayuso J 2020 Photocatalytic recycled mortars: Circular economy as a solution for decontamination. *Appl Sci* (Switzerland), **10**(20), 1–18

[27] Safranek M and Voznakova I 2019 *Proc. Int. Conf. on Metallurgy and Materials (Brno)* Use of slag in the construction industry, **1**, 2002-7

[28] Migliore M 2019 Circular economy and upcycling of waste and pre-consumer scraps in construction sector. The role of information to facilitate the exchange of resources through a virtual marketplace, *Environ Eng Manag J*, **18**(10), 2297–303

[29] Karayannis V G 2016 Development Of Extruded And Fired Bricks With Steel Industry Byproduct Towards Circular Economy *J Build Eng*

[30] Deselnicu D C, Militaru G, Deselnicu V, ZĂINESCU G and Albu L 2018 (October) *Int. Conf. on Advanced Materials and Systems* (Bucharest) Towards a Circular Economy– a Zero Waste Programme for Europe, 563–8

[31] Ford P, Santos E, Ferrão P, Margarido F, van Vliet K J and Olivetti E 2016 Utilizing economic value, resource availability, and environmental impact metrics to improve the weee and battery directives and promote alignment with the European Commission circular economy strategy. REWAS 2016: *Towar Mater Resour Sustain*, 289–295

[32] Parida V, Burström T, Visnjic I and Wincent J 2019 Orchestrating industrial ecosystem in circular economy: A two-stage transformation model for large manufacturing companies, *J Bus Res.*, **101**(June 2018), 715–25

[33] Kamble S S, Belhadi A, Gunasekaran A, Ganapathy L and Verma S 2021 A large multi-group decision- making technique for prioritizing the big data-driven circular economy practices in the automobile component manufacturing industry. *Technol Forecast Soc Change*, **165**, 120567

[34] McDowall W, Geng Y, Huang B, Barteková E, Bleischwitz R, Türkeli S and Doménech T 2017 Circular Economy Policies in China and Europe. *J Ind Ecol.*, **21**(3), 651–61

IOP Conf. Series: Materials Science and Engineering 1169 (2021) 012012 doi:10.1088/1757-899X/1169/1/012012

Oil condition monitoring, an AI application study using the Classification Learner Technics

G Grebenişan[1,3], N Salem[2], S Bogdan[1], D C Negrău[1]

[1]University of Oradea, Romania
[2]Zarqa University, Jordan

[3]grebe@uoradea.ro

Abstract. The present study uses an estimation and prediction method and characteristic techniques to predict the life of the lubricating oil based on data collected directly from the mechanical or hydraulic system. The collected data is part of a complex data set with 19 lubricating oil status parameters resulting from online measurements on an experiment stand built and operated under conditions similar to those in a mechanical machining company. The data set was collected during six months, continuously validating the data in several 258646 instances for 19 operating parameters. To predict the values of the next steps of a sequence, the Classification Learner Technics has been approached by support vector machines (SVM) models. The answers obtained characterize and equate the training sequences with values changed by a step of the time, and this means that at each stage of the input sequence, the data structure learns to predict the output value at the next time step. To prevent divergence of the forecast, it was necessary to standardize the training data so that it will achieve a zero mean and unit variance. Also, the test data set has been normalized in the same way as training data.

1. Introduction

Machinery and mechanical installations' condition monitoring has become a necessity with the rapid development of industrial automation. Due to the high cost of accidental interruptions generated by human errors or system failures, more intelligent and reliable manufacturing systems require related vehicle condition monitoring systems and the intervention decision to maintain and diagnose the malfunctions following based online measurements, dynamic response measurements periodically updated. In general, mechanical machining machines and operating conditions are monitored using vibration measurements generated during automated, technological processes or by online analyses of lubricating oils. The monitoring of lubrication conditions is recognized as an effective means of determining the health of gearboxes and transmission train components [1-6].

Mechanical vibrations occur due to imperfections and deviations generated both by the quality of the moving parts and lack of collinearity or coaxially. In addition, the degradation of lubricating oils is caused by altering the chemical or physical properties of these lubricants, Table 1, [7-13].

The role of an effective monitoring system, based on vibration intensity measurement and observation, is to provide singularities and abrupt increases/ decreases in vibration amplitudes in a certain period of observation and measurement. Undoubtedly, the signals contain the characteristics of the vibrations measured by several transducers or sensors, and it is infrequent to use a single transducer.

Therefore, such a system also involves numerical methods or algorithms for removing the noises from the signal components taken before the signal is used to process the information contained therein.

Table 1: Physical and chemical properties subject to degradation

Property	Definition of property
Viscosity at 40 and 100 °C	Contamination of the lubricant with other oils or oxidation thereof
Moisture, or water content	Verifying the presence of water
Insolubility	High-temperature indication in operation
Total Acid Number/ Total Base Number (TAN/TBN)	Acidity/ Alkalinity of the lubricant
Flashpoint	The presence of dissolved solvents and gases in the lubricant
Analyzes for the detection of wear results	Detection of wear-related elements in parts per million(ppm)
Particle counting (ISO count)	Detection of particle number, by size, in 100 cc

The success of a classification system depends significantly on the effectiveness of the extracted observation sequence to represent a particular condition or condition of the machine. Significant efforts have therefore been made in research works, the development of various techniques for extracting features, and monitoring the state of the machinery and installations. Characteristics extraction algorithms can be grouped, according to the basic methods used, into three main categories: the Principal Components Analysis, which is based on the identification of the axes on which data is most variable, the learning approach (Neural Networks, Pattern Classification, which have the advantages of superior learning, noise suppression, and parallel computational skills) and signal processing (Fourier transform, Hidden Markov Models or wavelet procedure, which allow the classification based on modulus maxima distribution), [14-26].

2. The Classification Learner application

This application allows you to approach the Feature Selection classification and establish Predictors Classes (in this study, the predictors are: column 16 (para_param: Param A used for oil degradation calculation), column 17 (oh_paramb: Param B used for oil degradation calculation) and column 18 (oh_paramc: Param C used for the oil degradation calculation) of the data file to be analyzed, see also Table 2, [27-29]:

Table 2- Parameters' definitions and symbols' significance

Crt. nr.	Parameter Symbol	• Significance of parameters
1	icm_rh	• Relative humidity of the oil (%)
2	icm_flow	• Oil flow (ml/min)
3	icm_temp	• Temperature of the sensor (°C)
4	icm_iso4	• Disaggregated ISO Cleanliness Code 4406:1999
5	icm_iso6	• Disaggregated ISO Cleanliness Code 4406:1999
6	icm_iso14	• Disaggregated ISO Cleanliness Code 4406:1999
7	icm_pc4	• Particle counter 4μ
8	icm_pc6	• Particle counter 6μ
9	icm_pc14	• Particle counter 14μ
10	fps_vcst	• Kinematic viscosity (cSt)
11	fps_v	• Dynamic viscosity (cPoise)
12	fps_density	• Density (g/cm^3)

13	fps_dielectric	• Dielectric
14	fps_temp	• Oil temperature (°C)
15	oh_temp	• The temperature of the sensor (°C)
16	oh_parama	• Param A used for oil degradation calculation
17	oh_paramb	• Param B used for oil degradation calculation
18	oh_paramc	• Param C used for oil degradation calculation
19	oh_od	• Oil degradation

From the same data file, column 19 (oh_od: Oil degradation, calculated using parameters oh_parama, oh_paramb, and oh_paramc, [27], [28]) is the basis of the learner classification response, used as the acceptance level of the application solution. The data contained in column 19, i.e., the values of the total oil degradation parameter were converted to binary values 0, corresponding to the values oh_od> 28 or 1, corresponding to the importance of oh_od <28, i.e., a degraded quality grade oil to facilitate the classification she predictors, depending on the target responses. The application was developed successively for six classification models (Tree, Linear SVM, kNN, Linear Discriminant, Quadratic Discriminant, and Quadratic SVM), validated with PCA modelling in all six models. Response Classes generated by the application were 64. The highest accuracy, 99.7%, was obtained for the k-nearest neighbour (kNN) model. This technique, kNN, is, in fact, a remote function, which searches in a set of data and allows the graphical display of the closest "neighbour" based on a required query. The kNN search technique and kNN-based algorithms are widely used as reference learning rules. In addition, the relative simplicity of the kNN search technology makes it easier to compare the results from other classification techniques with kNN results, [1], [30-33], figure 1:

Model 1: Trained

Results

Accuracy	99.7%
Prediction speed	~220000 obs/sec
Training time	21.699 sec

Model Type
Preset: Fine KNN
Number of neighbors: 1
Distance metric: Euclidean
Distance weight: Equal
Standardize data: true

Feature Selection
All features used in the model, before PCA

PCA
PCA is keeping enough components to explain 95% variance.
After training, 1 components were kept.
Explained variance per component (in order): 99.6%, 0.4%, 0.0%

Figure 1.- The General kNN model characteristics after classification run

IOP Conf. Series: Materials Science and Engineering 1169 (2021) 012012 doi:10.1088/1757-899X/1169/1/012012

In the Classification Learner, a particular problem is creating a proper separation. This approach can be made by graphical representation using a scatter plot through graphic representation in a pair of predictors before training a classifier. Figure 2 shown a complete plot scatter, for two pairs of predictors, consisting of the data columns: oh_parama versus oh_paramb; oh_parama versus oh_paramc; oh_paramb versus oh_paramc. Note that, in the figures, the data columns correspond to the data in the file containing the collected data set, with the following correspondence: column_1 = oh_parama, column_2 = oh_paramb, column_3 = oh_paramc. It is easy to see, analyzing the six representations, that the chosen predictors are valid for the data classification: separation of data is good, the concentration of data sets in the two classes being remarkable.

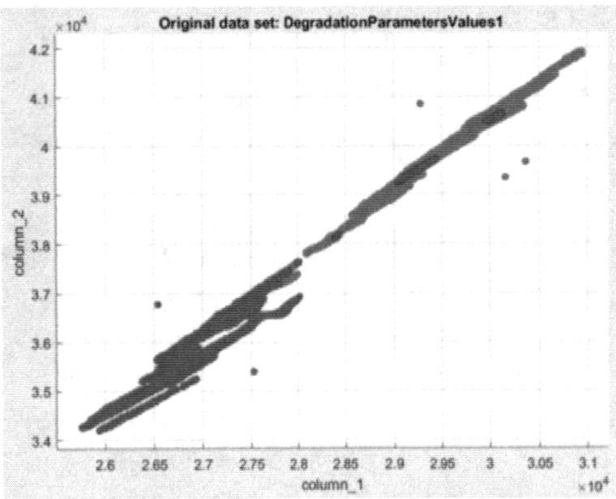

a)-oh_parama versus oh_paramb, data set

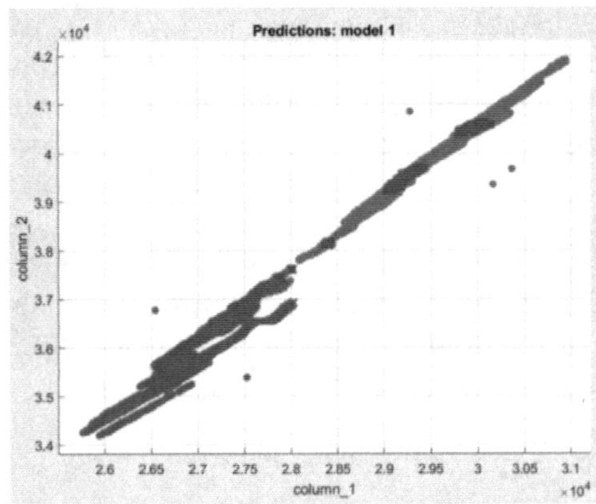

b)- oh_parama versus oh_paramb, model

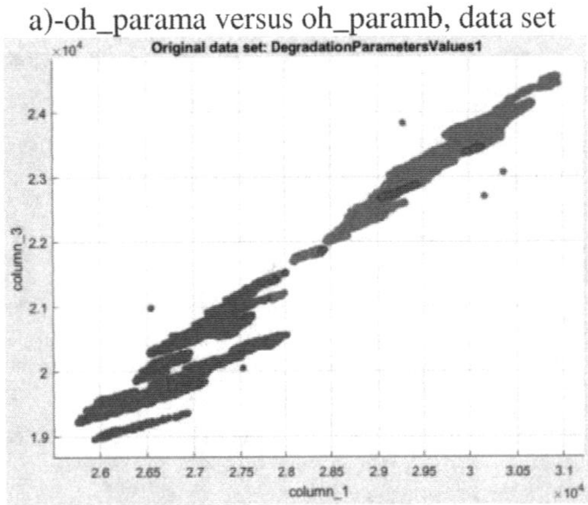

c)-oh_parama versus oh_paramc, data set

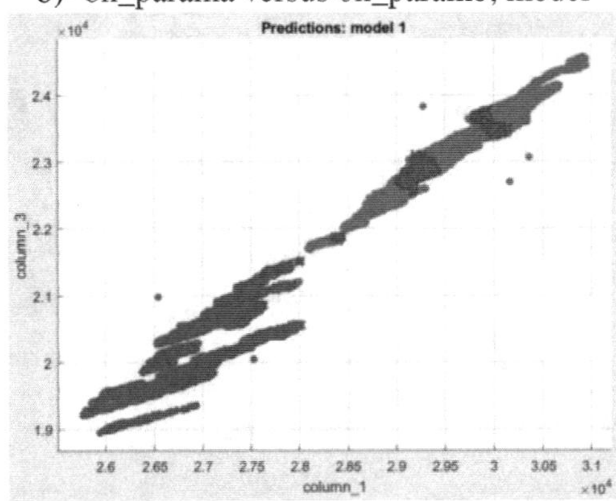

d)- oh_parama versus oh_paramc, model

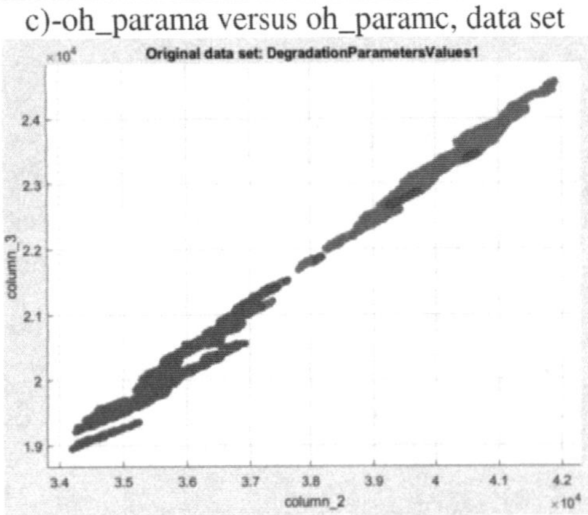

e)-oh_paramb versus oh_paramc, data set

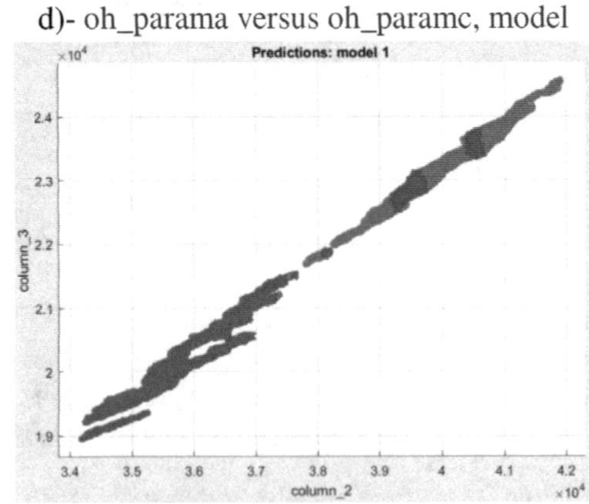

f)- oh_paramb versus oh_paramc, model

Figure 2-Scatter Plot for the data set and model predictions

IOP Conf. Series: Materials Science and Engineering 1169 (2021) 012012 doi:10.1088/1757-899X/1169/1/012012

The Confusion Matrix shown in Figure 3 and 4 describes a graphical method to help identify areas where the classifier had good results (diagonal area, green-colored):

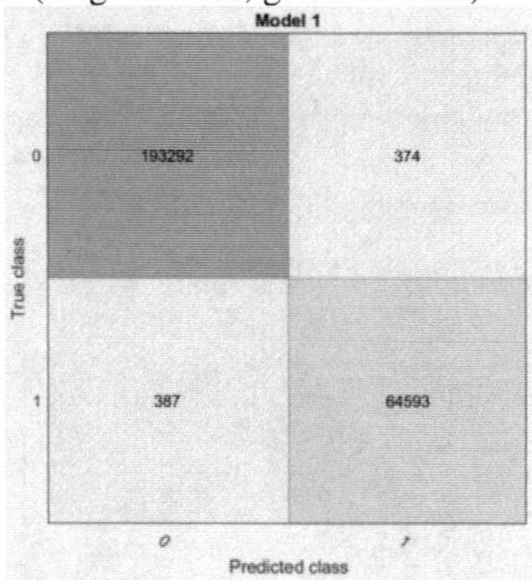

Figure 3- The confusion matrix, plotted using a number of observations

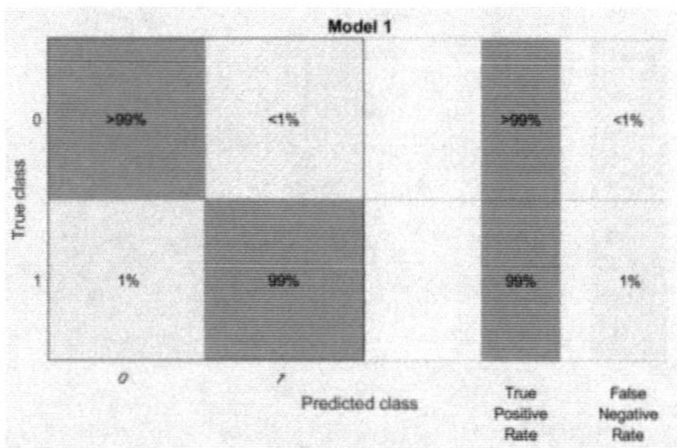

Figure 4- The confusion matrix, plotted using
True Positive Rate and False Negative Rate options

A simple method to show the true positive rate versus the false positive rate for the currently selected trained classifier consists of plotting the Receiver Operating Characteristic (ROC) curve for each of the two classes, figure 5. One can select different classes in the plot. The marker on the plot, in figure 5, a), (0.01, 1) shows the performance of the currently selected classifier (class 0). The marker shows the values of the false positive rate (FPR: 0.01) and the true positive rate (TPR: 1.00) for the classifier 0, in this case, and in figure 5, b), the marker shows the values of the false the positive rate (FPR: 0.00) and the true positive rate (TPR: 0.99) for the classifier 1, respectively.

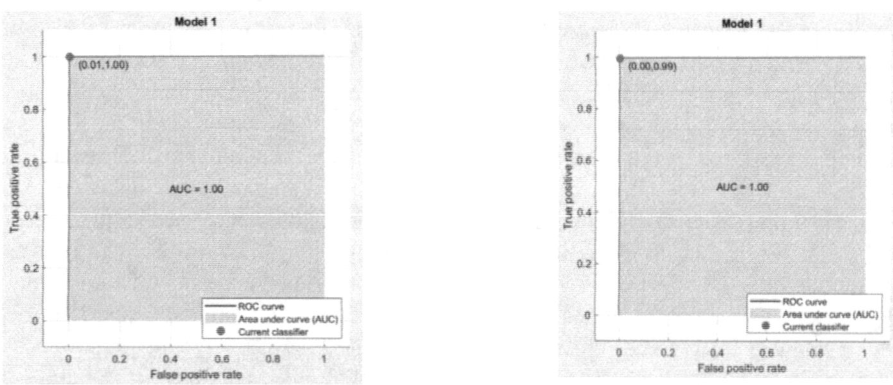

a)-Positive class: 0 b)-Positive class: 1

Figure 5- The receiver operating characteristic (ROC) curve for two classes

A decision, such as, which features to include in the classification, or which ones are excluded from the classification, respectively, can be determined by viewing the Parallel Coordinates Plot graphical representation. This graphical representation helps to understand the existing relationship between features and to identify useful predictors for different classifiers. You can thus view the training date and points unclassified or classified incorrectly. The latter is represented on the graph, always with dashed lines, figure 6:

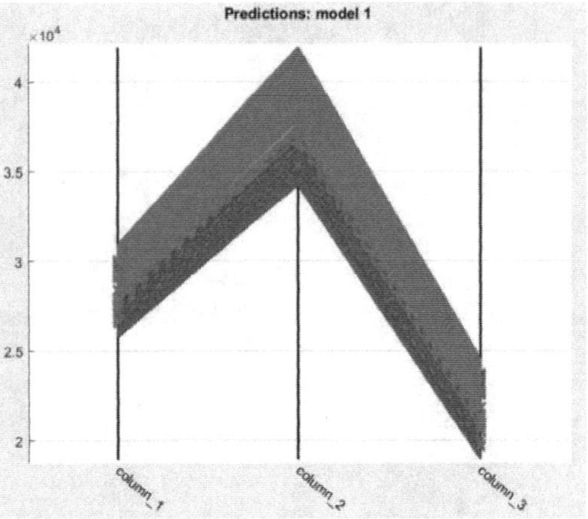

Figure 6- Parallel Coordinates Plot
(incorrectly classified points are represented with dashed lines)

3. Conclusions

Condition monitoring, diagnosis, and predicting the estimation of some output parameters for a technical system, respectively, can be framed in the "extrapolation" chapter as a mathematical, evaluation, testing and validation operation [27-29]. Doubtless, the time series for which mathematical operations of this kind can be applied are subject to errors due to the statistical nature of these mathematical classes. Therefore, a prediction is understood as an estimate of the values of a time function, based on values of a time series, values that can be, or can not be, affected by random errors. Thus, for example, a prediction problem could be expressed as follows: Given a series of time, $S(t)$, which consists of a set of values, and a random set of disturbing signals assimilated to a group of noise, $Z(t)$, it is proposed to estimate a future value, a prediction, therefore, $P(t+\tau)$, where τ is a positive constant, the forecast is also a continuous function of time.

This study was intended to show how to predict time series data using Artificial Intelligence techniques [3] on a set of data collected using direct/ indirect sensors and computational determinations based on empirical relationships. The data set contains 19 parameters, not all independent, whose values have been collected over six months or determined empirically. From this set, the data collected for three status parameters (which determines the degree of degradation of the lubricating oil) and a set of values corresponding to the set of target values were selected, the latter being used as a comparison element, training, learning, testing, and validation, respectively. The number of time steps is estimated at 258646 items. Outliers detection techniques have replaced missing or erroneous values by comparing the closest mean value set. The values with which these abnormalities have been replaced complies with the condition that they are interpolated linearly over a strictly determined range in the valid set of values.

To predict the values of the next steps of a sequence, the Classification Learner Technics has been approached by support vector machines (SVM) models. The answers obtained characterize and equate the training sequences with values changed by a step of the time. This means that at each step of the input sequence, the data structure learns to predict the output value at the next time step [1].

The prognosis process validating the time multiples values, the predictedAndUpdateState function in Matlab® [1] was used, which allows the anticipation of consecutive time steps to update the status of the network at each prediction. This experiment uses the data set collected within the National Program PN II project, ERA MANUNET: NR 13081221, [27], [28], [29]. These experiment models and analyzes at the same time the behaviour of a mechanical system, studying its behaviour based on data collected from applied sensors the hydraulic system of automatic machine lubrication. Computational determinations have been made for a small set of parameters to generate an estimate of the degradation state of the lubricating oil (A, B, C), respectively for the complete set of data (are used all 19 independent parameters, or not, collected during the experiment). The latter analysis required an excessive computational effort in the necessary resources, involving parallel computing techniques on an i7 computer with eight kernels. For this approach, it was also necessary to process the data provided by the last data column, which contains calculated estimations of the qualitative degradation of the oil (empirical determination, [34], [35]) so that the data column reflects the correct classification of the qualitative estimates as well as the values collected from the system sensors for the other parameters of the lubricating oil. In this sense, the last data column in the data set matrix contains binary values (0 and 1) for the two possible states of the oil in the plant: the logical value "1" corresponds to a degraded oil, and needs to be changed, respectively the logical value "0", represents a suitable quality oil, and can be used in the lubrication system. The data set is partitioned as follows: 70% of the data volume will be used for artificial network training, and 15% of the collected data set is used for network testing, and 15% of the dataset is used to validate the results. The working procedure states that the response values of the system are the network training data sequence [36]. Carefully prevent divergence of the forecast; it will be necessary to standardize the training data so that it has zero mean and unit variance.

Acknowledgment

This work is supported by Machine Intelligence for Diagnosis Automation (MIDA), Research Program, National Program PN II, ERA MANUNET: NR 13081221/13.08.2013

References

[1] Statistics and Machine Learning Toolbox™ User's Guide, https://www.mathworks.com/help/releases/R2018a/pdf_doc/stats/stats.pdf, Accessed in February 2021.
[2] Lihui Wang and Robert X. Gao (Eds.) Condition Monitoring and Control for Intelligent Manufacturing, 2006, Springer-Verlag.
[3] David Ortiz, Carl Byington, Romano Patrick, Cody Ture, John Farnach, John Moffatt, Combined Lubrication Monitor for On-Line Gearbox Health Assessment, **DOI:** 10.1109/AERO.2011.5747563, 2011, Aerospace Conference**, IEEE Xplore.**
[4] Henry S. Teng, Kaihu Chen and Stephen C-Y Lu, Adaptive Real-time Anomaly Detection Using Inductively Generated Sequential Patterns, Conference Proceedings on Research in Security and Privacy, 1990.., 1990, IEEE Xplore DOI: 10.1109/RISP.1990.63857
[5] Chandola, V., Banerjee, A., and Kumar, V. 2009. Anomaly detection: A survey. ACM Comput. Surv., 2009, 58 pages. http://doi.acm.org/10.1145/1541880.1541882
[6] Victoria J. Hodge and Jim Austin, A Survey of Outlier Detection Methodologies, Artificial Intelligence Review 22: 85–126, 2004, Kluwer Academic Publishers, DOI: 10.1023/B:AIRE.0000045502.10941.a9
[7] George E. P. Box, Gwilym M. Jenkins, Gregory C. Reinsel, Greta M. Ljung, Time Series Analysis: Forecasting and Control, Fifth Edition, 2015, John Wiley & Sons.
[8] G. E. P. Box & G. C. Tiao (1975) Intervention Analysis with Applications to Economic and Environmental Problems, Journal of the American Statistical Association, 70:349, 70-79, DOI: 10.1080/01621459.1975.10480264.
[9] Chang, Ih, George C. Tiao, and Chung Chen. "Estimation of Time Series Parameters in the

Presence of Outliers." *Technometrics*30, no. 2 (1988): 193-204. DOI:10.2307/1270165.

[10] Bovas Abraham, G. E. P. Box, "Bayesian Analysis of Some Outlier Problems in Time Series", 1979, Biometrika 66(2):229-236, DOI: 10.1093/biomet/66.2.229

[11] Peter J. Rousseeuw, Annick M. Leroy, "Robust Regression and Outlier Detection", 1987, John Wiley & Sons.

[12] Bernhard Schölkopf, Alex J. Smola, Klaus-Robert Müller, Nonlinear Component Analysis as a Kernel Eigenvalue Problem, Neural Computation 10(5):1299-1319, DOI: 10.1162/089976698300017467, 1998

[13] Ryohei Fujimaki, Takehisa Yairi, Kazuo Machida, An approach to spacecraft anomaly detection problem using Kernel Feature Space, Conference: Proceedings of the Eleventh ACM SIGKDD International Conference on Knowledge Discovery and Data Mining, Chicago, 2005, DOI: 10.1145/1081870.1081917.

[14] Daniel Pena,George C. Tiao, Ruey S. Tsay, A Course in Time Series Analysis, 2001, John Wiley & Sons, Inc., New York

[15] Basu, S. & Meckesheimer, Automatic outlier detection for time series: an application to sensor data, Springer-Verlag London Limited, M. Knowl Inf Syst (2007) 11: 137. https://doi.org/10.1007/s10115-006-0026-6

[16] Manahil Raza, Usman Qayyum, Classical and Deep Learning Classifiers for Anomaly Detection, 2019 16th International Bhurban Conference on Applied Sciences and Technology (IBCAST - 2019), DOI: 10.1109/IBCAST.2019.8667245

[17] Rolf Isermann, Process Fault Detection Based on Modeling and Estimation Methods A Survey, Published in Elsevier- Automatica, Volume 20, Issue 4, July 1984, Pages 387-404, DOI:10.1016/0005-1098(84)90098-0

[18] Gwidon W. Stachowiak, Andrew W. Batchelor, Engineering Tribology, Third Edition, 2005, Elsevier Butterworth-Heinemann

[19] BC. Sharma and O.P. Gandhi, Performance evaluation and analysis of lubricating oil using parameter profile approach, Industrial Lubrication and Tribology 60(3):131-137, DOI: 10.1108/00368790810871057

[20] Junda Zhu, David He, Eric Bechhoefer, Survey of Lubrication Oil Condition Monitoring, Diagnostics, and Prognostics Techniques and Systems, Journal of Chemical Science and Technology July 2013, Vol. 2 Iss. 3, PP. 100-115

[21] NK. Myshkin and LV Markova, Online Condition Monitoring in Industrial Lubrication and Tribology, Applied Condition Monitoring 8, Springer International Publishing AG, 2018, DOI 10.1007/978-3-319-61134-1_1

[22] Qingfei He, Guiming Chen, Xiaohu Chen, Chunjiang Yao, Application of Oil Analysis to the Condition Monitoring of Large Engineering Machinery, Published in 2009 8th International Conference on Reliability, Maintainability and Safety, IEEE *Xplore,* DOI: 10.1109/ICRMS.2009.5270033

[23] Junda Zhu, Online Industrial Lubrication Oil Health Condition Monitoring, Diagnosis and Prognostics, Ph.D. Thesis, the University of Illinois at Chicago, 2013.

[24] Saloka, G. and Meitzler, A., "A Capacitive Oil Deterioration Sensor," SAE Technical Paper 910497, 1991, https://doi.org/10.4271/910497.

[25] A.H. Sihvola, J.A. Kong, Effective Permittivity of Dielectric Mixtures, IEEE TRANSACTIONS ON GEOSCIENCE AND REMOTE SENSING. VOL 26, NO 4, 1988, pp 420-429, **DOI:** 10.1109/36.3045.

[26] B. Jakoby, M. J. Vellekoop, Physical sensors for water-in-oil emulsions, Sensors and Actuators A 110 (2004) 28–32, 2003 Published by Elsevier B.V., DOI:10.1016/j.sna.2003.08.005.

[27] G. Adriani, M. Campatelli, M. Paoli, -Monitoil® - *Online sensors for efficient and cost-saving oil conditions' monitoring*, Mecoil Diagnosi Meccaniche s.r.l., Florence, Italy, http://www.mecoil.net/wp-content/uploads/2013/04/20120914-Articolo-colonne_Monitoil_eng.pdf

[28] Grebenişan G., Salem N., and Bogdan S., *The lubricants parameters monitoring and data collecting,* MATEC Web of Conferences, **Volume** 126, 2017, https://doi.org/10.1051/matecconf/201818403008.

[29] Grebenişan G., Salem N., and Bogdan S. -*An approach of classification and parameters estimation, using the neural network, for lubricant degradation diagnosis*, MATEC Web of Conferences, **Volume** 184, 2018, https://doi.org/10.1051/matecconf/201818403009.

[30] Anveshan Bommareddi, An Engine Oil Life Algorithm, The Pennsylvania State University, Department of Mechanical and Nuclear Engineering, 2009

[31] Enrico Zio, Giovanni Peloni, Particle filtering prognostic estimation of the remaining useful life of nonlinear components, 2010 Elsevier Ltd, Reliability Engineering and System Safety 96 (2011) 403–409, DOI:10.1016/j.ress.2010.08.009

[32] Diego Galar Pascual, Artificial Intelligence Tools: Decision Support Systems in Condition Monitoring and Diagnosis, CRC Press Taylor & Francis Group, LLC, 2015

[33] Qiang Miao, Viliam Makis, Condition monitoring and classification of rotating machinery using wavelets and hidden Markov models, ELSEVIER, Science Direct, Mechanical Systems and Signal Processing 21 (2007) 840–855, DOI:10.1016/j.ymssp.2006.01.009.

[34] Qiang Miao, Hong-Zhong Huang, Xianfeng Fan, A Comparison Study of Support Vector Machines and Hidden Markov Models in Machinery Condition Monitoring, Journal of Mechanical Science and Technology 21 (2007) 607~615, DOI: 10.1007/BF03026965

[35] Norbert Wiener, Extrapolation, Interpolation, and Smoothing Of Stationary Time Series With Engineering Applications, The MII. Press Massachusetts Institute Of Technology Cambridge, Massachusetts, 1949.

[36] Lotfi A. Zadeh and John R. Ragazzini, An Extension of Wiener's Theory of Prediction, 1950, IEEE Xplore, Journal of Applied Physics 21(7):645 – 655, DOI:10.1063/1.1699725

Transforming 3D printed CNC prototype into a plotter

LI Culda[1], ES Muncut (Wisznovszky)[1], GM Erdodi[1]

[1]University "Aurel Vlaicu" of Arad, Department of AIITT B-dul Revoluției Nr. 77, 310130, România. P. O. BOX 2/158 AR,

lavyy_99@yahoo.com

Abstract. We often hear about numerical control or CNC abbreviated. With this type of control, we can automatically control various processing tools such as: drills, lathes, milling cutters, rapid prototyping printers, etc. The instructions they need to receive will be in the form of a Gcode sequential program generated by the CAD / CAM drawing and manufacturing program. In the case of 3D printers, the CAD / CAM information is passed through a slicer specific to that printer, which "cuts" the object into printing slices. CNC systems are used today for virtually any process that has movements and operations, including laser cutting, hole punching, fabric cutting, welding, etc. because most CNC systems built today are 100% electronic. 3D printing of a CNC was a real challenge that aimed to understand the principles of obtaining and operating a CNC. I found that we basically have 3 devices with the same basic principles: CNC milling cutter, 3D printer and plotter. So I created a CNC that I turned into a plotter just to highlight the above.

1. 3D CNC design

In the industry we often hear about numerical control or abbreviated CNC. With this type of control, we can automatically control various processing tools such as: drills, lathes, milling cutters, rapid prototyping printers, etc.

The command is coded in Gcode via a computer and has the advantage that we can process various materials: metal, plastic, wood, etc. without a human operator to perform the processing operation directly.

CNC control, even if it has become commonplace nowadays, brings a significant improvement in production compared to the classic processing method. The design of a part is done with CAD systems and then the overall data, processing speed, geometry, etc. are translated into values and commands necessary for manufacturing through the CAM software and loaded into the machine software.

The instructions they need to receive will be in the form of a Gcode sequential program generated by the CAD / CAM drawing and manufacturing program. In the case of 3D printers, the CAD / CAM information is passed through a slicer specific to that printer, which "cuts" the object into printing slices.

CNC systems are used today for virtually any process that has movements and operations, including laser cutting, hole punching, fabric cutting, welding, etc. because most CNC systems built today are 100% electronic.

Starting from these known data, we started to realize the concept of 3D printed CNC. The basics of a CNC lathe or 3D printer are the same significant difference being the mode of operation and control.

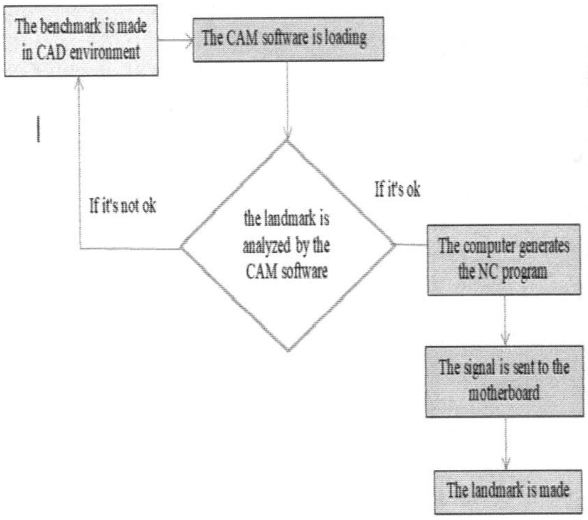

Figure 1.Logical scheme

Figure 2.Construction of 3D printed CNC

The movement I want to control is on the x and y axis (mass) and on the Z the tool. The position of the tool is driven by stepper motors with direct drive or actuators to provide precise movements, the control in the open loop works as long as the forces are kept small enough and the speed is not too high.

2. Making CNC components using 3D printing

To make the components of the CNC we used 3D printing with filament. Printing is done after the landmark is created in a CAD environment.

For my project I used CAD drawings made in Fusion 360 and I printed the parts on CTC printer.

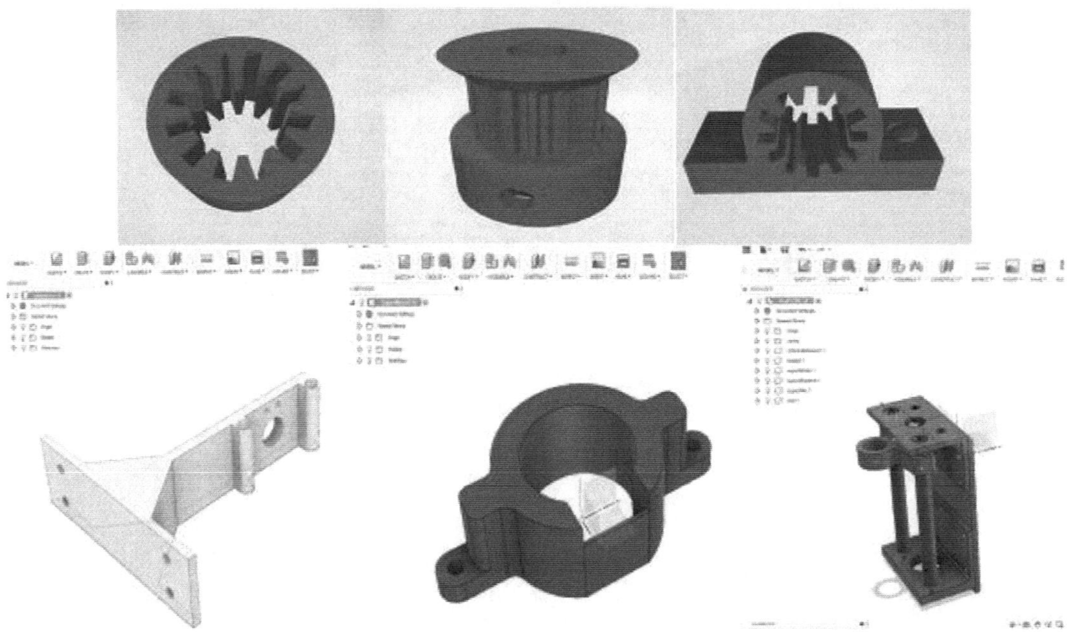

Figure 3.CAD drawings

IOP Conf. Series: Materials Science and Engineering 1169 (2021) 012013 doi:10.1088/1757-899X/1169/1/012013

To print them, each drawing was saved as an STL and passed through the CURA printer software

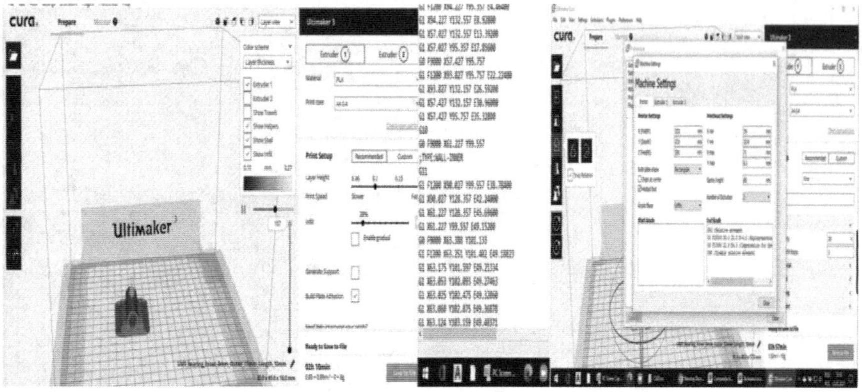

Figure 4.Cura (Cura printer software, 2021)

The Gcode is generated and the landmarks are printed

3. Realization of 3D printed CNC

After 3D printing of the components, the CNC was assembled.

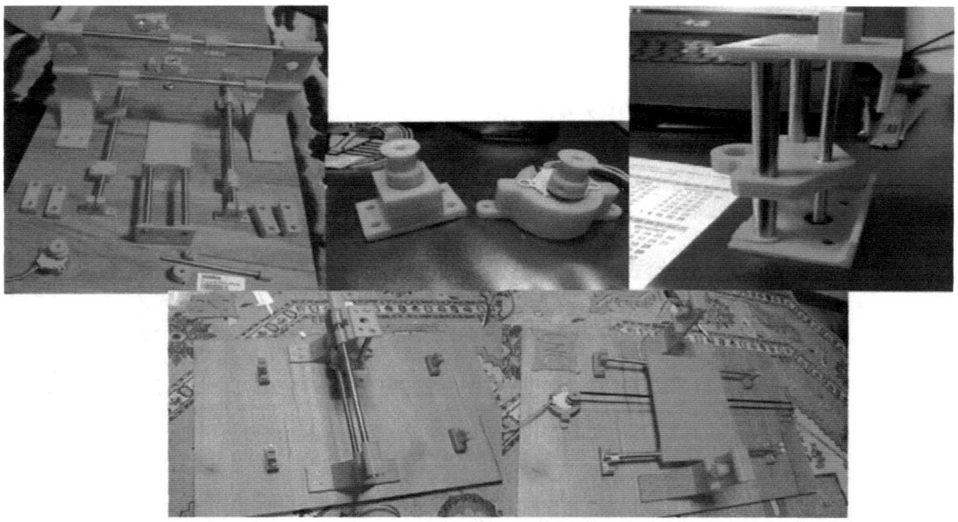

Figure 5. Realization of 3D printed CNC

The printed parts were used to mount the CNC prototype:

- Printed constructive landmarks
- Guiding and joining elements
- Stepper motors 28BYJ-48: are among the most used in various hobby-electronics applications, 3D printers, but also on certain industrial machines ex mini-conveyors, medical equipment, air conditioning systems, etc.

Figure 6.Stepper motors 28BYJ-48

- **Development board Arduino UNO.** The Arduino UNO microcontroller will represent the "brain" of this prototype. I chose the Arduino UNO version, because it is very different from the previous boards, in the sense that it does not use a USB-to-serial FTDI driver chip. Instead, it has the built-in Atmega8U2 microcontroller programmed as a USB-to-serial converter. The Arduino UNO module is a development board based on the ATmega328 microcontroller. Arduino UNO has 14 digital pins, of which 6 can produce pulse width modulation (PWM - Pulse with modulation) and 6 analog pins. Data transmission on the Arduino UNO board is done via a USB cable, connected between the board and the USB port on the computer.

Figure 7.Arduino UNO (Microcontroller Arduino Uno, 2021)

- **Arduino CNC Shield V3**. Arduino CNC Shield V3 is a platform specially designed for the Arduino UNO board that is to be used in the construction of CNCs with various applications of cutting, milling, drilling but it can also be adapted to marking CNCs , laser engraving or cutting, 3D printers, etc. This Shield is capable of supporting up to 4 drivers that can be used for 4 DC / Stepper motors and separately a servo motor, besides all this being a platform designed for CNC has adapted and the connection of mechanical limiters for each axis (+/- X, Y, Z), coolers, emergency button and supports a voltage supply between 12 and 36 V which gives it the possibility to be interconnected with a wide variety of motors in CNC applications. I chose this shield due to the fact that it was designed for Arduino UNO, giving it 100% compatibility in CNC applications that will be run through this shield, but also the ease of connecting the shield with Arduino UNO and drivers for engines. Basic knowledge of electronics is required to connect motors, mechanical limiters and certain safety buttons.

Figure 8 (Arduino CNC Shield V3, 2021)

IOP Conf. Series: Materials Science and Engineering 1169 (2021) 012013 doi:10.1088/1757-899X/1169/1/012013

- **Driver A4988**. The A4988 driver provides a motor drive solution for printers, scanners and other automated applications. The device has two integrated H-bridge drivers and a micro-step indexer and is designed to drive a bipolar stepper motor. The driver output block consists of MOSFETs - N-channel configured as complete H-bridges to drive the motor windings. A4988 is capable of conducting up to 2.5 A current on each output (with a corresponding cooling to 24V and 25 ° C). A simple STEP / DIR interface allows easy connection to control circuits. There is a reduced sleep mode, which stops the internal circuit to obtain a very low current draw. This standby mode can be set using the dedicated nSLEEP pin. Internal shut-off functions are provided for overcurrent, short circuit, voltage lock and high temperature. The fault conditions are indicated by the nFAULT pin.

Figure 9.Driver A4988 (Driver A4988- pine configuration, 2021)

For this project I chose the Arduino UNO microcontroller, together with a shield designed for CNCs capable of controlling up to 4 steering axes, Arduino CNC Shield V3, three stepper motors 28BYJ-48 controlled via the Arduino CNC Shield platform V3 driven by three drivers of A4988 motors and a scanner source 12V - 1.2A to supply the prototype, I also used four limiters for the X, Y axes, for the mechanical part of the X and Y axes we opted for guide axes and linear bearings of 8 mm. The rest of the mechanical parts were designed in Fusion 360 and printed in 3D. The stand that will serve as a foundation for the prototype will be built of chipboard (Particleboard).

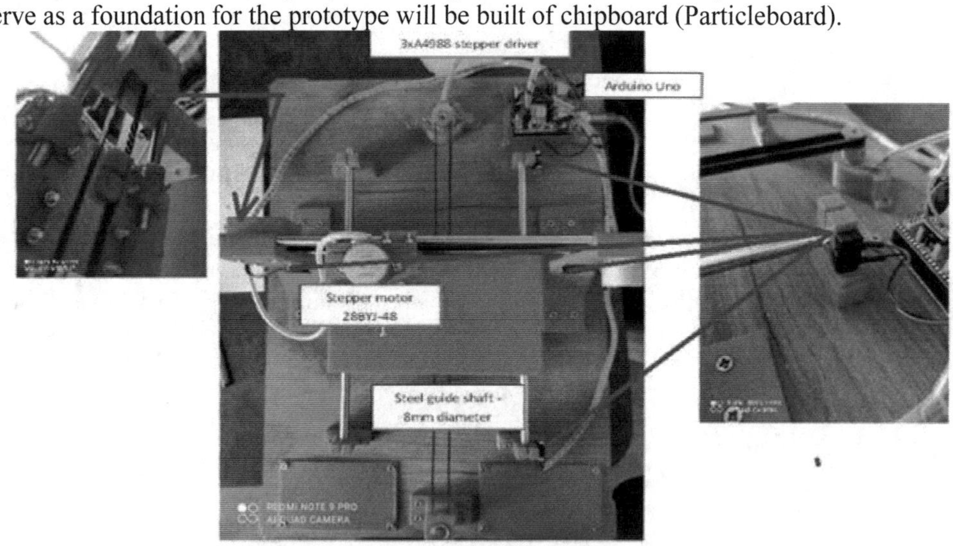

Figure 10.CNC prototype

IOP Conf. Series: Materials Science and Engineering 1169 (2021) 012013 doi:10.1088/1757-899X/1169/1/012013

4. Transformation of the CNC prototype into a Plotter

Because for CNC testing we did not want to attach tools, I transformed the CNC prototype into a plotter.

A plotter allows us to draw with a writing tool: pen, coloured pencil, pencils, markers, etc.

The drawing support can be on different materials: paper, cardboard and even glass. I opted for paper.

The attachment to the writing instrument is done as reliably as possible, so as not to disturb the printing. The writing object is with a small M3 screw.

Figure 11.Plotter (Howtomecatronics.com, 2021)

After successfully mounting the writing instrument on the printer, it is time to calibrate it and find the optimal print height. We installed the LaserGRBL v4.2.1 firmware's that our Arduino board can read the G code used by the created CNC prototype.

The choice of the printing model is made taking into account the fact that we will have to obtain a svg vector format. You can also convert JPEG and PNG images to SVG.

In the LaserGRBL software I had the advantage that no svg conversion was needed and I could use jpg images. We uploaded the Image file - Open File - Import Rate Image - we choose vectorization and Crop image to frame the image on the next page - we make the settings of the print mark - generated, at which point the plot code is generated. Choose Send to Machine and start plotting.

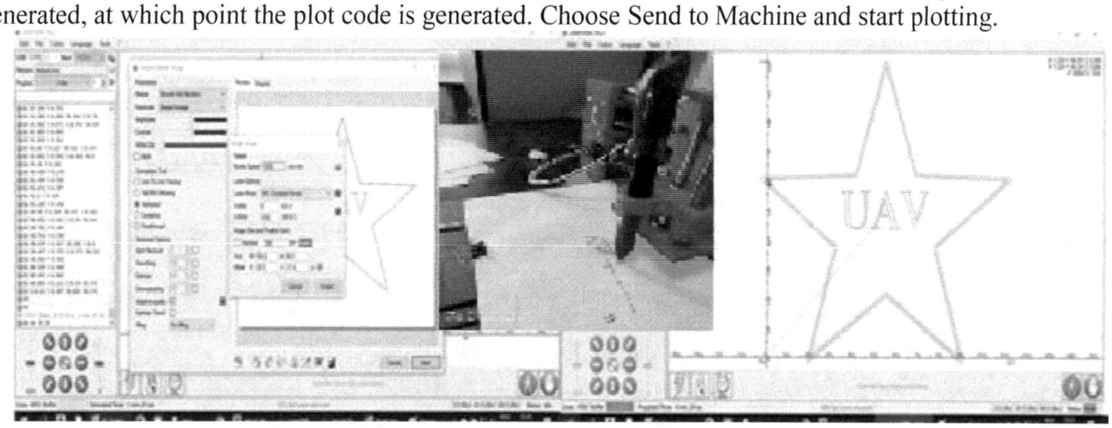

Figure 12.LaserGRBL software

5. Conclusion

3D printing is a new technology but with a very fast growth compared to traditional methods due to the relatively low costs. Although it cannot be used in series production, it allows a freedom of action, which makes it increasingly used in all fields.

The present project aimed to create from scratch a CNC prototype created, printed and assembled, which would allow us to test the degree of adaptability of our CNC to various transformations. Thus, the CNC prototype came to life thanks to the Arduino Uno boards and by mounting a writing device, it became a plotter. If we put a tool on him, he'll cut. So with low costs you can create various devices for teaching or research purposes. This method of obtaining the CNC prototype is recommended to be used in teaching so that engineering students can practically apply what they have learned in specialized courses.

References

[1] Mortoiu Doina, Săbăilă Lavinia, Babanatsas Theoharis, Gal Lucian, AutoCad 2006 Partea I – Modelarea 2D, Îndrumător pentu uzul studenţilor, 84 pagini, Editura Universităţii "Aurel Vlaicu",Arad, ISBN (10) 973 – 752-092-0, 2006

[2] Muncut Elena Stela; Erdodi Geza Mihai; Culda Lavinia Ioana; Komjaty Andrei; Sirghie Cecilia, Determination of speed and accelaration for a plane mechanism with the instantaneous rotation center method and euler's graf-analytical method using autocad mechanical, Source: eLearning & Software for Education., Vol. 2, p420-427. 8p., 2020

[3] Barbu Ionel; Komjaty Andrei; Fogorasi Magdalena Simona; Culda Lavinia; Bucevschi Adina; Kaminszky Robert, Teaching and e-learning about open-end spinning machine., Source: eLearning & Software for Education. Vol. 3, p285-293. 9p., 2020

[4] E S Muncut, L I Culda, G M Erdodi and G Sima, 8D complaint solving method in an automotive component processing company, Annual Session of Scientific Papers "IMT ORADEA 2019" IOP Conf. Series: Materials Science and Engineering568 (2019) 012020, IOP Publishing doi:10.1088/1757-899X/568/1/012020, 5pg

[5] Mortoiu D, Sima G, Săbăilă Lavinia. Holonic Manufacturing System for Safety Sistems Annals of DAAAM for 2009& Proccedings of the 20 th DAAAM International World Symposium, Austrian Society of Engineers and Architects –OIAV 1848, volume 20, nr.1,ISBN 978-3-901509-70-4,ISSN 1726-9679, pag.983-984,Vienna 2009

[6] T V Boshhenko and P V Chepur, Technology infusion of intellectual 3D printersbased prototyping of products into learning process, IOP Conf. Ser.: Mater. Sci. Eng. 327 022020, 2018

[7] *Arduino CNC Shield V3*. (2021). Retrieved from https://ardushop.ro/ro/electronica/148-cnc-shield-v3.html?gclid=Cj0KCQjwo-aCBhC-ARIsAAkNQisHGL1X8jQBPVyJPJc4Il4tuToO3CCz_B69ZDCtI27eTYphrFFSH0waAiVlEALw_wcB

[8] *Driver A4988- configurație pini*. (2021). Retrieved from https://www.sigmanortec.ro/Driver-Stepper-DRV8825-p125423334?gclid=Cj0KCQjwo-aCBhC-ARIsAAkNQiuuf-JhhiKChtQGWzTpx6sztpuZCZyOegiX0UuZnEIoW17lXlOAUBsaAonBEALw_wcB

[9] *Expeditor universal Gcode*. (2021). Retrieved from https://winder.github.io/ugs_website/#universal-gcode-sender

[10] *Howtomecatronics.com*. (2021). Retrieved from https://howtomechatronics.com/tutorials/how-to-setup-grbl-control-cnc-machine-with-arduino/

[11] *Microcontroller Arduino Uno*. (2021). Retrieved from https://ardushop.ro/ro/home/29-placa-de-dezvoltare-uno-r3.html?gclid=Cj0KCQjwo-aCBhC-ARIsAAkNQitnFZAg4TFZf006Id6QCWg90b9UWagG6-0WTKbM700VNEqp8qzoXwoaArD_EALw_wcB

IOP Conf. Series: Materials Science and Engineering 1169 (2021) 012014 doi:10.1088/1757-899X/1169/1/012014

Experimental studies regarding the thermal operating regime for SCR catalysts of heavy-duty commercial vehicles

D M S Capătă[1], M F Popa[1], T Dan[1], N Burnete[1]

[1]Technical University of Cluj Napoca, Romania

dorin.capata@gmail.com

Abstract. SCR, selective catalytic reduction, is the most efficient method for treating nitrogen oxides – one of the main pollutants of internal combustion engines. The SCR system involves the injection of a reductant (the solution AdBlue) in the exhaust gas flow, the reduction of the oxides being achieved in the catalyst's interior [1,4,6]. The operating limits of the SCR system are given by need to maintain the temperature conditions for the injected solution as well as the catalyst and its internal components (oxygen and temperature sensors, reductant injector) [2]. Overheating the AdBlue injectors may have a major influence for the operation of the SCR system, the deposits of the decomposed solution within the injectors leading to the system's out of use. Thus, maintaining an optimum thermal regime by cooling the AdBlue injectors becomes a required solution for a good functioning of the entire system. The paper presents experimental studies in relation to the thermal evolution of SCR catalysts on heavy-duty commercial vehicles in different turnover situations as well as an analysis of the measured values.

1. Introduction.

The heavy-duty commercial vehicles that fulfill the polluting norm Euro 6 use two combined methods for treating the evacuation gases; EGR for treating the nitrogen oxides and a complex system of catalysts which reduces the particles, carbon monoxide as well as nitrogen oxides, up to their entire neutralization. The two methods operate independently, EGR uses the evacuation gases that are partially collected and returned to the engine's admission gallery, while the catalysts system is set on the engine's evacuation path and treats those gases before they are sent out to the atmosphere [5,8,9].

The system for treating subsequent gases is made out of an assembly of catalysts: DOC, the oxidation catalyst that transforms CO in CO_2 respectively NO in NO_2; the particles filter that retains the particles from the evacuation gases; the SCR system that treats the emissions of NO_x, respectively the ammonia catalyst that prevents the evacuation of ammonia into the atmosphere [7,8,9].

The exhaust gas recirculation system, EGR, can be activated and becomes functional shortly after starting the engine, in comparison to the post-treating system that needs to reach a certain thermal regime to be activated. The starting temperature for the post-treating exhaust gas system is around 250^{0}C and the efficiency of the transformation processes within the catalysts improves with the increase of its internal temperature. There is also a superior limit for temperature, considered to be over 600^{0}C, a temperature at which the efficiency of the transformation in the catalysts is extremely diminished and the danger of the catalysts' deterioration becomes a real issue. Thus, we can actually talk about a working window of the catalysts between 250-450^{0}C, according to the materials used for

IOP Conf. Series: Materials Science and Engineering 1169 (2021) 012014 doi:10.1088/1757-899X/1169/1/012014

their manufacturing, an interval that insures maximum efficiency and a safe exploitation of their components [3].

The proposed experimental studies have been focused of the determination of the thermal operating regime of the catalysts in different exploitation conditions for the commercial trucks. Also, the way in which the temperature within the SCR catalysts might influence the thermal regime of the reductant injectors has also been pursued. Thus, there is a link between the temperature within the SCR catalyst and the deposits of the decomposed reductant on its injector as well as on the catalyst's interior walls. The partial obstruction of the reductant injector's nozzle supplies the perfect conditions for creating these deposits, and not cleaning them leads to the total blocking of the injector and the put out of use of the subsequent gases treating system.

2. The experimental installation, the equipment used and the progress of the experiments:

The thermal experimental determinations have been achieved on Volvo trucks, model FH(4), manufactured in 2018, endowed with 13 liters engines (D13K of 500CP). The catalysts group DOC, DPF, SCR is compact, while three temperature sensors placed on the direction of the exhaust gases flow monitor the thermal operating regime of these catalysts. The placement of the temperature sensors is being done before the oxidation catalyst, thus having information on the entrance temperature into the catalysts group for the exhaust gases; between the oxidation catalyst and the particles filter for information regarding the temperature of gases before the particles filter; between the particles filter and the catalyst of selective catalytic reduction for measuring the exhaust gases temperature at the entrance into the SCR catalyst.

The monitorization and registration of the temperature values obtained by the three sensors have been achieved with the diagnosis computer Volvo TechTool by selecting the proper parameters, thus obtaining temperature values for each second of the engine's operation.

A mixed route – highway-national road – has been achieved in order to simulate a transport for general goods, 28 km, respectively a route on a temporary work side in order to simulate a construction transportation regime, 7.5km.

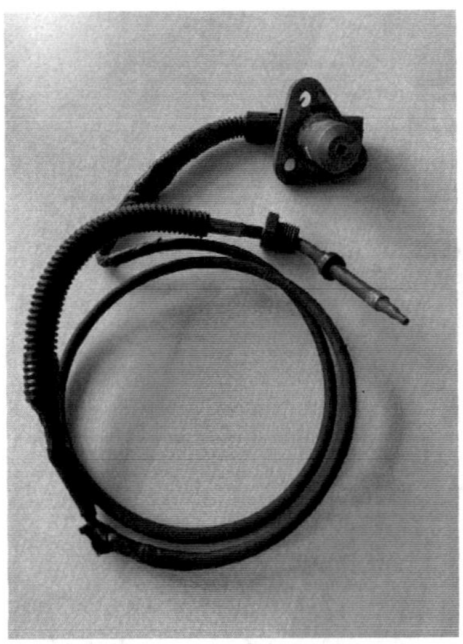

Figure 1. Montage temperature sensor

Figure 2. Montage assembly sensor on catalyst

The determinations for the temperature of the reductant injector have been achieved statically, by using a crafted montage and introducing a temperature sensor within the injector on the reductant's flow, as it can be noticed in figure 1. The information on the temperature within the injector have been

IOP Conf. Series: Materials Science and Engineering 1169 (2021) 012014 doi:10.1088/1757-899X/1169/1/012014

measured with the diagnosis computer Volvo TechTool, by connecting the temperature sensor to the electric installation of the truck – figure 2, instead of the temperature sensor placed at the entrance into the catalysts group.

Heating up the catalysts for the static determinations has been achieved with the regenerating tests SCR, the maximum temperature within the catalyst being 400^{0}C, followed by a slow cooling at 550rot/min, down to a temperature of 160^{0}C in the catalyst.

3. The analysis and the interpretation of the results:

In order to determine the temperature values recorded by the three sensors placed into the catalysts assembly for post-treating exhaust gases system, there have been achieved road tests in conditions closed to the exploitation ones.

Thus, the road tests for the general merchandise running mode has been accomplished at an average load of 15t, and the route included a mixed plan, including plan road and significant gradients of slope or ramp (figure 3)

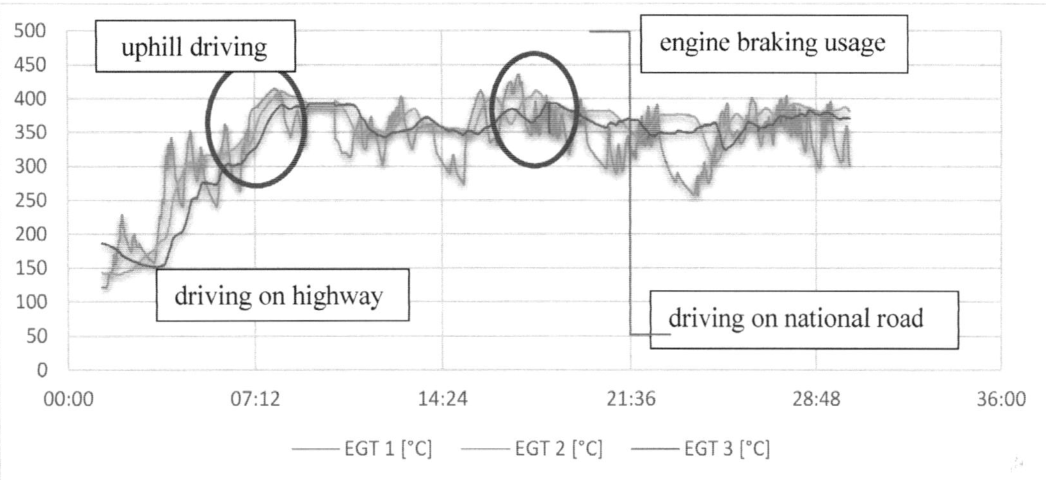

Figure 3. The registered temperatures for highway/national road

By analyzing the registered values for highway/national road, it can be noticed that there is a maintenance in temperature for the SCR catalyst between 350-380^{0}C, with tops reaching 400^{0}C when the engine is maximum loaded; when uphill driving the ramp or using engine brake. It can also be noticed the fact that the temperature within the catalysts (EGT2, EGT3) changes less with respect to the temperature of the exhaust gases measured at the catalysts' entrance (EGT1). The phenomenon is due to the fact that temperature is being kept within the catalysts group well enough, the heat loss through the thermal transfer towards the catalysts' walls or the exhaust gases within them being relatively small. The catalysts efficacy is due to their working thermal regime, so that the higher the temperature within the catalysts is (up to a limit), the ore transformations happen within them.

The decrease in temperature within the SCR catalyst are given by the free rolling of the vehicle or going down small slopes without using engine brake.

Repeated speeding periods, typical for driving on national road, lead to an increase in temperature on the SCR catalyst, its working value being much higher compared to driving on a highway.

The measurements for the construction sector have included highway constructions with loads and unloads at certain points and a mixed route of national road, off road, suppling for highway construction (figure 4). Loading the vehicles has been done at an average of 23 t.

IOP Conf. Series: Materials Science and Engineering 1169 (2021) 012014 doi:10.1088/1757-899X/1169/1/012014

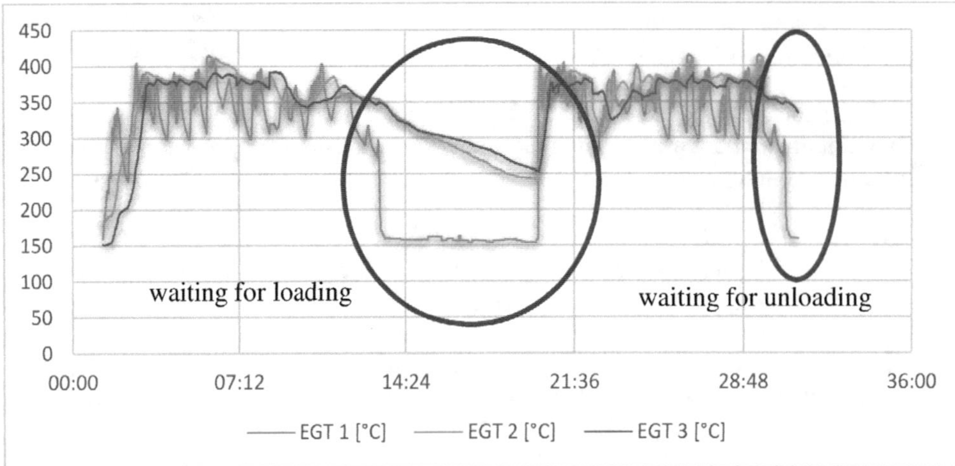

Figure 4. The registered temperatures for construction regime

The thermal operating regime for SCR catalysts for the construction segment is higher than driving on highway/national road, the average value being 370^{0}C. The major disadvantage is the jerky use of the vehicles, the alternation between driving and stops at the loading-unloading points having a major impact on the temperatures measured in the SCR catalysts.

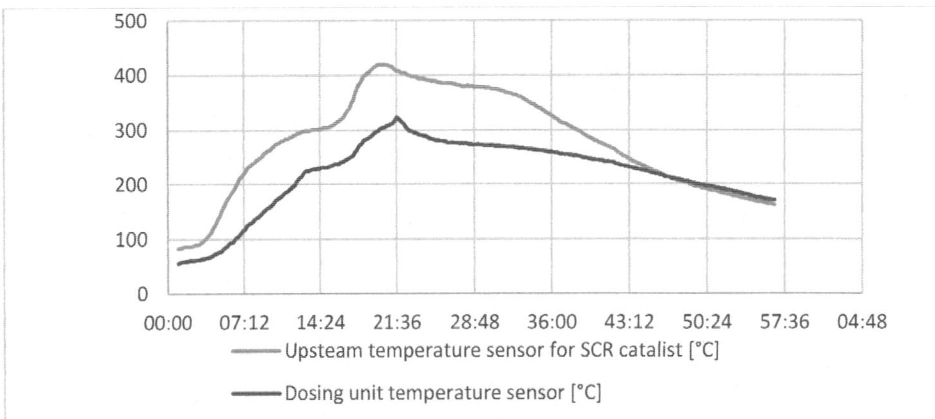

Figure 5. Static measurements of the temperature of the reductant injector

The measurements for the thermal regime of the reductant injector (figure 5) reveal a delay in its heating in comparison to the temperature of the SCR catalyst, a delay due to its placement outside the direction of the flow of the exhaust gases. It can be noticed a relatively constant difference, of about 80^{0}C, between the temperature of the SCR catalyst and the temperature of the injector during the heating phase, a difference that is being lost during the cooling phase. It can also be noticed that the temperature of the SCR catalyst decreases under the value that has been measured on the reductant injector at the end of the catalyst's cooling.

4. Conclusion regarding the achieved experimental studies:

Driving on highway or on national road offers a relatively constant plane for the temperature measured before the SCR catalyst, the average temperature being 375^{0}C. The top points correspond to driving on slopes and going down on ramps with using the engine brake when the temperatures of the SCR catalyst reach 400^{0}C.

The thermal stress on the reductant injector corresponds to the maximum temperature values registered in the SCR catalyst, the effect being more pronounced as the distance is higher.

There can also be issues of overheating the reductant injectors when there are stops in the vehicle's exploitation after an increased thermal exploitation of the SCR catalyst.

For the construction regime, an extreme variation of the thermal regime that has been measured in the SCR catalyst can be noticed. This variation is due mainly to the exploitation regime of the vehicles, thus longer the stop when loading/unloading are, higher is the possibility of creating decomposed reductant deposits around the injector. The situation can be worsened even more if the routes between the loading/unloading areas are short and the thermal regime in the SCR catalyst does not reach the required temperatures for the decontamination of the internal deposits.

Cooling the reductant injectors is extremely required for both regimes for commercial trucks, while active regeneration of the catalysts for driving in constructions needs to be done more often.

The protection of the reductant injector by placing it outside the exhaust gases flow, or by putting it into a thermal cover is a required condition in order to avoid the deposits on the reductant's nozzle.

References:

[1] Băţaga N, Burnete N, Barabas I, Căzilă A, Filip N, Naghiu A, Dan, F,Szasz, R, Karamusantas, D, *Motoare cu Ardere Internă, Combustibili, Lubrifianţi şi Materiale Speciale pentru Automobile Economicitate şi Poluare*, Editura U.T.Press, Cluj Napoca, 1999, ISBN 973-9471-20-X

[2] Capata D M S, Popa F M, Burnete N, *Aspects Regarding Fuel Consumption and the Pollutant Products of Internal Combustion Engines for Commercial Vehicles*, The 30th SIAR International Congress of Automotive and Transport Engineering, ISBN 978-3-030-32563-3, Springer Nature Switzerland AG 2020

[3] Cavina N, Mancini G, Corti E, Moro D, De Cesare M, Stola F, *Thermal Management Strategies for SCR After Treatment System*, SAE Technical Paper 2013-24-0153, 2013, 11th International Conference on Engines & Vehicles, ISNN 0148-7191

[4] Coli R, Ferguson and Allan T. Kirkpatrick, *Internal Combustion Engines: Applied Thermosciences*, Third Edition, John Wiley & Soons Ltd.,2016, ISBN 978-1-118-53331-4

[5] Klaus Mollenhauer, Helmut Tschoeke, *Hanbook of Diesel Engines*, Springer-Verlag Berlin Heidelberd, 2010, ISBN 978-3-540-89082-9

[6] Manea, Adriana T, Manea, LC, Sandu, V, *Motoare termice, procese poluare*, Editura Matrix Rom, Bucuresti, 2004, ISBN 973-685-731-X

[7] *** EATS – Euro 6 Volvo Trucks training material

[8] *** Mercedes-Benz Trucks, maintenance and repairs manuals

[9] *** Volvo Trucks, maintenance and repairs manuals

IOP Conf. Series: Materials Science and Engineering 1169 (2021) 012015 doi:10.1088/1757-899X/1169/1/012015

Quality improvement of the primary cable of the handbrake lever: a QFD approach

C F Baban[1], M Baban[1], I S Popi[2]

[1]University of Oradea, Faculty of Management and Technological Engineering, str.Universității nr.1, Oradea 410087, Romania
[2]S.C. Top Metal Factory SRL Oradea, Romania

E-mail:cbaban@uoradea.ro

Abstract. Quality function deployment is one of the most used tools for the quality improvement of products or services. According to the literature, manufacturing is one of the industries that has benefited most from its employment. Therefore, the purpose of this paper is to illustrate the usability of the quality function deployment in improving the quality of the primary cable of the handbrake lever. In this way, the technical priorities of this product were established by comparison with those of the main competitors in the market and the target values of each technical characteristic of the product were determined. In addition, the attention of the manufacturer of the primary cable of the handbrake lever can be also focused on the fulfillment of the customer expectations.

1. Introduction

Firms are competing today in a demanding economy, where quality is seen as one of the most important factors in addressing organizational success. To satisfy customers, firms have to improve their products to deliver greater quality characteristics than their competitors [1]. One of the main tools for achieving this goal is Quality Function Deployment (QFD), which has been the subject of much literature [2, 3]. Since its introduction in Japanese firms, it has been successfully employed around the world [4, 5]. This is also the case of a Romanian firm, code for confidentiality reasons as Our Company, which aims to improve the quality of its products using a QFD approach.

2. The Quality Function Deployment

A detailed literature review on QFD, based on different functional areas of its employment and considering both industrial and service applications, is shown in [3]. A comprehensive review of the application of QFD in different areas and industries/services was also presented in [6, 7].

According to existing studies, manufacturing is one of the industries that can benefit most from the employment of QFD. Our Company is a firm involved in manufacturing and has a wide range of products for the automotive industry. Among such products, the primary cable of the handbrake lever is one of the most manufactured products (figure 1). At the same time, the company aims to continuously improve its products, and QFD can be a solution to achieve this goal. Through the use of QFD, the customer needs can be translated into activities related to the development of the Our Company products [2].

IOP Conf. Series: Materials Science and Engineering 1169 (2021) 012015 doi:10.1088/1757-899X/1169/1/012015

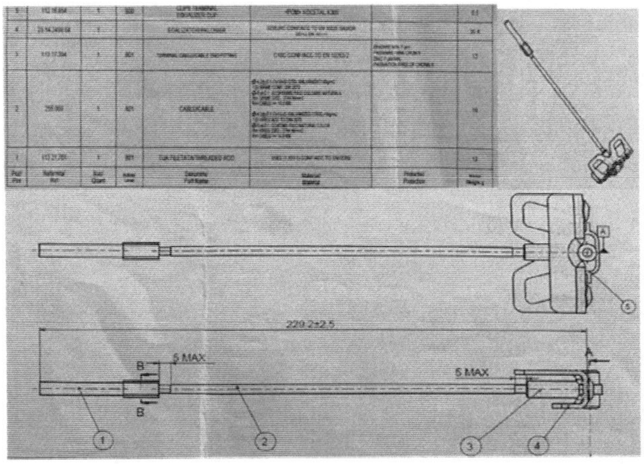

Figure 1. The primary cable of the handbrake lever

In conformity with [1, p.25], QFD involved a four-stage approach. Considering its importance to QFD this study addresses only the first stage-product planning, while the other three stages (part, process, and production planning) are somewhat similar to this first one. The main element of product planning is the House of Quality (HOQ), through which the customer requirements are correlated with the technical characteristics of the product to be developed. The framework of the HOQ and its elements are described in detail in [1, 8, 9].

3. The case study

The employment of the QFD through the HOQ in the case of the primary cable of the handbrake lever is presented below, considering the steps presented in [1, 9]. For confidentiality reasons, the product manufactured by Our Company was codified as Our Product, while the products of the main competitors in the Romanian market were codified as product Competitor A and product Competitor B, respectively.

Step 1) The identification of the customer needs (WHAT's List)

The customer needs have been identified by discussions and feedback from the clients. As a result, the following customer needs were established**:**

- conformity;
- warranty;
- price;
- quality of the packaging.

Step 2) The identification of the technical specifications (HOW's List)

The product improvement team has identified the following technical requirements:

- tensile strength (N);
- deviation from the length (mm);
- maximum deviation between axes (^{0});
- traceability.

Step 3) The planning matrix

The planning matrix includes the evaluation of Our Product compared with the product Competitor A and product Competitor B, based on the performance of the products in responding to the customer needs. For this purpose, the following steps have been completed (figure 2):

- the establishment of the degree of importance of each client need, using a five-point scale;
- the competitive evaluation against the main competitors, using a five-point scale;
- the establishment of the planned coefficient of each customer need;
- the computation of the improvement factor using the following relation [8, p.315]:

  Improvement factor = {(Planned coefficient - Coefficient of Our Product }*0.2}+1 (1)

IOP Conf. Series: Materials Science and Engineering 1169 (2021) 012015 doi:10.1088/1757-899X/1169/1/012015

- the establishment of the market leverage points (1.5, 1.2, 1), which correspond to (strong, moderate, no market points) [10].
- the computation of the global importance of each customer need using the relation [8, p.316]:

Overall importance=Importance of WHAT's* Improvement factor* Market leverage points (2)

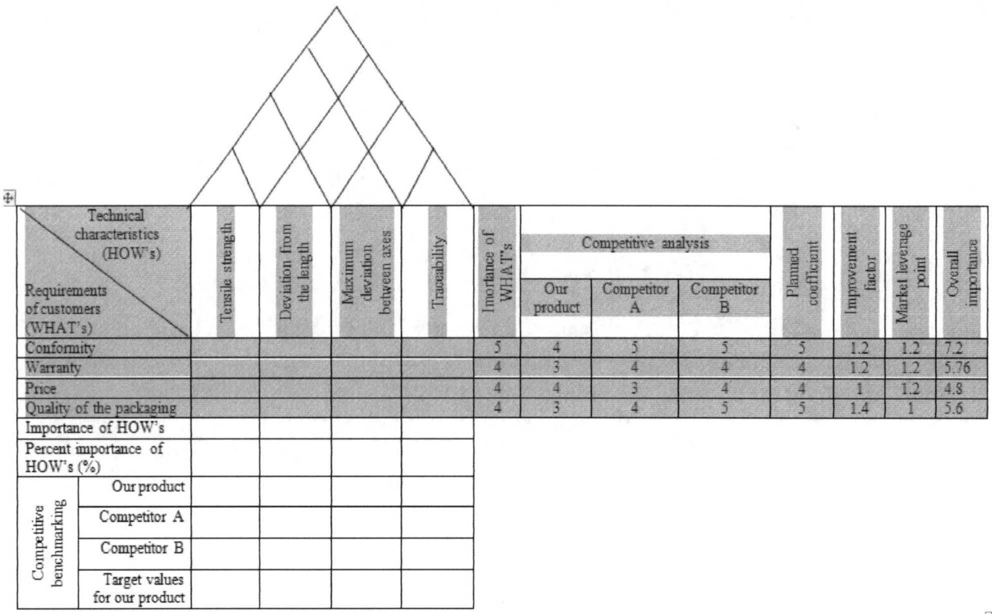

Figure 2. The WHAT's List, HOW's List and the planning matrix

Step 4) The relationship matrix

In order to establish the relationship between each customer need and each technical requirement (relationship WHATxHOW), a (9,3,1,0) scale was employed. The (9,3,1,0) scale corresponds to (strong, moderate, weak, no relationship) levels [1]. Figure 3 depicts the relationship matrix.

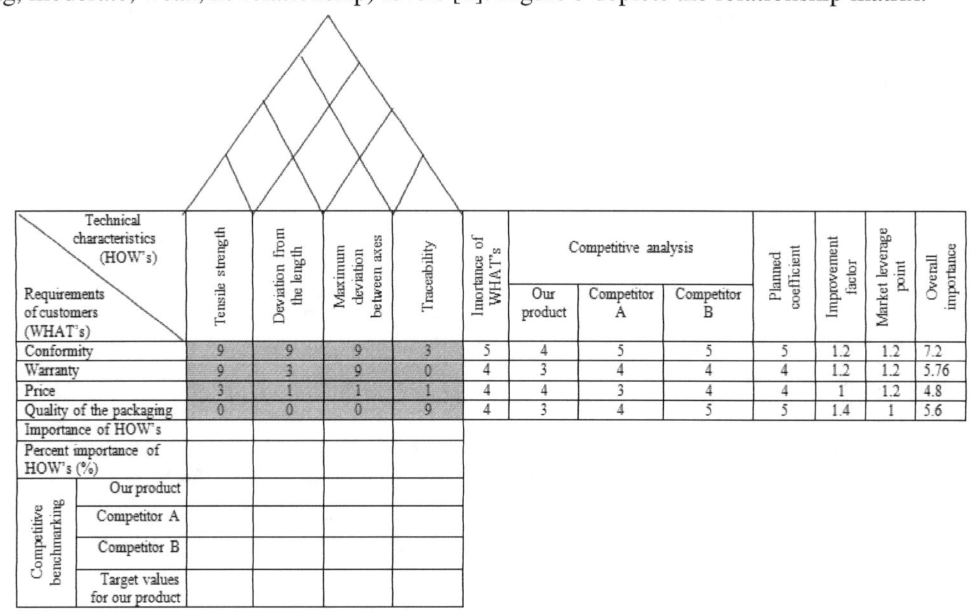

Legend
Relationship scale: 0=no relationship, 1=weak relationship, 3= moderate relationship, 9=strong relationship.

Figure 3. The relationship matrix

IOP Conf. Series: Materials Science and Engineering 1169 (2021) 012015 doi:10.1088/1757-899X/1169/1/012015

Step 5) The correlation matrix

The correlation matrix is used to assess the correlation between each two HOW's (+ represents a positive correlation, while – a negative one). Figure 4 depicts the correlation matrix.

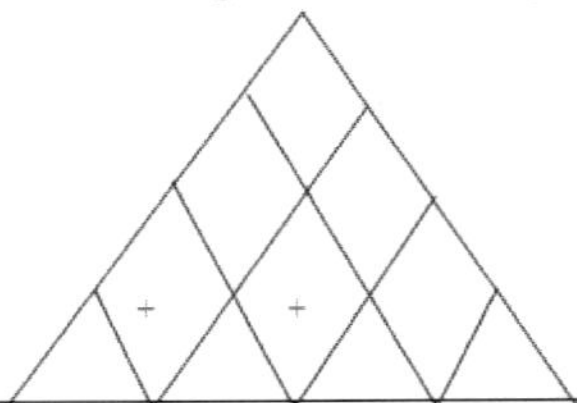

Technical characteristics (HOW's) / Requirements of customers (WHAT's)	Tensile strength	Deviation from the length	Maximum deviation between axes	Traceability	Imortance of WHAT's	Competitive analysis: Our product	Competitor A	Competitor B	Planned coefficient	Improvement factor	Market leverage point	Overall importance
Conformity	9	9	9	3	5	4	5	5	5	1.2	1.2	7.2
Warranty	9	3	9	0	4	3	4	4	4	1.2	1.2	5.76
Price	3	1	1	1	4	4	3	4	4	1	1.2	4.8
Quality of the packaging	0	0	0	9	4	3	4	5	5	1.4	1	5.6

		Tensile strength	Deviation from the length	Maximum deviation between axes	Traceability
Importance of HOW's					
Percent importance of HOW's (%)					
Competitive benchmarking	Our product				
	Competitor A				
	Competitor B				
	Target values for our product				

Legend
Relationship scale: 0=no relationship, 1=weak relationship, 3= moderate relationship, 9=strong relationship.
The correlation between the technical characteristics: +positive impact, -negative impact.

Figure 4. The correlation matrix

Step 6) The technical matrix

The last step in developing the HOQ is the technical matrix (figure 5).

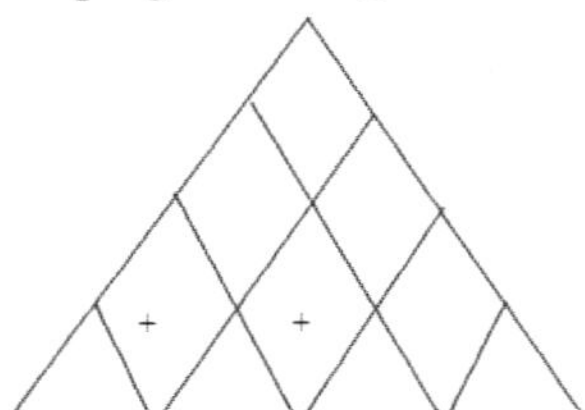

Technical characteristics (HOW's) / Requirements of customers (WHAT's)	Tensile strength	Deviation from the length	Maximum deviation between axes	Traceability	Imortance of WHAT's	Competitive analysis: Our product	Competitor A	Competitor B	Planned coefficient	Improvement factor	Market leverage point	Overall importance
Conformity	9	9	9	3	5	4	5	5	5	1.2	1.2	7.2
Warranty	9	3	9	0	4	3	4	4	4	1.2	1.2	5.76
Price	3	1	1	1	4	4	3	4	4	1	1.2	4.8
Quality of the packaging	0	0	0	9	4	3	4	5	5	1.4	1	5.6

		Tensile strength	Deviation from the length	Maximum deviation between axes	Traceability
Importance of HOW's		131.04	86.88	121.44	76.8
Percent importance of HOW's (%)		31.49%	20.88%	29.18%	18.45%
Competitive benchmarking	Our product	4500	±2	20	-
	Competitor A	5000	±2.1	20	Eticheta identif.
	Competitor B	4500	±2.1	20	Eticheta identif.
	Target values for our product	5000	±2	20	Eticheta identif.

Legend
Relationship scale: 0=no relationship, 1=weak relationship, 3= moderate relationship, 9=strong relationship.
The correlation between the technical characteristics: +positive impact, -negative impact.

Figure 5. The technical matrix

The importance of each HOW was computing using the following relation [1, p.31]:

$$\text{Importance of HOW}_i = \sum_{j=1}^{4}\left(\text{Relationship WHAT}_j \text{ xHOW}_i\right) * \left(\text{Overall importance of WHAT}_j\right) \quad (3)$$

$i = \overline{1...4}$

The competitive benchmarking was conducted by comparing the value of each HOW of Our Product with the corresponding values of product Competitor A and product Competitor B, respectively.

4. Conclusions

A QFD approach was used to correlate the client requirements with the technical possibilities of Our Company to achieve these requirements in the case of the primary cable of the handbrake lever. For this purpose, a HOQ has been developed based on the steps described in the literature. The advantage of this approach is the establishment of the technical priorities of the primary cable of the handbrake lever, by comparing each technical characteristic of the own product with those of the main competitors in the market.

References

[1] Chan L-K and Wu M-L 2002 Quality function deployment: a comprehensive review of its concepts and methods *Qual. Eng.* **15** 23-35

[2] Carnevalli J A and Miguel P C 2008 Review, analysis and classification of the literature on QFD-Types of research, difficulties and benefits *Int. J. Prod. Econ.* **114** 737-754

[3] Sharma J R, Rawani A M and Barahate Milind 2008 Quality function deployment: A comprehensive literature review *Int. J. Data Analysis Techniques and Strategies* **1** 78-103

[4] Akao Y and Mazur G H 2003 The leading edge in QFD: past, present, and future *Int. J. Q. Reliability Manage.* **20** 20–35

[5] Cristiano J J, Liker J K and White C C 2000 Customer-driven product development through quality function deployment in the U.S. and Japan *J. Prod. Innov. Manag.* **17** 286–308.

[6] Chan L-K and Wu M-L 2002 Quality function deployment: a literature review *Eur. J. Oper. Res.* **143** 463–497

[7] Lager T 2005 The industrial usability of quality function deployment: a literature review and synthesis on a meta-level *R&D Manag.* **35** 409-426

[8] Goetsch D L and Davis S 2013 *Quality management for organizational exellent: introducing to total quality* (Harlow: Pearson Education)

[9] Han S B, Shaw S K, Ebrahimpour M and Sodhi M S 2001 A conceptual QFD planning model, *Int. J. Q. Reliability Manage.* **18** 796–812

[10] Shina S G 1991 *Concurrent Engineering and Design for Manufacture of Electronics Products* (New-York: Van Nostrand Reinhold)

IOP Conf. Series: Materials Science and Engineering 1169 (2021) 012016 doi:10.1088/1757-899X/1169/1/012016

Financial performance management and economic cycle variations. Evidence for textile industry

M B Tudose[1] **and S Avasilcai**[1]

[1] Department of Engineering and Management, FDIMA, Gheorghe Asachi Technical University, 29 Dimitrie Mangeron Bd., Iaşi, 700050, Romania

brindusatudose@gmail.com

Abstract. The study analyses the dynamics of the financial performance of the companies in the textile industry in Romania and aims to identify its main determinants, corresponding to the different stages of the economic cycles. The selected sample was mixed; therefore, the econometric analysis (based on panel data) was performed on accounting data collected from secondary sources (for 2000-2019). Because some of the companies in the sample showed negative financial results, the financial performance was assessed based on sales' volume. The regression model used included seven determinants related to the company (internal determinants) and three external variables (GDP growth rate, inflation rate and unemployment rate). The results of the analyses indicated that (for the whole period) 51% of the sales dynamics can be explained due to the variation of the determinants included in the model. The analysis at the level of internal determinants indicated that all seven variables have a significant statistically influence on financial performance. From the perspective of macroeconomic determinants, the study showed that performance is negatively and significantly correlated with the inflation rate and unemployment rate. Applying the regression model to distinct periods (pre-crisis, crisis and post-crisis) provided very useful information from a practical point of view.

1. Introduction

The literature focused on research methodology in business and management highlights that financial performance is a more sensitive but fascinating topic [1]. Proof of this is the fact that the debate on performance has been and continues to remain in the attention of theorists and practitioners. The benchmark that fuelled researchers' interest was that, at the organizational level, the essence of performance consists in long-term growth and survival [2], and continuous performance improvement is an important goal. In the context of this research, performance has been accepted as an artifact based on which the success of an organization is appreciated, in the context of a free, competitive and globalized market [3].

The research topic of this article requires a theoretical, methodological and empirical assessment of the cause-effect relationship between financial performance and its determinants. This is because, depending on the stage of the economy, but also on their internal health, companies develop their own strategies to adapt to the economic and social context in which they operate. We consider at least three types of strategies: growth (adapting to the macroeconomic context), financing and performance strategies. The interdependencies between these strategies are very close. Moreover, these strategies have features specific to the fields of activity in which they are implemented. For this reason, this study focuses on the analysis of financial performance for a sample of companies in the textile industry.

The textile industry has an old tradition in Romania. Although the industry has focused on higher value-added production, recent official reports [4] indicate that identifying the determinants of modest performance in the textile industry (in the context in which it has made an important contribution to international trade) is a priority. Similar to the trends in other countries [5], from the 1990s to the present, the clothing industry in Romania has been characterized by an unfavourable evolution of financial performance indicators, respectively, by a decrease in the employed workforce. The main cause was the inability of national companies to cope with imports (especially those from China).

Starting from this identified problem, this study aims to assess the impact of the main determinants on the financial performance of companies in the textile industry. In order to achieve the assumed objective, the article was structured as follows. The next section presents a summary of the literature review on financial performance and its determinants. The third section presents the research methodology. The fourth section presents the analysis, interprets the results of performance measurement, details the impact of determinants and initiates discussions so that the analysis can be useful for increasing the performance of textile companies. The last section presents the conclusions and considerations regarding future research directions.

2. Literature review regarding to financial performance and its determinants

Over time, performance management has been assigned two important roles: a) to support organizations in formulating and assessing strategy, to motivate human resources, and to communicate the performance of all stakeholders [6,7]; b) to continuously measure the effectiveness and efficiency of capitalizing strategic and operational resources, human and information resources, marketing and financial perspectives [8,9].

The importance and necessity of measuring and assessing organizational performance was recognized early on. Since 1996, Sinclair and Zaire [10] have argued that measuring performance helps managers adopt long-term perspectives, develop more effective communication, motivate and encourage the adequacy of organizational behaviour, facilitate change management, allocate resources more efficiently, and adopt a more efficient operating, planning and control system. Later (in 1999), Wagoner, Neely and Kennerley [11] showed that measuring performance has multiple purposes: controlling performance and identifying areas that require special attention; increasing employee motivation; improving internal communication; strengthening the accountability of all stakeholders.

In general, financial performance was defined from the perspective of companies' financial results. Therefore, in order to admit that a company is financially successful, for a certain period of time, the income must be higher than the expenses of the income-generating activities. Financial performance was also defined from the perspective of the efficiency and effectiveness of the organization in achieving its objectives [12]. In other words, a superior performance is the proof of the proper management and use of the company's resources [13, 14], respectively, the proof of the effectiveness and efficiency of the management of resource use. Other authors [14] argued that financial performance reflects a company's ability to create economic value, respectively, to attract and generate profits for investors [15, 12].

From a macroeconomic perspective, performance reflects the ability of companies to contribute to a country's economy in general [16]. Achieving superior financial performance amplifies the positive influences of companies on society and the natural environment [17].

Studies that have focused on the analysis of financial performance have mainly used two tools: financial rates (which allow the diagnosis of the financial health of companies) and cash flow (which allows managers to manage liquidity - for operational, financial and investment activities - so as to ensure the sustainability of the business). Most studies of financial performance determinants use causal models between a dependent variable (financial performance) and one or more independent (explanatory) variables. The dependent variable (financial performance) is assessed on the basis of two categories of data: accounting and market. Accounting measures are calculated based on performance over time, such as return on assets (ROA), return on equity (ROE), return on sales (ROS), return on investment (ROI), earnings per share (EPS), net profit margin (NPM), volume of sales, operating cash flow and others. Market-based measures (TobinsQ value and the company's market value) assess the company's performance from a market perspective. Unlike market-based measures, accounting-based measures can reflect the internal decision-making process and performance of managers [18].

The determinants of financial performance are grouped into two categories: internal (specific to the company) and external factors (specific to industry and / or the national economy). Therefore, it is recognized that financial performance depends on micro and macroeconomic variables. Microeconomic variables (financing structure, product / service, costs of production factors, quality, organizational culture, etc.) determine a company's position in its competitive environment. The main macroeconomic variables with an impact on financial performance are: economic growth rate, inflation rate, unemployment rate, indebtedness of companies, etc.).

Pantea, Gligor and Anis [19] showed that concerns about performance determinants are divided into two levels: one based on economic tradition (emphasizing the importance of external factors, related to the market), and another one focused on behavioural and sociological paradigm (centred on organizational factors and their correlation with the environment). The authors also emphasize that little attention is paid to the company's competitive position, respectively, to the company's internal factors.

Dinu and Vintilă [20] grouped external factors on two levels: economic environment (customized by current legislation, market preferences and purchasing trends of the population) and the economic situation of the country (assessed by the cost of financing, inflation rate, capital market development, international trade relations, exchange rate, current account deficit / surplus, demand and supply of goods, prices of raw materials and utilities, income per capita). In the category of internal factors, the authors work with the following structure: factors related to the business itself, factors related to its operations, ownership, management, employees, products / services and strategy.

Capon, Farley and Henig [21] conducted extensive research to identify factors that affect financial performance. Of these, the most representative determinants with a positive impact on financial performance were: industry concentration; increase in assets and sales; market share; company size (measured as sales' volume); capital investment intensity. The authors also noted that most research focused on financial performance in terms of the impact of environmental factors (concentration, industry size, imports and exports, barriers to entry) and organizational strategy (growth, capital investment, advertising, market share, research and development, debt, diversification, product / service quality, vertical integration, corporate social responsibility). Identifying few studies that address the companies' problems, the authors argued that more attention is needed "for this general family of determinants of financial performance".

Starting from the hypothesis that belonging to a cluster generates valuable benefits for companies (with positive effects on performance), Pavelkova et al. [5] conduct an analysis of the financial performance of companies in traditional industrial sectors in the Czech Republic - plastics and textiles. They used financial indicators such as ROA, ROS, labour productivity and economic added value to measure financial performance. Although the results of their research did not confirm the hypothesis, their work was a reference for this research because it signals differences in the level of performance of companies in the textile industry and the level of performance of companies in other industries.

Analysing the impact of capital structure on the performance of listed textile companies (90 companies, 2008-2017), Ullah et al. [22] showed that financial performance (assessed by ROE) is influenced by capital structure (degree of indebtedness), company size, asset turnover and sales' growth. Specifically, the authors provided evidence on: the negative and significant influence of capital structure on financial performance; the positive and significant influence of sales' growth (including export sales) on financial performance.

The literature also notes that companies in the textile industry have a modest competitiveness due to low productivity and product quality, poor product diversification, high costs of production factors and underdeveloped infrastructure [23]. In order to assess the financial performance of companies in the textile industry, some authors have proposed [24] and applied [25] diagnostic methods and tools with several criteria associated with financial, marketing, technological and quality activities, general management and human resources.

3. Methodology

The literature [26, 27, 28] points out that organizational performance research is exposed to risks related to: the correct identification of determinants and their measuring instruments; choice of data sources (secondary sources provide historical information; primary data - based on observation - are not relevant for longer periods of time); representativeness of samples (lack of homogeneity limits representativeness; focusing on certain areas / industries limits generalization).

Taking into account these considerations, we constructed the sample based on the following criteria: object of activity (companies were selected according to CNEA 1413 - Manufacture of other clothing, excluding underwear); the volume of sales for the last year of the analysis period (companies with annual sales of at least 2 million euros were selected). From the total number of companies identified, 120 companies were selected, of which only 63 submitted data for the entire analysis period. The total sales of these companies (in 2019) were 665 million euros (representing 42% of total sales). Of the 63 companies, those with more than three years of losses were eliminated. The final sample was represented by 47 companies.

To track the impact of financial performance determinants in different phases of economic cycles, we opted for a 20-year analysis (2000-2019). The selected sample was mixed (comprising both large and medium-sized companies); therefore, the statistical analysis was based on accounting data collected from secondary sources (available on www.mfinante.ro). The usefulness of accounting information was also confirmed by [25], which analysed the performance of companies in the Romanian textile industry.

Because some of the companies presented (during the analysed period) negative financial results, the financial performance was assessed on the basis of sales' volume (considered a dependent variable in the regression model). The independent variables, their determination methodology and the predicted impact are presented in table 1. Due to the fact that some indicators are calculated in absolute terms, to obtain correct results in future analyses we calculated natural logarithm.

Table 1. Dependent and independent variables. Impact predicted on financial performance.

Variables	Symbol	Determination methodology	Estimated impact*
Sales' volume	*LN _S*	Natural logarithm of turnover	
Return on assets	*ROA*	(Gross profit / Total assets) x100	+
Return on equity	*ROE*	(Net profit / Equity) x100	+
Return on salles	*ROS*	(Gross profit / Turnover) x100	+
Share of trade receivables in turnover	*Tt*	(Trade receivables / Turnover) x100	+ / -
Level of indebtedness	*LI*	(Liabilities / (Liabilities + Equity) x100	+ / -
Level of liquidity	*Liq*	(Cash availability/Liabilities) x100	+
Labour productivity	*Lp*	(Turnover/No. of employees)	+
Logarithm of Lp	*LN _Lp*	Natural logarithm of LP	+
Firm size_1	*Ne*	Number of employees	
Logarithm of Ne	*LN _Ne*	Natural logarithm of Ne	+ / -
Firm size_2	*TA*	Total assets	
Logarithm of TA	*LN _TA*	Natural logarithm of TA	+ / -
GDP growth rate	*GDP*	-	+
Unemployment rate	*UR*	-	-
Inflation rate	*IR*	-	-

Note: * According to literature review. Source: Processed by authors.

In order to identify the interdependent relations between S and its determinants, we performed correlation and regression analyses. *Data Analysis* (from Excel) software was used to perform statistical analysis for the identified econometric model. Because the data in our sample consider a set of 12 indicators, for 47 companies, over a period of twenty years, the regression analysis was adapted to panel data. The general equation of the regression model applied is as follows:

$$X_{it} = Y_{it} \, x \, \beta_1 + Z_{it} \, x \, \beta_2 + u_{it} \quad (1)$$

where: *i* represents the companies included in the analysis, *t* is time (2000...2019); X_{it} is the dependent variable (in our case sales' volume); Y_{it} is the independent variables; Z_{it} is the control variables (in our case the size of the company, appreciated by total assets and number of employees); β_1 and β_2 represent the coefficients; u_{it} is the error term.

4. Results and discussions

As described in the previous section, for identifying the interdependent relationships between S and its determinants, we performed the correlation and regression analyses. Before performing these, we analysed the results obtained for descriptive statistics (see table 2). The results highlight the fact that the indicators considered in the analysis varied significantly during the analysed period, registering also important differences depending on the company and the period.

Table 2. Descriptive statistic.

	LN_S	*ROA*	*ROE*	*ROS*	*Tt*	*LI*	*Liq*	*LN_Lp*	*LN_Ne*	*LN_TA*	*GDP*	*UR*	*IR*
Mean	16.6	17.9	40.9	-84.6	120.1	42.1	99.1	10.9	5.7	16.3	4.0	6.5	9.6
Median	16.7	12.8	19.7	9.2	15.3	40.3	21.1	10.9	5.8	16.5	4.3	6.8	5.7
Standard deviation	1.5	18.0	258.2	2445.1	3073.0	27.1	187.4	1.3	1.3	1.4	3.7	1.0	11.6
Minimum	0.0	-48.4	-1887.0	-73000.0	0.0	1.8	-309.2	0.0	0.0	8.1	-5.5	3.9	-1.5
Maximum	19.8	94.1	7411.2	75.6	94233.3	146.1	1429.4	18.7	8.1	19.2	10.4	8.1	45.7
Count	940	940	940	940	940	940	940	940	940	940	940	940	940

Source: Processed by authors.

Usually, the lowest value of salles was recorded in the first years of the analysis, because some of the companies were precisely established in 2000. Through the sales' logarithm operation, these differences have been ameliorated. Rates of return (ROA, ROE) recorded average values of 17.9%, respectively,40.9%. The superiority of ROE over ROA indicated that the sampled companies used borrowed capital which led to increased efficiency in the use of equity. For this reason, these variables were considered to have a positive impact on sales' volume (S). The average value for ROS was negative, which indicated that, on the whole sample, sales did not cover the costs associated with the production of the goods sold.

The average *Tt* is higher than 100% (120.1% respectively), which indicates that many companies do not immediately collect the value of sales. The facilities granted to customers to pay invoices can have a positive (if customers' loyalty is achieved) or negative impact on sales (because they reduce the company's cash flow). The average level of indebtedness is below 42.1%, and indebtedness has a positive effect on ROE; the predicted sign for the impact of this indicator was a positive one. Holding cash facilitates the purchase of other factors of production, a fact for which we estimated that it has a positive impact on sales. Average labour productivity is 116 euro/ employee; this variable usually has a positive impact on performance. The company's size was assessed by two indicators: total assets and number of employees. Usually, larger companies have higher sales, which is why it was estimated to have a positive impact on performance (appreciated by sales' volume).

For the three macroeconomic variables (GDP, UR and IR) we estimated a different impact on performance. A positive GDP growth rate indicates an economic environment with increasing activity, which can be considered a premise for increasing the performance of companies. Instead, the unemployment rate and the inflation rate indicated certain macroeconomic imbalances, which can have a negative impact on financial performance of the companies. Correlation analysis indicated strong associations between some variables (TA and S, Lp and S; bolded in Table 3), which required their exclusion from the regression analysis.

IOP Conf. Series: Materials Science and Engineering 1169 (2021) 012016 doi:10.1088/1757-899X/1169/1/012016

Table 3. Correlation analysis.

	LN_S	*ROA*	*ROE*	*ROS*	*Tt*	*LI*	*Liq*	*LN_Lp*	*LN_Ne*	*LN_TA*	*GDP*	*UR*	*IR*
LN_S	1												
ROA	-0.17	1											
ROE	-0.14	0.18	1										
ROS	0.42	0.06	-0.01	1									
Tt	-0.33	-0.06	0.00	-0.23	1								
LI	-0.15	-0.12	0.12	-0.10	0.11	1							
Liq	0.11	0.00	-0.05	0.03	-0.02	-0.67	1						
LN_Lp	0.49	-0.09	-0.07	0.31	-0.24	-0.10	0.08	1					
LN_Ne	**0.56**	-0.10	-0.09	0.17	-0.14	-0.07	0.05	-0.34	1				
LN_TA	**0.88**	-0.30	-0.17	0.18	-0.19	-0.21	0.21	0.51	0.51	1			
GDP	-0.02	0.05	0.00	0.02	-0.01	0.06	-0.08	-0.09	0.07	-0.05	1		
UR	-0.25	0.23	0.06	-0.02	0.01	0.16	-0.13	-0.30	0.01	-0.29	-0.14	1	
IR	-0.51	0.25	0.17	-0.12	0.10	0.36	-0.27	-0.53	-0.05	-0.53	0.03	0.30	1

Source: Processed by authors.

In order to eliminate the problem of the existence of multicollinearity and to obtain the most accurate results, we formulated regression models as follows:

$$LN\_S_{it}=ROA_{it}\,\beta_1+ROE_{it}\,\beta_2+ROS_{it}\,\beta_3+Tt_{it}\,\beta_4+LI_{it}\,\beta_5+Liq_{it}\,\beta_6+LN\_Lp_{it}\,\beta_7+GDP_{it}\,\beta_8+UR_{it}\,\beta_9+IR_{it}\,\beta_{10}+u_{it} \quad (2)$$

The results of the regression analysis are centralized in Table 4. *R Square* indicates that 51.4% of the S variation can be explained by the variation of the independent variables included in the model. The model hypotheses were:

H0: $\beta_1 = \beta_2 = ... = \beta_{10} = 0$;

H1: there is at least a non-zero coefficient β_i.

Table 4. Test regression.

Regression Statistics		*ANOVA*					
							Significance
Multiple R	0.717		*df*	*SS*	*MS*	*F*	*F*
R Square	0.514	Regression	10	213.043	21.304	98.531	2E-138
Adjusted R Square	0.509	Residual	929	200.868	0.216		
Standard Error	0.464	Total	939	413.911			
Observations	940						

Source: Processed by authors.

The values of the significance test F provided non-zero values, indicating that there is at least a non-zero coefficient. The results obtained indicate that the regression equation is significant globally, noting that some coefficients may not be significant. Therefore, the null hypothesis is rejected in favour of the alternative hypothesis. The statistical assurance assumed for modelling was 95%, and the assumed significance threshold was 0.05.

The values obtained for the regression model coefficients (see table 5) indicate that - with one exception (GDP) - all independent variables have a significant impact (statistically) on sales (S). The free term indicates that there are other variables that influence the volume of sales, and whose influence is statistically significant (P-value <0.05). This information suggests that financial performance also depends on non-financial variables. The results of the regression analysis are summarized in Table 6.

Table 5. Regression analysis.

	Coefficients	*Standard Error*	*t Stat*	*P-value*	*Lower 95%*	*Upper 95%*
Intercept	5.9148	0.2202	26.8619	0.0000	5.4827	6.3469
ROA	-0.0027	0.0009	-2.8724	0.0042	-0.0045	-0.0009
ROE	-0.0002	0.0001	-2.5046	0.0124	-0.0003	0.0000
ROS	0.0001	0.0000	10.5778	0.0000	0.0001	0.0001
Tt	0.0000	0.0000	-7.6713	0.0000	-0.0001	0.0000
LI	0.0018	0.0007	2.4577	0.0142	0.0004	0.0033
Liq	0.0003	0.0001	2.9865	0.0029	0.0001	0.0005
LN_Lp	0.3446	0.0341	10.1165	0.0000	0.2778	0.4115
GDP	0.0017	0.0042	0.3950	0.6929	-0.0066	0.0099
UR	-0.0375	0.0164	-2.2867	0.0224	-0.0696	-0.0053
IR	-0.0152	0.0017	-8.7167	0.0000	-0.0186	-0.0118

Source: Processed by authors.

Table 6. Variables of regression model.

Variables	Symbol	Estimated impact*	Obtained results	Degree of significance	Hypothesis
Return on assets	*ROA*	+	-	*Significant* *P-value* < 0.05	Infirmed
Return on equity	*ROE*	+	-		Infirmed
Return on salles	*ROS*	+	+		Confirmed
Share of trade receivables in turnover	*Tt*	+ / -	+		Confirmed
Level of indebtedness	*LI*	+ / -	+		Confirmed
Level of liquidity	*Liq*	+	+		Confirmed
Labour productivity	*Lp*	+	+		Confirmed
Logarithm of Lp	*LN_Lp*	+	+		Confirmed
GDP growth rate	*GDP*	+	+	*P-value* > 0.05	Insignificance
Unemployment rate	*UR*	-	-	*Significant* *P-value* < 0.05	Confirmed
Inflation rate	*IR*	-	-		

Note: * According to literature review. Source: Processed by authors.

To identify the impact of independent variables on performance at different stages of economic cycles, we organized research over three periods: pre-crisis (2000-2008); crisis (2009-2015) and post-crisis (2016-2019). The regression model presented above was applied for each period. The results are summarized in Table 7.

Applying the regression model over the three periods (pre-crisis and post-crisis) indicated the following:

- for the period 2000-2008, 51% of the variation S can be explained by the change of the independent variables included in the model; for the other periods the representativeness of the model decreases to 49% during 2016-2019 and to almost 26% between 2000-2008;

- Significance F - for all three periods - has a lower value than the set significance threshold (0.05), which means that the regression equations are significant at the global level (noting that some coefficients may not be significant);

- for the three periods, the free term varies between 4 and 6, which indicates that there are other variables with significant influence on sales (because P-value <0.05);

- ROA variation has a significant and negative impact in the pre-crisis period (2000-2008); during the crisis (2009-2015) this variable changes its direction of influence (being positively correlated with

IOP Conf. Series: Materials Science and Engineering 1169 (2021) 012016 doi:10.1088/1757-899X/1169/1/012016

the volume of the sales);

- the variation of ROE is maintained with a negative influence, but it is only representative for the crisis and post-crisis periods;

- as expected, the share of receivables in total sales has a negative and statistically significant influence in the crisis and post-crisis periods; also, during this period, the degree of indebtedness has a positive and significant impact on sales;

- labour productivity is the only variable that remains positive and significantly correlated with the volume of the sales (the statistical significance of the results is confirmed both for the entire period, 2000-2019, and for the pre-crisis, crisis and post-crisis periods); by comparison,

- the GDP growth rate is variable without statistical significance (both at the level of the whole period and in the pre-crisis, crisis and post-crisis periods);

- labour market imbalances (assessed by the unemployment rate) do not have statistically significant influences on the volume of the sales;

- the change in the inflation rate is maintained with a negative and significant influence only in the pre-crisis period; during this period (2000-2008) the inflation rate varied between 45% and 79%; in the period 2009-2016, the inflation rate changes from positive values (6.5%) to negative values (-0.6), a variation that is not significant in the regression model run in this research; this variable remains statistically insignificant in the period 2016-2019 (time in which the transition from deflation to moderate inflation is marked with a growth rate of 3.8%).

Table 7. Regression analysis.

Regression Statistics	2000-2008	2009-2015	2016-2019	2000-2019
Multiple R	0.71	0.51	0.70	0.72
R Square	0.51	0.26	0.49	0.51
Adjusted R Square	0.50	0.24	0.46	0.51
Standard Error	0.55	0.37	0.32	0.46
Observations	423	329	188	940
Significance F	*2.31E-57*	*2.7E-16*	*3.49E-21*	*2E-138*

Regression								
	2000-2008		*2009-2015*		*2016-2019*		*2000-2019*	
Results	*Coefficients*	*P-value*	*Coefficients*	*P-value*	*Coefficients*	*P-value*	*Coefficients*	*P-value*
Intercept	6.035	0.000	4.937	0.005	5.000	0.000	5.9148	0.0000
ROA	-0.004	0.006	0.009	0.015	0.009	0.063	-0.0027	0.0042
ROE	0.000	0.094	-0.007	0.003	-0.007	0.020	-0.0002	0.0124
ROS	0.000	0.000	-0.005	0.126	-0.008	0.039	0.0001	0.0000
Tt	0.000	0.000	-0.003	0.033	-0.006	0.000	0.0000	0.0000
LI	-0.002	0.142	0.007	0.000	0.009	0.000	0.0018	0.0142
Liq	0.000	0.499	0.000	0.000	0.000	0.001	0.0003	0.0029
LN_Lp	0.299	0.000	0.262	0.000	0.411	0.000	0.3446	0.0000
GDP	0.004	0.730	0.007	0.518	0.004	0.832	0.0017	0.6929
UR	-0.005	0.905	0.156	0.546	0.047	0.705	-0.0375	0.0224
IR	-0.012	0.000	-0.032	0.126	0.019	0.640	-0.0152	0.0000

Source: Processed by authors.

The results of the econometric analysis are of great practical use because they provide important benchmarks in substantiating long-term organizational strategies. Companies' managers in the textile industry must take into account that: in pre-crisis and crisis periods the goal of increasing sales is more important than the goal of maximizing earnings; credit sales (which generates trade receivables) does not contribute to the increase sales' volume; labour productivity is one of the important factors of performance (appreciated by sales' volume); significant monetary imbalances have a negative impact on sales.

5. Conclusions

This study focuses on the analysis of financial performance for a sample of companies in the textile industry which has an old tradition in Romania. The identified research problem (the unfavourable evolution of the financial performance in the last 20 years) determined us to identify which are the factors that set up this evolution.

The bibliographic research carried out indicated that the identification of the determinants of financial performance is a current and important concern. This is because performance management aims to support companies in formulating and assessing the strategy, taking into account both the microeconomic and macroeconomic context.

The empirical research aimed to determine the meaning and intensity of the influences of the determinants of financial performance (appreciated by the volume of sales). The statistical analysis performed on a sample of companies in the textile industry (inhomogeneous sample in terms of company size) provided representative information both for the period 2000-2019 and by subdivisions of this period (pre-crisis, crisis and post-crisis). The regression model used included ten internal determinants (ROA, ROE; ROS; share of receivables in total sales; labour productivity, indebtedness and liquidity) and three external variables (GDP growth rate, inflation rate and unemployment rate).

The results of the analyses performed for the whole period (20 years) indicated that all ten internal variables have a significant influence (from a statistical point of view) on the financial performance. From the perspective of macroeconomic determinants, the study showed that performance is negatively and significantly correlated with the inflation rate and the unemployment rate. The results of the analyses performed over shorter periods (pre-crisis, crisis and post-crisis) indicated that most of the determinants analysed had a different impact. The only variable that keeps the same impact on performance is labour productivity (with the stipulation that the intensity of the impact increases in post-crisis periods). Debt has a positive and significant impact only for the crisis and post-crisis periods.

In the literature review we did not identify assessments of the impact of performance determinants on different economic phases (growth, crisis or decline). Therefore, we admit that this study fills the existing research gap in the field of performance management and presents evidence for companies in the textile industry in Romania. The elements that give originality to the research are: the selected sample (which ensures the representativeness of the results at the level of the textile industry); the analysed period (extended for 20 years, so as to include pre- and post-crisis periods) and the results obtained (which can be very useful in substantiating organizational strategies).

The study has some limitations. Data were collected from simplified annual financial statements, which reduced the number of variables with a potential impact on performance. In the regression model applied, the free term varies between 4.9 and 6.03, which indicates that there are other variables with a (statistically significant) impact on financial performance. In future research we intend to identify these factors, so as to expand knowledge about the determinants of financial performance of companies in the Romanian textile industry.

6. References

[1] Saunders M, Lewis P and Thornhill A 2009 *Research methods for business students* (Italy: Rotolito Lombarda)

[2] Bercovitz J and Mitchell W 2007 When is more better? The impact of business scale and scope on long-term business survival, while controlling for profitability *Strat Manag J* **28**(1) pp 61–79

[3] Tudose M B and Avasilcai S 2020 A Review of the Research on Financial Performance and Its Determinants *Innovation in Sustainable Management and Entrepreneurship (Springer Proceedings in Business and Economics)* ed G Prostean, J J Lavios Villahoz, L Brancu and G Bakacsi (Cham: Springer) chapter 4 pp 229–44

[4] BNR 2019 Annual report, available at https://www.bnr.ro/Raport-anual-2019-21552.aspx accessed on 5 Mart 2021

[5] Pavelkova D, Zizka M, Homolka L, Knapkova A and Pelloneova N 2021 Do clustered firms outperform the non-clustered? Evidence of financial performance in traditional industries *Economic Research-Ekonomska Istraživanja* DOI: 10.1080/1331677X.2021.1874460

[6] Kennerley M and Neely A 2002 A framework of the factors affecting the evolution of performance measurement systems *Int J Oper Prod Manag* **22**(11) pp 1222–45
[7] Frolick M and Ariyachandra T 2006 Business performance management: one truth *Inf Syst Manag* **23**(1) pp 41–49
[8] Melchert F and Winter R 2004 The enabling role of information technology for business performance management *Decision Support in an Uncertain and Complex World: The IFIP TC8/WG8* pp 535–46
[9] Chvatalova Z and Koch M 2015 Optimizing of information systems in companies: support of sustainable performance *Procedia Soc Behav Sci* 213 pp 842–47
[10] Sinclair D and Zairi M 1996 Assessing the effectiveness of performance measurement systems: A case study *Total Quality Management* **7** pp 367–78
[11] Waggoner D, Neely A and Kennerley M 1999 The forces that shape organizational performance measurement systems: An interdisciplinary review *Int J Prod Econ* **60**(1) pp 53–60
[12] Kurniaty S, Handayani S R and, Rahayu S M 2018 Stock return and financial performance as moderation variable in influence of good corporate governance towards corporate value *AJAR* 2443-5175 pp 18–34
[13] Mahrani M and Soewarno N 2018 The effect of good corporate governance mechanism and corporate social responsibility on financial performance with earnings management as mediating variable *AJAR* **3**(1) pp 41–60
[14] Orozco L A, Vargas J and Galindo-Dorado R 2018 Trends on the relationship between board size and financial and reputational corporate performance: The Colombian case *Eur. J. Manag. Bus.* **27**(2) pp 183–97
[15] Al-Sa'eed M A A 2018 The Impact of Ownership Structure and Dividends on Firm's Performance: Evidence from Manufacturing Companies Listed on the Amman Stock Exchange *Australas. Account. Bus. Finance J.* **12**(3) pp 107–26
[16] Egbunike C F and Okerekeoti C U 2018 Macroeconomic factors, firm characteristics and financial performance: A study of selected quoted manufacturing firms in Nigeria, *AJAR* **3**(2) pp 142–68
[17] Belás J and Gabčová L 2016 The Relationship among customer satisfaction, loyalty and financial performance of commercial banks *E a M*: *Ekonomie a Management* **19**(1) pp 132–47
[18] McWilliams A and Siegel D 2000 Corporate social responsibility and financial performance: Correlation or misspecification? *Strateg Manag J* **21** pp 603–09
[19] Pantea M I, Gligor D and Anis C 2014 Economic determinants of Romanian firms' financial performance *Procedia Soc Behav Sci* **124** pp 272–81
[20] Dinu A T and Vintilă N 2017 Determinant factors for the performance of companies *ABC-UL LUMII FINANCIARE* WP **5** pp 1–22
[21] Capon N, Farley J and Hoenig S 1990 Determinants of financial performance: a meta-analysis. *Management Science* **36**(10) pp 1143–59
[22] Ullah A, Pinglu C, Ullah S, Zaman M and Hashmi S H 2020 The nexus between capital structure, firm-specific factors, macroeconomic factors and financial performance in the textile sector of Pakistan *Heliyon* **6** pp 2–10
[23] Shetty R, Kiran KB and Dash M 2013 A Study on the Impact of Quota Removal on the Textile and Clothing Industry in Karnataka State *Int J of Econ Management and Soc Sci* **2**(6) pp 277–84
[24] Alexa V, Kiss I and Ratiu S 2013 Diagnostic analysis – tool in research and innovation *Annals of Faculty Engineering Hunedoara* **11** pp 321–24
[25] Hada T, Bărbuţă-Mişu N and Avram T M 2019 Evaluating firms' financial performance in textile industry of Romania *Ind Textila* **70**(3) pp 272–76
[26] Hamann M, Schiemann F, Bellora L and Guenther T 2013 Exploring the Dimensions of Organizational Performance A Construct Validity Study *Organizational Research Methods* **16**(1) pp 67–87
[27] Lee J 2009 Does size matter in firm performance? Evidence from US public firms *Int J Econ Bus* **16**(2) pp 189–203
[28] March J and Sutton R 1997 Crossroads-Organizational Performance as a Dependent Variable *Organ Sci* **8**(6) pp 698–706

IOP Conf. Series: Materials Science and Engineering 1169 (2021) 012017 doi:10.1088/1757-899X/1169/1/012017

Leadership approach towards Agile, Waterfall and Iterative implementation of the software development products

O Popa[1], C. Mihele[1], C Făgărăşan[1]and A. Pîslă[1]

[1] Design engineering & Robotics Department, Machine Building Faculty, Technical University of Cluj-Napoca, 103-105 Muncii Avenue, Cluj-Napoca, Romania

[1]ou.popa@gmail.com

Abstract. The automotive industry is one of the most dynamic markets, gathering around many different sectors from material extractions to robotics and cloud computing. The current research analyses the influences and particularities of leadership and identifies the challenges and advantages of the two principal methodologies in software development projects within the automotive industry, Agile and Waterfall. Some aspects of reducing methodological disadvantages are suggested and presented. The aim is to arrive at a sustainable solution that can ensure the delivery of reliable and flexible software products. The impact of these methodologies on the quality standards and project documentation is analyzed, which is a vital part of automotive software development. By influencing the product development lifecycle, there is an opportunity to develop a structured framework that can be adopted to build software products in a more flexible and agile way.

1. Introduction

It is impossible to ignore the global value of the automotive industry that employs directly around five million people and indirectly many times more. Only in South Korea, this industry employs over 1.4 million people. Besides that, the automotive sector is one of the most dynamic fields reporting continuous and sustainable growth. It gathers around different other industries from raw materials extraction, materials development, design, machine-tools and robotics, manufacturing process, technologies development, cabling, testing, maintenance, ensuring, dealership, planning, programming, AI and cloud computing, bringing to the market more than 60 million cars and trucks in a single year.

The automotive industry contribution to technology and value creation acts as a pillar of the global economy and the central driver of macroeconomic growth and stability using innovation and technological advancement in developed and developing countries, with a high percentage in the national GDP, representing 3% of the global GDP. It is known that governments can have a direct impact on automotive manufacturers through the regulations introduced. As an effect, the production costs may be increased, and limitations may be added to the way cars are sold and marketed. Therefore, the governments and the industry's key stakeholders must closely collaborate to build a shared strategy.

Car manufacturing implies many engineering specializations. At the end of the XX century, the automotive industry was dominated by the physical suggested disciplines, directly contributing to automotive manufacturing, the so-called "conventional engineering". The XXI century is greatly influenced by the fourth industrial revolution, which promotes the automation of traditional manufacturing and industrial practices, going fully digital towards building an automotive "brain" and "sensorial" system.

The boost for innovation needs substantial Research and Development investment by automakers, driven by the consumer demands for more product variety, better performance, improved safety and higher emission standards, all together at lower costs. Automobile companies are in third place on the list with the most significant investments in Research and Development. In the first place is the pharmaceuticals companies with $120 billion (as reference 100%), then the technology companies with $111 billion (92.5%) and the automotive industry with $108 billion (90%). The essential automotive investing countries that cover 71% of the automotive Research and Development investment are Germany (33%), Japan (20%) and South Korea (18%). Furthermore, the automotive industry's Research and Development and innovation can benefit other sectors, such as insurance or automotive telematics, the communication technology for the automobile industry based on information flowing from vehicles via wireless networks.

The core automotive industry, the vehicle manufacturers and parts makers, will continue to support a wide range of upstream and downstream business segments with adjacent sectors. This will continue to lead to a multiplier effect for growth and economic development. The automotive industry remains at the forefront of cutting-edge manufacturing technology that spread to other sectors. Leading and production processes within automotive (lean manufacturing, assembly lines, supply chain integration and modular sourcing) are standard for many different industries. This sector was a pioneer in using robots and other automation solutions, and now the Robotics business is an over $25 billion industry.

Recognizing it as a forefront means that learning from the global automobile industry is more than beneficial, especially for the fact that project management is a vital component as is also defined in DIN 699901 [1] by the core characteristics: novelty, unicity, complexity, and nevertheless interdisciplinary, interlacing different resources from different fields that depend on one another. By focusing on kernel processes, automakers have improved profitability and served niche markets more efficiently. Therefore, the interdisciplinary projects involving hardware and software development for the automotive industry face three significant challenges.

1.1 The organization leadership best implementations strategy

By definition, the Agile methodology can be considered a product of successful leadership. The Agile methodology's implementation in any organization automatically determines an increase in the leadership quality level [2]. The fusion of core values, concepts, and behaviour promotes highly-rated products that consistently exceed customer expectations and increase the project team's potential. As a result, it can be affirmed that efficient leadership and agile methodology can be considered interdependent components of the organization strategy in project management [3].

In this complex equation, other critical factors can lead to project success when Agile or Waterfall methodologies are applied. They include a brief definition of scope and objectives, increasingly focused on adding value to the organization, implementing a functional management structure, pushing its members to adopt a new mindset that can lead to determination, proactive and efficient communication. In the Agile context, the project attributes tend to be amplified, and their impact extrapolated.

The transition from Waterfall to Agile needs to be evaluated as a complete change inside the organization because of the significant impact on the project management methodology that grants success. During waterfall project applicability, the project's constraints are high, and the software development lifecycle is sequential; a phase needs to be finalized and accepted before the next one can be started. Regarding the scope, it is defined at the beginning of the project, and once it is approved, there are rigorous processes to follow to change the agreed scope baseline. The Agile methodology was introduced to promote change and welcome early feedback in the software development life cycle to respond to customer requests continually. The budget and time may remain fixed, depending on the existing agreement.

The Agile transformation should start with understanding the organization's desire to change the standard approach by adopting Agile, as each organization might have a different one. The most apparent reasons are improving delivery, increasing the quality of the products and improving the time to market. For situations where no development delivery method was officially instated, the Agile methodology represents a suitable solution.

The agile framework offers the possibility of fragmented product increments delivered in short periods through development iterations. Decoupling big projects in fragments represents a solution for blocking stops in the cascade approach. In the automotive industry, the quality of final delivered products may not correspond to customer expectation because it might have changed between the project's start-up till the end of the project. Still, the Agile methodology is preventing such situations by constant reviews with customers and continually adapting the product according to their vision.

The approach toward Agile strategy needs to address the organizational culture and management style, on top of their critical constraints [4]. Culture is one of the foundation pylons for the business strategy, and the most popular culture constraints are the management style, trust, acceptance of change and intense process focus. Therefore, awareness about the existing organizational structure is crucial as long as that will reflect the adopted business strategies.

The automotive industry has more cascade-type organizational structures as a typology that helps in the tasks and responsibilities distribution in an organized way. The delegation is transparent, and there are predetermined decision processes. Hence, the automotive industry relies on three types of organization diagrams: Functional, Project-based, Matrix-based.

The most traditional structure is the Functional organization, represented in Figure 1, consisting of divisions and departments oriented on specialization, product development or different manufacturing facilities. A project may be split into phases from various departments and divisions. As an example, this is highlighted in grey in Figure 1. Such structure is proven to favour flexibility towards the clients demanded products, with the capability that some manufacturing technology can be imposed. In such an organization, the procedures, communication and responsibilities are very well-defined. As a disadvantage, this system is reluctant to change, and usually, there is a lengthy process through which organizational changes are considered.

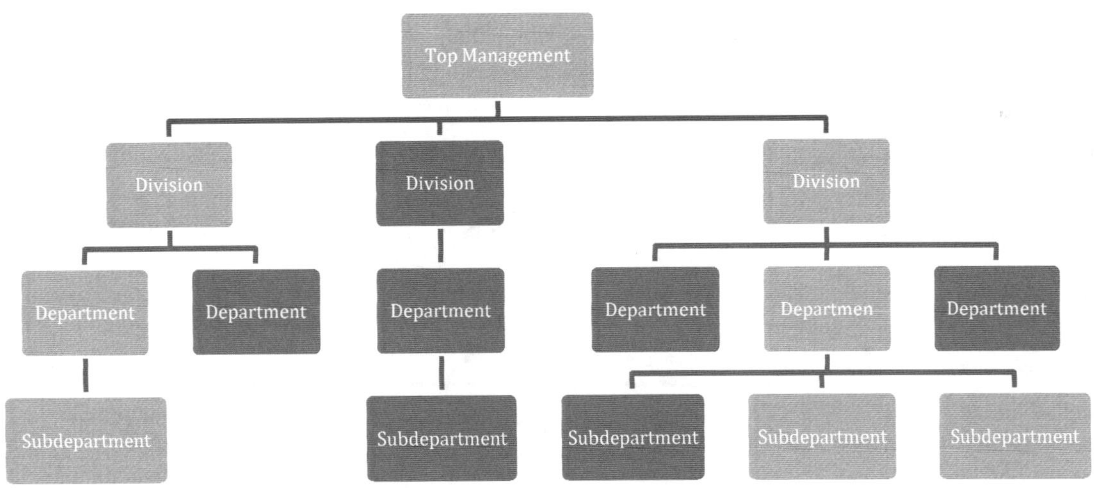

Figure 1. Functional organization [16]

Figure 2 displays a project-based organization, a typical structure for the organizations that offer professional services, including the product design or product development companies, facilitating the cooperation within the company members of different specializations, and having the same membership structure an entire project. There is a risk here that can result in technical capping and frustration caused by isolation; other teams have nothing in common despite being company member.

In this structure, the project manager will be responsible for team management and motivation (organizational and functional), including all project management activities.

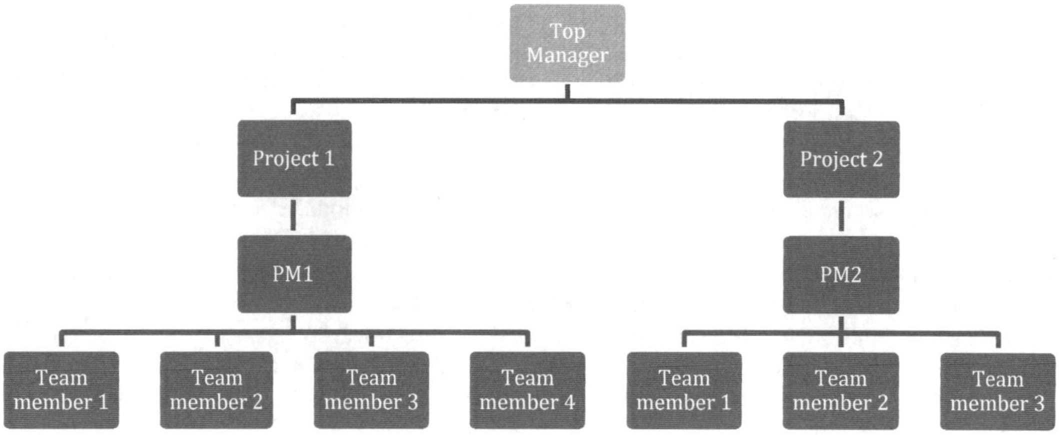

Figure 2. Project-based organization [16]

The structure presented above has its disadvantages because it implies that companies have a continuous flux of projects. There will be periods of peak times during which the teams are stressed due to a lack of personnel to deliver the higher flux of projects. On the other hand, when the company will have a lower flux of projects, the company's budget will be at risk. The situation presented led to a derivate version of the project-based organization. As described in Figure 3, in this hybrid project-based structure, the project managers may form their teams from a single "human resources pool", offering each individual the chance to work in different groups and gather novelty and knowledge. This structure introduces flexibility and promotes teamwork within the organization.

The big challenge of the hybrid structure may be the projects prioritization and the adjacent sharing of the human resources, with less personal or emotional involvement leading to ownership loss.

Figure 3. Derivate project-based organization [16]

Figure 4 reveals the Matrix-based organization. This structure is recognized for counteracting most of the disadvantages of the previous structures, offering a balanced environment in which the Project Managers (PM) are organized in a single department. The Project Management Office administrates all the projects, led by a Team Leader (TL), who is usually in charge of distributing the projects within the team. The project managers are at the same level within the department, there is no natural hierarchy, and they are reporting to the Team Lead of the Project Management Office or the Top management.

In this case, the main challenge is the projects distribution and the project managers allocation. The excellent cooperation between the Team Lead and the Project Managers is the key to getting increased department efficiency and project success. The project managers are directly responsible for their assigned projects and their completion. The project managers team has a consultative role but is under the subordination of the Team Lead.

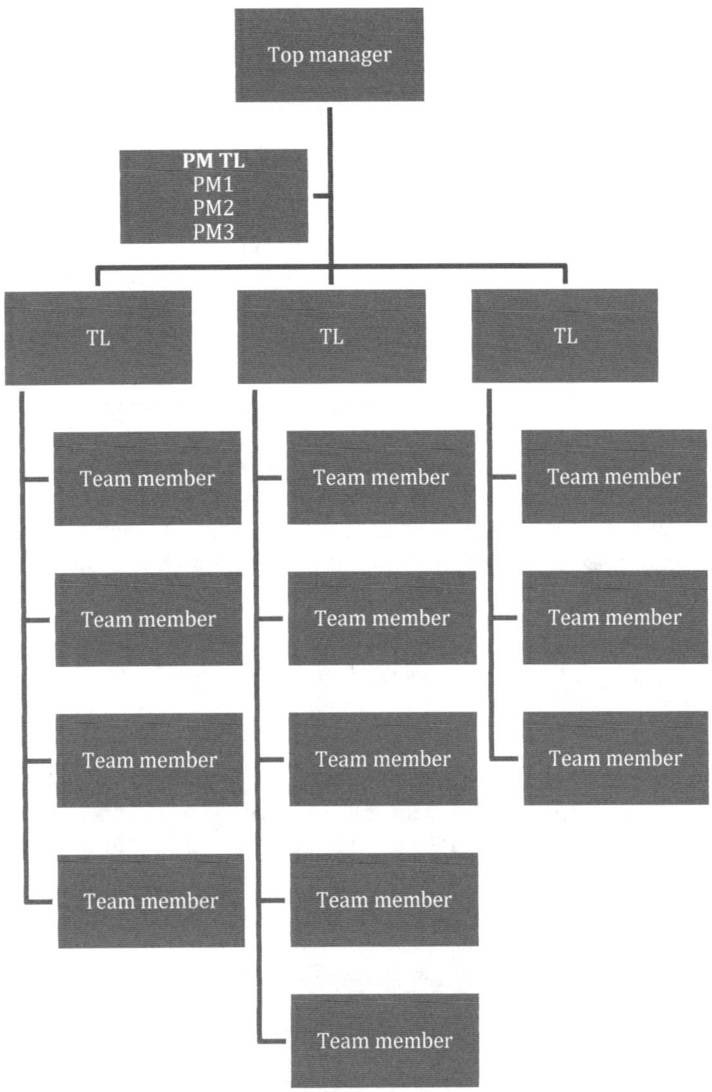

Figure 4. Matrix-based organization [16]

The second challenge of such an organization is the two heads' impressions and the danger of redundancy in documentation, processes and reports. The Agile management style can help overcome this challenge through collaboration and decentralized decision-making. If the organization management style is directive, the managers need to learn to encourage cooperation, delegate and allow their subordinates to make decisions.

Another cultural constraint is the resistance to change, which needs to be reduced. Willingness to embrace the change is a determinant factor for agile implementation success. If the organization culture is reluctant to change, the agile transformation process needs to incorporate existing change leaders or external leadership representatives on board [5]. The managers need to prepare for the transition and learn about the advantages of the agile methodology. Furthermore, trust is an essential element for Agile growth, facilitating the leadership changed approach [6].

A successful agile methodology implementation is always the result of the team's collaboration and effort and should start with a pilot project. Developing a group of agile supporters that understand the methodology and are interested in testing and promoting it represents another pylon of organization agile transition. A team composed of people with knowledge about agile who are also willing to contribute can ensure the pilot project success. Having a team of representatives in each project area with an agile mindset provides the basis for the pilot project's success.

The agile methodology needs supporters in all layers of the organization, from executive levels to individual contributors. One major step in the agile transformation process is to designate agile influencers on the executive company level.

After the pilot project and pilot team are defined, the visual presentation becomes the next major step in implementing the agile methodology. From the project's initiation phase, everyone needs to be informed, and the communication channels must remain open and transparent. The executive support must be demonstrated within the organization and is needs to be visible, starting with the pilot project presentation, available to every member of the organization.

I. The developed software components can be swift and appropriate into the hardware component that requires a longer development process and comes with constraints in the flexible software development.

II. Totally qualification to standards. The hardware-software components are in direct dependency on passenger safety. For these reasons, the industry adopts plenty of quality standards. The software projects should qualify to standards such as ASPICE [7], ISO 9001 [8], ISO 26262 etc. It is necessary for these met standards, but it must be taken into considerations also the generate challenges. For example, a significant impact is the level of detail in documentation, which is very helpful for traceability, planning, and development history that is slowing down the development process and requires a significant level of resources.

1.2 Leadership issues in the automotive industry

In the automotive industry, leadership is much connected with project management and human interaction [9], [10], [11, [12], [13], as the first and necessary more vital element that face the four forces of the change: Political, Economic, Social and Technological. In case of failure, managers are considered the ones responsible. In case of success, the sums of money leaders are paid in the automotive industry each month is justified.

Thus, the organization leader and the methodologies adopted by him are unique factors for success or failure. Consequently, besides the knowledge, skills, competencies, and the employees' involvement, the result is defined by the role of the leader to align all of the above and obtain sustainable and long-lasting competitive advantages. It is essential that the new leadership model and culture work well over time [14].

IOP Conf. Series: Materials Science and Engineering 1169 (2021) 012017 doi:10.1088/1757-899X/1169/1/012017

1.3 Project management methodologies in the automotive industry

The planning of the automotive software projects is a genuine challenge when part of the project scope is to ensure a specific quality for this industry. The development process should be competitive and recognized in the software development world.

A classic waterfall methodology would be disadvantageous for software development because it is very rigid, as described in [15]. This effect can produce significant gaps in the performance of the product's milestones, and customer satisfaction can be seriously impacted at the end of the product development. Incremental development can increase these key performance indicators because the product is developed in an active and dynamic synchronization with the customer needs. On the other side, Agile development with an applied incremental delivery for mechanical/hardware parts can increase significant development costs or be predictable for robotic development.

Figure 5 reveals the hybrid methodology designed for the automotive industry. Taking into consideration all hybrid scenarios [16], [17], the advantages and the disadvantages of this methodology were evaluated [18], with the purpose of reducing the sides effects of this in a mapped methodology with the project's specific needs.

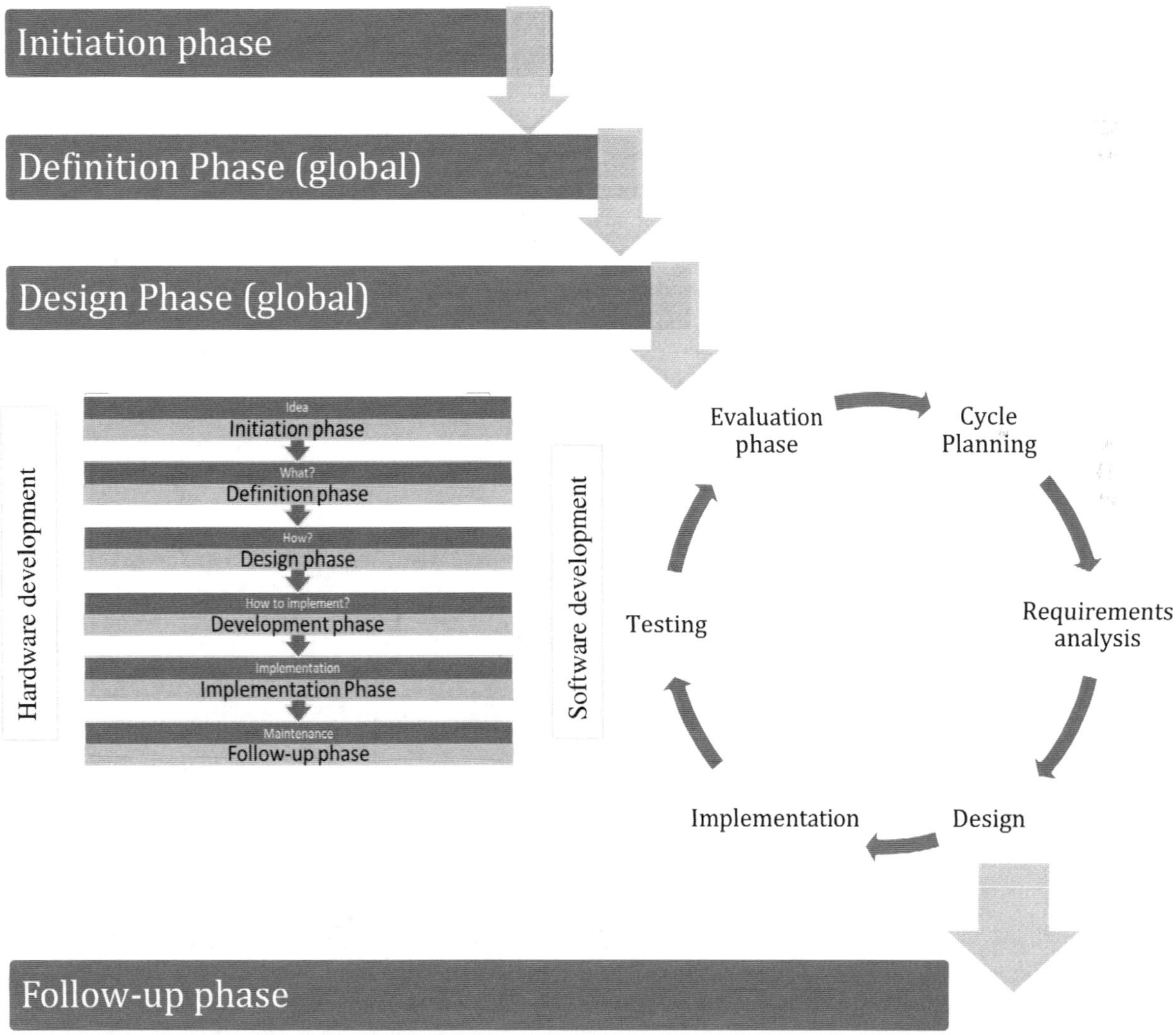

Figure 5. Hybrid methodology in automotive [16]

IOP Conf. Series: Materials Science and Engineering 1169 (2021) 012017 doi:10.1088/1757-899X/1169/1/012017

2. Software component development in automotive industry development

The development workflow was applied to an automotive software development project for the infotainment system centred on the central media system from a car console. For the example below, the project scope is to deliver the product just with unit and integration tests.

As it is shown in Figure 6, the customer requirements play an essential role at the beginning of the project. It is necessary to elicit clear and precise customer requirements. The client (PAG) raises the business requirements with the company's implementation partner, internal or external. The Agile teams need to be cross-functional; they need to have all the skills necessary to deliver increments of working software. Usually, the development team has a business analyst who takes the client requirements and evaluates them together with the team. In this phase, the team collaborates with the client to refine and clarify the business requirements and transforms them into functional requirements.

After the client has approved the specifications, a more detailed analysis occurs, and the final software requirements are defined. After they are revised and validated, the application architecture is developed, which must be designed, reviewed and approved.

The next step is for the development team to create a detailed design of the application. After this phase is complete, the unit test specifications description follows based on the design specifications. In correlation with the defined architecture, the integration tests specifications are prepared. After the acceptance criteria are validated during the testing phase, the final version can be delivered to the client.

The customer may have his own processes to validate the acceptance criteria. For big projects, the companies have acceptance team which verify that the functionality provided is according to the business requirements and the implementation satisfies the business need. After this validation, the customer is doing qualification and validation tests followed by the approval.

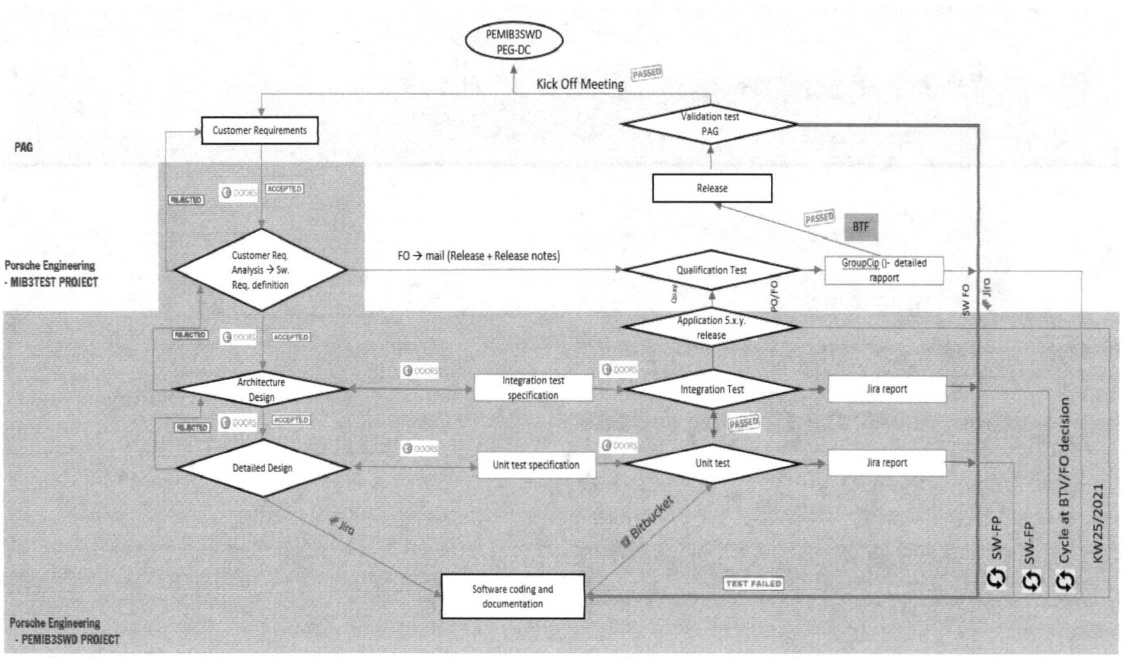

Figure 6. Development workflow for a software component used in the infotainment system

As it can be observed, a V-Model methodology is used for the development of the software component. However, the application can be developed in more iterations by leveraging the implementation methods of the Agile methodology. The most popular ones are Scrum and Kanban, which are well-known for their success rates and easiness of usage. Project iterations are called sprints in the Scrum framework. The agile development process was adjusted to create a new working methodology, as shown in Figure 7.

IOP Conf. Series: Materials Science and Engineering 1169 (2021) 012017 doi:10.1088/1757-899X/1169/1/012017

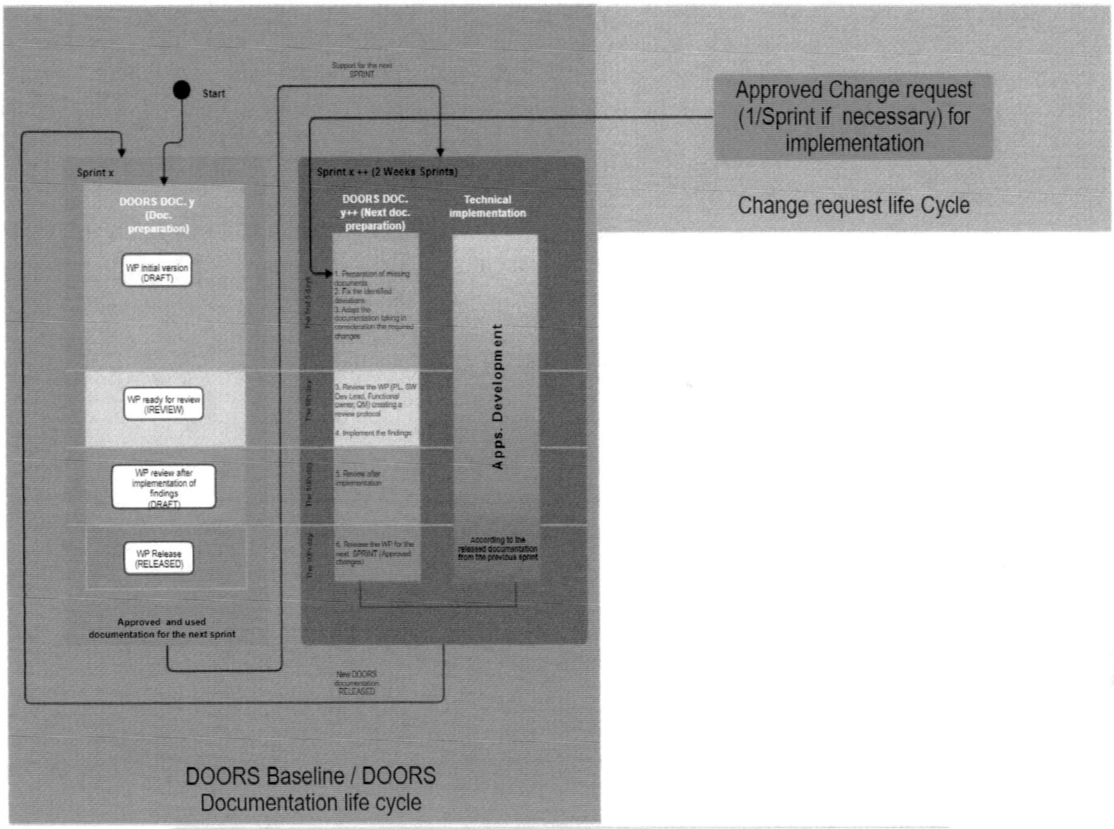

Figure 7. Iterative methodology integrated into V-Model development workflow

All projects have kick-offs during the initiating phase. After this formal meeting with the project team, as Figure 7 shows, Sprint x is started, which is the first sprint of the development phase. In this iteration, the first version of the project's documentation is produced, and any technical project setup can be done. Once the documentation is revised and approved during the sprint review meeting, the priorities for the next iteration are discussed. The application development can begin based on the approved documentation from the previous sprint in the second sprint, even if it is not entirely validated.

As part of the new iteration, the documentation is further developed by adjusting it or introducing any missing elements. Furthermore, in the first week of the current sprint, the approved change requests can be submitted in the documentation so the client can adapt the requirements according to the dynamic changes or needs. Once the second sprint is finalized, a new version of the documentation is approved. This documentation will become the base of the application development in the third sprint.

Through this interactive method, the project can permanently be flexible related to new requirements of the client through a Change Request process, taking the overall picture of the product development. It can be observed that the V-Model allows systematic documentation and planning to the automotive industry.

3. Conclusions

The project management methodologies have the main scope to help the projects in every domain perform and compete in the market. The automotive industry has procedures and standards, which are well-defined. Still, the sector should inspect and continually adapt the development procedure to be faster, more flexible, and incorporate the Agile mindset. Despite the fact that the hardware development and the standard's requirements come with many limitations, the results can be significant if the agile spirit is adopted in all industry branches.

For the automotive software development area, it was shown that a combination of the Agile software development lifecycle and a V-Model documentation process adapted in short, incremental releases can be a suitable solution in this dynamic and rigorous context.

Acknowledgement: This paper was supported by Porsche Engineering Romania.

4. References

[1] Die DIN 69901:2009-01 "Projektmanagement - Projektmanagementsysteme", January 2009, replacing the former valid norms DIN 69901:1987-08, DIN 69902:1987-08, DIN 69903:1987-08; DIN 69904:2000-11 and DIN 69905:1997-05, (2009).
[2] Gary Yukl, Leadership in Organizations (7th edition), Prentice-Hall (2009).
[3] Editors: Jay A. Conger & Ronald E. Riggio, The Practice of Leadership, John Wiley & Sons Inc, ISBN13 (EAN):9780787983055, (2007).
[4] Edgar H. Schein, Organizational Culture and Leadership, (4th edition), Jossey-Bass Business & Management, (2010).
[5] Richard L. Hughes, Leadership: Enhancing the Lessons of Experience. (7th edition). Ginnett, & Curphy. McGraw-Hill/Irwin, (2010).
[6] James MacGregor Burns, Leadership, New York: Harper Collins, (1978).
[7] Automotive SPICE, Process Reference Model, Process Assessment Model, Version 3.1, Author VDA QMC Working Group 13 / Automotive SIG (2021).
[8] International standard ISO 9001, Quality Management System-Requirements (2021).
[9] Narcis Popescu, Project Management MZT815 Materiale Auxiliare, (2021).
[10] Adrian Danet, Managementul Proiectelor, editura Disz Tipo, Brașov (2001).
[11] Project Management, Olad Passnheim & Ventus Publishing ApS, (2009).
[12] R. N. Lussier, Human Relations in Organizations: Applications and Skills Building, (1999).
[13] David V. Day & John Antonakis, The Nature of Leadership (2nd edition), SAGE press, (2011).
[14] Bruce J. Avolio, Full Range Leadership Development, (2nd edition) SAGE press, (2010).
[15] Kelley O'Connell, Transitioning from Waterfall to Agile Project Management, (2015).
[16] Project Management Handbook, Wouter Baars, (2006).
[17] Bernard M. Bass with Ruth Bass, Bass's Handbook of Leadership (4th edition), New York: Free Press, (2008).
[18] Management 3.0, Leading Agile developers, developing Agile leaders, Jurgen Appelo, (2011).

IOP Conf. Series: Materials Science and Engineering 1169 (2021) 012018 doi:10.1088/1757-899X/1169/1/012018

Productivity assessment of the Romanian construction industry using Malmquist Productivity Index

C Cristea[1], M Cristea[1], F M Şerban[1], C Făgărăşan[1] and C E Stoenoiu[1]

[1] Technical University of Cluj-Napoca, 28 Memorandumului Street, Cluj-Napoca, Romania

[1]ciprian.cristea@emd.utcluj.ro

Abstract. The construction industry is one of the largest economic sectors worldwide. It is essential that the firms operating in the construction industry to evaluate accordingly their productivity change both in economic depression and when there is registered an expansion in the economy. This paper investigates the productivity changes of the Romanian enterprises from construction industry during the period of 2006 through 2019 by using the Malmquist Productivity Index. Four input variables and two output variables have been employed for measuring the change of construction productivity. The findings indicated that Malmquist Productivity Index can be effectively employed for assessing the productivity changes of companies. The results indicated that the majority of the Romanian companies exhibited better performance in terms of technical change while efficiency change regressed. This investigation provides valuable information for decision makers in order to enhance the performance of the Romanian construction industry organizations.

1. Introduction

The construction industry plays a major role in the European Union (EU) economy, making up to about nine percent of the EU's Gross Domestic Product (GDP) and, at the same time, providing direct jobs for eighteen million people [1]. This industry was severely affected by the economic and financial crisis, between 2007 and 2013, its value added has fallen by about a fifth [2]. The construction industry plays an important role in economic development, and although is confronting with significant transformation processes, 1324 billion Euros were invested in this sector in 2019 across EU [3].

The construction industry is a strategic sector in Romania's national economy. Between 2006 and 2019, this sector's contribution to GDP ranged between about five and ten percent [4]. The development of infrastructure for the other national sectors is sustained by the construction industry. This industry provided total employment ranging between about six and eleven percent of the Romanian workforce during the period of 2006 through 2019 [4].

Productivity points out the relationship between goods or services produced and the specific resources used to obtain them. Construction productivity evaluates the efficiency by comparing inputs and outputs [5]. For gaining and holding competitive advantage, the firms operating in the construction industry use several methods to assess their performance [6]. Malmquist Productivity Index (MPI) represents one of the most valuable methods used for assessing the productivity. One of the main benefits of the aforementioned method is the possibility to be broken down into separate elements that can provide useful information regarding the resources that influence both positively and negatively the productivity.

IOP Conf. Series: Materials Science and Engineering 1169 (2021) 012018 doi:10.1088/1757-899X/1169/1/012018

Over the last years, the productivity assessment of the companies activating in the construction industry using MPI has attracted the attention of different scholars. Li and Liu [7] evaluated the total factor productivity changes for the Australian construction industry from eight states during the period 1990-2007 by using MPI. The findings indicated a slow improvement of the productivity exemplified by the model composed of two input variables and one output variable. Hung Chiang et al. [8] evaluated the efficiency of twenty construction firms listed in China Mainland and Hong Kong during the period of 2004 through 2010 by employing MPI and found that China Mainland contractors had lower efficiency scores compared to those in Hong Kong. The managerial and strategic capabilities have been identified as the principal strength of Hong Kong companies. Horta et al. [6] used MPI for assessing the productivity change of construction firms from several continents. The results revealed that the efficiency of Asian and European companies is lower than that of companies in North America. Nabavieh et al. [9] employed MPI for investigating the productivity of the construction companies listed in the Tehran Stock Exchange during the period 2006-2012. The results indicated a declining of the companies' productivity levels during the analysed period. Also, the findings revealed that, on average, the productivity and flexibility of the Iranian construction enterprises is improving as they decline in size.

The main objective of this paper is to assess the productivity changes of the Romanian companies that activates in the construction industry listed on the Bucharest Stock Exchange for a period of fourteen years from 2006 to 2019 using MPI. No paper, to the best of our knowledge, has investigated the levels of productivity for the Romanian companies operating in the construction industry using MPI.

The rest of the paper is organized as follows: second section describes the methodology, the results are presented in the third section and the final section summarizes the conclusions of this study.

2. Methodology

In this paper, for evaluating the productivity changes of the Romanian construction companies, 35 decision making units (DMU) have been taken into account. The investigated construction firms are presented in table 1.

Table 1. Investigated construction companies.

DMU No.	Company name	DMU No.	Company name
1	Cemacon	19	Galați Railway Construction
2	Constanța Construction-field Assembly Company	20	Helios Astileu
3	Condmag	21	Concifor Buzău
4	Energopetrol Company	22	Simbeton Oradea
5	Impact Developer & Contractor	23	Iași Hydro-technical Construction
6	Prefab	24	Elco Electro Construction Alba-Iulia
7	Transilvania Construction Company	25	Electromontaj Carpați Sibiu
8	Prebet Aiud	26	ATM Construction Ploiești
9	Napoca Construction Company	27	Comtram Sibiu
10	Electroconstrucția Elco Suceava	28	Duplex Făgăraș
11	Icsim București	29	Crișeni Construction-field Assembly
12	Concivia Brăila	30	Admet Galați
13	Scut Bacău	31	Prima Construct Timișoara
14	ICMRS Galați	32	Mindo Dorohoi
15	Imotrust Arad	33	Sinter Ref Azuga
16	Buzău Complex Construction	34	Concas Buzău
17	Macofil Târgu Jiu	35	Conex Prahova București
18	Bihor Oradea Construction Company		

MPI has been used for determining the productivity change of construction firms. It has been assumed that DMUs use inputs $x^t \epsilon R_+^m$ for producing outputs $y^t \epsilon R_+^m$. $D^t = (x^t, y^t)$ represents the relative distance that defines the technical efficiency of DMU in the period *t* while $D^{t+1} = (x^{t+1}, y^{t+1})$ is the technical efficiency's measurement for DMU in the time interval of *t+1*. Equation (1) show the way MPI in period *t* is determined [10].

$$MPI^t = \frac{D^t(x^{t+1}, y^{t+1})}{D^t(x^t, y^t)} \quad (1)$$

where x_j, $(j = 1, 2 \ldots, n)$ – input vector of DMU_j; y_j, $(j = 1, 2 \ldots, n)$ – output vector of DMU_j; $(t = 1, 2 \ldots, T)$ – time intervals; D – distance function.

The MPI in the time interval of *t+1* is defined by equation (2) [10].

$$MPI^{t+1} = \frac{D^{t+1}(x^{t+1}, y^{t+1})}{D^{t+1}(x^t, y^t)} \quad (2)$$

Fare et al. [10] proposed equation (3) for measuring output oriented MPI between the time interval of *t* and *t+1*.

$$MPI = \mathrm{EC} \cdot TC \quad (3)$$

where *EC* – efficiency change; *TC* – technical change.

If the value of MPI is greater than unity, then the productivity of DMU is improving compared to the initial period of time. If MPI value equals to zero or is sub unitary, then the productivity remains unchanged or is declining.

EC represents the relative change in efficiency between the time intervals and is defined by equation (4).

$$EC = \frac{D^{t+1}(x^{t+1}, y^{t+1})}{D^t(x^t, y^t)} \quad (4)$$

If the value of EC is less than unity, then the efficiency of the DMU is regressing compared to the initial period. If EC value equals to zero or is over unity, then the efficiency remains constant or is enhancing.

Equation (5) indicate the way TC is computed. This index is used for measuring the frontier shift between the time interval of *t* and *t+1*.

$$TC = \left[\frac{D^t(x^{t+1}, y^{t+1})}{D^{t+1}(x^{t+1}, y^{t+1})} \cdot \frac{D^t(x^t, y^t)}{D^{t+1}(x^t, y^t)}\right]^{\frac{1}{2}} \quad (5)$$

where $D^t(x^{t+1}, y^{t+1})$ – the technical efficiency's measurement for DMU when DMU data in the time interval of *t* is replaced with those in the time interval of *t+1*; $D^{t+1}(x^t, y^t)$ – the technical efficiency's measurement for DMU when DMU data in the time interval of *t+1* is substituted with those in the time interval of *t*.

If the value of TC is greater than one, then the technology of DMU is increasing with respect to the initial period of time. If TC value equals to zero or is sub unitary, then the technology frontier remains unchanged or is regressing.

It is very important that proper data are selected for MPI model. For determining the potential input and output variables, a literature-based review has been conducted. The variables that have been chosen in this paper were consistently used in previous investigations. Number of employees, total assets, fixed assets and operating expenses has been chosen as input variable while total revenues and operating revenues are the output variables.

The data were collected manually from the financial statements and annual reports of construction companies from Romanian capital market during 2006-2019, thus making the sample of 490 observations.

The monetary data have been adjusted to remove the effect of the inflation by using data provided by the National Institute of Statistics [4].

IOP Conf. Series: Materials Science and Engineering 1169 (2021) 012018 doi:10.1088/1757-899X/1169/1/012018

3. Results

A summary description of the chosen input and output variables employed in this study are presented in table 2. It can be noticed that the investigated enterprises are diverse following the standard deviation values.

Table 2. Descriptive statistics.

Period	Variables	Mean	Std. Deviation	Period	Variables	Mean	Std. Deviation
2006	Total revenues*	24.245	28.307	2013	Total revenues*	21.595	26.362
	Operating revenues*	27.294	29.577		Operating revenues*	24.013	28.474
	No. of employees	269.2	257.21		Number of employees	128.09	155.01
	Total assets*	37.434	68.526		Total assets*	60.074	82.039
	Fixed assets*	18.949	27.299		Fixed assets*	39.173	58.972
	Operating expenses*	25.481	27.029		Operating expenses*	24.889	30.560
2007	Total revenues*	30.074	39.922	2014	Total revenues*	22.096	31.471
	Operating revenues*	37.194	47.135		Operating revenues*	25.037	34.389
	No. of employees	262.46	253.66		No. of employees	109.4	118.07
	Total assets*	57.929	102.729		Total assets*	60.258	89.600
	Fixed assets*	31.606	50.799		Fixed assets*	38.573	60.504
	Operating expenses*	34.284	43.821		Operating expenses*	24.394	34.123
2008	Total revenues*	32.046	36.495	2015	Total revenues*	20.776	25.563
	Operating revenues*	38.639	47.009		Operating revenues*	23.193	31.262
	No. of employees	249.91	242.67		No. of employees	100.4	99.19
	Total assets*	67.084	110.116		Total assets*	60.410	92.813
	Fixed assets*	39.236	59.193		Fixed assets*	38.315	59.648
	Operating expenses*	35.522	42.699		Operating expenses*	23.561	31.278
2009	Total revenues*	27.554	39.418	2016	Total revenues *	21.884	31.747
	Operating revenues*	29.609	40.699		Operating revenues*	25.187	38.452
	No. of employees	197.2	194.9		No. of employees	95.86	94.54
	Total assets*	67.572	102.102		Total assets*	62.055	97.366
	Fixed assets*	42.54	63.025		Fixed assets*	37.043	56.101
	Operating expenses*	27.966	36.912		Operating expenses*	23.844	33.352
2010	Total revenues*	27.106	39.347	2017	Total revenues *	19.627	27.559
	Operating revenues*	32.161	41.952		Operating revenues*	24.311	39.717
	No. of employees	179.37	195.71		No. of employees	88.46	89.33
	Total assets*	68.539	98.536		Total assets*	63.308	115.332
	Fixed assets*	41.181	59.604		Fixed assets*	39.455	64.450
	Operating expenses*	30.699	37.896		Operating expenses*	22.497	29.697
2011	Total revenues*	23.079	25.771	2018	Total revenues*	19.724	26.495
	Operating revenues*	27.503	29.998		Operating revenues*	23.854	30.220
	No. of employees	167.54	186.76		No. of employees	83.4	88.78
	Total assets*	66.308	89.676		Total assets*	61.486	112.728
	Fixed assets*	45.258	67.594		Fixed assets*	37.933	67.174
	Operating expenses*	28.128	30.265		Operating expenses*	21.772	25.115
2012	Total revenues *	22.815	25.244	2019	Total revenues *	24.476	36.125
	Operating revenues*	25.925	27.655		Operating revenues*	30.138	50.348
	No. of employees	145.4	169.41		No. of employees	84.43	94.17
	Total assets*	70.784	98.593		Total assets*	69.798	136.836
	Fixed assets*	46.359	69.604		Fixed assets*	43.863	87.686
	Operating expenses*	25.546	27.256		Operating expenses*	22.530	27.678

* – in mil. RON.

IOP Conf. Series: Materials Science and Engineering 1169 (2021) 012018 doi:10.1088/1757-899X/1169/1/012018

Table 3 presents the MPI results for the examined construction companies. The results indicate the changes in productivity of DMUs that are operating in the Romanian construction industry during the investigated period.

Table 3. Results of MPI.

DMU No.	2006-2007	2007-2008	2008-2009	2009-2010	2010-2011	2011-2012	2012-2013	2013-2014	2014-2015	2015-2016	2016-2017	2017-2018	2018-2019
1	1.215	1.041	0.922	0.941	0.94	1.014	1.202	1.052	1.007	1.117	1.006	0.989	1.083
2	1.04	1.034	0.85	0.936	1.022	1.12	0.982	1.023	0.688	0.645	1.055	1.557	0.338
3	1.243	0.928	1.096	0.973	0.886	0.796	1.062	1.284	0.427	0.677	0.629	2.343	1.228
4	1.036	0.999	1.044	0.981	1.046	0.95	0.684	0.958	1.288	0.894	1.133	1.684	0.498
5	1.033	1.091	0.689	0.947	0.768	1.255	0.823	1.283	1.431	1.446	0.811	0.868	1.652
6	0.958	0.999	1.054	0.976	0.96	1.043	0.982	1.006	0.997	0.984	1.003	1.049	0.918
7	1.187	0.949	0.94	0.953	1.039	1.215	1.122	1.082	1.042	0.902	0.991	1.135	1.238
8	0.947	0.96	1.07	0.884	1.046	1.024	0.898	1.053	0.944	1.049	0.875	1.128	1.075
9	1.053	0.978	0.918	0.917	0.812	2.429	0.996	0.988	0.865	1.072	0.946	0.93	1.038
10	0.881	0.913	0.972	1.004	1.029	0.922	0.945	1.034	1.068	1.038	0.92	1.021	1.093
11	0.694	1.14	1.137	1.175	0.935	0.966	1.18	1.007	0.871	0.924	1.019	1.164	0.674
12	1.068	1.035	0.969	0.974	0.992	1.001	0.954	0.86	1.228	0.855	1.057	1.064	1.256
13	0.991	0.995	0.963	1.012	1	1	1	1.097	0.888	1.018	0.901	1.077	1.003
14	1.027	0.95	0.642	1.009	1.206	0.89	1.076	0.953	1.212	1.161	0.994	1.01	0.982
15	2.281	0.499	1.234	2.192	0.54	2.017	1.26	0.842	1.274	0.569	0.209	1.803	2.171
16	1.025	1.161	0.836	1.006	0.86	1.044	1.204	0.887	1.136	1.072	0.959	0.806	1.229
17	0.927	0.983	0.973	1.107	0.912	0.844	1.21	1.03	1.113	0.951	0.956	1.026	1.149
18	1.161	1.213	0.797	0.955	1.088	0.978	0.885	1.235	0.937	0.962	0.975	0.971	1.039
19	1.116	1.116	0.912	1.024	0.977	0.712	1.49	0.985	0.995	1.099	0.911	1.019	1.162
20	1.039	1.03	1.095	0.943	0.985	1.052	1.014	1.023	1.04	0.954	1.082	0.869	0.93
21	1.03	0.944	0.946	1	0.866	1.029	1.069	0.923	1.019	0.959	1.05	1.644	2.699
22	0.888	0.804	1.134	1.2	1.046	0.576	1.526	0.891	1.253	0.991	1.021	0.984	1.173
23	0.936	0.979	1.269	1.128	0.839	0.889	1.005	0.983	1.003	1.007	0.962	0.994	0.925
24	0.942	1.171	0.914	0.929	0.971	0.974	0.895	0.868	1.018	1.001	1.017	1.006	0.963
25	1.038	0.939	1.059	0.997	0.974	0.998	0.992	0.638	1.182	1.354	1.156	1.027	1.068
26	1.058	0.953	0.96	0.987	0.929	1.214	0.929	1	0.883	0.984	1.086	1.034	1.02
27	1	0.96	1.057	0.967	1.029	0.9	0.941	0.933	1.044	0.985	0.899	0.964	1.018
28	0.971	0.989	0.965	0.972	0.993	0.535	1.532	1.462	0.869	1.002	0.881	1.075	1.064
29	0.661	1.229	0.978	0.963	0.684	1.396	1.01	1.006	0.57	1.037	0.539	1.085	0.502
30	1.177	0.973	0.278	0.06	2.05	1.466	1.202	1.234	1.127	0.978	0.858	0.614	0.905
31	0.53	1.333	1.253	0.757	0.83	0.949	1.046	0.952	0.984	0.715	0.825	1.976	0.814
32	1.006	1.065	0.843	1.382	1.015	0.91	1.169	1.147	0.897	1.086	1.141	1.029	0.957
33	1.01	1.078	0.873	0.931	0.871	1.325	1.628	0.477	1.019	0.939	0.817	0.794	1.226
34	1.057	1.158	0.989	1.226	0.982	1.019	0.895	0.887	1.045	0.749	0.794	1.207	1.139
35	1.936	0.621	0.713	0.826	0.882	0.736	1.701	0.526	0.55	1.327	1.978	0.822	0.238

It can be noticed a downward trend of the construction companies from 2007 through 2011. During the aforementioned period, between 20 and 23 firms experienced decline in the level of their productivity.

For the examined firms that activate in the Romanian construction industry, the period between 2017 and 2019 is on average the most favourable, because between 22 and 23 enterprises experienced improvement productivity.

The results presented in table 3 indicate that several companies experienced longer periods of productivity improvement during 2006-2019. The productivity of DMU1, DMU7 and DMU30 has grown for at least five consecutive years during 2011-2017, 2010-2015 and 2010-2015. On the other hand, the productivity of DMU24 and DMU28 decreased for six consecutive years during 2008-2014 and 2006-2012.

Figure 1 shows the average efficiency scores of Romanian construction companies for the interval 2006-2019. The results indicate that 16 out of the investigated firms increased the level of productivity during this period.

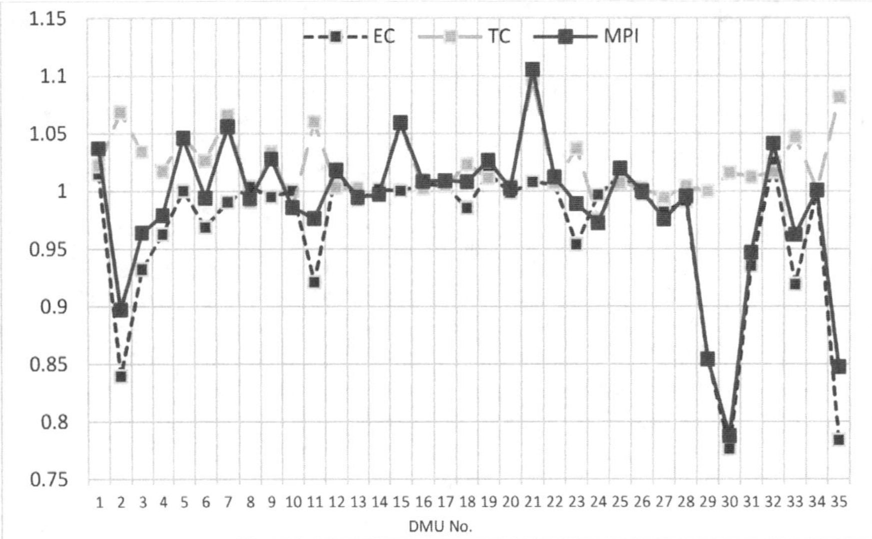

Figure 1. Average EC, TC and MPI of DMUs.

DMU21 experienced the highest growth rate of about 11 percent. On the other hand, DMU30 registered the highest decrease of productivity of over 21 percent. It can be noticed that the productivity enhancement was mainly due to technology improvement.

Figure 2 illustrates the annual means of efficiency scores for the investigated construction firms. The annual means of MPI indicate that the productivity of Romanian construction industry has grown in four years.

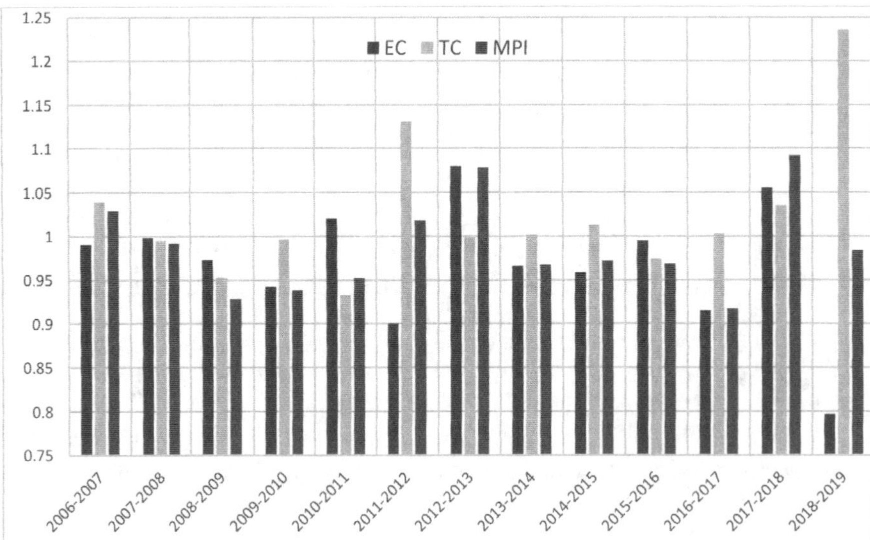

Figure 2. Annual means of EC, TC, and MPI.

IOP Conf. Series: Materials Science and Engineering 1169 (2021) 012018 doi:10.1088/1757-899X/1169/1/012018

The annual means of MPI reflect that the productivity in the construction industry has declined in 2007-2009, 2013-2014, 2015-2017 and 2018-2019 mainly due to the poor performance of efficiency. The best construction productivity enhancement has been registered in 2017-2018, acquiring a rate of over 17 percent growth while the largest regress in productivity has been noticed in 2013-2014 with about 11 percent decrease.

4. Conclusions

The productivity evaluation of the construction firms represents one of the most valuable sources of information for enhancing their performance in time. This study investigates the productivity evolution of the Romanian companies, from the construction industry, listed on the Bucharest Stock Exchange from 2006 to 2019 by applying MPI. Over the period of this investigation, the geometric mean of MPI was 0.986, meaning that the average total productivity of construction enterprises declined slightly by more the one percent. The geometric means of EC and TC were 0.9658 and 1.021, meaning that the technological progress instead of efficiency was the main driver of construction productivity enhancement. The results provide valuable information for policy making and suitable strategies for productivity growth in the Romanian construction industry.

References

[1] European Commission 2016 The European Construction Sector *Eur. Union.* **16** http://ec.europa.eu/growth/sectors/construction/, accessed on 30.04.2021
[2] Horta I M, Camanho A S, Johnes J and Johnes G 2013 *J. Product. Anal.* **39** 89
[3] European Construction Industry Federation, Construction Activity in Europe, https://fiec-statistical-report.eu/, accessed on 5.05.2021
[4] National Institute of Statistics, www.insse.ro, accessed on 6.05.2021
[5] Allmon E, Haas C T, Borcherding J D and Goodrum P M 2000 *J. Constr. Eng. Manag.* **126** 97
[6] Horta I M, Camanho A, Johnes J, Johnes G 2013 *J. Product. Anal.* **39** 89
[7] Li Y and Liu C 2010 *Constr. Manag. Econ.* **28** 933
[8] Hung Chiang Y, Li J, Choi TN Man KF 2012 *J. Facil. Manag.* **10** 179
[9] Nabavieh A, Cyrus K M, Gholamiangonabadi D and Gholamveisy S 2019 *Proc. Int. Conf. on Industrial Engineering and Operations Management* March 5th -7th (Bangkok, Thailand) pp 376-387
[10] Färe R, Grosskopf S, Norris M and Zhang Z 1994 *Am. Econ. Rev.* **84** 66

IOP Conf. Series: Materials Science and Engineering 1169 (2021) 012019 doi:10.1088/1757-899X/1169/1/012019

Insular genetic algorithm for operational management

A Vîlcu[1], I Herghiligiu[1], I Verzea[1] , M Pîslaru

[1] "Gheorghe Asachi" Technical University of Iasi, Department of Engineering and Management, Mangeron Blvd. 28, TEX1, 700050, Romania

adrian.vilcu@academic.tuiasi.ro

Abstract. Problems in the operational management that involve the calculation on graphs are numerous, being encountered in many economic fields: transport problems, problems of resource allocation, logistics problems. Within these problems are found the problems of partitioning graphs with partitions that meet different partitioning criteria. The paper presents a case study for the problem of partitioning graphs - organizing the graph, practical applicability of the problem, partitioning function, method of solving the problem by using three genetic algorithms working individually on islands of problem solutions, changing synchronously with each other solutions, and the application of the algorithm on a numerical example. Finally, the paper proposes conclusions and future directions of development.

1. Introduction

The stage of optimizing the flow of activities by balancing is an integral part of any technological process [1][2]. The importance of this stage increases in direct proportion to the increasing complexity of technological systems, the decrease of time for their design and the introduction of new restrictions [3][4]. Given the increased interest in this stage of optimization, numerous algorithms have been proposed that can be classified into two categories: exact algorithms provide optimal solutions for balancing problems in an exponentially time that exponentially increasing with problem size and class of approximate algorithms (heuristic and evolutive algorithms) which provides good quality solutions in a time independent of the size of the problem [5][6].

The objective of this paper is to balance a technological line for a textile product by developing an algorithm that takes into account the dynamic characteristic of the processing line with a new objective function that simulates a restriction of supply process line.

2. Material and method

2.1. The problem of partitioning graphs

Having in execution phases of a technological process, a graph of precedence restrictions and a technological limitation: each grouping of phases must contain at most L entries. Thus, the study problem in this paper is to determine the minimum number of groups of execution phases of a technological process that respects the supply restriction. The L integer is a grouping parameter.

With the new elements introduced, the problem is to partition a graph into (unfixed) subgraphs, each of which has at most L input elements.

Graph partitioning involves cutting arches; thus, each time an arc is cut, a primary pseudo-input (pseudo-input) and a primary pseudo-output (pseudo-output) are created. It is necessary for each subgraph to respect the condition, where it represents the number of entries and the number of pseudo-entries.

2.2. Objective function

The objectives that are considered in partitioning a graph are:

• the total number of connections between subcircuits (number of cut arcs) should be as small as possible;

• the number of subcircuits should be minimal.

These two objectives are contradictory and all the algorithms proposed in the literature try to optimize one objective to the detriment of the others. The objective function used in the present model is to minimize the number of subcircuits.

The first component of each list contains the number of predecessors of the vertices.

2.3. Graph representation

The graph is represented by adjacency lists (figure 1). These are massive ones that contain the ends of V chained lists, one for each node, each list containing all the predecessors of the node.

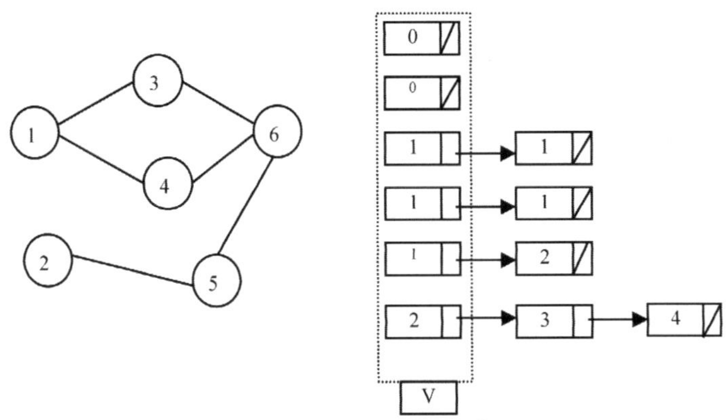

Figure 1. Graph representation

2.4. Graph partitioning model (GPM)

GPM consists in partitioning the set of vertices X into $X_1,..., X_j,..., X_k$ subsets (k not fixed) and the graph G into subgraphs $G_1,..., G_j = (X_j, A_j),..., G_k$ so that:

i) $X = \bigcup_{j=1}^{k} X_j$, $X_j \bigcap X_i = \phi$ $\quad\forall i \neq j \; i, j = 1,..., k$

ii) $X_j \bigcap (P \bigcup M) \neq \phi$ $\quad\forall \; j = 1,..., k$

iii) $n_j + c_j = \left| X_j \bigcap E \right| + \left| \varpi^-(X_j) \right| \leq L$ $\quad\forall \; j = 1,..., k$

where: M - the set of primary inputs, P - the set of intermediate nodes, n_j – the set primary inputs of the each group, c_j

iv) let $G_j = (X_j, A_j)$ the associated subgraph is connected $\forall \; j = 1,..., k$ or G_j is compound from H disjoint graph $G_1,..., G_h = (X_h, A_h),..., G_H$ where $X_h \bigcap (P \bigcup M) \neq \phi$ and G_h is connected $\forall \; h = 1,..., H$

It was noted with $\varpi^-(X_j)$ - the set of neighbors - the set of vertices that do not belong to the set and that have at least one successor in X_j: $\omega^-(X_j) = \{v \in X / v \notin X_j, si \; \exists \, arc(v, w) \; pentru \; w \in X_j\}$

GPM is equivalent to the problem of partitioning graphs for the balancing assembly lines with supply constraints [7][8].

2.5. Genetic algorithm for GPM
The developed genetic algorithm respects the structure of the standard genetic algorithm through the presence of the three modules: the initial population module, the evaluation module and the recombination and modification module [9][10][11].

2.5.1. The fitness function. The fitness function determines the number of subgraphs, the number of cuts and the class membership of the genes of a chromosome. The following Greedy method was used: on a generated permutation, the procedure with the first gene of the chromosome is started by calculating the number of primary entries and pseudo-entries of the gene sequence in the order of position. When the restriction is violated, increase the number of subcircuits, assign the current partition vertices, and continue the procedure.
Let graph from figure 2.

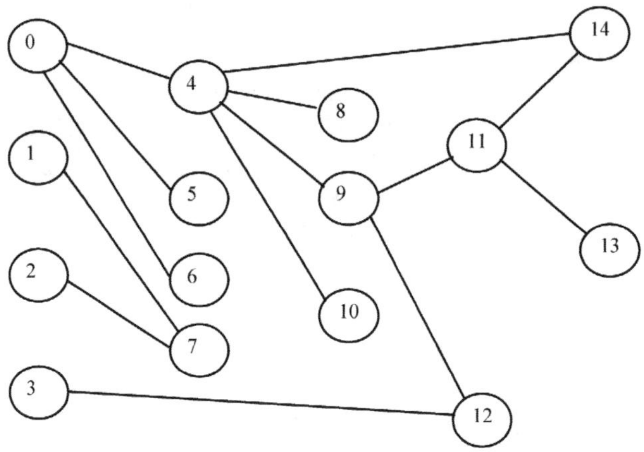

Figure 2. Graph example for the evaluation of the fitness function

For $L=3$ and for the generated solution, sol[1]=[14, 11, 4, 1, 9, 10, 3, 13, 0, 6, 5, 8, 2, 7]
The value of the fitness function is 2 (table 1).

Table 1. The fitness function for the example shown in figure 2

Group	Vertices													Inputs	Cuts
1	14	11	4	1	9	10	12	3	13	0	6	5	8	3	0
2	2	7												2	1

2.5.2. Population management. At the end of each stage of evolution all new chromosomes create a separate list. This list is interclassed but the initial population and the best chromosomes are retained for the next stage of evolution in the limit set for the population. This elitist selection focuses the search on the most promising regions in the solution space.

2.5.3. Genetic operators. The role of these operators is to introduce new individuals in the population, thus resulting in a search in the entire space of solutions.

The mutation operator - it prevents premature convergence, generating an additional diversity of individuals in a population. This diversity allows exploration of larger areas in the search space.

A mirroring mutation operator with a self-adaptive adjustment was used: the cutting edges are randomly generated according to the histogram of the parent chromosome - the number of elements in each class. Thus, the number of elements between the two cutting fronts is greater than the maximum number of elements in a partition. This self-adaptation of the mutation operator will avoid the generation of offspring identical to the parent chromosomes.

The crossover operator - is responsible for increasing diversity in a given population, combining what is good in the population and proliferating the most promising sequences of peaks. The offspring obtained by crossing will have the characteristics of both parents. This is a variant of the OX operator [12][13] adapted to GPM:

- 2 random positions are generated by extracting 2 chromosomes;
- 2 cutting fronts are generated for each parent chromosome so that the restriction is observed:, where it represents the maximum number of elements in a class;
- move by a circular permutation all the elements between the 2 fronts at the beginning of each descendant and fill in the remaining positions with unassigned tips from the other parent.

2.6. Insular genetic algorithm (IGA) for GPM

2.6.1. The structure of the insular genetic algorithm. During each stage of evolution of the IGA, each of the three GA contains mutation and crossing operations on the island of solutions given by the initial population, locally improving the value of its objective function. After the elitist selection is applied on each solution island of the three GAs, the migration operation is performed (figure 3). The policy is to make a circular infusion of solutions from one island to another in order to avoid premature achievement of a local optimum of the value of the estimation function.

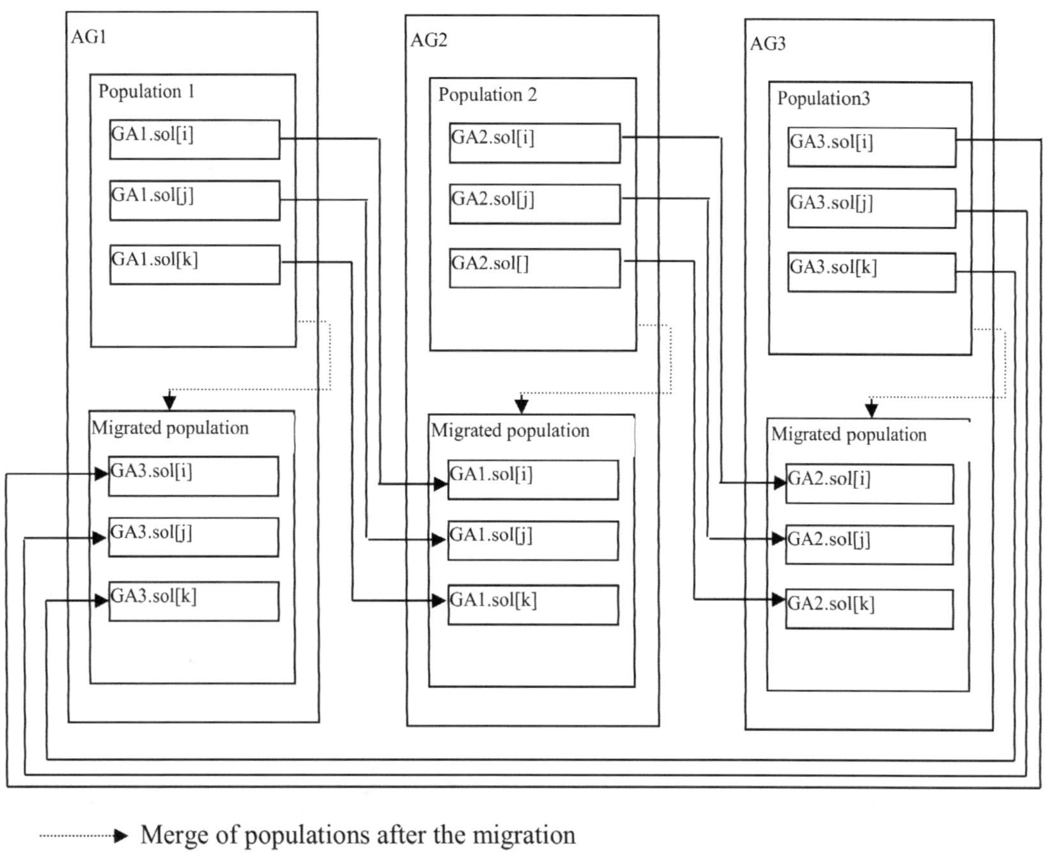

Figure 3. A stage of evolution of IGA

This operation will result in a list of individuals (GPM solutions) copies of the individuals selected for migration.

IOP Conf. Series: Materials Science and Engineering 1169 (2021) 012019 doi:10.1088/1757-899X/1169/1/012019

2.6.2. The migration operator. Steps of the migration procedure of individuals from:

• random numbers are generated between 0 and 1, where it represents the size of the population set by the court of the problem;

• for each random number with the property where (migration probability) - parameter provided by the problem instance, one is generated and it will indicate the individual that will migrate;

2.6.3. Master program function - global control. The evolution of the entire IGA is controlled by the master program function by:

• establishing the probabilities of mutation, crossing and migration;

• synchronization of processes after each stage of evolution;

• global stop criterion:

The global stop criterion. For each GA at the end of each stage of evolution, the average value of the fitness function of the individuals in the population marked with is calculated. The evolution of the process stops when the relationship becomes true for each of the GA, where there is a prescribed tolerance.

3. Experimental results

Let be a graph with 100 vertices, 4 primary inputs and *L=3*. The graph is shown in the figure 4.

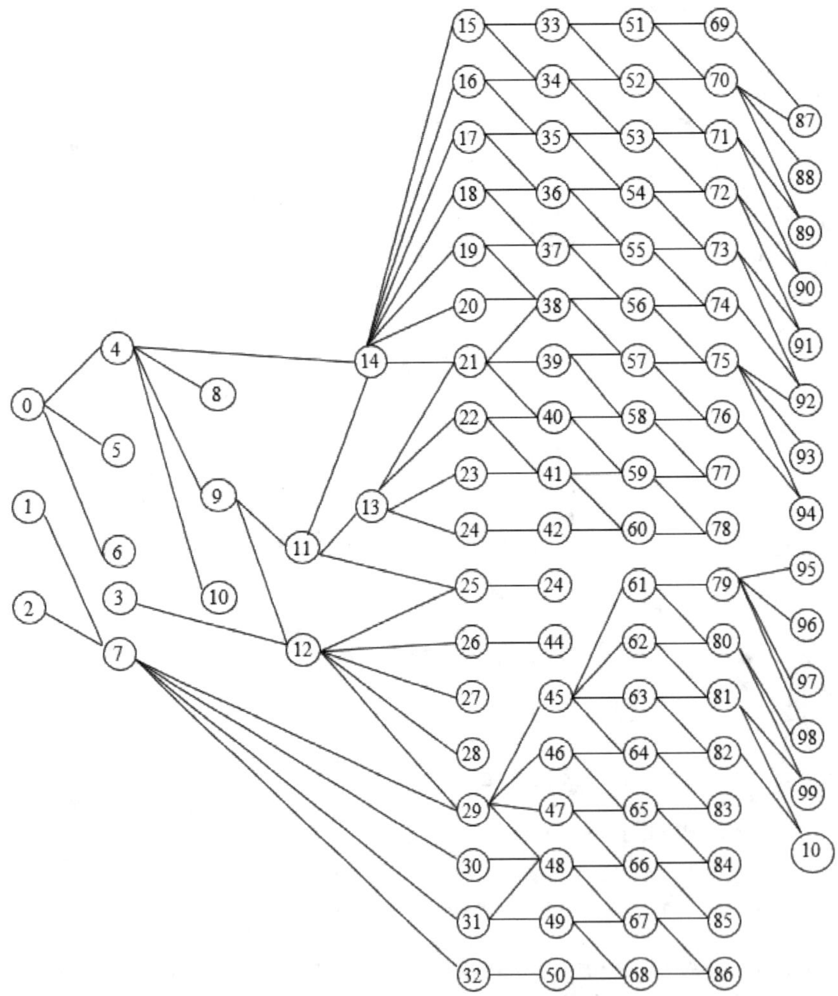

Figure 4. Graph for the studied technological process

In table 2 is presented the adjacent list for the problem graph, for each of the vertex the number of predecessors and their list are generated.

Table 2. Adjacent list for the graph

Vertex	No	List of predecessors			Vertex	No	List of predecessors			Vertex	No	List of predecessors		
0	0				36	2	18	17		72	2	53	54	
1	0				37	2	18	19		73	2	54	55	
2	0				38	3	19	20	21	74	2	55	56	
3	0				39	1	21			75	2	56	57	
4	1	0			40	2	21	22		76	2	57	58	
5	1	0			41	2	22	23		77	2	58	59	
6	1	0			42	1	24			78	2	59	60	
7	2	1	2		43	1	25			79	1	61		
8	1	4			44	1	26			80	2	61	62	
9	1	4			45	1	29			81	2	62	63	
10	1	4			46	1	29			82	2	63	64	
11	1	9			47	1	29			83	2	64	65	
12	2	3	9		48	3	29	30	31	84	2	65	66	
13	1	11			49	1	31			85	2	66	67	
14	2	4	11		50	1	32			86	2	67	68	
15	1	14			51	1	33			87	2	69	70	
16	1	14			52	2	33	34		88	1	70		
17	1	14			53	2	34	35		89	2	70	71	
18	1	14			54	2	35	36		90	2	71	72	
19	1	14			55	2	36	37		91	2	72	73	
20	1	14			56	2	37	38		92	3	73	74	75
21	2	13	14		57	2	38	39		93	1	75		
22	1	13			58	2	39	40		94	2	75	76	
23	1	13			59	2	40	41		95	1	79		
24	1	13			60	2	41	42		96	1	79		
25	2	11	12		61	1	45			97	1	79		
26	1	12			62	1	45			98	2	79	80	
27	1	12			63	1	45			99	2	80	81	
28	1	12			64	2	45	46		100	2	81	82	
29	2	7	12		65	2	46	47						
30	1	7			66	2	47	48						
31	1	7			67	2	48	49						
32	1	7			68	2	49	50						
33	1	15			69	1	51							
34	2	15	16		70	2	51	52						
35	2	16	17		71	2	52	53						
36	2	18	17		72	2	53	54						

Population size was set at 100 individuals and tolerance at $2 * 10^{-7}$. For each instance of the problem (including probability values) a group of 10 runs of the GA as performed. The following characteristics were calculated for each group:

- BGP - the best graph partitioning
- ABGP - the average of the best partitions of the graph
- LABGP - the average of the best partitions of the graph in the last stage of evolution;

IOP Conf. Series: Materials Science and Engineering 1169 (2021) 012019 doi:10.1088/1757-899X/1169/1/012019

• ANI - the average number of iterations made, until the stop condition is verified;
• ANC - average number of cuts

We will consider the following values for the probabilities of crossing, mutation and for the migration operator: $pM = 0.2, 0.4$; $pC = 0.4, 0.6$, $pMi = 0.0, 0.2$.

The value 0.0 for means that the migration operator was not considered and GPM is solved independently of the three GA.

The results for each of the groups of 10 runs considering all combinations of probabilities were centralized in the following table 3:

Table 3. Results of IGA

No	pM	pC	pMi	GA1					GA2					GA3				
				BGP	ABGP	LABGP	ANI	ANC	BGP	ABGP	LABGP	ANI	ANC	BGP	ABGP	LABGP	ANI	ANC
1	0.2	0.4	0.0	5	5.7	6.2	70	12	6	6	6	70	14	5	5	6	70	12
2	0.2	0.6	0.0	5	5.5	6.1	85	11	5	5.7	5.6	85	12	4	5.5	6	85	11
3	0.4	0.4	0.0	4	5	5.5	95	9	3	3.5	3.5	95	8	5	5	5.5	95	9
4	0.4	0.6	0.0	3	4	4	107	5	3	3	3.7	107	5	**2**	3	3	107	4
5	0.2	0.4	0.2	3	3.4	3.4	98	5	3	3	3.3	98	5	3	3.4	3.4	98	5
6	0.2	0.6	0.2	**2**	3	3	95	4	**2**	2	3	95	4	**2**	3	3	95	4
7	0.4	0.4	0.2	**2**	2	2	97	1.5	**2**	2	2	97	2	**2**	2	2	97	**1**
8	0.4	0.6	0.2	**2**	2	2	102	**1**	**2**	2	2	102	**1**	**2**	2	2	102	**1**

The role of mutation and crossover operators is observed in the evolution of the number of partitions of the graph, which decreases with increasing probabilities of mutation and crossover.

The introduction of the migration operator has the following effects:

• standardization of the solutions provided by the three genetic algorithms;
• improving the solution, Gr. no. 8 providing the optimal solution to the problem: 2 partitions and a single cutting operation.
• the number of iterations does not show a significant increase compared to the execution groups when the other probabilities remain constant.

4. Conclusions and future directions of investigations

As a conclusion, the implementation of a genetic algorithm for GPM using the island model with migration generates solutions of a quality close to the optimal ones, with a shorter execution time than the exact and pseudo-exact methods.

Future directions of investigation will be in hybridizing the algorithm, implementing and executing it on a distributed system using the MPI function library and performing a statistical analysis of the results obtained after running the program on a graph with a number of vertices of thousands.

References

[1] Vîlcu Adrian 2012 *A hybrid genetic algorithm for a combinational circuits partitioning problem* Buletinul Institutului Politehnic din Iaşi Automatic Control and Computer Science Section Volumul 3 Tome LVIII (LXII) fasc.3 p 43-51

[2] Vîlcu Adrian 2013 *Double hybridized embryonic genetic algorithm for optimal delivery routing* Scientific Studies and Research Series Mathematics and Informatics Bacau Vol. 23 No.1 p 213

[3] Adrian Vîlcu, Ion Verzea, Marius Pislaru & Ionut-Viorel Herghiligiu 2019 *Fuzzy modeling of dependability optimization for supporting the production-quality strategies - case study in technical field* ModTech International Conference Modern Technologies in Industrial Engineering June 19-22 2019 Iasi Romania

[4] Marius Pîslaru, Ramona-Diana Leon, Vîlcu Adrian 2018 *Using a fuzzy expert system for service quality improvement. the case of a car wash station* STRATEGICA International Academic Conference -Sixth Edition-Bucharest, Romania, October 11-12, 2018 Challenging the Status Quo in Management and Economics p 490-500

[5] Vîlcu Adrian, Ion Verzea, Catalin Vilcu 2016 *Heuristic algorithm for optimization of manufacturing process by balancing a production line from a textile company*,16th Romanian Textiles and Leather Conference, CORTEP, Iasi, 27-29.10

[6] Ion Verzea, Gabriel-Petru Luca, Rachid Chaib, Vîlcu Adrian 2016 *Conventional work units breakeven point assessment in the garment Industrie*, 16th Romanian Textiles and Leather Conference–CORTEP, Iasi, 27-29.10.2016

[7] Labontu Oana 2014 *Constructive aesthetic design of a blouse model for women and optimization of the manufacturing process through balancing* bachelor's thesis DIMA Faculty, Iasi

[8] Verzea I and Luca GP 2009 *Production and maintenance management methods*

[9] Aminur Rahman Khan, Adrian Vilcu, Nahid Sultana, Syed Sabbir Ahmed 2015 *Determination of Initial Basic Feasible Solution of a Transportation Problem: A TOCM-SUM Approach* Buletinul Institutului Politehnic din Iaşi Technical University of Iasi, Tomul LVIII (LXII) Fasc. 1 Automatic Control and Computer Science Section

[10] Adrian Vilcu 2012 *A hybrid genetic algorithm for a combinational circuits partitioning problem*, Buletinul Institutului POLITEHNIC Technical Unversity of Iasi Tomul LVIII (LXII) Fasc. 3 Automatic Control and Computer Science Section

[11] M Davis-Moradkhan 1993 *Problemes de partitionnement dans la Technologie des VLSI* These de Doctorat Universite Paris VI

[12] D Dumitrescu 2000 *Genetic algorithms and evolutionary strategies - applications in artificial intelligence and related fields* Ed. Albastră Cluj Napoca

[13] Michalewiz Z 1994 *Genetic Algorithms + Data Structures = Evolution Program,* 2nd ed. Artificial Intelligence Springer Verlag Berlin

IOP Conf. Series: Materials Science and Engineering 1169 (2021) 012020 doi:10.1088/1757-899X/1169/1/012020

A comparative experimental study for dry and wet collisions

F-C Ciornei[1], S Alaci[1], P Bulai[1] and I-C Romanu[1]

[1]Mechanics and Technologies Department, „Stefan cel Mare" University of Suceava, Suceava, Romania

E-mail: florina.ciornei@usm.ro

Abstract. The collision between polyurethane prismatic foam and a dropping steel ball is investigated experimentally. Because the foam presents large deformations, an analytical model is not available. The impact is studied on a test rig from laboratory, for dry foam and for the mineral oil wetted foam. The post impact velocity is found measuring the flight time and the coefficient of restitution and lost energy are estimated. For the dry impact case the damping effect is smaller than for the case of oil-wetted target.

1. Introduction

The mathematical models that describe the behaviour of a system with percussions have continuously evolved. A characteristic of these systems is the sudden variation of the kinematical parameters of first order without any change in the system's configuration.

The simplest models study systems made of two bodies with unidirectional motion and accept as hypothesis that percussion is an instantaneous phenomenon [1-3]. In order to describe the change of the kinematical state of the system, a series of coefficients, expressing the ratio between the values of the considered parameter post and before impact, are used. Amid these parameters, the most important is the coefficient of restitution, defined by Newton as the ratio with changed sign between the postimpact relative velocity and the relative velocity before impact:

$$e=-\frac{v'_2-v'_1}{v_2-v_1} \tag{1}$$

According to equation (1), the values of the coefficient of restitution may vary from $e=0$ which is characteristic to plastic collision, when due to impact the system loses the entire kinetic energy, to $e=1$ which is the case of elastic collision, when the kinetic energy of the system is maintained. The requirement of modelling more complex dynamical systems lead to friction consideration, as shown in [4]. Kane [5] analysed the impact with friction for a pendulum and using the equation (1) for the coefficient of restitution reached a paradoxical conclusion, that the final energy of the system is greater than the initial one. To overcome this situation, it is necessary to modify the definition of the coefficient of restitution, equation (1). To this purpose, the notion of percussion is introduced:

$$\boldsymbol{P}=\int_0^t \boldsymbol{F}(t)dt \tag{2}$$

where $\boldsymbol{F}$ is the impact force between the two bodies and t is the time variable. It is accepted that the collision period $[0,t_f]$, is divided in two phases: the compression, during $[0,t_c]$, where t_c is the instant corresponding to the maximum approach between the bodies and the restitution phase, during the interval $[t_c, t_f]$, where t_f is the final time, when the bodies separate. According to [6], Poisson introduced the following expression for the coefficient of restitution:

$$e = \frac{\boldsymbol{n} \cdot \int_{t_c}^{t_f} \boldsymbol{F} dt}{\boldsymbol{n} \cdot \int_0^{t_c} \boldsymbol{F} dt} = \frac{P_{nr}}{P_{nc}} \tag{3}$$

where $\boldsymbol{n}$ is the versor of the normal to the surfaces in the theoretical point of contact. Thus, the coefficient of restitution is the ration between the normal percussion from the restitution phase and the normal percussion from the compression phase.

Wang and Mason elaborated a classification of plane percussions with dry friction and made a foremost observation as conclusion, that is, unlike forces, where the presence of friction forces requires the existence of normal forces, the tangential percussions may exist independent of the normal percussions. Based on this remark, Wang and Mason [1] emphasize that when studying a problem of impact with friction, the principle of linkages formulated by Kilmister and Reeve [7] must be considered; *the constraints should be maintained using forces as long as possible and only on contrary situation, the percussions should be used.*

One of the main drawbacks of the instantaneous impact hypothesis consists in the impossibility of estimation of the contact forces that develop during the impact process. The requirement of correct design of the regions of different mechanical parts coming into contact compelled the decline of instantaneous collision and the acceptance of the hypothesis of finite duration collision. Timoshenko [Ti] considers the perfect elastic impact between two spheres under the assumption of Hertz type impact forces and determines the values of the maximum impact force and of the total collision time.

The centric impact of two viscoelastic spheres was analysed by Lankarani [9] who concluded that the only equation provided by the dynamic theorems [10] is the momentum theorem. To solve the problem, Lankarani used an energy hypothesis and accepts that the quantities of energy lost by friction during the two phases of impact, compression and restitution, are equal and describes the impact force using the equation:

$$F = \boldsymbol{K}x^{\alpha}\left[1 + \frac{3(1-e)^2}{4}\frac{\dot{x}}{v_0}\right] \tag{4}$$

where K is a coefficient depending on the elastic and geometrical characteristics of the bodies [11], the exponent takes the value $\alpha = 3/2$ for a point Hertz contact type and v_0 is the initial relative velocity. The model proposed by Lankarani describes accurately the quasielastic collisions for which $e > 0.9$. Flores, [12-13], makes the assumption that the characteristic point describes arcs of ellipse in the phase plane and improves the Lankarani model proposing the following equation for the impact force:

$$F = Kx^{\alpha}\left[1 + \frac{8(1-e)}{5e}\frac{\dot{x}}{v_0}\right] \tag{5}$$

The equation (5) can be applied for any type of collision, regardless of the values of the coefficient of restitution.

As it can be observed form equation (4) and (5), even for describing the behaviour of the simplest systems, nonlinear differential equations appear. It is expected that when considering more complex systems, with the aim of an accurate description of their behaviour, more intricate differential equations will be used. Additionally, the equations describing the models may lead to the bifurcation phenomenon that makes impossible reaching a conclusion regarding the evolution of the system for a longer period.

2. Statement of the problem and proposed solution

For the situations mentioned above, one of the solutions is to construct the actual device and to elaborate the conclusions based on an experimental study. In this category is framed the present paper that studies the effect the moist upon the damping characteristics of a material. The bodies used in the experiment are a bearing ball as the projectile and a polyurethane parallelepiped, the target. The experiments were conducted in two situations: first, for dry polyurethane foam and second, for wet polyurethane obtained after the foam was soaked in mineral oil and then squeezed. A theoretical model is difficult to elaborate for lubricant-wet body since even for the dry body, none of the equations (4) or (5) is applicable because the exponent α takes different values for the compression and restitution phase [14] and moreover, the hypothesis of small deformations required by the Hertzian theory is not valid. The principle scheme of the test rig is presented in figure 1.

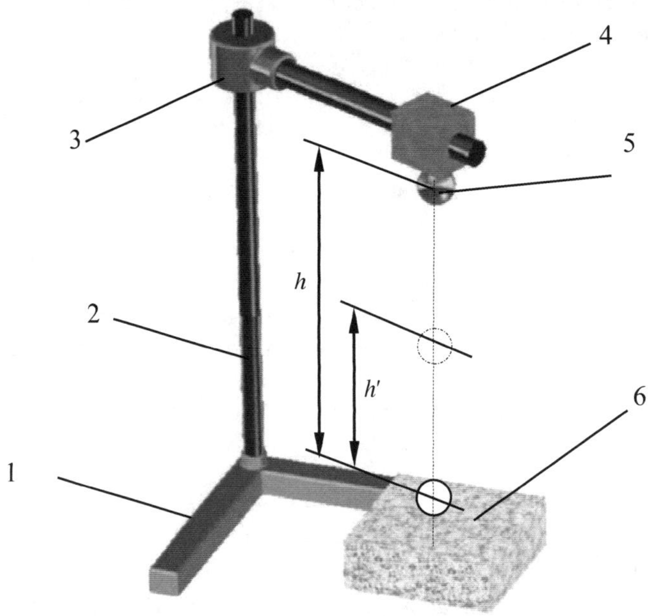

Figure 1. The principle scheme of the test rig

The base 1 supports the cylindrical vertical rod 2. The part 3 can glide along the rod 2 and it can be blocked with a screw in the needed position. An horizontal rod is fixed to the part 3 and a prismatic body 4 can slide on it; the body 4 is set on the required position with a screw. A conical hole is made in the inferior side of the part 4. The ball 5 is brought into contact to the conical hole and thus the same launch position h is ensured. When the ball is set free, it falls and collides the prismatic polyurethane foam 6 and rebounds up the height h'.

The velocity of the ball when it collides the surface of the polyurethane foam is:

$$v = \sqrt{2gh} \tag{6}$$

The post impact velocity of the ball is:

$$v' = \sqrt{2gh'} \tag{7}$$

IOP Conf. Series: Materials Science and Engineering 1169 (2021) 012020 doi:10.1088/1757-899X/1169/1/012020

The distance h from equation 6 is well determined because is constructive parameter of the test rig, but the height h' is difficult to determine. To carry out this, the experiment is filmed with a camera at $240\,frames\,/\,sec$. The film is split into frames and the precise duration t' of the bounce of the ball between the first and the second impact can be found. During this period, the ball reaches the height h' and then falls and impacts the target again:

$$v' = g\frac{t'}{2} \tag{8}$$

With known values of the velocity before and after impact, the coefficient of restitution and the energy lost during collision can be found.

3. Experimental results. Discussions

A question that was raised during the experimental tests concerns the estimation of the flight time of the ball, specifically the instant when the flight time t' starts. As it can be observed form figure 2, during the impact period, the polyurethane target presents large deformation, comparable to the dimensions of the ball, and it also moves. A simple calculus shows that the ball is in contact with the target for a duration of $(1581 - 1573)/240 = 0.033\,sec$. The polyurethane is free on the board of the test-rig and because there is possible that it will deform differently for two successive launchings, the time of the flight was calculated between the instant when the ball reaches the target for the first time and the instant when it detaches completely from the target, after the second impact.

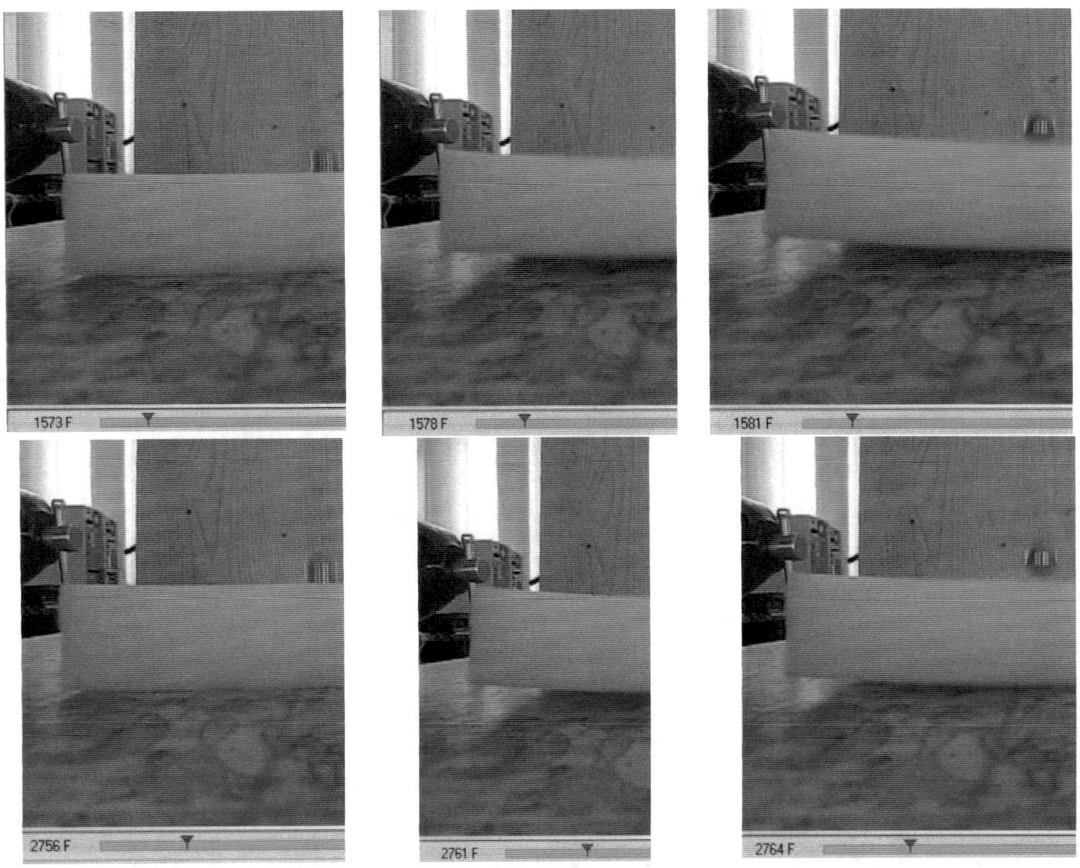

Figure 2. Images of the ball during the contact period for two successive launchings

IOP Conf. Series: Materials Science and Engineering 1169 (2021) 012020 doi:10.1088/1757-899X/1169/1/012020

In figures 3 and 4 there are presented two film frames with the ball at the moment when it detaches from the surface of the target and at the moment when it contacts again the foam after the flight time.

Figure 3. The ball at the first impact with the target

Figure 4. The ball at the second impact with the target

From each figure, in the lower left side it can be noticed the index of the frame that permits calculation of the time of flight:

$$t' = \frac{f_{new} - f_{old}}{240 \frac{fr}{sec}} = \frac{1663 - 1573}{240} sec = 0.375\, sec \tag{9}$$

For each of the two cases, dry and wet foam, 12 launches were made. The results are presented in figure 5, where there are also shown the mean values of the flight time for the launchings: $t'_{dry} = 0.376\, sec$ and $t_{wet} = 0.365$.

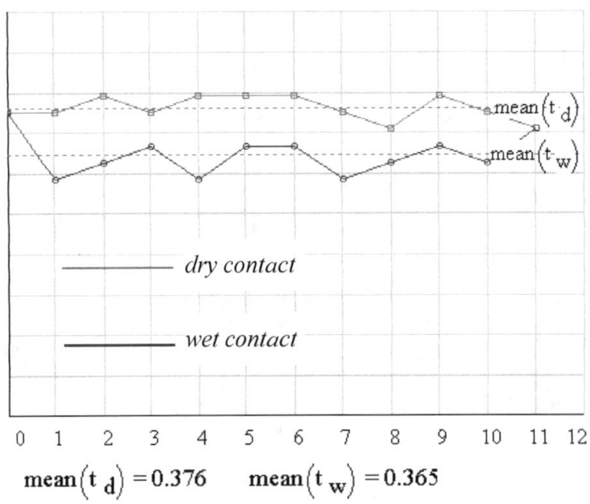

Figure 5. The values of the time of flight for the two cases

It is observed that the time of flight of the ball after the impact with the oil-wet foam is smaller and from here, the conclusion that the post impact velocity of the ball is smaller for the oil-wet contact than for the dry contact.
In the experiments made, the launching height of the ball was $h=0.38m$ and thus the impact velocity with the target, according to equation (6), is $v=2.73m/sec$. Based on the relation (8), the post impact velocities are:

$$v'_{dry}=1.844m/sec\ ,\quad v'_{wet}=1.788 \tag{10}$$

And the corresponding coefficients of restitution are:

$$e_{wet}=0.655\ ,\ e_{wet}=0.676 \tag{11}$$

The ratio of the energies lost during impact is:

$$\frac{\Delta E_{c\_wet}}{\Delta E_{c\_dry}}=\frac{\frac{mv^2}{2}-\frac{mv_{wet}^2}{2}}{\frac{mv^2}{2}-\frac{mv_{dry}^2}{2}}=\frac{1-\frac{v_{wet}^2}{v^2}}{1-\frac{v_{dry}^2}{v^2}}=\frac{1-e_{wet}^2}{1-e_{dry}^2}=1.05 \tag{12}$$

In order to estimate the magnitude of the maximum impact force, it is accepted that during the contact period, the impact force has a variation of the form:

$$F(t)=F_0\left[1-\cos\frac{2\pi t}{t_f}\right] \tag{13}$$

where t_f is the total contact time, $t_f=0.033\,sec$. The impact force is plotted in figure 6.

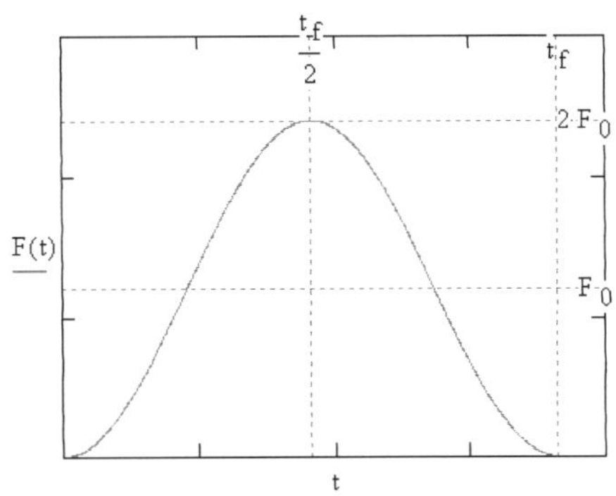

Figure 6. The accepted variation of the impact force

The mean value of the impact force during the collision is:

$$F_{med}=\frac{\int_0^{t_f}F(t)dt}{t_f}=F_0 \tag{14}$$

On the other side, the mean impact force can be calculated using the relation:

$$F_{med}=\frac{m(v-v')}{t_f} \tag{15}$$

Since for the accepted variation of the force the maximum force is:

$$F_{max} = 2F_0 \tag{16}$$

It results

$$\frac{F_{max}}{2} = \frac{m(v - v')}{t_f} \tag{17}$$

The relation (17) permits an estimation of the maximum impact force:

$$F_{max} = \frac{2m(v - v')}{t_f} \tag{18}$$

The ball used in the experiments has the diameter $d = 3/4'' = 19mm$ and the mass $m = 0.028kg$. Applying the relation (18) for the dry impact, for which there were determined $v = 2.73$ m/s, $v' = 1{,}844$ m/s, it is obtained $F_{max} \approx 1.5newton$.

In literature, [15-17], tests performed on porous materials or on materials capable of absorbing liquids, conclude that the wet materials behave considerably different from the dry ones both in quasi-static and dynamic tests: the stiffness, strength and the elasticity modulus decrease but the lost energy increases as more liquid is absorbed by the material. This is in agreement with the results of the present work, given that the damping for the tests on oil-wet foam was greater than for the dry case.

4. Conclusions

The paper presents an experimental study concerning energy aspects for the impact between a ball and a polyurethane target, for two situations: the dry polyurethane and the oil-wet polyurethane.

The test rig was made in the laboratory and ensures the same conditions for ball launching. The impact velocity is found by calculus, knowing the height from which the ball falls and the post impact velocity is found by using a video camera, based on the duration of the post impact flight time. During the tests and recordings it was noticed that the deformations of the target are comparable to the dimensions of the ball and thus the theoretical models from literature cannot be applied.

For the situations considered there were determined the flight times as mean value of the times obtained experimentally. Using the experimental data, the values of the coefficient of restitution and the losses of energy were found. One of the main conclusions is that the losses in the case of lubricated impact are greater than for the dry impact. The energy lost in the oil-wet target case is with 5% greater than for the dry target impact.

Finally, accepting for the impact force a variation similar to the ones from technical literature, an estimation of the maximum impact force is made.

References

[1] Wang Y and Mason MT 1992 Two dimensional rigid-body collisions with friction *Journal of Applied Mechanics* **59** pp 635–642
[2] Brach R M 1989 Rigid body collision *Journal of Applied Mechanics* **56** pp 133–138
[3] Brach R M 1991 *Mechanical Impact Dynamics, Rigid Body Collisions* (New York Wiley)
[4] Routh ET 1905 *Dynamics of a system of rigid bodies* (Macmillan London)
[5] Kane T R 1985 A Dynamic Puzzle *Stanford Mechanics Alumni Club Newsletter* pp 6-10.
[6] Goldsmith W 2001 *Impact, The Theory and Physical Behaviour of Colliding Solids* (Dover Publication)
[7] Kilmister C W and Reeve J E 1966 *Rational Mechanics* (American Elsevier New York).
[8] Timoshenko S P and Goodier J N 1970 *Theory of elasticity* (McGraw-Hill New York)
[9] Lankarani H M and Nikravesh P E 1990 A contact force model with hysteresis damping for impact analysis of multibody systems *J. Mech. Design* **112** pp 369-376.
[10] Ardema M D 2005 *Newton-Euler Dynamics (*Springer-Verlag US) p 316
[11] Johnson K L 1985 *Contact Mechanics* (Cambridge University Press)

[12] Flores P, Ambrosio J, Pimenta Carlo J C and Lankarani H M 2008 *Kinematics and Dynamics of Multibody Systems with Imperfect Joints, Models and Case Studies* (Springer)
[13] Flores P, Machado M, Silva M T and Martins J M 2011 On the continuous contact force models for soft materials in multibody dynamics *Multibody Syst Dynamics* **25** pp 357-375.
[14] Ciornei M C, Alaci Stelian, F C Ciornei and I C Romanu 2017 Use of loading-unloading compression curves in medical device design *IOP Conf Series MSE* **227** 012026
[15] Limmahakhuna S, Oloyede A, Sitthiseripratip K, Xiao Y and Yan C 2017 3D-printed cellular structures for bone biomimetic implants *Additive Manufacturing* **15** pp 93–101
[16] Mohamed M, Huo Z, Hawkins S, Chandrashekhara K, Birman V and Volz J 2013 Moisture effects on performance of polyurethane composite sandwich panels manufactured using VARTM, *Int SAMPE Technical Conference*, Society for the Advancement of Material and Process Engineering (SAMPE) https://scholarsmine.mst.edu/mec_aereng_facwork/698/
[17] Kanyanta V and Ivankovic A 2010 Mechanical Characterisation of Polyurethane Elastomer for Biomedical Applications *J Mech Behavior of Biomedical Materials* **3**(1) pp 51-62

IOP Conf. Series: Materials Science and Engineering 1169 (2021) 012021 doi:10.1088/1757-899X/1169/1/012021

Analysis of driver's seat rigidity in the case of rear-end collisions using a virtual multibody model

A I Radu[1]

[1] University "TRANSILVANIA" of Brasov, Brasov, Romania, 150000

alexansru.radu@unitbv.ro

Abstract. The aim of the paper was to analyse the injury potential of the occupant in the case where the backrest of the seat had lower value than normal simulating a broken seat for a rear-end collision. The study was conducted using a virtual environment and a multibody model for the occupant. Two test scenarios were taken into account, a rear end collision at 50 km/h with the normal seat and the second where the seat had a lower rigidity value. The results of the simulation show that while the head acceleration values were similar, due to the "whiplash "phenomenon when the seat had a lower rigidity, the injury potential for the occupant was two times higher compared to the normal seat.

1. Introduction

Worldwide, rear-end collisions are the most common type accident between two motor vehicles [1]. The cause of this type of collision is due to the fact that most drivers don't maintain a safety distance between them and the car ahead [2]. This is one reason why automotive manufacturers implemented the adaptive cruise control and the autonomous emergency braking systems as standard in all new vehicles. They can reduce the collision rate by up to 81% [3, 4, and 5]. Rear end collisions have a high injury potential for the occupant's neck and head due to the "whiplash" phenomenon. This type of injury is significant in rear end collisions because it causes long term health problems for the cervical spine, leading to pain and muscle tightness [6]. Over 10 million euros were paid, in Europe, by insurance companies to people who reported symptoms of whiplash after rear-end accidents [7, 8]. This is why this study is conducted in order to obtain valid results regarding the injury potential of the occupant in these types of accients using a virtual model.

The probability of "whiplash" can be predicted using a correlation model between the head-neck and torso rotation angle. A model was obtained by NHTSA(National Highway Traffic Safety Administration) from experimental tests done in a controlled environment and it's presented in figure 1 [9].

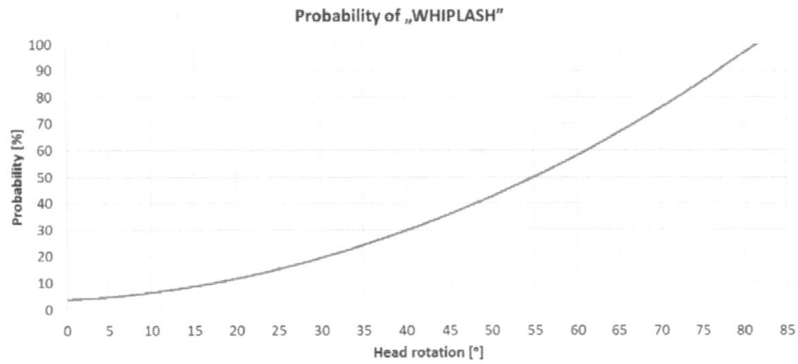

Figure 1. Probability of "whiplash" in accordance with the head and torso rotation angle [9]

For a whiplash injury of 100%, the rotation angle is around 84 degrees. This can cause injuries such as muscle tear, vertebra dislocation and fractures. These injuries are not life threating but they can create long term symptoms and limitations such as pain, dizziness, anxiety, muscle stiffness and head rotation turn limit [10].

2. Conducting the simulation

This study was conducted in a virtual environment using the accident reconstruction software called PC-Crash 12.0. The occupant was a multibody mathematical model included in PC-Crash and is presented in figure 2 [11].

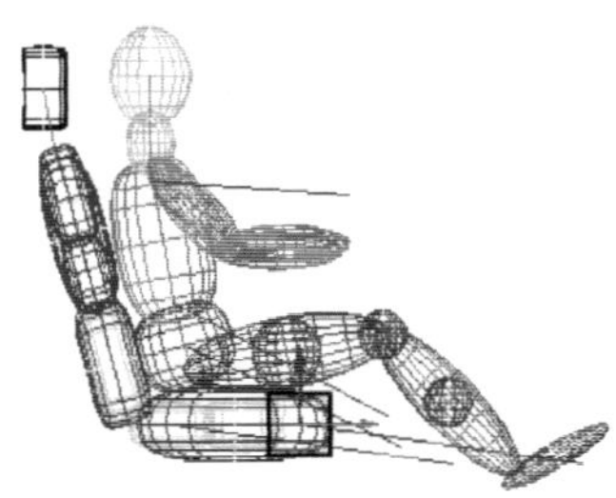

Figure 2. Occupant multibody model

The model consists of the different body parts, connected with joints and hinges in order to simulate a real human movement. Also, the seat has 3 components (lower seat, backrest and headrest).
In order to simulate a low rigidity backrest, a lower stiffness value was used in the hinge between the lower seat and the backrest. This can be observed in figure 3.

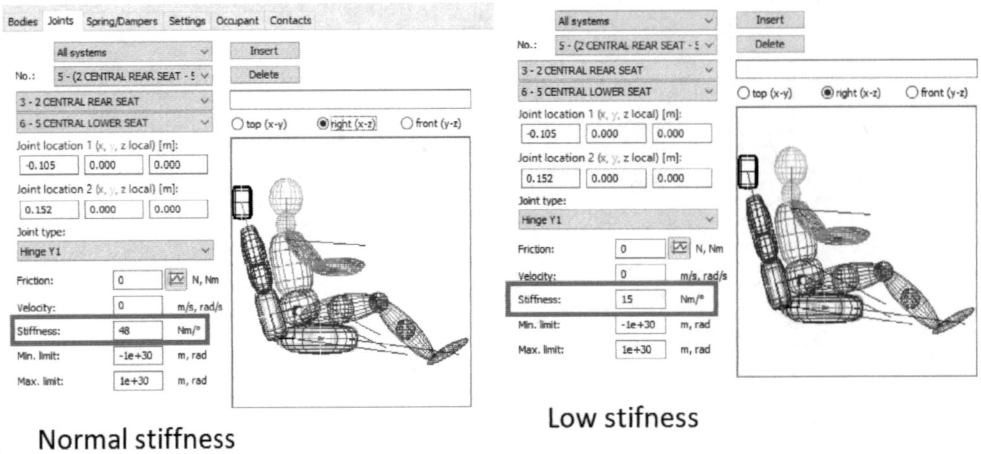

Figure 3. Seat rigidity values

For this study, a rear-end collision was simulated at a velocity of 50 km/h and the occupant model was positioned in struck vehicle. The vehicle impact position is presented in figure 4.

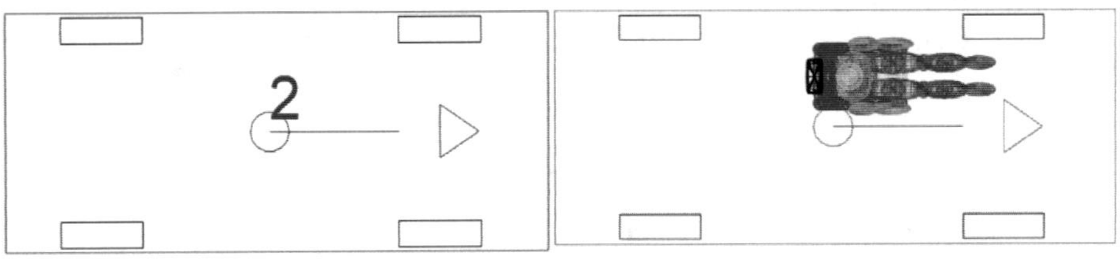

Figure 4. Vehicle impact position

Vehicle 1 is the struck vehicle (with the occupant) and vehicle 2 is the striking vehicle. The velocity of both vehicles during the collision is presented in figure 5.

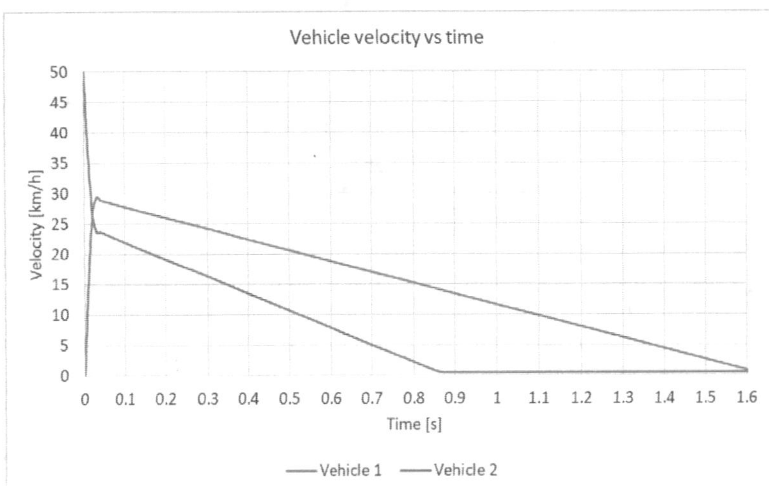

Figure 5. Vehicle's velocities during the collision

The striking vehicle velocity drops from 50 km/h to 24 km/h during the impact phase (in the first 50 ms) while the struck vehicle's velocity reaches 29 km/h during that same phase. At 26 km/h we can see that the velocities of the two vehicles are equal.

IOP Conf. Series: Materials Science and Engineering 1169 (2021) 012021 doi:10.1088/1757-899X/1169/1/012021

3. Results obtained from the simulation

The first result is the occupant kinematic during the collision for both situations for normal and low rigidity of the backrest. The kinematic comparison is presented in figure 6.

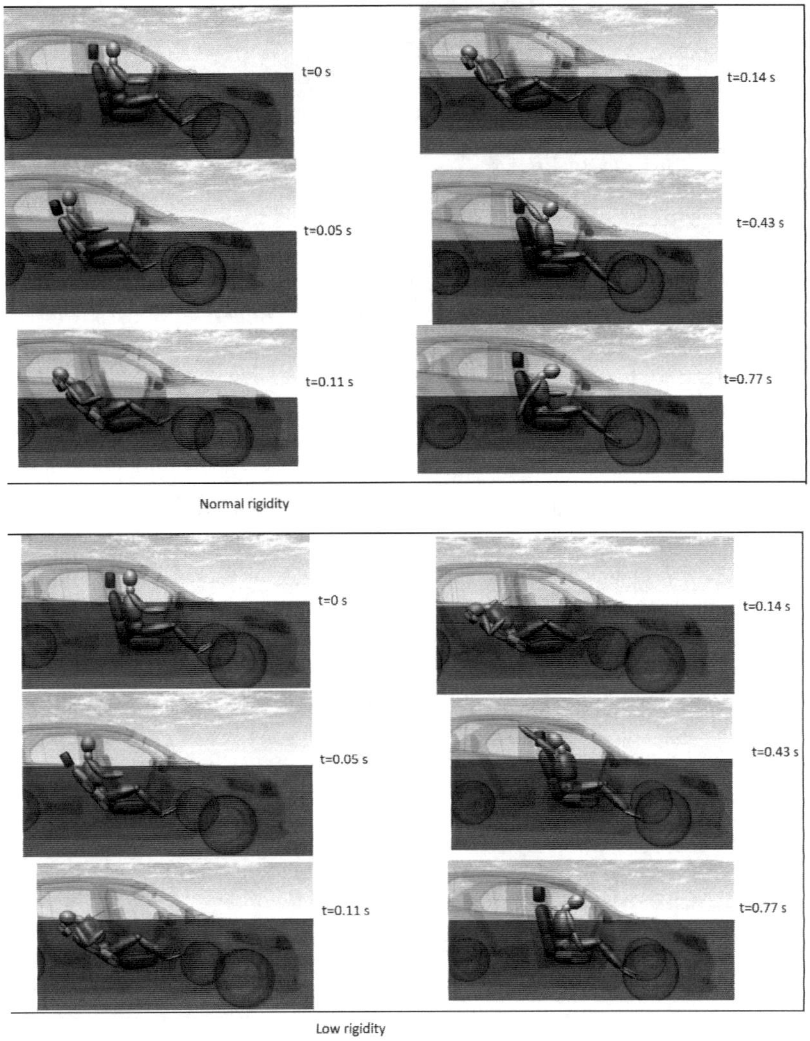

Figure 6. Occupant kinematic comparison between normal and low rigidy backrest

In can be observed that the case when the backrest has a lower rigidity, the torso has a longer displacement and also the head and neck move at a dangerous angle at 0.14 s during the impact. The maximum angle measured for the head and neck is presented in figure 7.

IOP Conf. Series: Materials Science and Engineering 1169 (2021) 012021 doi:10.1088/1757-899X/1169/1/012021

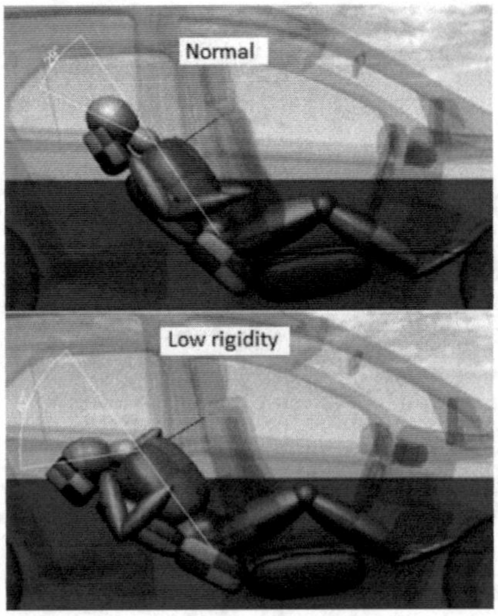

Figure 7. Maximum angle of the head and neck during the collision

If we observe the maximum angle of neck and head, relative to the torso during the simulation, we obtain a value of 29 degrees in the case of the normal rigidity seat and 62 degrees for the low rigidity case. These value can be correlated with the diagram from figure 1 and a probability of whiplash can be predicted. In the case of the normal seat, a 29 degree angle has a probability of whiplash of 19% while in the other case, 62 degrees corresponds to a probability of whiplash of 63%. The higher probability of whiplash can lead to whiplash related injuries such as neck pain and muscle tightness.

Another result we can focus on is the head acceleration value during the collision. A comparison was achieved for the two cases and presented in figure 8.

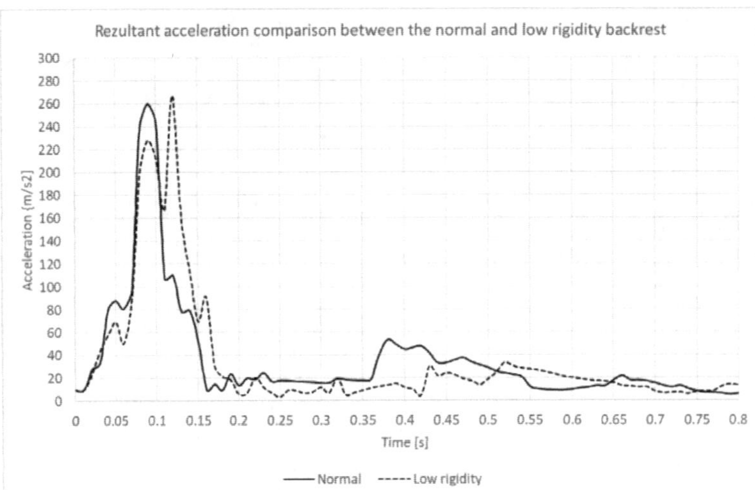

Figure 8. Head acceleration comparison

The resultant acceleration is presented for both situation and we can observe that the peak value is 260 m/s^2. In the low rigidity case of the backrest, two peak values were observed, the first (230 m/s^2 at 0.1 s) represents the initial deceleration of the head, neck and torso in first stage of the collision and the second peak (260 m/s^2 at 0.14 s) was caused by the head hitting the headrest. From this point of view, the maximum peak acceleration of 260 m/s^2 (approx. 26 g's) is not a threat to the occupant due to the fact that studies show the maximum head tolerance limit is 40 g's for more than 50 ms [12].

The final result presented is the contact force between the head and the headrest. This is shown as a comparison graph, similar to the acceleration graph, in figure 9.

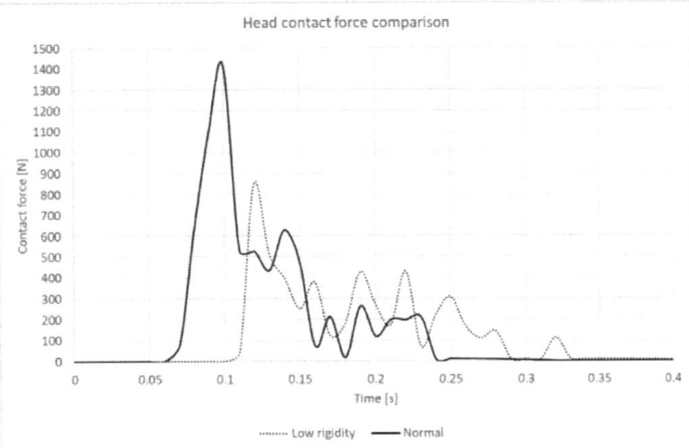

Figure 9. Head contact force comparison

In the figure we can see that in the normal seat rigidity, the contact force between the head and the headrest was around 1400 N, while in the second case, the contact force was 850 N. Also, in the low rigidity case we can observe a delay of 50 ms caused by the larger displacement of the backrest. These results show that if the backrest displacement increases, the contact forces are lower because of energy absorption. This phenomenon was observed by Jakobson that developed an anti-whiplash system, installed on Volvo cars called WHIPS (Whiplash Protection System). The system secures the occupant and moves the backrest more than in normal vehicles to absorb a part of the energy [13, 14]. In our case, while the forces are lower, the whiplash probability is higher.

4. Conclusions

The simulation shows that given the case where the backrest has a lower rigidity, the injury potential is higher because of the "whiplash "phenomenon. This is caused by the larger movement of the backrest along with the occupant's body during the collision and generating a high neck angle (of 62 degrees) relative to the torso. Compared to the angle when the seat has normal rigidity (29 degrees), the low rigidity value is double and corresponds to a probability of "whiplash" of 63%.

Head acceleration values for both cases show a similar value of 260 m/s^2 and do not suggest a threating situation for the occupant. An interesting result is the head contact force with the headrest. In the lower rigidity case is was 40% lower because of the higher movement of the occupant. Because the backrest had a larger displacement, a part of the energy was absorbed and so the contact force had a lower value. Even thou this result would suggest otherwise, the case where the backrest rigidity is lower can present a high risk of injury for the occupant because of the "whiplash" phenomenon and how it can injure the human body.

References

[1] IIHS Statistics. 2019. http://www.iihs.org/iihs/topics/t/general-statistics (accessed 2020).
[2] Jones WD. Keeping cars from crashing. IEEE spectrum. 2001 Sep;38(9):40-5.
[3] Xiao L, Gao F. Practical string stability of platoon of adaptive cruise control vehicles. IEEE Transactions on intelligent transportation systems. 2011 May 31;12(4):1184-94.
[4] Fildes B, Keall M, Bos N, Lie A, Page Y, Pastor C, Pennisi L, Rizzi M, Thomas P, Tingvall C. Effectiveness of low speed autonomous emergency braking in real-world rear-end crashes. Accident Analysis & Prevention. 2015 Aug 1;81:24-9.

[5] Cicchino JB. Effectiveness of forward collision warning and autonomous emergency braking systems in reducing front-to-rear crash rates. Accident Analysis & Prevention. 2017 Feb 1;99:142-52.

[6] Ono K, Kaneoka K. Motion analysis of human cervical vertebrae during low-speed rear impacts by the simulated sled. Traffic Injury Prevention. 1999 Sep 1;1(2):87-99.

[7] Cassidy JD, Carroll LJ, Côté P, Lemstra M, Berglund A, Nygren Å. Effect of eliminating compensation for pain and suffering on the outcome of insurance claims for whiplash injury. New England Journal of Medicine. 2000 Apr 20;342(16):1179-86.

[8] Foreman SM, Croft AC. Whiplash injuries: the cervical acceleration/deceleration syndrome. Lippincott Williams & Wilkins; 2002.

[9] Asada H, Nawata K. Study on static and quasi-dynamic evaluation method for assessing whiplash-associated disorders in rear impacts. ESV, Paper. 2007(07-0264).

[10] Chen HB, King HY, Wang ZG. Biomechanics of whiplash injury. Chinese Journal of Traumatology (English Edition). 2009 Oct 1;12(5):305-14.

[11] Teng TL, Chang FA, Liu YS, Peng CP. Analysis of dynamic response of vehicle occupant in frontal crash using multibody dynamics method. Mathematical and Computer Modelling. 2008 Dec 1;48(11-12):1724-36.

[12] Kramer M. Safety properties of laminated glass panes with changing pane temperatures (Sicherheitseigenschaften von verbundglasscheiben bei wechsel nden scheibentemperaturen). AUTOMOBILTECHN. Z.. 1972 Oct;74(10).

[13] Jakobsson L, Lundell B, Norin H, Isaksson-Hellman I. WHIPS–Volvo's whiplash protection study. Accident Analysis & Prevention. 2000 Mar 1;32(2):307-19.

[14] Lundell B, Jakobsson L, Alfredsson B, Lindström M, Simonsson L. The WHIPS seat-a car seat for improved protection against neck injuries in rear-end impacts. InProc. 16th ESV Conference, Paper 1998 May 31 (Vol. 98, p. S7).

IOP Conf. Series: Materials Science and Engineering 1169 (2021) 012022 doi:10.1088/1757-899X/1169/1/012022

Study of the clamping force in a pneumatic wedge device

P D Tocuț[1], Stanasel I[1], F T Avram[1]

Faculty of Managerial and Technological Engineering, University of Oradea, Romania

tocutpd@yahoo.com

Abstract. The continuous development of computing technology and design and manufacturing software applications have contributed to improving clamping devices used in manufacturing processes, which led to an increase in the quality of cut parts. This paper presents a study on the variation of the clamping force depending on the constructive and exploitation parameters of a pneumatic wedge device.

1. Introduction

In the context of companies' competition on the market, continuous improvement of products requires preparation and realization of manufacturing in a short period of time and with minimal expenses. One of the main ways to increase quality, productivity, and reduce expenses is to realize rationally designed and built technological equipment. Compared to the technological equipment used in manufacturing processes, the clamping devices used in manufacturing have the largest share.

Starting from a variant of a manually operated clamping device intended for small series production, an improved variant intended for large series production is presented, in which the action is carried out by a linear pneumatic cylinder with double action. The paper addresses an important topic in clamping device design, namely the study of the variation of the clamping force according to the constructive and operational parameters of the device.

In literature, numerous studies address different aspects in the field of technological devices.

Article [1] presents an algorithm for computer-aided design of a device for parts with freeform surfaces. Based on the algorithm, the minimum number of clamping elements is determined, as to maintain the stability of the part.

Paper [02] presents the design of mechanical clamps with extra-long wedge grips for static and fatigue testing of composite materials in tension and compression.

In thesis [03], the study of a device layout and clamping force for a disk in a numerically controlled vertical lathe were based on factorial experiments and it was established the shape (layout) of the clamping device. By finite element method, the displacements for the range of variation of the clamping force.

By using finite element method for the dynamic analysis of the workpiece clamping system in [04], the clamping force was optimized as to result in the minimal deformation of the workpiece.

Also, by using finite element method, [05] presents the study of a device for "thin-walled curved surface parts" and analyzes the deformation of clamped parts.

In [06], a study on the recent developments on computer-aided device design presents a design solution by using artificial intelligence and proposes to integrate the design steps in order to obtain an effective device.

IOP Conf. Series: Materials Science and Engineering 1169 (2021) 012022 doi:10.1088/1757-899X/1169/1/012022

2. Fixture design

Wedge clamping mechanisms are characterized by the existence of one or more wedges in their structure, which clamp the workpiece by the action of external forces. The transmission of external forces can be done directly on the blank or indirectly through intermediate elements. The device is intended for clamping cylindrical parts, in order to process the front surface.

The first variant of the manually operated device is shown in figure 1. The clamping force is achieved by rotating the screw (1), which moves the wedge clamp (2) and acts on the mobile clamp (3).

Figure 1. Manually operated device.

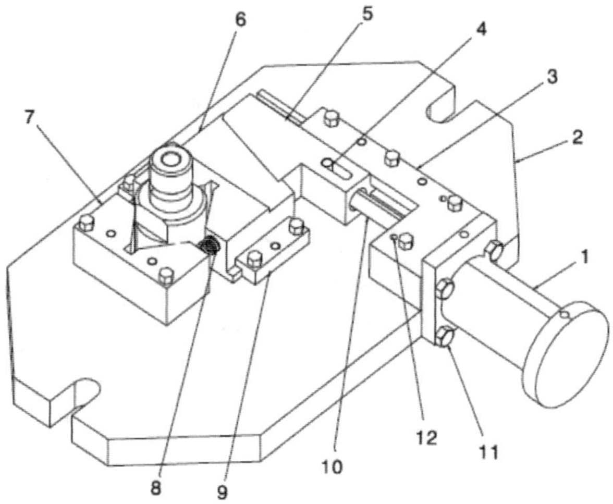

Figure 2. Pneumatic operated device.

For large series production, a variant of pneumatic drive device shown in figure 2 has been proposed. On the device, the acting force of the wedge mechanism is achieved by the double-action pneumatic actuator (1), fixed in horizontal position on the structure of the device base plate, as well as on the guiding system of the mobile wedge (3).

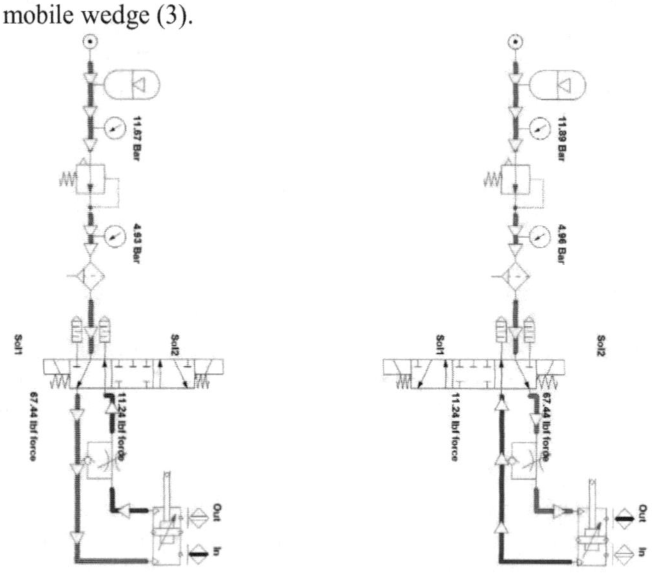

a. b.

Figure 3. Pneumatic diagram designed for device operation.
a. for locking; b. for unlocking

IOP Conf. Series: Materials Science and Engineering 1169 (2021) 012022 doi:10.1088/1757-899X/1169/1/012022

The pneumatic cylinder rod ensures that through the cylindrical bolt action (4), the connection between the movable wedge (5), which acts on the movable prism (6) is realized, and which clamps the part by moving the wedge.

The pneumatic diagram designed for device operation is shown in figure. 3.a, b, the electromagnetically controlled position of the type 5/3 pneumatic distributor at the clamping of the blank in the device, indicates the compressed air circuit, which determines a pushing force on the wedge which by linear motion actuates the fixing system.

According to the pneumatic diagram shown in figure 3.a, it can be observed that the wedge motion (5) is done at lower speed, which is achieved by pushing a linear stroke, which determines the motion of prism (6) from right to left, achieving blank clamping and orientation. The track drosel restricts the compressed air flow through the sense valve, thus achieving a lower travel speed. In the pneumatic scheme shown in figure 3, b it can be seen that when the blank is unscrewed from the device, the motion of the mobile wedge is made in reverse with higher speed by pulling. In this case, the track drosel does not restrict the compressed air flow through the sense valve, thus achieving a higher retraction speed, favoring a hydraulic shock, so that it does not remain stuck.

3. Clamping force

The clamping forces are calculated by considering the wedge in Figure 4., on which the outer force Q acts in the wedging direction. Due to the action of external force Q on the inclined and horizontal plane, the normal force reactions N1 and N2, respectively the friction forces F1 and F2 result. By combining these forces, the resultant forces R1 and R2 appear, which are decomposed by horizontal and vertical directions, so forces P and S are obtained.

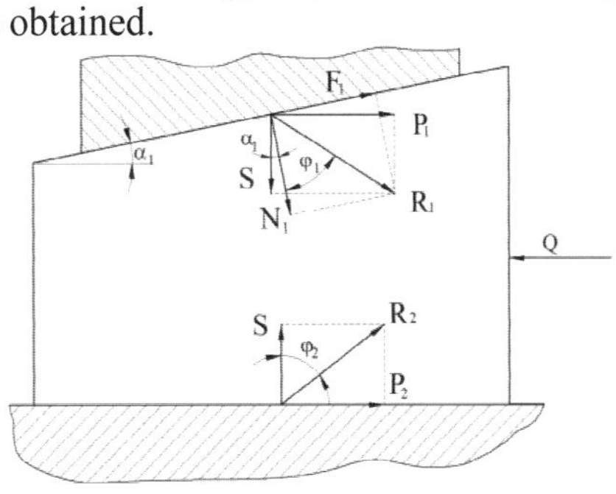

Figure 4. Sketch for calculating the clamping forces.

Knowing the acting force Q, the clamping force "S" can be determined using the diagram in figure.4.

$$Q = P_1 + P_2 \tag{1}$$

$$P_1 = S \cdot tg\left(\alpha_1 + \varphi_1\right) \tag{2}$$

$$P_2 = S \cdot tg\left(\varphi_2\right) \tag{3}$$

α_1 –the wedge angle

φ_1 *şi* φ_2 – are the friction angles on the two sides of the wedge

$$\varphi_1 = arctg\,\mu_1 \;;\; \varphi_2 = arctg\,\mu_2 \tag{4}$$

μ_1 şi μ_2 are the friction coefficients on the two sides of the wedge

After replacement results:

$$S = \frac{Q}{tg(\alpha + \varphi_1) + tg\varphi_2} \tag{5}$$

The clamping mechanisms must comply with the self-braking condition in order to keep the blank in the device after the external force ceases to act,

The self-braking condition involves the determination of the α angle, so that it will not self-operate when the drive Q equals 0.

By approximating $\varphi_1 = \varphi_2 = \varphi$, after calculations, the relationship (5) becomes:

$$tg(\varphi - \alpha) + tg\varphi = 0 \tag{6}$$

in order to have self-braking, must

$$\alpha < 2\varphi$$

Considering that the surfaces of the feathers are small roughness, the friction coefficients μ = 0,1……0,15, respectively $\varphi = 5^0 43'$, where we get

$$\alpha < 10^0 ......12^0 \text{ if } \mu = 0.1, \text{ and } \alpha < 16^0 ......18^0 \text{ if } \mu = 0,15$$

4. Numerical study

The purpose is to determine the variation of the clamping force depending on the working pressure and the angle of inclination of the wedge and the diameter of the piston. The pressure parameters and the angle of inclination considered have a continuous variation between the following limits:

$$p: 04....0.8 \text{ MPa}; \ \alpha: \ 8°...18°; \ \ D: \ 10,12, \ 16,20,25,32 \text{ mm}$$

From figure 5 result in how the three parameters influences the clamping force. It can be seen that the clamping force varies in proportion to the pressure and the diameter of the pneumatic actuator and inversely proportional to the angle of the wedge.

In figure 6 is shown the 3D variation of the clamping force depending on the air pressure and the wedge angle.

Parallel Plot Representation offers a better understanding of dependencies between variables, constraints, and system response. Each variable and system response is represented on a vertical axis and has its own representation scale. The combinations between variables and system response are represented as a series of lines connected to each axis. Using the parallel plot representation, one can determine the parameters that can produce a particular clamping force. For example, the parameters that produce a clamping force contained in the interval 600-700 N are those connected with the blue lines in figure. 7.

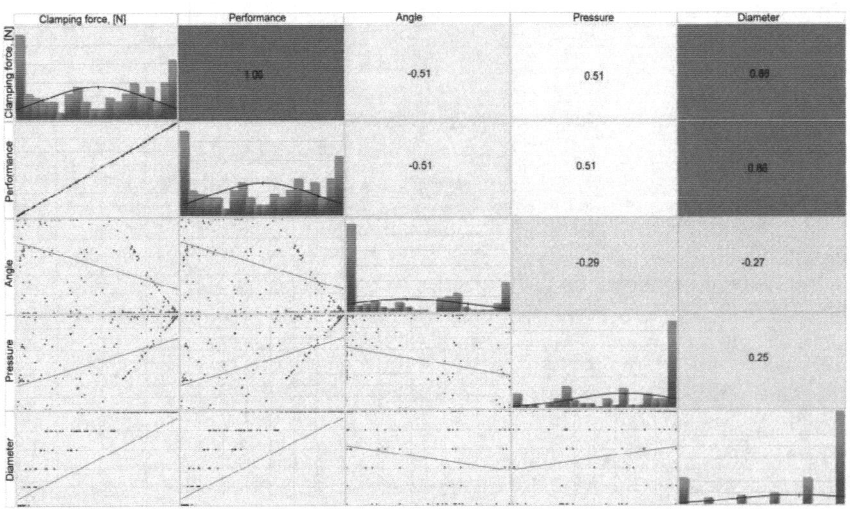

Figure 5. Correlation plot.

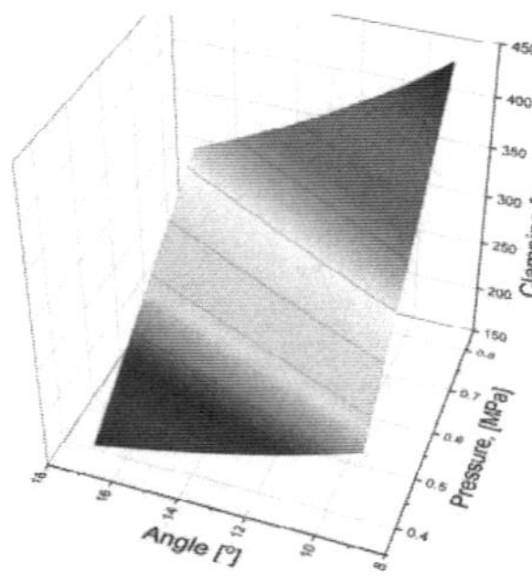

Figure 6. Clamping force vs. air pressure and wedge angle.

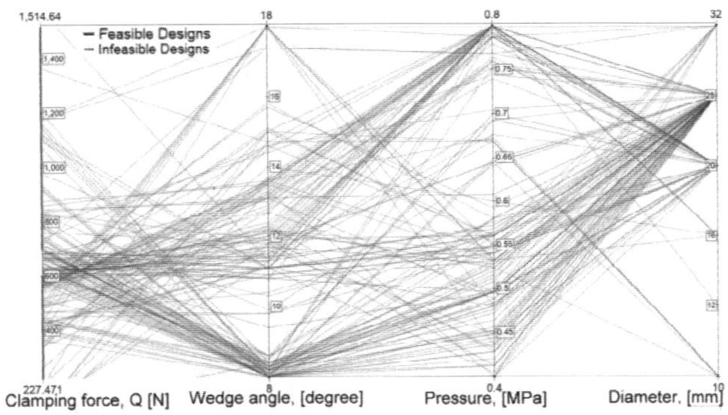

Figure 7. The parallel plot.

5. Conclusions

A quick-acting wedge device driven by an actuator has been designed to reduce the clamping time. According to these parameters, it was determined the relationship of calculation of the clamping force according to the constructive parameters and the air pressure at the compressed air network,

Numerical research was carried out on the variation of the clamping force., and it was shown that the clamping force increases when increasing the pressure and diameter of the pneumatic cylinder and when decreasing the angle of the drive wedge.

References

[1] M. J. Nategh, H. Parvaz 2018 Development of computer aided clamping system design for workpieces with freeform surfaces, *Science Direct*, Volume 95, p.52-61.

[2] I. De Baere, W. Van Paepegem, and J. Degrieck 2008 Design of mechanical clamps with extra long wedge grips for static and fatigue testing of composite materials, *Tension and compression, experimental techniques*, Society for Experimental Mechanics.

[3] M. Fang 2009 A Study of Fixture Layout and Clamping force for a Ti-6Al-4V Disk în a Vertical Turning Lathe, master thesis, Rensselaer Polytechnic Institute, Hartford.

[4] S. Selvakumar, K.P. Arulshri, K.P. Padmanaban and K.S.K. Sasikumar 2010 Clamping Force Optimization for Minimum Deformation of Workpiece by Dynamic Analysis of Workpiece-fixture System, *World Applied Sciences Journal* 11 (7) p. 840-846,

[5] L. Qu, and H. Liu 2012 Clamping deformation analysis research based on a flexible clamping fixture of thin-walled curved surface parts, *Advanced Materials Research* Vols. 476-478 p. 2028-2031, Trans Tech Publications, Switzerland,

[6] H. Hashemi, A. M. Shaharoun, S. Izman, and D. Kurniawan 2014 Recent Developments on Computer Aided Fixture Design: Case Based Reasoning Approaches, *Advances în Mechanical Engineering*, Hindawi Publishing Corporation.

IOP Conf. Series: Materials Science and Engineering 1169 (2021) 012023 doi:10.1088/1757-899X/1169/1/012023

Designing a vacuum chamber and substrate positioning system for magnetron sputtering deposition applications

D.I. Țarcă[1], R.V. Ghincu[1], D. Crăciun[2]

[1]Doctoral School of Industrial Engineering, University of Oradea
[2]Faculty of Managerial and Technological Engineering, University of Oradea

tarca.danioan@student.uoradea.ro

Abstract. A new custom vacuum chamber with a substrate positioning system in our SMARTMAT laboratory is needed, as the previous one that we installed did not suit our needs. Based on our experiences with the old sputtering system, we designed a smaller vacuum chamber that could be attached to our existing load-lock system, and that could have a substrate positioning system suited for our magnetrons which would work with the load-lock. The new system replicates our initial experiments and is compatible with our experimental data acquisition system. We mounted the sensors and vacuum equipment on the chamber, and initial vacuum tests were a success, which validated our design.

1. Introduction

After installing a high-capacity sputtering system in our SMARTMAT [1] laboratory, we had the necessary expertise to elaborate a design that would better suit our needs [2-5]. The new chamber has some old parts, such as the rough and turbopump, the vacuum sensors, and the gas inlet flange, but the rest are new and specific to our needs. While the existing vacuum chamber could have been modified to our needs, the process would have been too expensive and would require more moving parts inside the main compartment, which could have malfunctioned and required a lot of design work. So we opted instead for a simple solution with as few moving parts as possible while still getting the job done. We followed some basic guidelines set by previous researchers [6-11] while designing the chamber and positioning system, with the goal of making an experimental setup that is easier and faster to operate.

2. Vacuum chamber

We started designing the chamber represented in Figure 1. by having some minimal requirements and left some unused flanges for further expansion if needed.

We needed three DN40CF flanges on the cylindrical shape: two for measuring the pressure (one for rough vacuum, one for high vacuum) and one flange for the gas inlet. The gas inlet flange has no importance where it is situated. Still, the flanges where the measuring equipment will be mounted are best positioned in a spot with no direct line of sight with the magnetron as there will be less cleaning needed – so we put them underneath the magnetron gun. Although some scattered electrons may hit the measuring equipment, the chances are significantly reduced. Also, on the cylindrical shape, we have three DN 100CF flanges: one for the turbomolecular pump (which we already had) and two for seeing inside the chamber. Just one viewing flange would be insufficient as we discovered with the old

IOP Conf. Series: Materials Science and Engineering 1169 (2021) 012023 doi:10.1088/1757-899X/1169/1/012023

equipment that we need a source of artificial light to see inside the section, and shining a light through the same viewing port has some drawbacks such as reflection. At this time, as we are experimenting with an uncalibrated and self-developed system, it is essential to visually confirm that the substrate is in the correct position and that we do not hit the positioning system with the load-lock rod. Therefore, one DN160CF flange is needed for the already built and tried load-lock system.

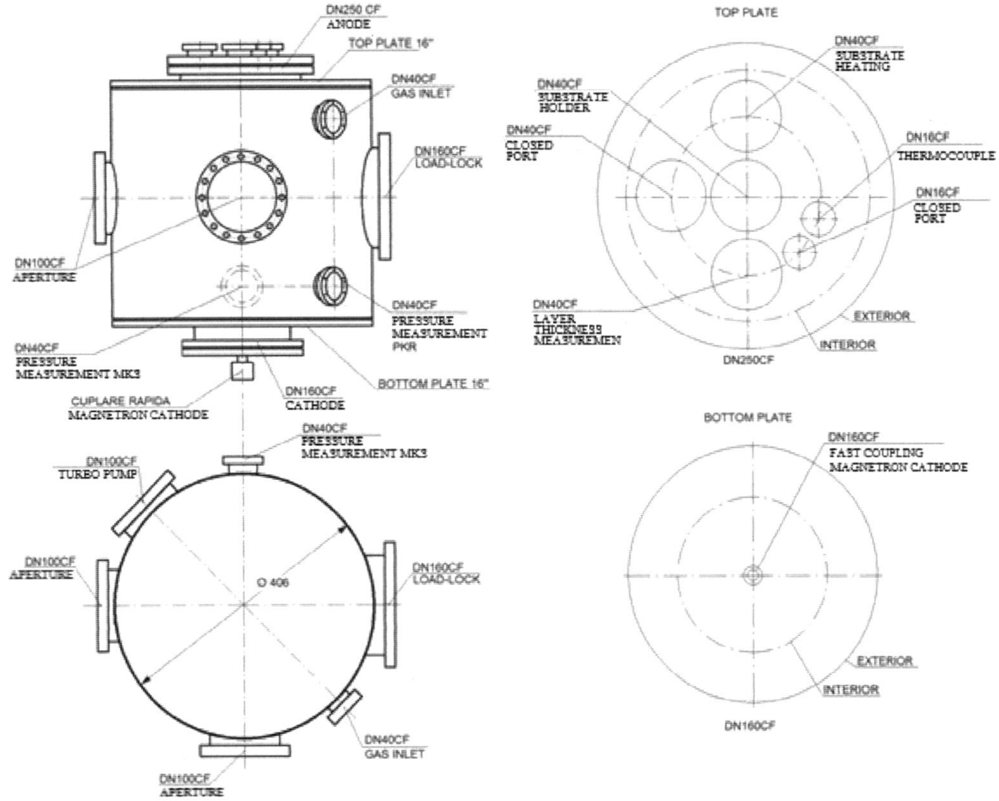

Figure 1. Vacuum chamber design.

On the top of the chamber, we have six flanges: two DN16CF and four DN40CF. One of the two DN16CF flanges will be used as a feedthrough for a thermocouple, and the other one is a spare. The four DN40CF will be used as follows: one for measuring the thickness of the deposited layer, one for heating the substrate, one for the substrate positioning system also designed by us and covered in this paper, and one is a spare. In order to measure the thickness of the deposited layer, we are also developing a piezoelectric sensor that will be mounted on this system.

There is only one DN160CF flange for the sputtering gun that can accommodate both old and new guns on the bottom of the chamber. It has a central position to be in line with the opposite counterpart, the substrate positioning system.

The top plate with the six flanges is an independent part of the vacuum chamber and can easily be removed. A Viton synthetic rubber makes the seal for easy access inside. The bottom plate is welded on the vacuum chamber.

The whole chamber is made of aluminum. However, it is not the best material as diffusion is faster because it is not as dense as steel, and some targets can react with the aluminum and contaminate the chamber. Still, on the other hand, it is cheaper, lighter, and easier to machine than steel. In addition, aluminum is non-magnetic, while for non-magnetic steel, the price is significantly higher than standard steel.

Price and execution time was an essential part of deciding if the vacuum chamber should be made of aluminum or steel. However, given that this is an experimental setup, and we already have a steel chamber if needed for specialized applications, choosing aluminum was a natural choice. Also, we now can experiment with different types of materials used in building vacuum chambers and how it affects the vacuum, deposition quality etc.

3. Substrate positioning system

The design of the vacuum chamber revolves around the substrate positioning system and the gun. Therefore, it is essential to have a positioning system that can hold the substrate and position it closer to the magnetron gun (target). Also, it has to have a substrate heating system that can reach about 300° Celsius. The heater is essential because the electric charge can heat the target. Electrons emitted by it do not adhere well to the substrate if there is a significant temperature difference. Therefore, by reducing the difference in temperatures, the deposited layer adheres better to the substrate.

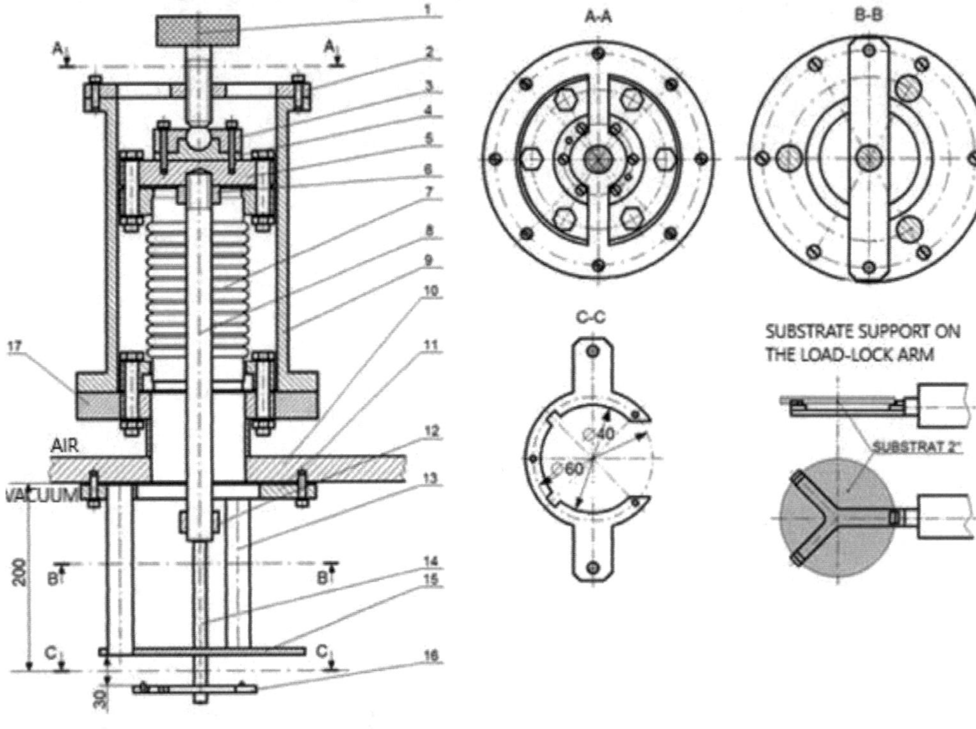

Figure 2. Substrate positioning system: 1 - Drive screw; 2 – Lid; 3 – Spherical join connection element; 4 - Spherical joint base plate; 5 – Blank flange DN40CF; 6 – Lock-nut; 7 – Bellows DN40CF; 8 – Central rod; 9 – Cylindrical shield; 10 – DN250CF flange; 11 - Fixed rod support ring; 12 – Yoke for moving rods; 13 – Fixed rods; 14 – Moving rods; 15 – Substrate base plate; 16 – Substrate support. 17 – Flange. C-C represents a top view for the substrate support (16), matching with the substrate support on the Load-Lock arm.

The substrate positioning system design presented in Figure 2 is based on similar designs, with some components already available on the market – such as the bellows or the spherical joint connection. The whole operation is airtight, with no direct contact with the exterior, as it is magnetically driven. By rotating the drive screw (1), the moving rods (14) are pushing or pulling the substrate support (16). The Y-shaped substrate support on the load-lock arm is accommodated by the substrate support (16) from the substrate positioning system thanks to the two notches. As the load-lock arm positions the substrate on the substrate support (16), the rods (14) are pulling up the substrate and fixing it against the substrate base plate (15). Thanks to the notches, the Y-shaped substrate support on the load-lock arm stays at the same vertical level and can be retracted. After a successful deposition, the moving rods (14) lower the substrate support (16) by turning the drive screw (1). The load-lock arm waits in position, right below the substrate support (16). As the substrate support (16) lowers, the substrate will be placed on the load-lock arm, which can be retracted to evacuate the substrate from the vacuum chamber.

4. Results

An image with the whole system is presented in Figure 3.

Preliminary results in vacuum measurements show that the new system is 3-4 times faster than the old one in terms of hitting 10^{-5}mbar, in large part because it is a smaller chamber, and the vacuum pump is oversized. More measurements will be made when the positioning system will be installed, and the whole deposition system will be mechanically complete.

As shown in Figure 3, the whole vacuum deposition system is placed on a steel frame because we want it to be a mobile system. The primary pumps are mounted inside the frame while the turbomolecular and the load-lock rod are exceeding the frame. As it is an experimental setup, by the time it will be complete, we will need to find a solution to protect both the pumps and the rod. However, the rest of the equipment can easily fit inside the frame.

Figure 3. Vacuum chamber with sensors, load-lock, and vacuum pump

5. Conclusion and further work

The design activities are concluded, and tests are yet to come once the system is mechanically complete, the only piece missing being the positioning system.

Regarding the positioning system, moving from the design phase to the execution phase should not be difficult as the main components are readily available to be purchased from well-established retailers. The main challenge will be installing and adjusting the system so that it can correctly pick up the substrate and that it will also be at a minimal distance from the sputtering gun.

The next phase is to connect the magnetron to the power transformer and calibrate the gas intake. Once this step is finished, more calibrations will be needed on the whole deposition system before depositing some metal targets.

References

[1] https://erris.gov.ro/Research-Laboratory-for-Adva-1
[2] Dan-Ioan ȚARCĂ, Moldovan Octavian ALIN, Remus GHINCU, Moldovan Ovidiu GHEORGHE, Traian COSTEA, Polojințef Corbu NICOLAE, Installing a high capacity sputtering system at University of Oradea, *Electrical Engineering and mechatronics conference EEMC'19, University of Debrecen, Faculty of Engineering,* May 2019
[3] D I Țarcă, T. Costea, I. Moga, Custom vacuum load-lock for the sputtering vacuum chamber,

loading/unloading and locking automatisation mechanisms, *IOP Conf. Ser.: Mater. Sci. Eng.* ***898***, 2020, doi:10.1088/1757-899X/898/1/012046

[4] Moga, I., Ghincu, R., Costea, T., Tarca, D., and Moldovan, O. A. (2018). Some experiments regarding magnetron sputtering deposition with small capacity devices. *Nonconventional Technologies Review* 22(4), Romanian Association of Nonconventional Technologies, Dec 2018

[5] D I Țarcă, L Csokmai, O Moldovan, R Ghincu and D Potroviță, Experimental data collection system for reading pressure levels in a vacuum environment, *IOP Conf. Ser.: Mater. Sci. Eng.*, 2019, DOI: 10.1088/1757-899X/568/1/012078

[6] K. Jousten, Handbook of Vacuum Technology, Second, Completely Revised and Updated Edition, Wiley-VCH, 2016.

[7] J. Moore, C. Davis, M. Coplan și S. Greer, Building Scientific Apparatus, Fourth Edition ed., Cambridge: Cambridge University Press, 2009.

[8] D. M. Hoffman, B. Singh și I. John H. Thomas, Handbook of Vacuum Science and Technology, Academic Press, 1998

[9] J. F. O'Hanlon, A User's Guide to Vacuum Technology, Third Edition, Hoboken, New Jersey: John Wiley & Sons, 2003.

[10] K. Wasa, I. Kanno and H. Kotera, Handbook of Sputter Deposition Technology; Fundamentals and Applications for Functional Thin Films, Nanomaterials and MEMS, 2 ed., Elsevier, 2012.

[11] K. Seshan, Handbook of Thin-Film Deposition Processes and Techniques, 2 ed., New York: William Andrew Publishing, 2002.

IOP Conf. Series: Materials Science and Engineering 1169 (2021) 012024 doi:10.1088/1757-899X/1169/1/012024

Development of a data acquisition system for a vacuum thin film deposition equipment

R. V. Ghincu[1], D.I. Țarcă[1], O. A. Moldovan[1]

[1]Doctoral School of Industrial Engineering, University of Oradea, Oradea, Romania

remusghincu@yahoo.com

Abstract. Digital data acquisition is a must for precise data collection, and monitoring systems should be present on all modern equipment. This article focuses on improving the vacuum deposition systems already present in the SMARTMAT laboratory at the University of Oradea. Previous advancements have been made in installing vacuum chambers and digitally monitoring the vacuum level with the final objective of having a complete and versatile vacuum deposition system: magnetron sputtering, RF sputtering, and resistive heating. This article proposes a complex monitoring system based on modern sensors, data acquisition modules, and a processing and storage unit capable of monitoring all the vacuum deposition steps regarding the resistive heating system. The monitoring system can show real-time charts and store all the data in a database to be further used in modeling. As we already have a digital monitor system for the vacuum level, we have developed a monitoring system for temperature, voltage, and current, thus completing the design and controlling the whole evaporation process.

1. The work so far

The present work described in this article completes the vacuum thermal evaporation deposition system [1-4] and puts it in line with modern systems capable of being monitored electronically. It is clear that in a research environment, the work never stops, but further improvements to the system would be just small increments, and further automatization of the processes will be intensely dependent on the system created so far.

By giving the vacuum deposition system updated features, such as electronic monitoring and control, which has an already sound mechanical and electrical system, we are upcycling old technology and giving it a new life in our SMARTMAT [5] laboratory.

2. Data collection and monitoring system

The schematic of the data collection and monitoring system is presented in Figure 1.

The data collection and monitoring system presented in this paper is part of a complex monitoring system and focuses on measuring and logging data from the two Thermocouples K and R and from the two digital Voltmeters and Amperemeters.

Well-established books [6-8] but also recent articles [9-11] all have in common the same descriptions for data collecting when it comes to vacuum systems: pressure, temperature, and power management (voltage and current intensity).

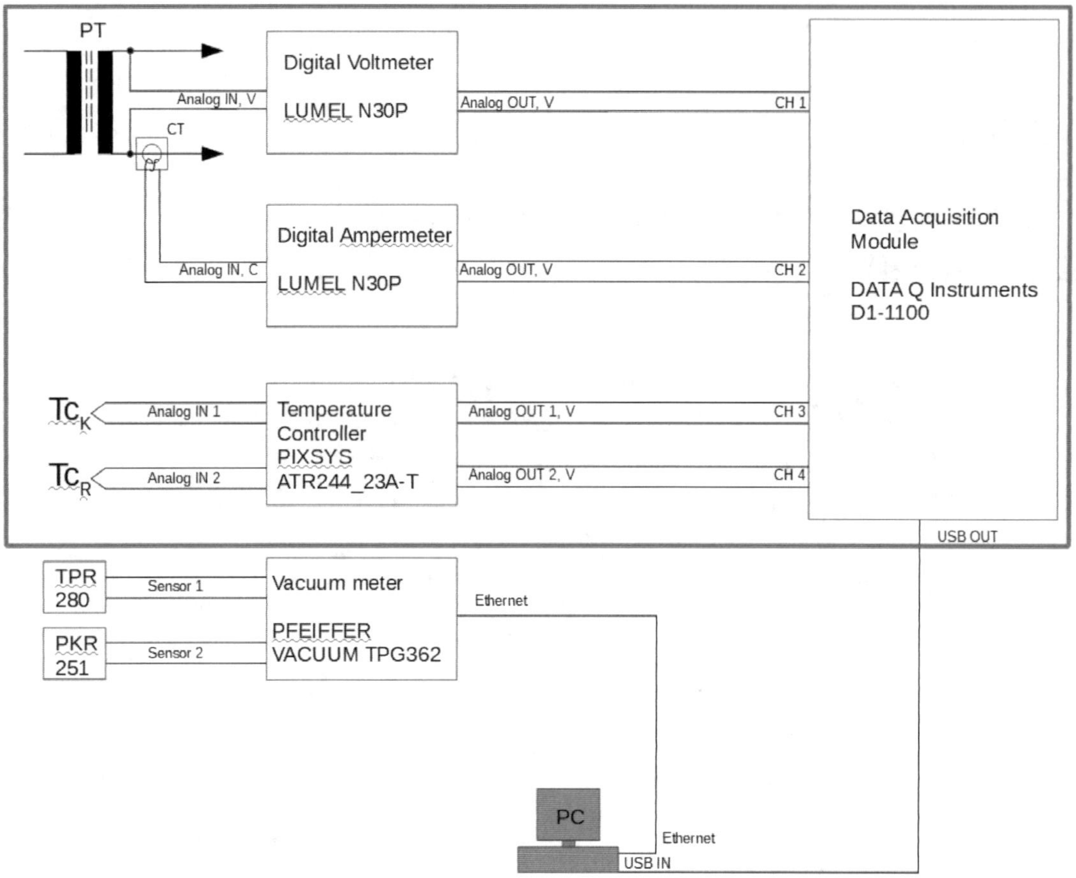

Figure 1. Diagram of the entire data acquisition system. The red rectangle shows the system described in this paper. PT = Power Transformer 3kW; CT=Current Transformer; TC_K = Thermocouple K; TC_R = Thermocouple R; TPR 280, PKR 251 = Vacuum sensors

2.1. Equipment and sensors used.

All the equipment and sensors used are readily available on the market and can be purchased from specialized retailers.

On the baseplate of the vacuum chamber we installed two thermocouple feedthroughs, individually for each sensor. On the outside, the feedthroughs are connected to the temperature controller PIXSYS [12]. On the inside, we connected the temperature sensors. We need two types of sensors because there will be two different ranges of temperatures recorded. The K-type thermocouple [13] has a limit of 1260° Celsius, while the Type R [14] has a maximum temperature range of 1760° Celsius. Because of the two different types of temperature sensors, we had to program the PIXSYS temperature controller with different settings. The calibration of the two thermocouples is made using an already calibrated sensor, as shown in Figure 4. It is essential to have two temperature sensors inside the vacuum chamber, as we need to monitor both the substrate temperature and the evaporation boat temperature. Without the sensors, we cannot accurately replicate different evaporation procedures.

The digital voltmeter was put in parallel on the output lines of the power transformer. The voltage to be measured is the supply voltage for the heating boat. It is an AC voltage with the frequency of 50Hz, adjustable in the range 0…15VAC. The heating current is an AC current, with a frequency of 50Hz, and a value in the range 0…300AAC To measure the heating current intensity, we used a current transformer connected to the digital amperemeter. The current transformer has a primary current nominal value of 300A and a secondary current nominal value of 5A, accuracy class 1.0. Both measuring devices are the same type LUMEL N30P [15], a digital panel meter capable of measuring both voltage and current but with different parameter settings. Unfortunately, one device cannot measure both at the same time. Calibration was needed, and it was done with a pre-calibrated true RMS digital multimeter.

Figure 2. Actual image of the whole setup (above); Previous configuration (below)

For the data acquisition platform, we opted for a DATAQ data acquisition kit with four channels. In retrospect, we should have chosen one with more channels to read the pressure sensors as well, but then again, they are directly connected to our PXI [3].

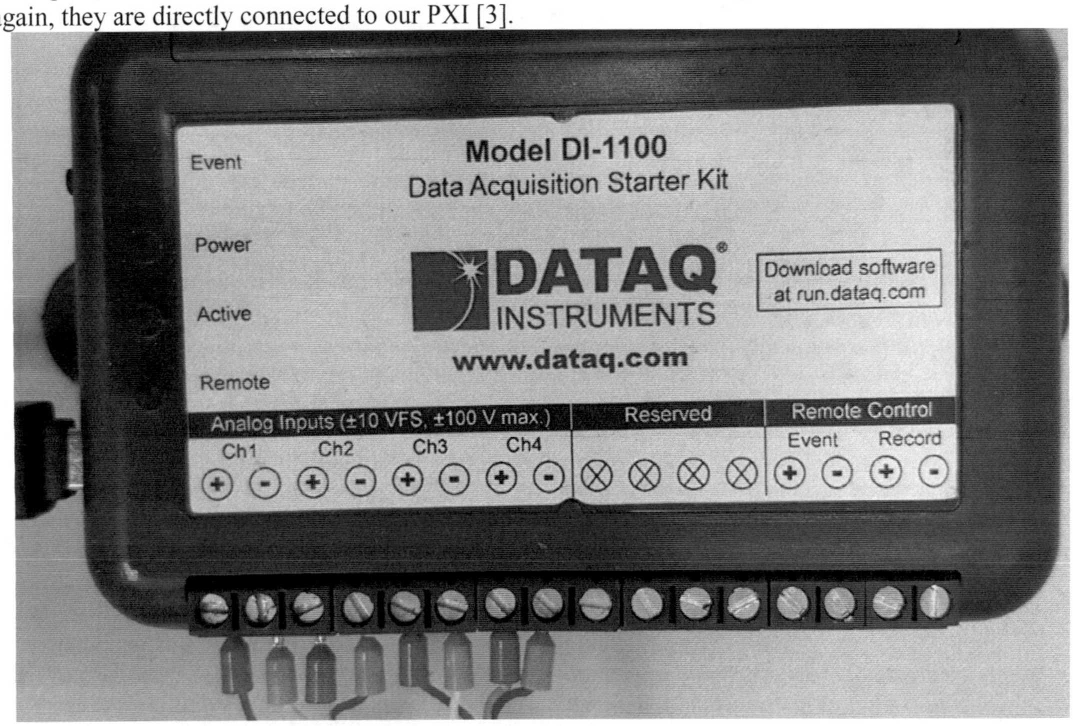

Figure 3. DATAQ DI-1100 data acquisition module

IOP Conf. Series: Materials Science and Engineering 1169 (2021) 012024 doi:10.1088/1757-899X/1169/1/012024

The DATAQ instrument [16] presented in Figure 3 is connected to our PXI via USB, and all the information is stored in a SQL database with timestamps. For each sensor (temperatures, voltage, current intensity), we take a reading each second. There is no need for a greater speed, as the thermocouples do not react instantly to the changes in temperature. Readings in voltage and current are only for reference as the most critical parameter, after the pressure, in a vacuum evaporation deposition system, is the temperature of the boat, but still plays an essential part in replicating the same technique and for validating that the system works correctly.

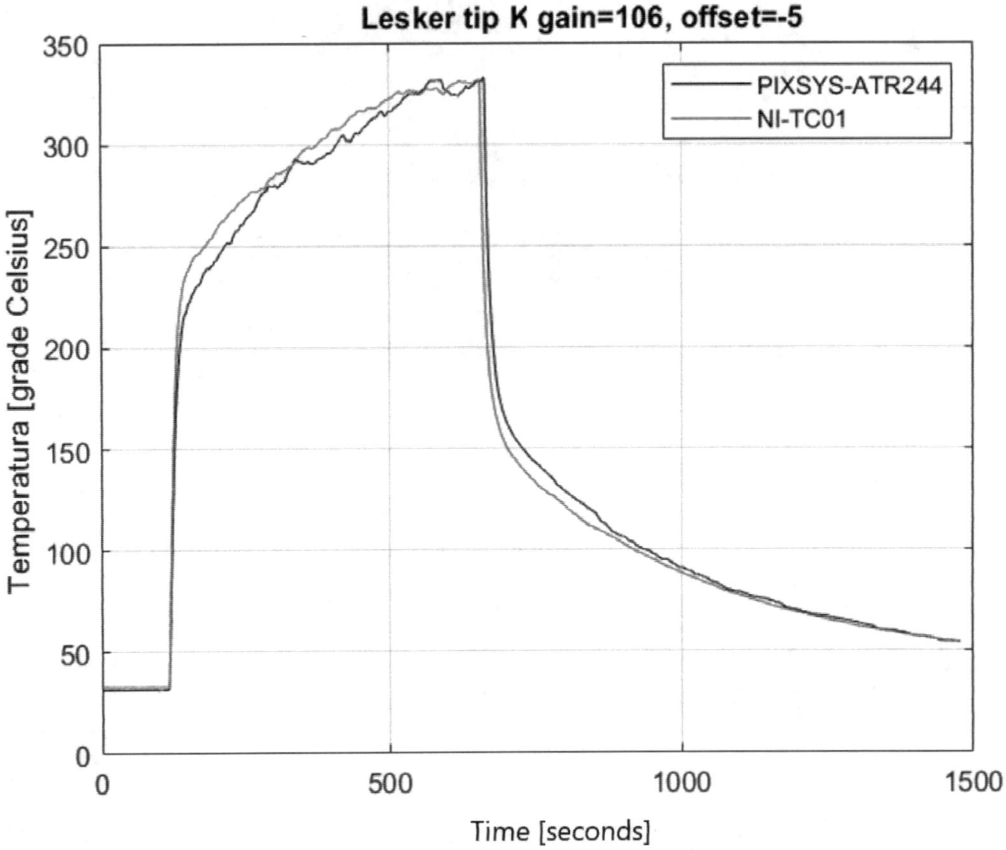

Figure 4. Calibration of type K thermocouple. Black (PIXSIS) – our device; Red (NI) – control device

3. Results and interpretation

Figure 5 presents a test made with both thermocouples R and K in a vacuum environment. The boat was not heated to the operating temperature, but a clear result was obtained. The sensors are working correctly, and as was expected, the R thermocouple is more dynamic in registering temperatures, as its wires are thinner and connection junction is smaller.

Of course, slight differences also appear because the sensors were placed on opposite ends of the boat, and the connection was not perfect.

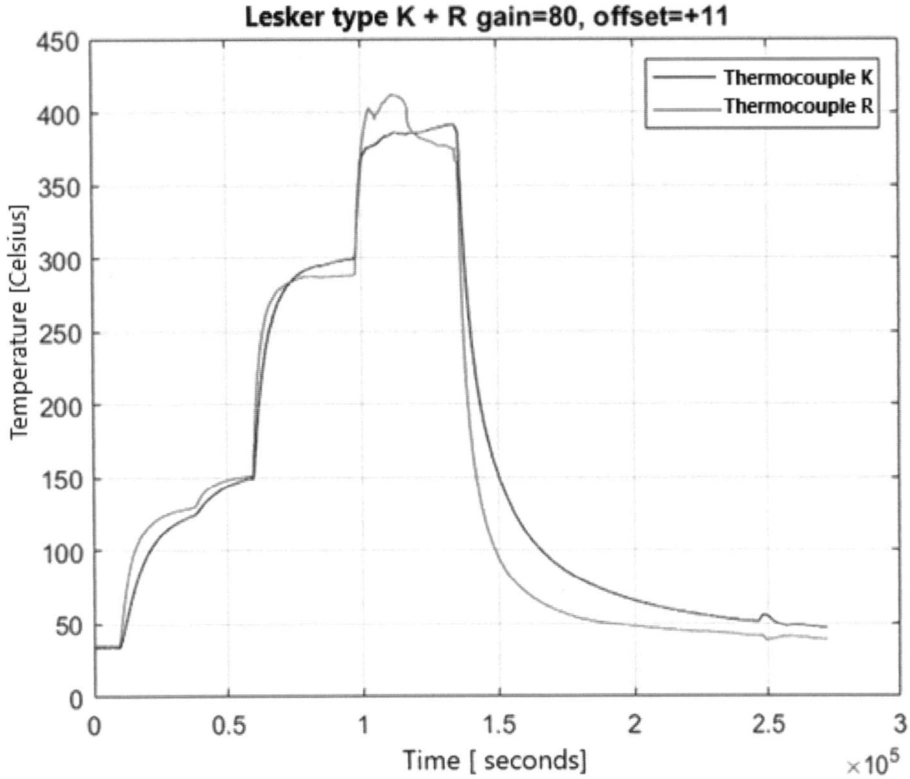

Figure 5. Testing both thermocouples K and R on the boat

4. Conclusions and further work

Given that this is an experimental setup, and optimizations are needed, the results are encouraging, and the data collection system is now complete. Furthermore, it can be translated or replicated in other systems. Since this is a developed system, we now have the know-how and skill-sets to implement such a complex data collection system in similar environments.

A correlation between all parameters will be made to better understand the interdependency between pressure, temperature, voltage and current, but that will be the subject of future work. Optimizations will be made on the system based on these measurements. Also, an electronic adjustment system for the current that is supplied to the boat will be installed in order to have more precise adjustments and, in the end, to have a system that auto-corrects itself and keeps either the temperature constant, either the voltage or current.

References

[1] Moga, I., Ghincu, R., Costea, T., Tarca, D., and Moldovan, O. A. (2018). Some experiments regarding magnetron sputtering deposition with small capacity devices. *Nonconventional Technologies Review* 22(4), Romanian Association of Nonconventional Technologies, Dec 2018

[2] Dan-Ioan ȚARCĂ, Moldovan Octavian ALIN, Remus GHINCU, Moldovan Ovidiu GHEORGHE, Traian COSTEA, Polojințef Corbu NICOLAE, Installing a high capacity sputtering system at University of Oradea, *Electrical Engineering and mechatronics conference EEMC'19, University of Debrecen, Faculty of Engineering,* May 2019

[3] D I Țarcă, L Csokmai, O Moldovan, R Ghincu and D Potroviță, Experimental data collection system for reading pressure levels in a vacuum environment, *IOP Conf. Ser.: Mater. Sci. Eng.,*

2019, DOI: 10.1088/1757-899X/568/1/012078
[4] D I Țarcă, T. Costea, I. Moga, Custom vacuum load-lock for the sputtering vacuum chamber, loading/unloading and locking automatisation mechanisms, *IOP Conf. Ser.: Mater. Sci. Eng.* ***898***, 2020, doi:10.1088/1757-899X/898/1/012046
[5] https://erris.gov.ro/Research-Laboratory-for-Adva-1
[6] K. Jousten, Handbook of Vacuum Technology, Second, Completely Revised and Updated Edition, Wiley-VCH, 2016.
[7] J. Moore, C. Davis, M. Coplan și S. Greer, Building Scientific Apparatus, Fourth Edition ed., Cambridge: Cambridge University Press, 2009.
[8] N. Yoshimura, Vacuum Technology Practice for Scientific Instruments, Berlin: Springer, 2008.
[9] AGARWAL, Reema, HIMANSHU, PATEL, S.L., VERMA, M., CHANDER, S., AMETA, C. and DHAKA, M.S.. Tailoring the physical properties of titania thin films with post deposition air and vacuum annealing. Optical Materials. 2021. Vol. 116, p. 111033. DOI 10.1016/j.optmat.2021.111033.
[10] GHOSH, Joydip and GIRI, P.K.. Vacuum deposited PbI2 film grown at elevated temperatures for improved efficiency of CH3NH3PbI3 based planar perovskite solar cells. Materials Research Bulletin. 2021. Vol. 139, p. 111255. DOI 10.1016/j.materresbull.2021.111255.
[11] LI, Zhiming, QIAN, Shiqiang and WANG, Wei. Characterization and oxidation behavior of NiCoCrAlY coating fabricated by electrophoretic deposition and vacuum heat treatment. Applied Surface Science. 2011. Vol. 257, no. 10p. 4616–4620. DOI 10.1016/j.apsusc.2010.12.097.
[12] https://www.pixsys.net/en/products/pid-controllers/atr244 Pixsys Electronics, ATR244 Controller user manual
[13] https://www.thermocoupleinfo.com/type-k-thermocouple.htm REOTEMP Instrument Corporation, Type K thermocouple reference table
[14] https://www.thermocoupleinfo.com/type-r-thermocouple.htm REOTEMP Instrument Corporation ,Type R thermocouple reference table
[15] https://www.lumel.com.pl/resources/Pliki%20do%20pobrania/N30P/N30P_service_manual_rev_B.pdf Lumel SA, Digital Panel Meter N30P user manual
[16] https://www.dataq.com/products/di-1100/ DataQ Instruments, Data Acquisition USB DAQ DI-1100

IOP Conf. Series: Materials Science and Engineering 1169 (2021) 012025 doi:10.1088/1757-899X/1169/1/012025

Agile, waterfall and iterative approach in information technology projects

C Fagarasan[1], O Popa[1], A Pisla[1] and C Cristea[2]

[1] Design engineering & Robotics Department, Machine Building Faculty, Technical University of Cluj-Napoca, 103-105 Muncii Avenue, Cluj-Napoca, Romania
[2] Electrical Machines and Drives Department, Faculty of Electrical Engineering, Technical University of Cluj-Napoca, 26-28 G. Baritiu Street, Cluj-Napoca, Romania

[1]office.cristif@gmail.com

Abstract. The aim of this paper is to analyse the Agile methods implementation capability within software projects as a more viable alternative to the waterfall model and determine the key factors that guarantee the continuous delivery of software to market at the right and agreed quality level. Two Agile implementation methods were considered: Scrum and Kanban, which were evaluated based on factors like predictability, practicability and complexity. The conclusions show the selected methodologies advantages and disadvantages, and project key attributes can be used to choose the suitable methodology for a project. Regarding the Agile implementation methods, the most straightforward and predictable one was proven to be Scrum. This framework supports frequent delivery of software products that can address complex problems and adapt to the rapid market changes.

1. Introduction

The Information Technology (I.T.) industry is continually growing and expanding, forcing the people who work in this domain to enhance their workflow and improve the software delivery methods, even those considered to be well-known within the industry constantly adapting through learning.

In conformity with the report, edited by the United States Bureau of Labor Statistics [1], the software industry is one of the largest in the world economy in 2019, displaying the most significant increase anticipating that until 2022 will perform a 20% increase in the industry working places.

Considering the I.T. industry in Romania, there are about 17.000 active companies with over 100.000 registered employees and an annual income of over 4.8 billion Euros, as is presented in a study run by the Romanian Association for Electronic Industry and Software (ARIES) [2]. The analysis presented in the paper is interested in the development of I.T. processes and sustainability of the software companies, starting from now when this industry is in constant growth. It is essential to achieve a certain level of standardisation at the organisational level that enables to align the software products delivery process at the desired quality, and in the meanwhile to be able to measure the performance of each version that is accepted to be directed to the production teams. In this sense, various measuring instruments and methodologies like Life Cycle Management Index Tool (LY-MIT) [3] are developed.

It is accepted that complex software projects may take years to complete, so the tendency in the last twenty years was to migrate from a waterfall mindset, which means to deliver a working product at the end of the project, to an Agile mindset which implies to provide small increments of working software through time-boxed iterations. In [4], it is stated that the waterfall method is suitable for a predictable environment, whereas Agile works better in a change-driven climate, given the fact that the

IOP Conf. Series: Materials Science and Engineering 1169 (2021) 012025 doi:10.1088/1757-899X/1169/1/012025

software projects are continually evolving and the final shape of the product is practically far from clear in the beginning. Agile seems a rational option when talking about the implementation of software projects.

In [5], it is presented how the Lean methodology inspired the Agile methodology creation. It is argued that through this method, customer value is created by eliminating waste, amplifying learning, empowering the team, delivering fast and optimising the product as a whole. A known implementation method used in I.T. that derived from Lean is Kanban, a Japanese system that helps to visualise the workflow, maximise efficiency and continually improve. Another well-known methodology is the Agile implementation methods, detailed in [6], where the terminology and processes belonging to each one is described. The methodologies that organisations use and which are worldwide spread are Scrum, Kanban and Extreme Programming.

Starting from Peter Drucker's saying, "What gets measured, gets managed" [7], when a new software project is initiated, the project team, together with the project manager, choose between different methodologies available, deciding about the one that will be used for the project implementation. This happens during the discovery phase that is part of the project planning stage.

After the methodology is selected, the team might explore the various implementation methods frequently used to develop software projects and choose the best one to fulfil the project's needs [8]. Understanding the complexity of the future project and the environment in which the project will be situated will help the project manager select the potential implementation methods that can deliver the project's product. As mentioned before, this is done during the planning stage. However, there are a total of five stages through which each project will find itself in a moment in time: initiating, planning, executing, monitoring and controlling, and closing, according to [8].

These stages are usually included in traditional projects, and only when the objectives of a phase are achieved, the next one is started.

When an Agile methodology or iterative implementation drives the projects, most of the stages are repeated in each iteration. An advantage in applying this method is that the project's scope can be adapted during the implementation phase based on early feedback from customers or end-users. If the made assumptions during the requirements specification stage (identification stage) prove faulty, the adaptive implementation methods introduce flexibility into the delivery model, enabling the development teams to reprioritise and introduce changes that can increase the final product's value. Depending on the used framework, a different number of rules and principles will have to be applied, in correspondence and proportional with the project's agility level.

This paper aims to compare different methodologies used to deliver information technology projects. Moreover, based on [5], [9], and [10], Scrum and Kanban frameworks have been analysed with an emphasis on the differences between the frameworks and the key points that can help software development teams to select the best implementation method for their projects by leveraging the foreseen advantages and disadvantages.

2. Agile and Waterfall methodologies

Agile software implementation methods become three times more prevalent than Waterfall. In the last twenty years, statistics show that they have three times as many successful outcomes. That does not mean Waterfall is outdated, but particular projects may call for different solutions. Alternative methodologies need to be well-understood to make the right choice, so an appropriate decision is taken when starting a software project. The Waterfall model was named from the analogy to the finish-to-start relationship that exists between the project phases. Each stage involves different skills and has a formal acceptance and approval at the end. Figure 1 presents the Waterfall software development lifecycle.

When talking about Agile methodology, the product scope is divided into smaller shippable increments delivered at the end of each iteration. As presented in Figure 2, the phases are adapted at a smaller scale in the Agile methodology and are executed for each product increment. The increments are then split over iterations and then implemented. At the end of each iteration, will be delivered a functional product increment. Agile software teams must possess all the skills necessary to produce a working product at the end of each sprint, a term used in Scrum, and it refers to one iteration.

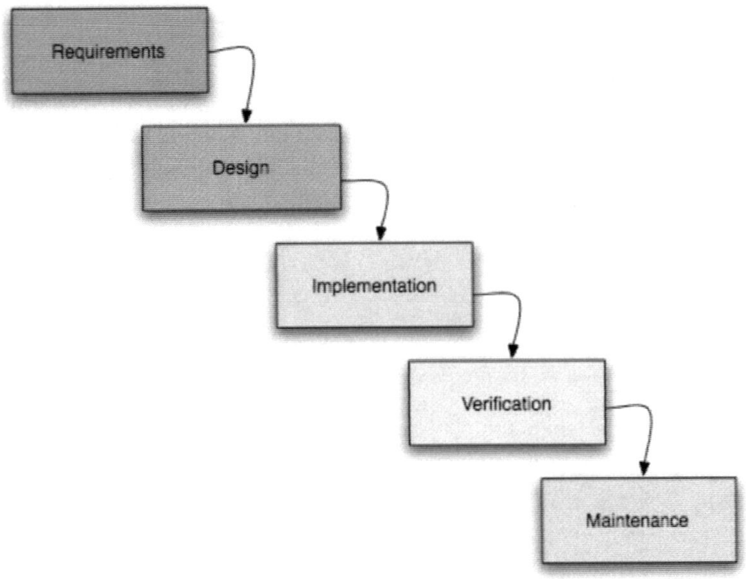

Figure 1. Waterfall software development lifecycle

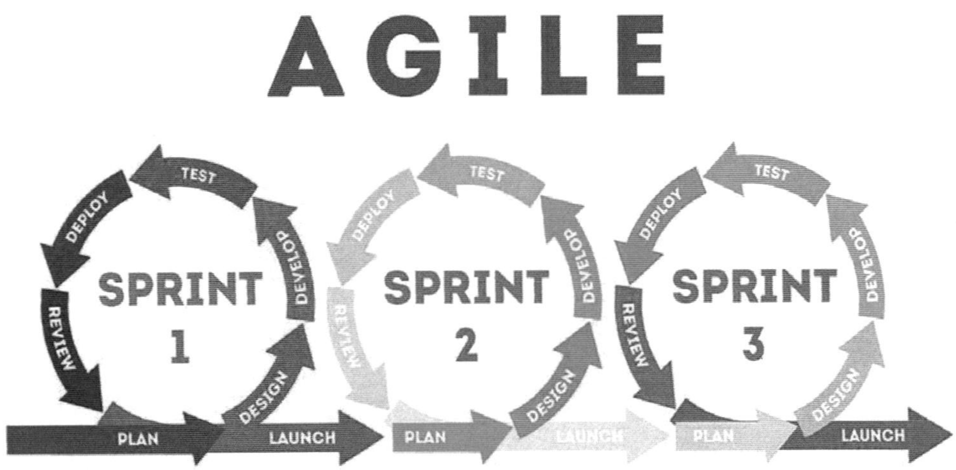

Figure 2. Iterative software development in the Agile methodology

In Table 1, some of the key metrics were taken as an example to explain the differences between these two software implementation methodologies. Figure 3 shows how using the iterative implementation reduce the overall project risk.

The metrics are related to the project management triangle: scope, time and cost. Other factors like time-to-market and return of investment (ROI), project risk, flexibility focus, and adaptability to change were considered. By following these metrics, the differences between Agile methodology and Waterfall methodology are more evident and can help make an informed decision when starting a new software development project. Choosing the suitable methodology for a project is crucial for its success.

The metrics presented can be used as a checklist in the project initiating phase. They can help the project managers understand the advantages and disadvantages of the analysed methodologies.

IOP Conf. Series: Materials Science and Engineering 1169 (2021) 012025 doi:10.1088/1757-899X/1169/1/012025

Table 1. Comparison between the Waterfall and Agile methodologies

Metric	Waterfall methodology	Agile methodology
Time spent on planning the project	Long. Scope and processes need to be defined before starting the actual product development	Short. There is limited time spent on planning before the first iteration is started
Time needed to get the implementation started	Long. The initiating and planning stages are formal and take a lot of time to complete	Short. Agile values customer collaboration over contract negotiation
Development direction	Plan-driven	Change-driven
Built-in flexibility to discover problems	None	Possibility to adapt during and after each iteration
Built-in flexibility to discover problems	None	Possibility to adapt during and after each iteration
Delivery time for a usable product	Long, at the end of the project	Short, at the end of each iteration
Time-to-market and Return of Investment	Fixed, at the end of the project	Adjustable, early increments might start bringing revenue
Project Implementation Cost	Higher due to fixed-price contracts, which can include premium risks	Time and Materials contracts usually offer more flexibility and can be value-dependent
Requirements	They are defined at the beginning of the project. These are unlikely to change during the implementation phase	Likely to change during the implementation phase based on end-user feedback
Scope changes	There is a lengthy process to follow if scope changes are introduced during execution	Scope changes are welcomed
Project delivery	One delivery, at the end of the project	Incremental deliveries, at the end of each iteration
Project Risk	High at the end of the project since this is the point when the product is delivered to the market.	Low at the end of the project because a shippable increment is delivered with each iteration.
Project Focus	Processes	End-user value
Collaboration between the customer and product development teams	Low	High, they can be part of the project team
Communication modes	With the Project Manager	With the product development team
Customer-vendor relationship	Contract negotiation before anything else	Customer collaboration before anything else

In the Waterfall methodology, the final product is delivered to the customer at the end of the project. The product development team does not receive any feedback during the implementation stage, and the focus is only on the requirements established in the planning stage. In contrast, in Agile methodology, the customer receives a working product increment after each iteration, and the customers' feedback can be implemented in the subsequent iterations. This is the main reason why most software organisations prefer the Agile methodology for software products development.

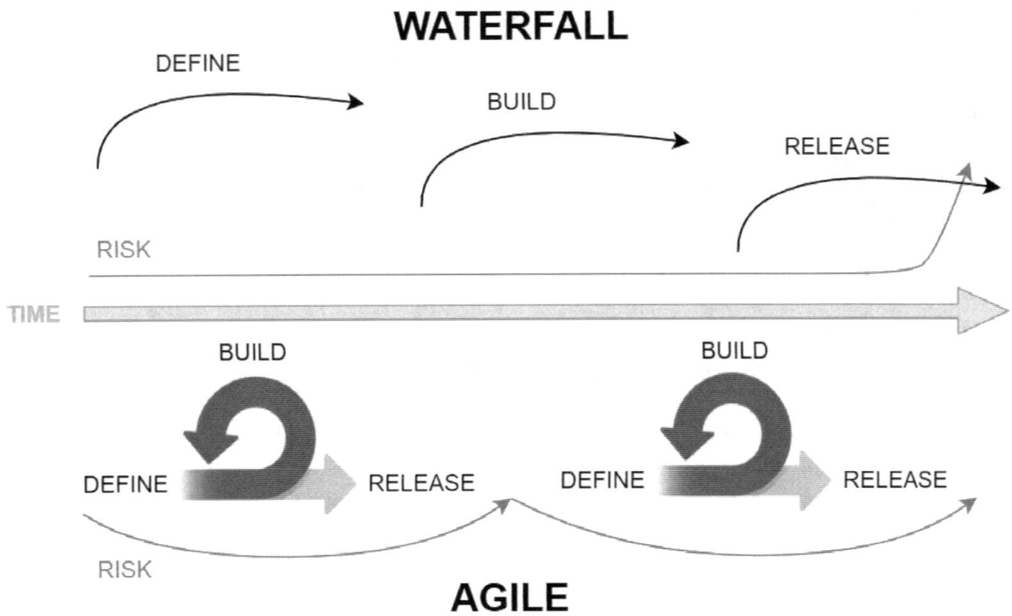

Figure 3. The project risk in Agile and Waterfall methodologies

As shown in Figure 3, by delivering software projects using the Waterfall methodology, the risk is high at the end of the project when the delivery of the project's outcome happens. During the project, the team focuses on implementing the requirements established during the planning phase. The risk is lower at the end of the project when using the Agile methodology because a shippable increment is delivered with each iteration. The customers' feedback can be implemented in subsequent iterations. However, some key project attributes can be evaluated to determine how they comply with the two methodologies presented in Table 2.

Table 2. Project attributes in the Waterfall and Agile methodologies

Project Attribute	Waterfall methodology	Agile methodology
Magnitude	Small or Medium-sized projects.	Any size projects.
Requirements	Clear from the beginning of the project, no room for change.	Not clear, they will become clearer after analysing the feedback received on the implemented increments.
Scope	Fixed scope, mentioned in the contract.	Likely to change during the implementation of the software product.
Budget	Fixed-budget mentioned in the contract. It has to be approved before the project is started.	The project can be started if a rough order of magnitude budget exists.
Delivery	The product has to be delivered in one piece.	The product can be delivered through iterations
Success	It is determined by the project plan implementation.	It is determined through customer feedback and end-user value.

3. Software projects development lifecycle – in Agile methodology

Figure 4 presents the popular Agile iterative development lifecycle that has been adapted for the software industry, given the flexibility needed during the software projects' implementation stage. Every organisation that is active within the I.T. industry can have specific processes and procedures. Generally, the same stages are present in the Agile methodology project lifecycle, as displayed in Figure 4. The project initiation phase, also known as the discovery phase, happens before starting the project. Usually, the team that will complete this stage is not the same as the one that will implement the project. Software development organisations have a separate department, called the "Presales department", which handles all business initiatives.

After the presales department receives a request for the proposal, a small team of experts is gathered for a short time to elicit the requirements, estimate the effort, and negotiate a contract. The period dedicated to this activity may vary for different reasons, but it is usually between two to four weeks.

After a contract is signed, a project manager is assigned to the project, and the planning stage is started. The first iteration is light in terms of scope because, during this iteration, the team will begin to set up the project, refine the requirements for the next iteration and build the backlog. Once these elements are completed, the development work can be started. In the meantime, the project manager will secure the necessary allocations, gather the project team and starts with the next iteration. The product development team should have all the skills required to deliver the first product increment.

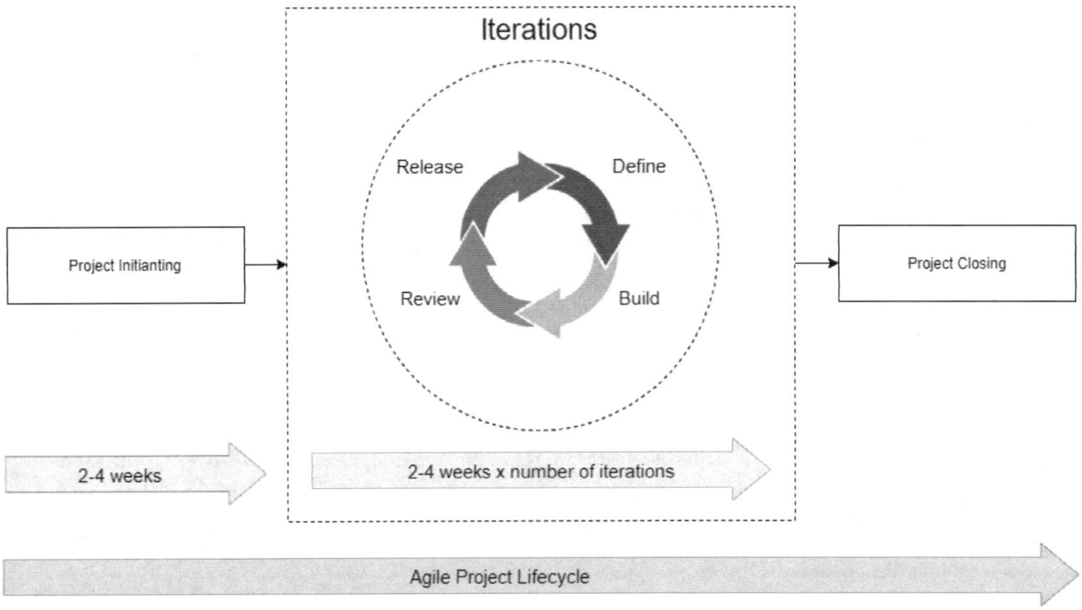

Figure 4. Agile software development lifecycle

The cycles Define, Build, Review, and Release will continue until all product increments are delivered and the scope of the project has been completed. The project will find itself in the Project Closing stage, where the attention is focused on the formal acceptance and the reallocation of team members to other projects. However, even if the review and retrospective sessions are organised within each iteration, it is recommended to schedule a "lessons' learned meeting" in which the product development team discusses the project challenges and how they were overcome.

4. Agile implementation methods: comparison between Scrum and Kanban

As it was mentioned, no implementation method will guarantee the project success. These are frameworks that can support the product development team, and their tools can be leveraged to perform at the highest efficiency. Both of the considered methods have their roots in Agile methodology, and they focus on delivering frequently and continuous improvement [12]. Their best practices are similar; the total scope is divided into small manageable pieces and delivered. Incrementally, a synthetic comparison between them is depicted in Table 3.

Table 3. Comparison between Kanban and Scrum

Kanban	Scrum
There are no pre-defined project roles, and the project manager is optional.	All roles are pre-defined, and a Scrum Master role is required.
The project tasks are shared by the team as a whole.	The project tasks have dedicated owners as part of the product development team.
Timelines are not fixed; they evolve on a need basis, making it difficult to predict a product release.	Timelines are time-boxed into sprints.
Changes can be made at the customers' discretion.	Changes can only be made when a new sprint starts.
Cannot guarantee the meeting of deadlines due to the lack of predictability.	It has a more organised structure with defined procedures that can keep the product development on track.
Focused on changing expectations of the client, the team can adapt during the product iterations.	It is focused on customer-driven development and shipping value at the end of each iteration.
Project priorities might change daily.	Project priorities might change with every iteration.
It works better for large and distributed teams.	A scrum team should have a maximum of 9 members, and they should be collocated.
The project work can start immediately.	A Scrum master is needed to coach the team in order to make sure that the Scrum framework is understood and implemented.
Used for small pieces of work, defect fixes or enhancement requests.	It is used for feature development tasks with clear goals.
It is concentrated on work items.	It is concentrated on iterations.
The productivity is measured through cycle time.	The productivity is measured through team velocity.
The project team is specialised, and its team members are specialised human resources.	The project team is cross-functional and has all the skills to deliver a working increment.

It may seem that the Waterfall methodology can deliver the final product of the project faster, specifications of the project being detailed before the executing phase starts. In contrast, in Agile methodology, there is a tendency to go in a spiral, given the high number of changes that can appear. Despite that, the product starts bringing revenue earlier by implementing Agile methodology due to the significantly reduced time-to-market. Even if it has a minimum set of features, the product brings value to the customer in the beginnings. It is developed in parallel with end-user feedback, which can be incorporated in future product increments.

According to the statistics presented in the Annual State of Agile Report [13], developed by Digital.ai, 58% of the software projects that were implemented by using the Agile methodologies chose Scrum as an implementation method. In contrast, only 7% of the total software projects were implemented using the pure Kanban development method, as shown in Figure 6. Nevertheless, this statistic should not influence a project managers' decision when choosing the suitable implementation method.

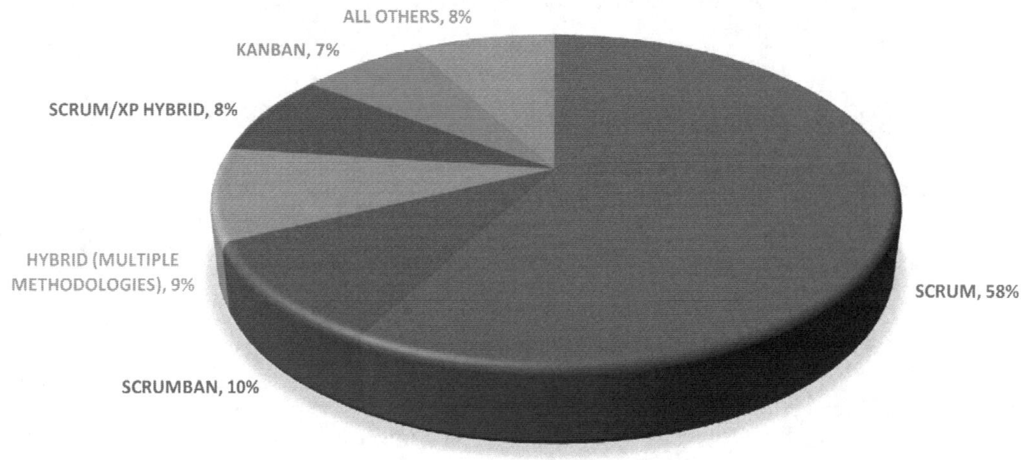

Figure 5. Agile methodologies used for implementing software projects

5. Case study: delivering a small-sized project by using Scrum

This paragraph presents the use of the Scrum methodology applied to deliver a small-sized project with a three-month estimated timeline case study. There are several steps to follow in order to reach a project timeline. Firstly, the client's requirements need to be clarified to ensure that the scope is understood and the questions that the product development team raised are entirely answered.

Secondly, a breakdown is created, containing all the features and functionalities that the customer wants to include in the product's first version. Those are negotiated until the team and the customer reaches an agreement on the scope, timeline and budget allocated for the first delivery. These features are estimated, and after considering the relationships and dependencies between the project activities and project team, a high-level timeline is created, as shown in Figure 6. This timeline is used as a baseline, together with the breakdown structure that contains detailed information about the features, documented risks, assumptions and constraints.

The dependencies play a crucial role when creating a new project roadmap. They need to be assed, and any risks or issues should be flagged and documented in a registry so they can be tracked.

In this case study, two-week iterations (sprints) were considered. These are short periods, which are time-boxed, during which a product increment is created [11]. As presented in Figure 6, the first delivery is scheduled at the end of Sprint 2 when the first product increment is deployed on a User Acceptance Testing environment (UAT) and presented to the customer to decide if the product is ready to be released to a broader audience or not. Even if an increment is delivered at the end of each sprint, this does not mean that the product will be made available to the public by deploying it in a production environment. This decision lies with the Product Owner, which usually is the customer or a customers' representative.

The deliverable of Sprint 1 is a backlog of work items that can be picked up in Sprint 2 by the product development team and the project setup activities, which enable the team to start the work on the project. During the discovery phase and Sprint 1, the project manager will ensure that all needed project physical resources and project allocations are secured so that the work can be started immediately. At the end of the project roadmap, there is a formal acceptance phase for the product as a

IOP Conf. Series: Materials Science and Engineering 1169 (2021) 012025 doi:10.1088/1757-899X/1169/1/012025

whole. This usually happens in the User Acceptance Testing environment, and after the sign-off is received, a final deployment of the product in the production environment is made. Usually, after this point, the project goes into a maintenance phase.

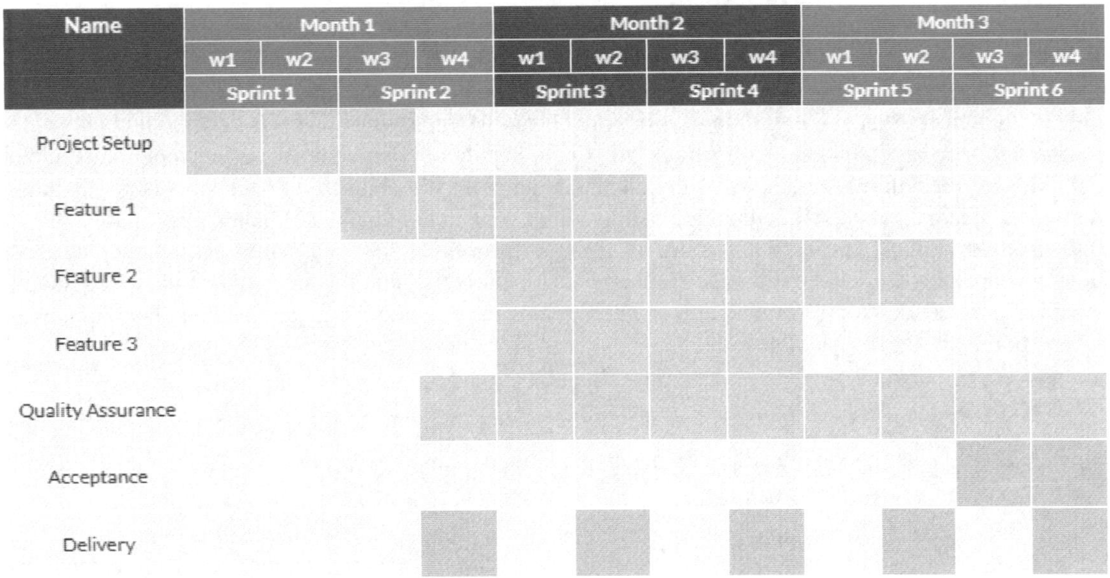

Figure 6. High-level timeline for an Agile project delivery

6. Conclusions

There can be as many delivery methods as project types, which are called hybrid. It is known that the applied methodology can help deliver a project, but depending on the projects' specificity, the delivery model can be adapted to work for each specific project. The frameworks and methodologies available are seen as guidelines that can be used as tools by the project managers and product development teams to deliver the project's desired outcome.

Regarding the Agile methodology implementation methods and the Agile Toolkits [14], Scrum has proven to offer a stable framework that product development teams can quickly adapt if a Scrum Master exists to coach and offer support during the implementation process. It has been established that Scrum can offer predictability while leveraging flexibility regarding scope changes during the project executing stage. There are hundreds of projects that are implemented through hybrid models, influenced by both Scrum and Kanban. The project management software systems have incorporated these methodologies, and they can help get any project started. The templates provided can be adjusted by learning after several iterations. Regardless of the selected implementation method, the product development team should stick with it for several iterations. The implementation process should be adapted at the teams' discretion through the inspect and the adapt sessions. The Agile methodology is a practical implementation approach that offers efficacious instruments which can improve the performance of any software organisation's delivery model. Its principles can be adopted within the Waterfall projects. By leveraging the iterative implementation concept, the phases of a project can be delivered incrementally rather than having just one delivery at the end of each stage.

A typical hybrid model encountered within the information and technology industry is when the initiating and planning phases are implemented using the Waterfall approach. However, the implementation activities are done through the iterative delivery model, which Agile promotes. There are two ways in which it can be used and adopted at an organisational level so that the delivery practices can be improved:

- For junior organisations, which is new within the software delivery market, it can be quickly implemented by following the methodology's principles and adopting the Agile mindset at the organisation level to define the delivery model.
- For existing corporations, which already have a determined delivery model, a more hybrid model is recommended to blend the traditional approaches that made the organisation successful to date with the Agile principles, enhancing the value of the delivered software products.

Although the Agile methodology started to become the standard methodology in software projects implementation, several organisations are currently employing Waterfall methodology because it simply works, and it has a proven track record. Using this system, the managers feel comfortable knowing that everything is extensively documented and there is a centralised decision-making process that reduces human error. All companies will become software companies as technology advancement is focused on digitalising and automating all business processes. The performing companies need to have the capability to deliver software at a large scale. Currently, the digital age is at its turning point and it is recommended for the organizations that are using the existing Waterfall software delivery model to incorporate the Agile principles.

7. References

[1] U. S. Bureau of Statistics 2021 *Labor f. stat. f. the cur. pop. sur.* (https://www.bls.gov/cps/cpsaat18.htm)
[2] ARIES 2019 *Market survey*
[3] Niemann J and Pisla A 2018 *Sust. poten.and risks assess in autom. and robot. using the l. c. manag. ind. tool—LY-MIT* **10** 21 (https://doi.org/10.3390/su10124638)
[4] Kisielnicki J and Misiak A M 2017 *Effect. of A. comp. to W. implem. meth. in I.T. proj. anal. bas. on b. i. proj.* **3** 14
[5] Llahm O, Faraj B, Dalla and Llahm O B D 2020 *Lean soft. dev. prac. and princ. in ter. of obs. and evol. meth. to incr. work env. produc.* **7** 24
[6] El-Sseid A M, El-Sseid H and Llahm O B D 2020 *Scrum Dev. Proc.* **13** 31
[7] Prusak L 2010 *What can't be measured* (https://hbr.org/2010/10/what-cant-be-measured)
[8] PMI *A Guide to the Project Management Body of Knowledge*, 6th Edition
[9] Schwaber K and Sutherland J 2020 *The Scrum Guide* **3-7** 14
[10] Beck K, Beedle M, Bennekum A, Cockburn A, Cunningham W, Fowler M, Grenning J, Highsmith J, Hunt A, Jeffries R *et al* 2001 *Manifesto for Agile Software Development* (https://agilemanifesto.org/)
[11] Mahnic V and Drnovscek S 2005 *Agile soft. dev. with scrum* **1-4** 7
[12] Moniruzzaman A B M and Hossain S A 2013 *Comp. st. on a. soft. dev. meth.* **2-4** 15
[13] Digital.ai *The 14th Annual State of Agile Report* 2020 (https://stateofagile.com/#ufh-i-615706098-14th-annual-state-of-agile-report/7027494)
[14] Poppendieck M and Poppendieck T 2003 *Lean Software Development: An Agile Toolkit* (Addison-Wesley Professional) chapter 2 pp 15-45

IOP Conf. Series: Materials Science and Engineering 1169 (2021) 012026 doi:10.1088/1757-899X/1169/1/012026

Experimental study on the settings of Delta and Cartesian 3D printers for samples printing

M Ratiu[1], D M Anton[2] and D C Negrau[2]

[1] Department of Mechanical Engineering and Automotive, University of Oradea, 410087 Oradea, Romania
[2] Doctoral School of Engineering Sciences, University of Oradea, 410087 Oradea, Romania

E-mail: mratiu@uoradea.ro

Abstract. This paper presents the first part of the experimental procedures developed at the beginning of a larger research, consisting of printing, on two 3D printers, one parallel and one cartesian, of some test specimens, in different working conditions and with different settings of some parameters, and then investigation of the tensile and compressive behavior of the test specimens. The research was carried out on two sets of standardized samples, printed with three types of the commonly used and low-cost material, Polylactic acid (PLA), Carbon fiber PLA (CF-PLA), and Polyethylene terephthalate glycol (PETG). The purpose of the study is to create a database with the results obtained, which will be a working basis for future printings with these printer models, available in our laboratory, in order to understand the 3D-printed material behavior under certain available working conditions.

1. Introduction

In the last decades, due to multiple advantages and lower costs, the use of 3D printing in manufacturing processes is in a more and more quickly growing. At present, there are a large range of additive manufacturing technologies (AMT) or 3D printing processes, the most popular of them, together with the most common applications and materials, being presented in [1]. The most widely used 3D Printing technology, belonging to the category Material extrusion, of AMT, is considered Fused Deposition Modeling (FDM), or Fused Filament Fabrication (FFF), in special regarding its quickly and cost-effective possibility of producing plastic prototypes. But, this type of additive manufacturing technology is used, at present, not only to fabricate diverse and complex plastic prototypes but also to produce more and more final products, in different industrial sectors and in the most diverse areas.

The advantages of 3D printing over traditional manufacturing techniques (speed of fabrication, single-step manufacture, cost, risk mitigation, complexity and design freedom, customization, ease of access and sustainability), and the growth of the number of printers sold are exposed in [2]. Choosing the most suitable 3D printing process has to be made from different perspectives (like required material, characteristics of the end parts or certain process capabilities)as presented in [3]. If, initially, the most used 3D printers were that with a cartesian configuration, in recent years, Delta parallel printers have appeared and are widely used, considered more advantageous, from several points of view [4].

A large range of materials can be used for FDM printing, and other new materials appear as this technology becomes more and more widespread. The most popular and used is Polylactic Acid (PLA), along with many others as Acrylonitrile Butadiene Styrene (ABS), Polycarbonate (PC), Polyethylene Terephthalate (PET), Polyethylene Terephthalate Glycol (PET-G), Polyether Ether Ketone (PEEK) or High-Impact Polystyrene (HIPS), base materials or with various inserts, such as carbon fiber, to improve the properties of the filament. A comprehensive review on the characterization of the mechanical properties of the structures and materials used in the FFF process is presented in [5] considering an extensive number of experimental, computational and theoretical approaches.

With the increasing interest and use of 3D printing process, a lot of experimental research has been done, and numerous scientific articles have been published, each addressing, in a different manner, the 3D printing and testing of specific standardized test specimens mainly at traction and compression. An extensive state-of-the-art review regarding the optimization of the FDM additive manufacturing process in order to achieve the maximum tensile strength, through a comparative study with PLA, PC, PEEK and ABS is presented in [6].

Considering the results of the previous studies, as well as the available material capacities, we resorted to our own research, by using three types of filaments, various printing parameters and different working conditions, in order to establish the best 3D printing variants in our laboratory.The experimental study regarding the settings of Delta and Cartesian 3D printers, for printing two types of test specimens, is further presented in this paper. Their testing, on the tensile and compressive, and the results obtained, will be presented in another paper.

2. Technical specifications of 3D printers and software used

The two models of 3D printers, available in our laboratory, are part of the medium printing category, one being the FLSun QQ-S PRO Delta model, which guarantees speeds between 60-120mm/s on an area with a diameter of 260mm and a maximum height of 360mm, and secondly, the Cartesian model TEVO Tornado, which guarantees speeds between 60-150mm/s, with the printing platform of 300x300mm and height 400mm. These two printers have a 0.4mm extrusion nozzle or nozzle. The Delta printer has a much better print resolution than the Cartesian one, several upgrades on the controller side as well as motor drivers, which makes its motors more precise, but also much quieter during printing.

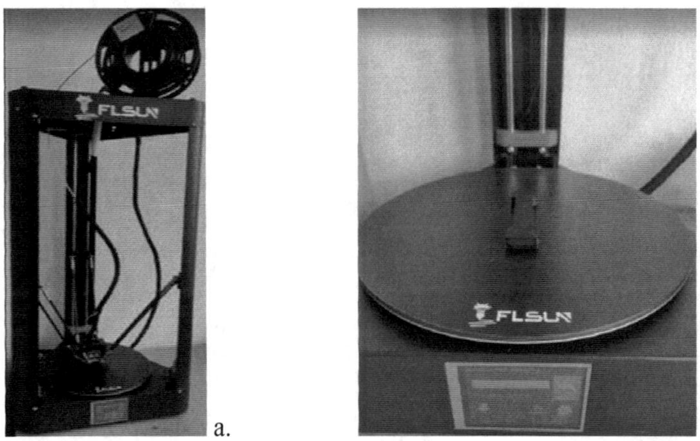

a. b.

Figure 1. Delta FLSun QQ-S PRO 3D printer and PLA sample.

The Delta FLSun QQ-S PRO 3D printer, shown in figure 1, during the printing of the sample, in large view in figure 1.a and with the sample printed in figure 1.b, allows you to select the speed during printing, increase or decrease the print speed, or increase or decrease the filament feed speed, if certain inconveniences occur, such as the inability to insert both filaments, as long as the nozzle allows its extrusion. The problem that may arise is with the filament insertion motor, the motor starting to jump steps. The controller also allows the temperature to rise or fall during printing, which allows us to increase the temperatures. The controller also allows us to raise or close the z axis at the beginning of printing to the bed on which it is printed, when the first layer is made, if the nozzle is too close to the bed, it can also rise 0.05 mm on the z axis whenever this is desired until a layer equal to the original layer in the printer settings is created. There is also the possibility to adjust the z-axis with the sensor in the printer, which can be easily mounted on the print head and to select the print bed height check setting in 21 points from the printer menu. This printer allows building layers of at least 0.05 mm, each such layer being created due to the trapezoidal screw and the motor that allows this printing resolution.

The Cartesian printer model Tevo Tornado, shown in figure 2, during the printing of the sample, in large view in figure 2.a and with the sample printed in figure 2.b, has the capacity of a layer thickness of 0.1mm; this resolution is given by the motor driver controllers, it can not be lower, only with an upgrade to TMC 2208 drivers that allow the movement of 256 microsteps/revolution . The controller of this printer does not allow us to change the temperature, speed or other details after printing, all remain exactly as in the program used.

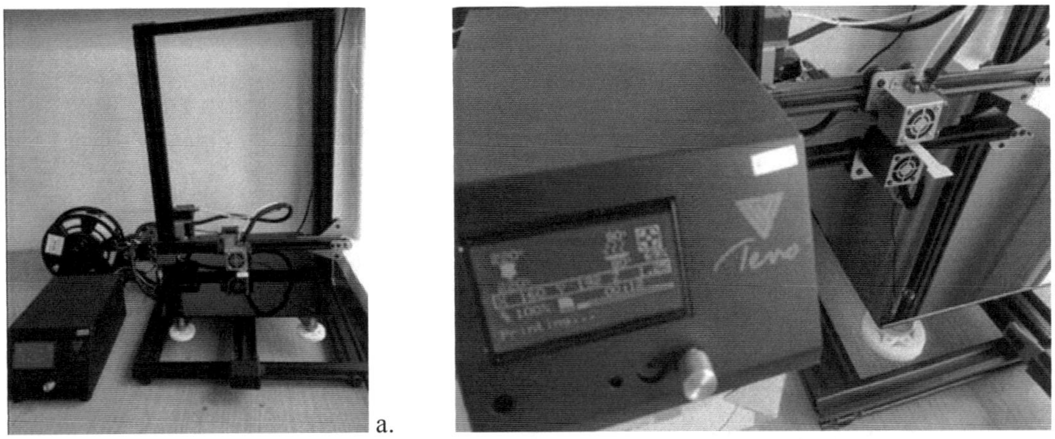
a. b.

Figure 2. Tevo Tornado Cartesian 3D Printer and PET-G sample.

For both printers, the gcode creation software Ultimaker Cura 4.7.0 is used, the version used for its very good statistics, in which most of the gcode generation problems or other printing problems have been solved.

3. 3D printing settings

The work profiles for these two printers were set over approximately 6 months of use, with these profiles using approximately 80% of its print capacity for the Delta printer and 60% of its capacity for the Cartesian printer. These limitations have been introduced because, for higher speeds, to both models, a box for keeping temperature must be used for the printed filament, as it contracts and lifting off the printing bed. This problem leads, at best, to raise the corners of the part by about 0.5 mm, or, in the worst case, to detach it completely. Problems with filament contractions occur when any airflow enters the printing area, the filament tending to often detach from the Cartesian printer because it has a glass bed. The good adhesion temperature of the PLA filament is at least 60°C, and for PLA with carbon insert and PET-G of 80°C.

Often the layers at the beginning come off, as presented in figure 3, even due to the low temperature of the bed, because they suffer a rapid contraction due to the ventilator mounted to cool them, the propagation of airflow becoming much more pronounced in areas of angles below 90°. This problem is often solved by raising the bed temperature, but it can also be solved by adding the "brim" function to the program. This creates successive layers in the first layer, and they attach the parts much better to the printing bed.

The filaments used for printing the samples are PLA, PET-G and PLA with 20% carbon fiber insertion. These filaments are from the printing area of these printers, so for the PLA filament the printing temperature is between 190-220°C, each increase in temperature by 5°C leading to an increase in speed by about 15 mm/s, all depending on the ambiental temperature in which are printed the samples.

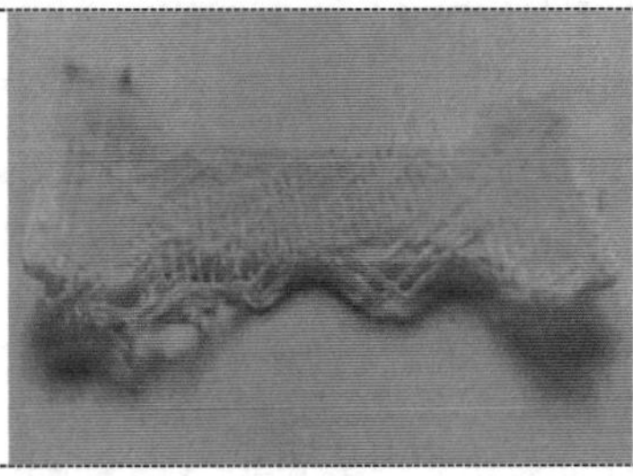

Figure 3. Parts detached from the printing bed.

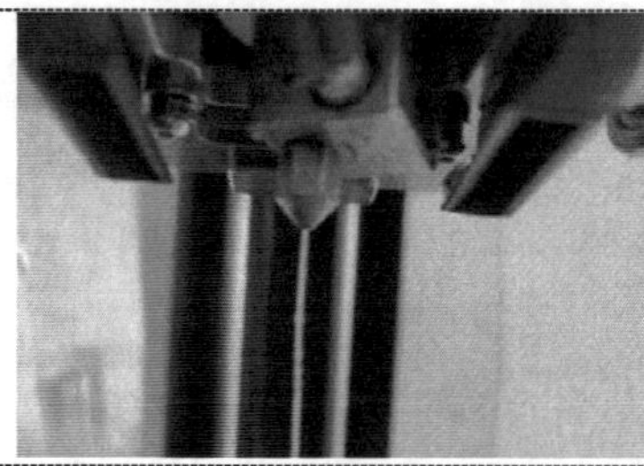

Figure 4. Lumpy on the filament.

In the settings for the Delta printer, with the use of PLA filament, a printing temperature of 225°C was reached, a set speed of 85mm/s for the initial samples, the printing temperature for bed of 60 °C, and for the first layer was set the temperature of 210°C, because the speeds at this layer are up to 24.5mm/s.

For the Cartesian printer, the settings are slightly different, in this case, the printing temperature was 220°C, the printing speed of 65 mm/s for all samples, the temperature for the printing bed of 60°C, and the speed at the first layer of 20 mm/s.

The adhesion of the filament on the printer bed is different for the first layer, because the Delta has a layer of matte foil with holes pierced on the glass surface, for very good adhesion and preventing lifting up during printing, and on Cartesian is printed directly on the heated glass.

Both printers have, for both the nozzle heating resistance and the bed heating resistance, the PID use functions, these helping to keep the temperature around ±0.3 °C, which is very important because it does not allow the production of very high temperature variations during printing. If it were a PWM type system, it, depending on the supply frequency, would act much too much above the set temperature, both during heating and when it is off and too cold. The same happens in the case of simple on-off hysteresis controllers, these being two-stage or on-off controllers, they have feedback that switches between two states, so there is no middle state, these controllers being systems with variable structure.

All samples were printed at an indoor temperature of 28°C, except for the last cube sample, sample 3 from PLA CF, which was printed at an indoor temperature of 25°C. It is very important that the print nozzle temperature to be stable, because any sources of air currents can cool the nozzle top from brass, copper, or steel, and may even cause the filament blocking in the extruder.

It was observed that the PET-G filament, after a period of printing time at an ambient temperature of 28°C, absorbs moisture from the air, so that when the filament melts, the moisture in the filament reaching the extruder block, reaches the boiling point as a result, when it is pushed, it explodes into small particles, and the printed filament is no more uniformly applied.

For printing the second type of test specimen, the 20x20x20 mm cube-type model, was chosen the printing of a single piece on the bed, in order to avoid the passage between the pieces and the appearance of filament threads on the outside of the piece, but also to avoid changing the settings while printing the piece. The printing settings were determined according to the printing capacity of the filament used, for each filament the temperature was changed, this being that differentiates all the pieces from each other, because the composition of the material for the three types of filaments differs greatly. Therefore, this second type of specimen was also printed using the temperatures specified above, in the case of printing the first type of specimen, the standard settings for all three materials being chosen.

The layer height for all the samples used was chosen at 0.2 mm, a height that is widely used for the good adhesion of the layers with the previous ones, and also very good for the fluidity of the filament, and for the faster printing time. For a height of 0.1 mm, time is doubled, which is not always to the advantage of the operator, and for a height of 0.3 mm it is already enough that the next layer does not adhere properly to the previous one, and, also, the resolution grows very much, visibly, the extruded filament sometimes generating filament bubbles due to the pressure exerted on the 0.4 mm nozzle, as appears in the figure 4. The problems of avoiding the bubbles that appeared in the printed piece can be solved by simply reducing the flow by 5%. If its decrease does not help, then it can be reduced by another 5% or even 10%.

When printing the cubes, 2 exterior walls were used for the first five samples and 5 exterior walls for the following ones. When determining the exterior walls, the function of printing 0.5 mm walls with a 0.4 mm nozzle can be used, as this function increases the flow of filament, creating a much deeper joint of the layers.

When choosing the parameters for these cubes, a filling percentage of 100% was established for two models. In the first case, using the line filling model, with the thickness of each printed line of 0.5 mm, it is obtained, for each layer, on the diagonal of the square, a passing model next to each other, this model being the only one that intertwines. The other models are different and are created according to the preset parameters in the software. For the second case, with the filling setting of 100% with the cubic subdivision model, simple division networks are created inside the part, these forming a network of equilateral triangles which, in 3D view, on the z axis, have an angle of 60° compared to the base.

In the next figures, two sets of test specimens, printed from different filaments and with various printing settings and working conditions, are presented. In figure 5 are presented samples that will be tested to tensile, and in figure 6 samples for compressive testing. Their experimental testing at the tensile and compressive, and the results obtained, will be done and presented in future work.

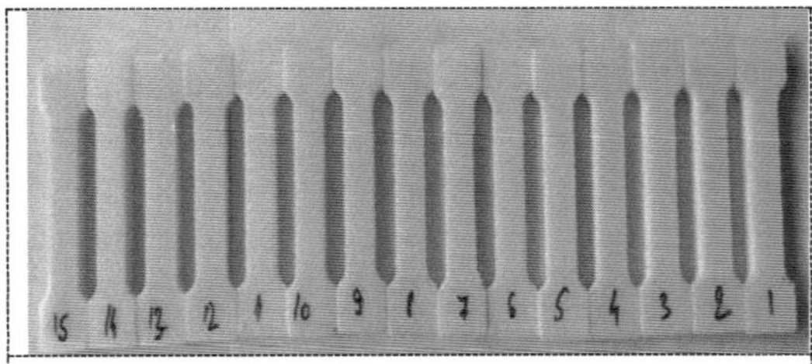

Figure 5. Samples printed for tensile testing.

Figure 6. Samples for compressive testing.

4. Conclusion

Considering the wide interest regarding the 3D printing process, the previous studies, as well as the available material capacities and the principle of the minimum costs, an experimental study on the finding the optimal settings of Delta and Cartesian 3D printers, for printing two types of test specimens, was developed and presented in this paper. The testing on the tensile/compressive strength of the printed samples will be done further and the results obtained will be exposed in a future paper. The purpose of the study is to create a database with the results obtained, which will be a working basis for future printings with these printer models, available in our laboratory, in order to understand the 3D-printed material behavior under certain available working conditions.

References

[1] Redwood B Additive manufacturingtechnologies: An overview https://www.3dhubs.com 4/7/2021
[2] Redwood B The advantages of 3D printing https://www.3dhubs.com 4/8/2021
[3] Varotsis A B Selecting the right 3D printingprocess https://www.3dhubs.com 4/8/2021
[4] Schmitt B M et all 2017 A comparative study of Cartesian and Delta 3D printers on producing PLA parts *Material Research* **20** (2)
[5] Cuan-Urquizo E et all 2019 Characterization of the mechanical properties of FFF structures and materials: A review on the experimental, computational and theoretical approaches *Materials* **12**, 895
[6] Gordelier T, Thies P R.Johanning L and Turner L 2019 Optimising the FDM additive manufacturing process to achieve maximum tensile strength *Rapid Prototyping Journal* 2019

IOP Conf. Series: Materials Science and Engineering 1169 (2021) 012027 doi:10.1088/1757-899X/1169/1/012027

Parametric study of air-cooled TEG heat exchanger design for waste heat recovery in heavy-duty vehicle

NV Burnete[1*], F Mariasiu[1], D Moldovanu[1], N Burnete[1], D Capata[1], B Jurchis[1]

[1]Technical University of Cluj-Napoca, Automotive Engineering and Transports Department, Bd. Muncii 103-105, Cluj-Napoca, Romania

E-mail: nicolae.vlad.burnete@auto.utcluj.ro

Abstract. This study focuses on the influence of different heat exchanger (HX) design parameters like the fin height, fin number, and fin thickness on the performance of an air-cooled exhaust gas thermoelectric generator (TEG). The TEG is considered to be installed on a heavy-duty truck traveling at 85 km/h on the highway thus ensuring the required cooling air for the cold side HX. It was found that the number of fins has the highest influence on the amount of extracted heat and power output of the system, but it also negatively affects the pressure drop. Increasing the fin height improves the pressure drop, but the performance increase is limited. Similarly, increasing the fin thickness provides only limited output improvements, but it also affects the pressure drop. Finally, compared to the hot side HX, the cold side HX has a limited influence on TEG output, but its design is not restricted by the pressure drop.

1. Introduction

Current environmental protection requirements are directly linked to the need for new solutions to reduce pollutant emissions caused by the transport sector. Of all possible ways to reduce pollutant emissions, the currently preferred solution is to employ electric motors and batteries for vehicle propulsion, but current technology still has a major disadvantage in terms of range [1]. However, the main energy source of vehicles (especially heavy-duty vehicles used in freight transport) is still the internal combustion engine (ICE) due to its characteristics and functional performances. If the immediate advantage of using ICEs is given by the range of the vehicles, the big and immediate disadvantages are the low thermal efficiency and the pollutant emission. Consequently, it is necessary to continue researching and developing solutions to increase the efficiency and reduce the environmental impact of ICEs. Since the hot exhaust gases expelled in the atmosphere contain approximately 30 to 40% of the thermal energy resulting from the combustion of the air-fuel mixture, it follows that there is a significant amount of wasted thermal energy which could be harnessed and converted into other types of energy used in the vehicle, thus improving its global energy efficiency [2]. Such a solution is the use of thermoelectric generators (TEG) to directly convert the otherwise wasted thermal energy into electrical energy. The operating principle of TEGs is based on the Seebeck effect which represents the electromotive force generated when a temperature gradient is applied across a thermocouple (TC). A TEG system used for ICE waste heat recovery (WHR) usually consists of several thermoelectric modules (TEMs), arranged in a series or parallel electrical configuration. These systems are usually installed in all ICE areas where heat losses occur, like the exhaust system [3], the exhaust gas recirculation (EGR) system [4], the radiator [5] and the oil pan [6]. Depending on the number and characteristics of the TEMs used to form the TEG, installation location, design etc., the power output values can vary significantly and, to the knowledge of the authors, has not exceeded approximately 1.5 kW. The TEG thermal to electrical energy conversion performance also depends on the temperature difference between the hot and the cold side (temperature gradient) and on the properties of the thermoelectric materials used to manufacture the TEM. Therefore, to achieve high power outputs, researchers usually prefer the combination of high-temperature TEMs with liquid cooling, but this solution has the disadvantage of a more complex and heavier construction. In contrast, this study focuses on the use of an air-cooled (due to vehicle travel) TEG installed on a heavy-duty vehicle traveling at highway speed. A Simulink TEG model (developed and validated in a previous effort of the authors [7]) is used to highlight the influence of the hot and cold side heat exchanger (HX) design parameters on the performance of the device.

2. Methodology

To highlight the influence of various heat exchanger parameters (fin height, fin number and fin thickness) on the performance of the TEG, a numerical study was performed using a model developed by the authors in a previous effort [7]. Compared to the initial model, a slight change has been done to the hot side heat exchanger to allow for the mass flow as an input value instead of the flow velocity. The HX basic design, material (aluminium) and other data remain unchanged. The TEG under study is considered to have a single row of TEMs (perpendicular to the flow direction), electrically connected in series, with the main characteristics given in table 1. With respect to the input data for the hot exhaust gas (mass flow of 0.25 kg/s and temperature of 250 °C) and cooling air (flow velocity of 23.6 m/s and temperature of 20 °C – which is equivalent to a heavy-duty truck traveling on the highway at a speed of 85 km/h and an outside temperature of 20 °C), these were determined from the research of Heber et al. [8]. For the parametric study, two situations were considered: a) the cold side HX design is fixed, while the hot side HX parameters vary according to the data presented in table 2 and b) the hot side HX design is fixed, while the cold side HX parameters vary according to the data presented in table 3.

Table 1. Datasheet of the TEM [9].

Type	GM250-127-28-10
Number of TCs	127
Peak power	28.3 W
Maximum operating temperature	Hot side – 250 °C Cold side – 175 °C

Table 2. Test cases for analysis of hot side heat exchanger design.

	Case 1	Case 2	*Case 3*	Case 4
Fin height (mm)	50	20	35	65
Fin number (-)	80	40	60	100
Fin thickness (mm)	1	0.5	2	3

Table 3. Test cases for analysis of cold side heat exchanger design.

	Case 1	Case 2	*Case 3*	Case 4
Fin height (mm)	60	70	85	100
Fin number (-)	200	150	250	300
Fin thickness (mm)	1	1.5	2	2.5

3. Results and discussions

The simulation results for the hot side HX parametric analysis (table 1) are presented in figures 1 to 10, while for the cold side HX they are presented in figures 11 to 16. In all figures, the naming is formed from the parameter name, the variable symbol (fin height – H, fin number – FN, and fin thickness – Ft) and the variable value (table 1 or 2). Figures 1 to 4 highlight the influence of both the fin height and load resistance ratio on various parameters like the TEG power output (Pel), hot (T_h) and cold (T_c) side temperatures of the TEMs, TEM voltage (V) and TEM current (I). Analysing the results, it was found that increasing the fin hight the TEG output improves. However, the improvement is not linear, and a limit appears to be reached for a fin height of 65 mm. An additional benefit of a larger fin height is a larger hydraulic diameter which leads to a lower pressure drop across the TEG. Considering that manufacturers usually set a limit of 10 mbar or lower for the pressure drop [8], that the TEG design should be lightweight and that the 65 mm fins provide only a slight performance improvement, the optimum fin height appears to be 50 mm. Furthermore, the taller fins also cause a reduction in the HX efficiency because the heat transfer area increase is not proportional with the amount of extracted heat.

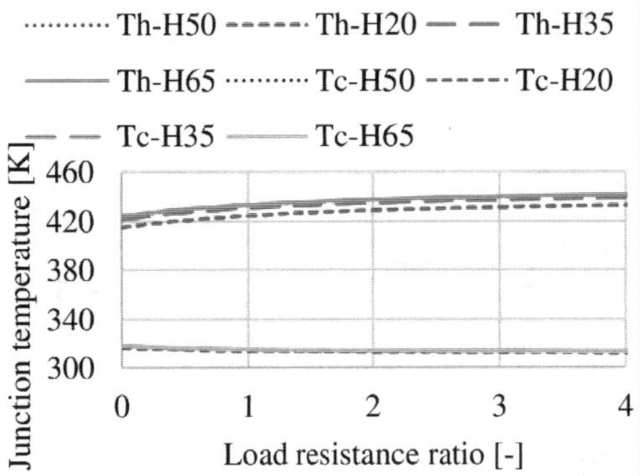

Figure 1. Influence of the fin height on the hot and cold side temperatures of the TEMs.

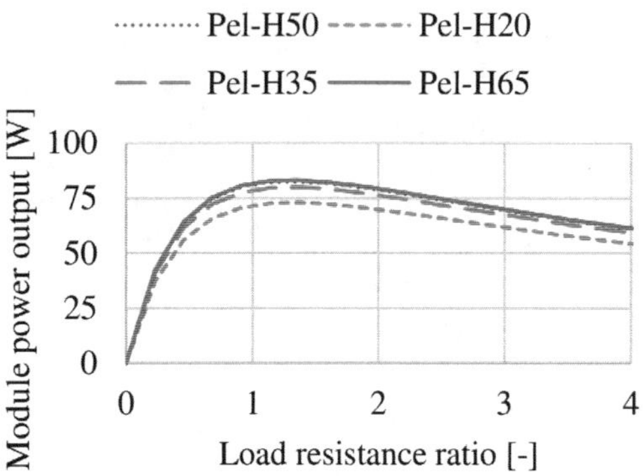

Figure 2. Influence of the fin height on the power output of the TEMs.

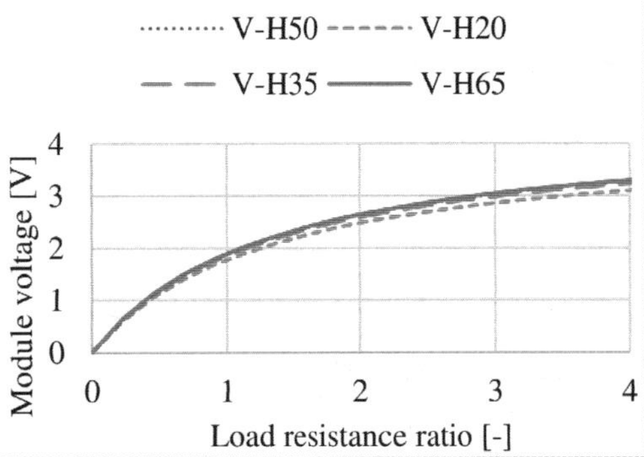

Figure 3. Influence of the fin height on the voltage of one TEM.

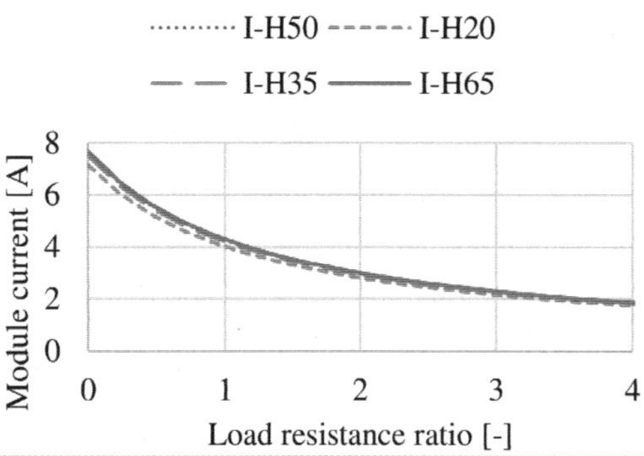

Figure 4. Influence of the fin height on the current of all TEMs.

Compared to the fin height, the number of fins shows a higher influence on the TEG performance. Increasing the fin number from 40 to 100 causes the power output to double, from 47.5 W to 95.4 W, but it also increases the pressure drop from 2.6 mbar to 17.3 mbar. A higher number of fins leads to a higher heat transfer surface, thus benefiting the amount of extracted heat, but a smaller fin spacing (since the HX diameter is fixed) which negatively affects the viscous drag and consequently the pressure drop. As a result, the optimum fin number was found to be 80. Similarly, the fin thickness also affects the heat transfer surface and the fin spacing. However, compared to the number of fins, increasing the fin thickness has a higher impact (negative) on fin spacing than on the area available for heat transfer. Consequently, even if the TEG performance increases with the fin thickness, so does the pressure drop (from 7.6 mbar for a 0.5 mm fin thickness to 32.1 mbar for a 3 mm fin thickness). Therefore, for the present study, the optimum fin thickness was considered 1 mm.

IOP Conf. Series: Materials Science and Engineering 1169 (2021) 012027 doi:10.1088/1757-899X/1169/1/012027

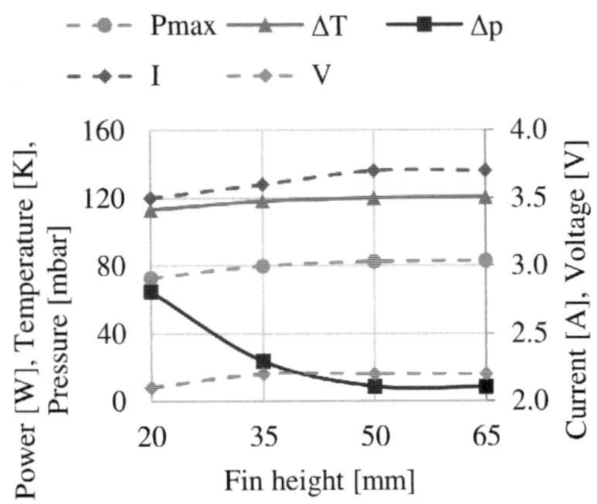

Figure 5. Influence of hot side HX fin height on TEG performance (power (P_{max}), temperature gradient (ΔT), pressure drop (Δp), and TEM current (I) and voltage (V)).

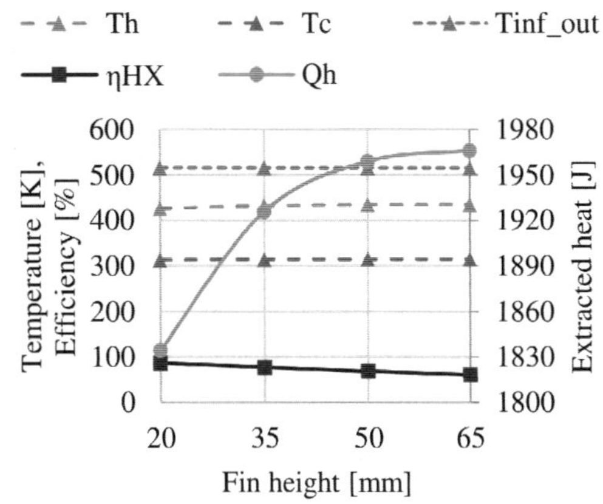

Figure 6. Influence of hot side HX fin height on hot (T_h) and cold (T_c) side temperatures, hot side outlet temperature (T_{inf_out}) and HX performance (extracted heat Q_h and efficiency η_{HX}).

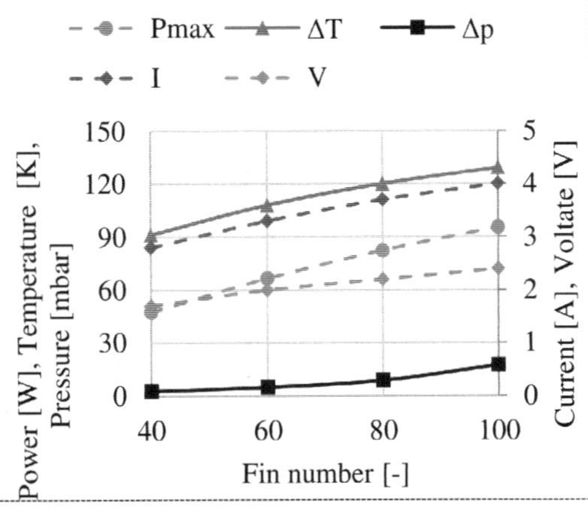

Figure 7. Influence of hot side HX fin number on TEG performance.

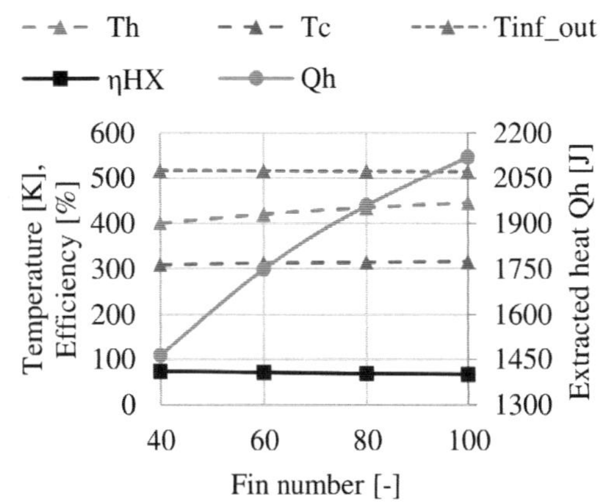

Figure 8. Influence of hot side HX fin number on hot and cold side temperatures, hot side outlet temperature and HX performance.

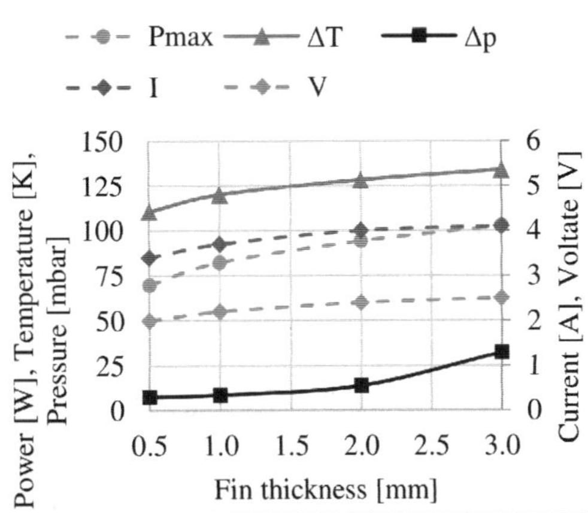

Figure 9. Influence of hot side HX fin thickness on TEG performance.

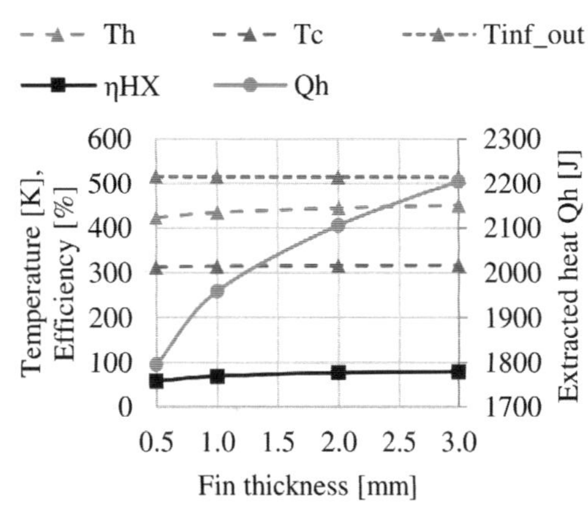

Figure 10. Influence of hot side HX fin thickness on hot and cold side temperatures, hot side outlet temperature and HX performance.

IOP Conf. Series: Materials Science and Engineering 1169 (2021) 012027 doi:10.1088/1757-899X/1169/1/012027

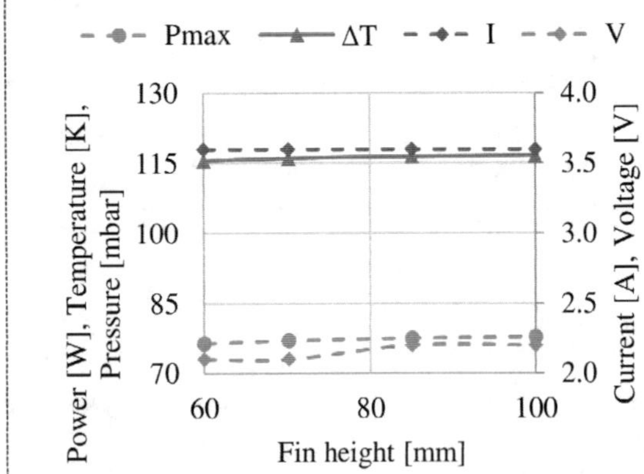

Figure 11. Influence of cold side HX fin height on TEG performance (power (P_{max}), temperature gradient (ΔT), and TEM current (I) and voltage (V))

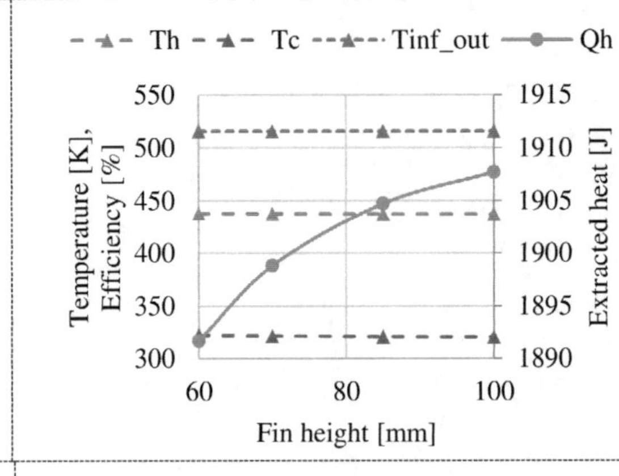

Figure 12. Influence of cold side HX fin height on hot (T_h) and cold (T_c) side temperatures, hot side outlet temperature (T_{inf_out}) and HX performance (extracted heat Q_h)

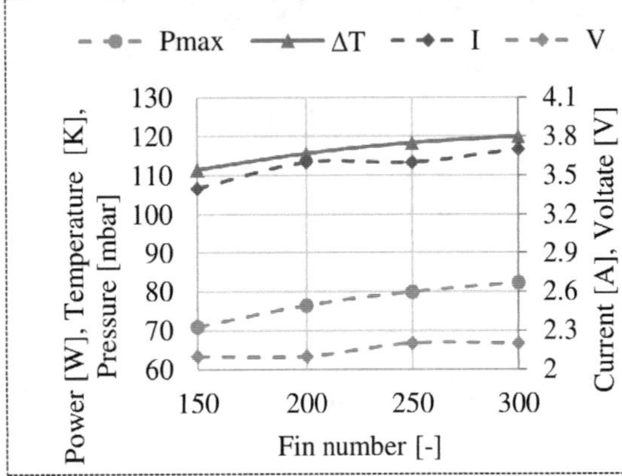

Figure 13. Influence of cold side HX fin number on TEG performance.

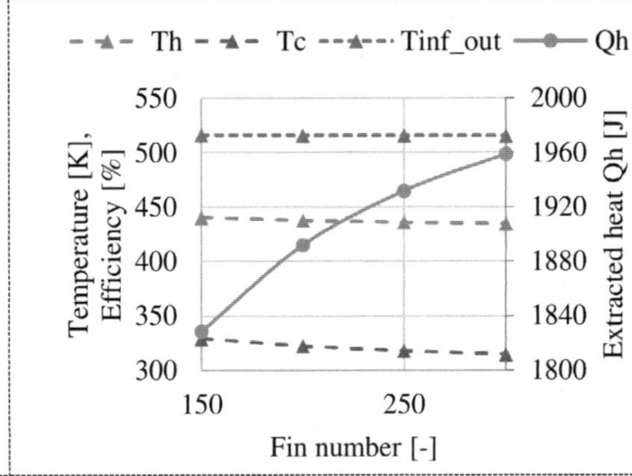

Figure 14. Influence of cold side HX fin number on hot and cold side temperatures, hot side outlet temperature and HX performance.

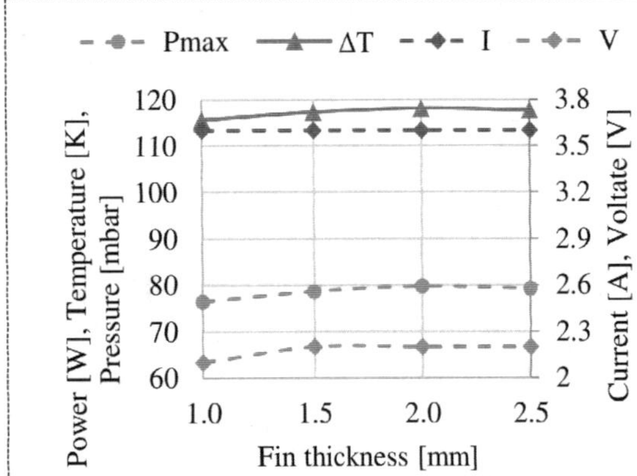

Figure 15. Influence of cold side HX fin thickness on TEG performance.

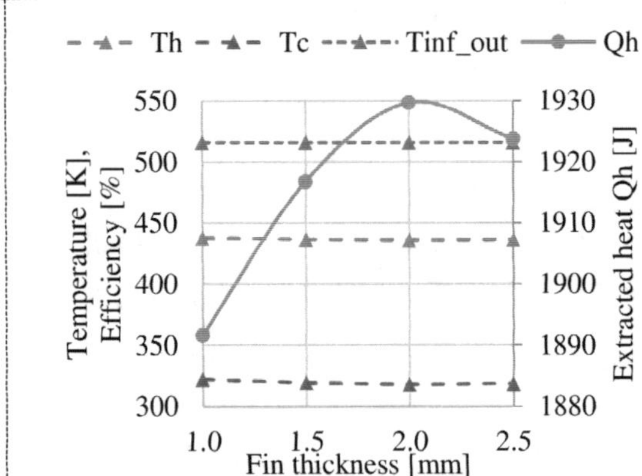

Figure 16. Influence of cold side HX fin thickness on hot and cold side temperatures, hot side outlet temperature and HX performance.

A second set of results (figures 11 to 16) was obtained for a variable cold side HX design, while the hot side HX parameters were kept constant. The analysis of the obtained results has shown that varying the fin height, fin number and fin thickness leads to similar trends to those obtained for the hot side HX. However, the impact of the different designs is significantly smaller. For example, increasing the fin number from 150 to 300 leads caused an increase in the amount of extracted heat of approximately 130 J. In comparison, increasing the fin number from 40 to 100 leads an additional 650 J of extracted heat. Because the cold side HX is not subjected to same restraints as the hot side HX in terms of pressure drop, the design can be optimised for high heat rejection performance. However, care must be taken when opting for design changes that affect the fin spacing (like the fin number and fin thickness) since, at very low fin spacing values the flow is restricted and as a result, the amount of rejected heat drops, thus reducing performance.

4. Conclusions

The present study focused on analysing the influence of the hot and cold side HX design parameters on the performance of an exhaust gas TEG for heavy-duty vehicles, that uses air as cooling fluid. The design variables under consideration were the fin height, fin number, and fin thickness. The analysis of the simulation results has shown that the number of fins has the highest influence on the amount of extracted heat and power output of the system, but it also negatively affects the pressure drop. Increasing the fin height also increases the TEG output and improves the pressure drop, but after a certain value, it reaches a plateau. Of the three variables, fin thickness has the smallest influence on the output, but the highest influence on the pressure drop. Finally, it was found that the cold side HX design has only a limited influence on the TEG output compared to the hot side HX, but, since it is not directly affecting the ICE, it is subjected to fewer design restrictions.

5. References

[1] Varga B O, Sagoian A and Mariasiu F 2019 Prediction of Electric Vehicle Range: A Comprehensive Review of Current Issues and Challenges *Energies* **12**

[2] Deng Y D, Hu T, Su C Q and Yuan X H 2017 Fuel Economy Improvement by Utilizing Thermoelectric Generator in Heavy-Duty Vehicle *J. Electron. Mater.*

[3] Shen Z-G, Tian L-L and Liu X 2019 Automotive exhaust thermoelectric generators: Current status, challenges and future prospects *Energy Convers. Manag.* **195** 1138–73

[4] Lan S, Yang Z, Chen R and Stobart R 2018 A dynamic model for thermoelectric generator applied to vehicle waste heat recovery *Appl. Energy* **210** 327–38

[5] Abderezzak B and Randi S 2020 Experimental investigation of waste heat recovery potential from car radiator with thermoelectric generator *Therm. Sci. Eng. Prog.* **20** 100686

[6] Aljaghtham M and Celik E 2020 Design optimization of oil pan thermoelectric generator to recover waste heat from internal combustion engines *Energy* **200** 117547

[7] Burnete N V, Mariasiu F, Moldovanu D and Depcik C 2021 Simulink Model of a Thermoelectric Generator for Vehicle Waste Heat Recovery *Appl. Sci.* **11**

[8] Heber L, Schwab J and Friedrich H E 2019 Design of a Thermoelectric Generator for Heavy-Duty Vehicles: Approach Based on WHVC and Real Driving Vehicle Boundary Conditions BT - Energy and Thermal Management, Air-Conditioning, and Waste Heat Utilization ed C Junior and O Dingel (Cham: Springer International Publishing) pp 206–21

[9] European Thermodynamics Ltd. *https://www.europeanthermodynamics.com/products/datasheets/GM250-127-28-10-v2.pdf*

Acknowledgements

This paper was supported by the Project POCU/380/6/13/123927 "Entrepreneurial competences and excellence research in doctoral and postdoctoral programs-ANTREDOC", project co-funded by the European Social Fund.

This research was possible due to the MATLAB-SIMULINK campus-wide license of the Technical University of Cluj-Napoca.

Modeling and building a 3D print head

D C Negrau[1,2] , G Grebenişan[1] ,T Vesselenyi[1] ,D M Anton[1] , C I Indre[1]

[1] University of Oradea, Faculty of Managerial and Technological Engineering, Universităţii Street No. 1, Oradea, Romania

[2] dan.negrau@yahoo.com

Abstract. The paper presents 3D modeling and the construction of a printing head for plastic, a head that will be used on a CNC milling machine. The component parts were modeled with Solid Edge software, and for manufacturing the parts we use a CNC milling machine and one conventional lathe. Following the manufacture of the parts, they were mounted together. After that, the mechanical parts was connected together with the electrical components and the result is a functional head that uses fused deposition modeling (FDM) technology.

1. Introduction

Additive manufacturing (AM) is a fast and efficient method of manufacturing parts [1], manufacturing based on technologies (figure 1) such as fused deposition modeling (FDM) [2], a technology that is widely used in the production of parts in various fields. [3], of high complexity and technical quality [4],[5], parts that are built layer by layer [6], based on a 3D model saved in stl format.

The generated STL file is loaded into the machine's built-in software, which divides it into individual slices and as sign attributes, such as fill percentage and wall thickness, and as basic printing materials using plastic filaments, such as be PLA, carbon fiber, ABS and other plastics [7].

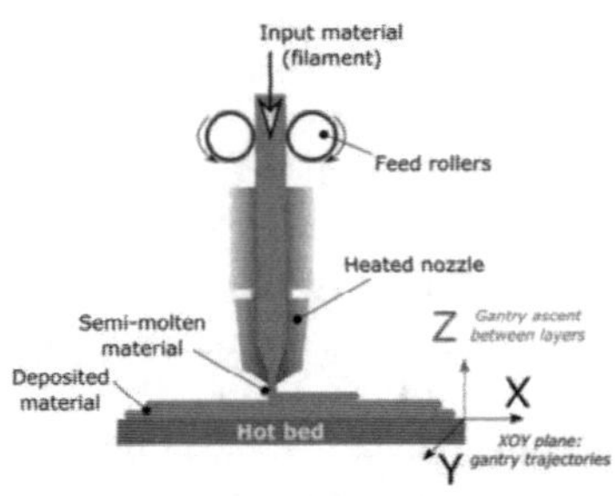

Figure 1 Schematic representation of FDM system [6]

There are also other companies that during this period do research and produce additive printheads such as: HURCO, FASTRAX, MELTIO ENGINE, 3D HERNDON.

- **HURCO** in 2014 he began to explore ways to combine subtractive and additive technologies thus managing to produce a printhead (figure 2) to be used on the same CNC machine together with cutting tools without the need for repeated configurations of the CNC machine (setting print head such as a cutting tool-offset tool)[8], concluding that such operation is much

IOP Conf. Series: Materials Science and Engineering 1169 (2021) 012028 doi:10.1088/1757-899X/1169/1/012028

cheaper than purchasing a complete system, and at the same time the manufacturing process is much shorter and cheaper (figure 3).

Figure 2. Print head HURCO [8]

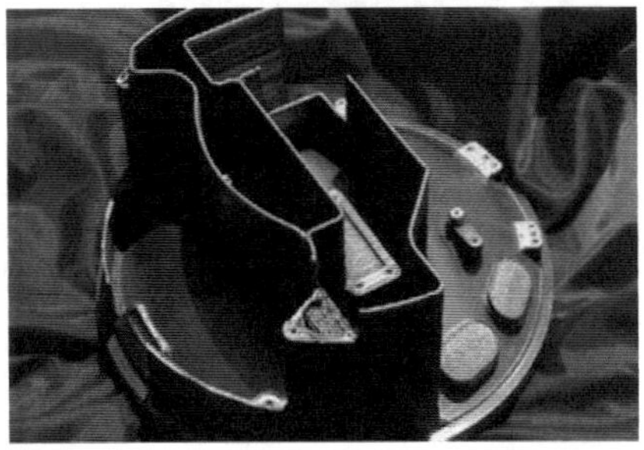

Figure 3. Part obtained by printing + milling in 3 weeks compared to 52 weeks with another method [8]

- **FASTRAX** also a print head (figure 4), which was designed and built by Dante Dudy for personal use, and later being produced as an industrial product. This printhead can be adapted to any CNC machine (figure 5), having the control part separately from the CNC PLC, having as control panel a touch screen [9] .

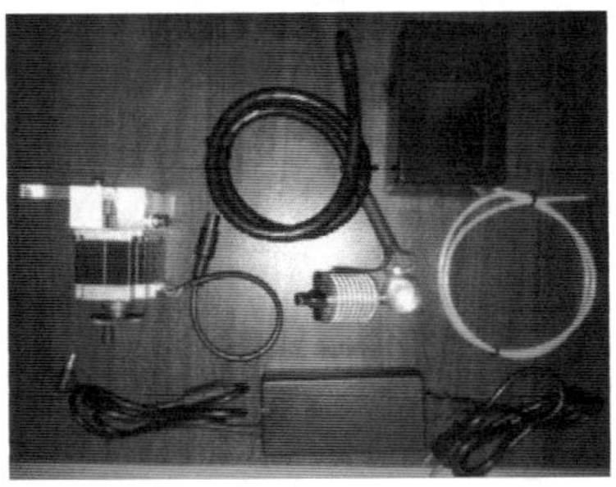

Figure 4. Print head FASTAX [9]

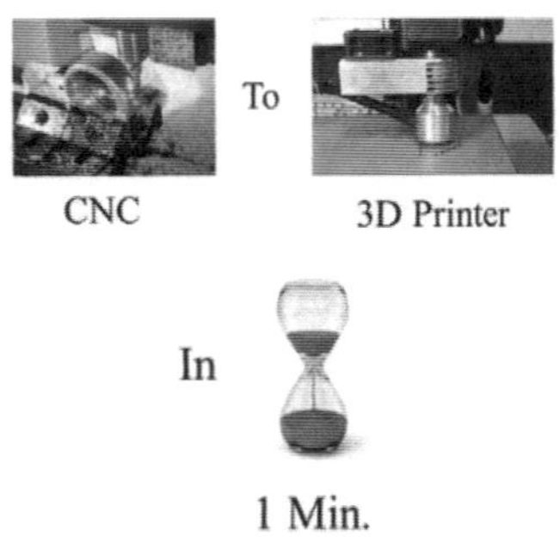

Figure 5. Steps for assambly on the CNC- FASTAX head [9]

- **MELTIO ENGINE** manufacture a 3D laser deposition print head that can be adapted to almost any CNC machine (figure 6). This head is precise and suitable for 5-axis milling machines and also suitable for mounting it on robots (figure 7), to produce a part from "zero" or to load with material a part that has wear and then milling it [10].

Figure 6. Meltio Engine print head fixed on a CNC [10]

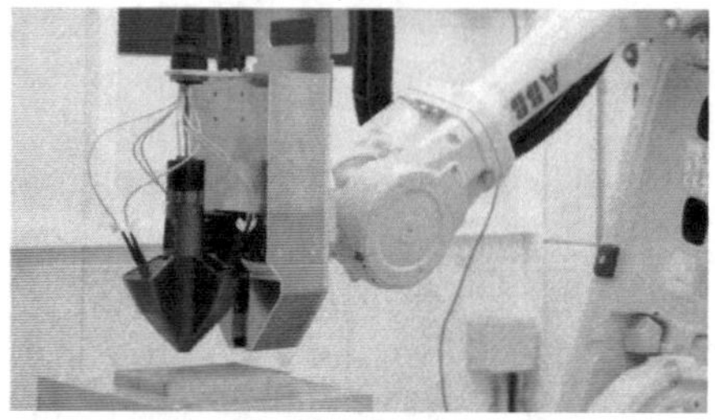

Figura 7. Mounting the Melted Engine head on an ABB robot [10]

As a price it is around 90.000 EURO, a relatively low price compared to a mixed car that reaches 1.000.00 EURO. Such processes are used to repair large and expensive parts, such as injection molds, car assembly stations and in the aeronautical industry [1].

2. 3D modelling and Finit Element Analysis (FEA) print head

Starting from the idea of "TRANSFORM" the CNC + TMA-AL-550, in a mixed machine, we mechanically designed the extruder head using the software Solid Edge ST7[11], software that allows both 3D modeling of parts and the creation of 2D models based on 3D ones, which will later be used in the production area for the manufacture of 3D printhead parts.

The printhead is made of both standardized elements (electrical resistance, thermocouple, stepper motor…) and parts that are not stas (cooler, heater block, mounting bracket, adapter for CNC mounting),

elements that can be changed very quickly and easily, so that this head (figure 8) can be dimensionally modified, as well as its parameters (melting power mounting a higher electrical resistance, filament supply + change of filament supply motor) depending on the requirements of a possible customer, who has a certain milling CNC (vertical, horizontal) (table 1).

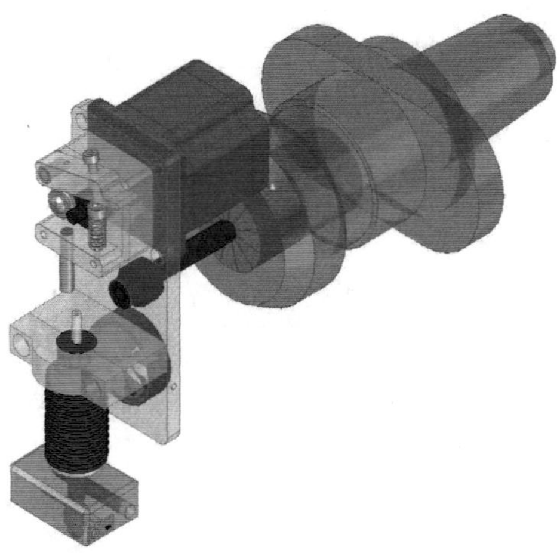

Figure 8 Print head for horizontal CNC machine

The machine on which we mounted the printhead is a 5-axis CNC milling machine (TMA-AL-550)[12], and the printhead (figure 8) has been designed so that it can be mounted on this horizontal CNC, and also can operate in the mode of working with 5 consecutive axes of the machine (figure 9).

IOP Conf. Series: Materials Science and Engineering 1169 (2021) 012028 doi:10.1088/1757-899X/1169/1/012028

Table 1 The parts of print head

Nr. Crt	(a) Support plate	(b) Heater block	(c) Supports heat exchanger	(d) Cartridge electrical resistance Ø6,5x40 mm, 160W, 230V, Thermocouple type J, S5435/3	(e) Stepper motor Nema 23 Filament pull assembly	(f) Screws and cooling cooler
Name picture	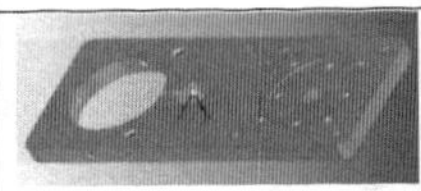			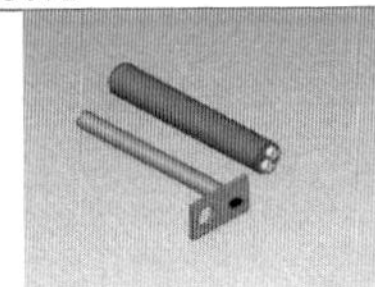	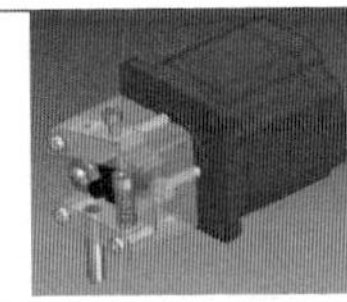	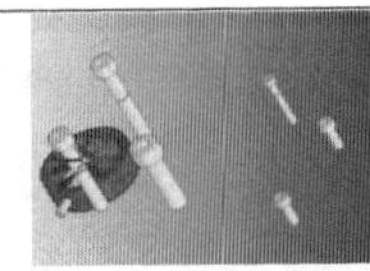
Description	Supports all printhead components and connects to the CNC Made of EN AW-7075 ALZN5 on CNC with an accuracy of 0.02m	In this heater block is mounted the electric resistance cartridge, thermocouple J, printing nozzle, heat exchanger Made of EN AW-7075 ALZN5	They have the role of supporting the heat exchanger, which in turn supports the print head Made by plastic 3D printing: PLA + carbon fiber	The resistor has the role of heating the heater block, and the thermocouple J monitors temperatures and transmits the information to the temperature controller Delta DT3 STAS Parts	The two components assembled together have the role of feeding the extruded head with filament. STAS Parts	The cooler has the role of cooling the heater sink so that the print head temperature can be maintained at the set value and at the same time the heat is not transferred by conduction to the neighboring parts. STAS Parts

IOP Conf. Series: Materials Science and Engineering 1169 (2021) 012028 doi:10.1088/1757-899X/1169/1/012028

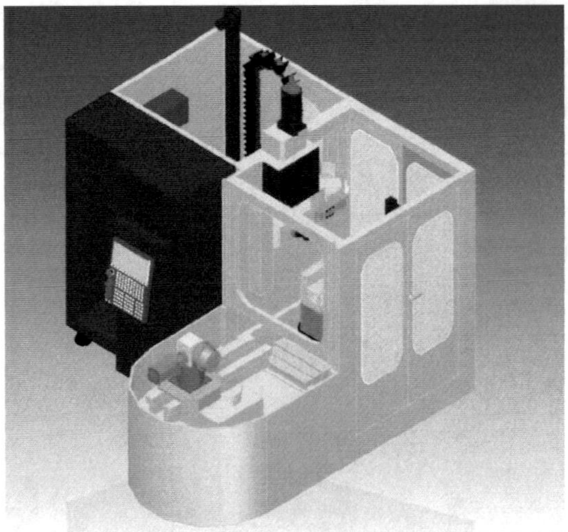

Figure 9 CNC machine TMA-AL-550 [12]

For the dimensioning of the heater block, of the cooler, the choice of the materials from which to build the components of the printhead, we used the analysis with finite element [13], [14], and based on the obtained results we found that the filament melts before it enters the heater block. In this analysis (figure 11) is presented the section of the 3D model of the extruder assembly (figure 10) with its components and the temperature of each component. With the help of FEA we modified both the design of the printhead and the addition of new components (teflon spacer) to prevent heat transfer from the hot end to the cooler, so that the temperature of the cooler is maintained at 50 degrees.

Teflon tube from the inside of cooler,hot end an nozzle, plays a double role, it is used primarily to direct the filament to the nozzle, because if it were transferred directly through the metal, due to the high roughness, the filament could become blocked.

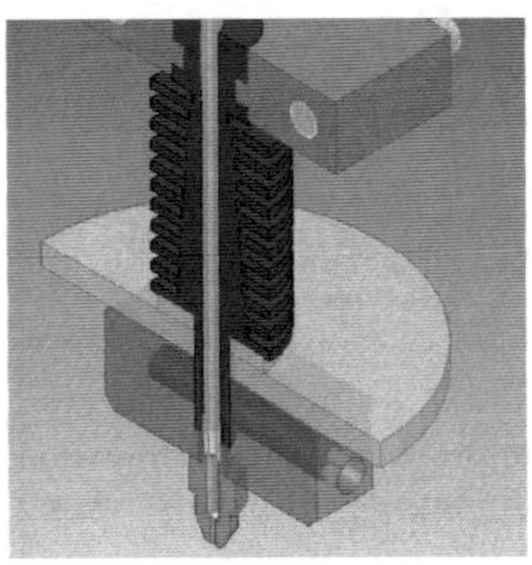

Figure 10 Print head modify after FEA

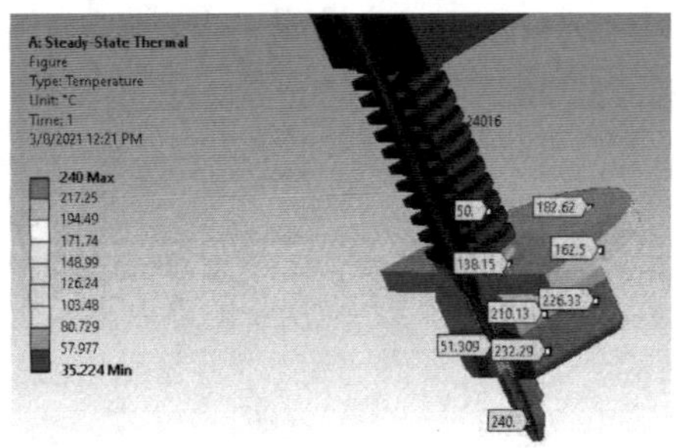

Figure 11 Analysis FEA

3. Conclusion

Additive manufacturing (AM) is a new industrial revolution, and the main machine tool manufacturers are developing this technology and with the help of finite element design and analysis software, an optimized print head can be built.

Using the printhead presented in this paper, he transformed the TMA-AL-550 CNC milling machine into a hybrid CNC machine. The electrical design of its printhead, as well as its control and programming, will be presented in other papers.

4. References

[1] J. Gausemeier, N. Echterhoff, and M. Wall, "*Thinking ahead the Future of Additive Manufacturing – Innovation Roadmapping of Required Advancements*," Univ. Paderborn Direct Manufacuring Res. Cent., p. 110, 2013, [Online]. Available: http://www.hni.uni-paderborn.de/en/pe.

[2] V. Shanmugam et al., "*Fatigue behaviour of FDM-3D printed polymers, polymeric composites and architected cellular materials,*" Int. J. Fatigue, vol. 143, no. October 2020, p. 106007, 2021, doi: 10.1016/j.ijfatigue.2020.106007.

[3] L. Suárez and M. Domínguez, "*Sustainability and environmental impact of fused deposition modelling (FDM) technologies,*" Int. J. Adv. Manuf. Technol., vol. 106, no. 3–4, pp. 1267–1279, 2020, doi: 10.1007/s00170-019-04676-0.

[4] F. Rayegani and G. C. Onwubolu, "*Fused deposition modelling (fdm) process parameter prediction and optimization using group method for data handling (gmdh) and differential evolution (de),*" Int. J. Adv. Manuf. Technol., vol. 73, no. 1–4, pp. 509–519, 2014, doi: 10.1007/s00170-014-5835-2.

[5] D. C. Negrau, G. Grebenian, and C. Gherghea, "*A brief overview of Additive Manufacturing,*" IOP Conf. Ser. Mater. Sci. Eng., vol. 898, no. 1, 2020, doi: 10.1088/1757-899X/898/1/012029.

[6] U. K. uz Zaman, E. Boesch, A. Siadat, M. Rivette, and A. A. Baqai, "*Impact of fused deposition modeling (FDM) process parameters on strength of built parts using Taguchi's design of experiments,*" Int. J. Adv. Manuf. Technol., vol. 101, no. 5–8, pp. 1215–1226, 2019, doi: 10.1007/s00170-018-3014-6.

[7] Y. Hu, R. B. Ladani, M. Brandt, Y. Li, and A. P. Mouritz, "*Carbon fibre damage during 3D printing of polymer matrix laminates using the FDM process*," Mater. Des., vol. 205, p. 109679, 2021, doi: 10.1016/j.matdes.2021.109679.

[8] "Adding and Subtracting to Make the Next Generation of Hybrid 3DCNC Machines - ASME." https://www.asme.org/topics-resources/content/adding-subtracting-make-next-generation-hybrid (accessed May 04, 2021).

[9] "CNC Mill into a Hybrid 3D Printer - 3D Printing Industry." https://3dprintingindustry.com/news/turn-your-cnc-mill-into-a-hybrid-3d-printer-57755/ (accessed May 04, 2021).

[10] "Meltio Engine transforms any CNC machine into a hybrid 3D print system." https://newatlas.com/3d-printing/meltio-engine-hybrid-manufacturing-system/ (accessed May 04, 2021).

[11] D. Santoro, "Solid Edge 2021 | Siemens Digital Industries Software - Ingenuity for Life," Accessed: Apr. 26, 2021. [Online]. Available: https://solidedge.siemens.com/en/solutions/products/complete-product-development-portfolio/whats-new-in-solid-edge-2021/ (accesed in May 04, 2021)

[12] T. D. E. Doctorat, "*Contributions to the Concept of Processing Applications in 5 Axis CNC Rotary Tipper Table with 2 Removable Axle DOMENIUL : INGINERIE INDUSTRIALĂ LA CONCEPTUL DE PRELUCRARE IN 5,*" 2014.

[13] "*An optimization approach by Finite Element Analysis, using Design of-*Negrău_2019_IOP_Conf._Ser.__Mater._Sci._Eng._568_012065.pdf." .

[14] "*A neural networks approach of process fault diagnosis using time series-*Grebenişan_2019_IOP_Conf._Ser.__Mater._Sci._Eng._568_012079.pdf." .

Acknowledgments
Published with the support of the University of Oradea because it provided me with scientific resources and with the support of the ID 123008 - POCU/380/6/13 project, - SMARTDOCT – Programe de înaltă calitate pentru doctoranzii și cercetătorii postdoctorat ai Universității din Oradea pentru creșterea relevanței cercetării și inovarii în contextul economiei regionale.

Control system for FDA deposition using a CNC milling machine

D C Negrau[1,2] , G Grebenişan[1] ,M Ratiu[1] ,D M Anton[1]

[1] University of Oradea, Faculty of Managerial and Technological Engineering, Universităţii Street No. 1, Oradea, Romania

[2] dan.negrau@yahoo.com

Abstract. This paper presents the command and control part of a 3D printhead, the command and control that is made by combining the PLC of a CNC milling machine with the electrical assembly of an extruded head for FDA(Fusion Deposition Adhesion) processing.
The command and control part of an extruder head is presented. On the same, the paper shows how to connect to the PLC of a CNC and presents the use and operation of an extruder block for FDA machining on CNC

1. Introduction

Regarding the increase in performance, it is expected that the price-performance ratio will improve, in favor of performance, but not as spectacular as it was in previous years. The physical and mechanical properties of the materials used have seen significant improvements and the accuracy has improved from ± 0.25 mm to ± 0.075 mm [1],[2]. . The most important improvements have been, and will be, in rapid prototyping (PR) programming systems, which are easier to use and smarter. The result is a reduction in working time, an improvement in the quality of machined surfaces and finally a reduction in the cost of parts made with the help of these technologies. Thus, if in 1992 a piece cost 15 units of price, today the same piece costs 2 units (fig. 1) [3].

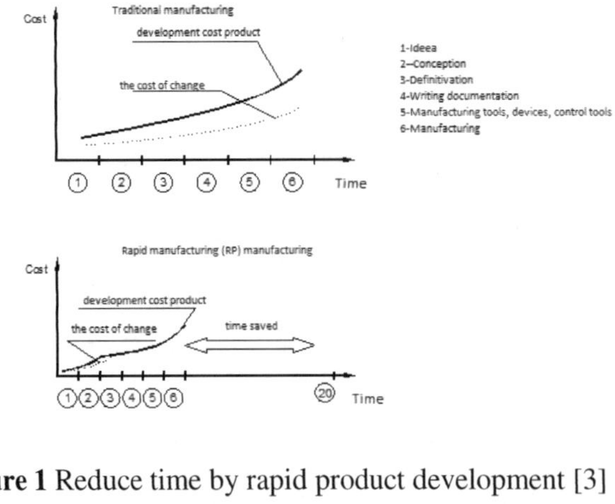

Figure 1 Reduce time by rapid product development [3]

Combining the technique of 3D mechanical design [4], with electrical design [5], or rather the formation of a team containing mechanical engineers and electronic engineers, leads to the manufacture of new equipment and the discovery of new technologies [6],[7] which make it possible to reduce manufacturing costs and increase the performance of the products obtained [1], [8], [9],[10], [11].

2. PLC of CNC and electrical components of FDA system

This paper present the control and command part of the print head assembly has been developed, which will be connected to a TMA-AL-550 CNC (figure 2) [6]. The control panel for FDA filing has several components, starting with the temperature controller that was chosen based on several models, each bringing pluses and minuses, because they were not compatible with either the heating resistance of the extruder block or the correct reading in real time temperature, or with CNC.

For this CNC model, a DELTA DTK 4848 temperature controller was chosen, with the desired reading specifications for the J-cartridge type temperature thermocouple.

During this time, the CNC was also modified on the control side (figure 3), here it was added to the Fanuc PLC to control both start / stop of the filament motor advance, start / stop of the heating resistance, and as input the signal reading temperature when it has reached the temperature set on the controller (figure 4).

Figure 2 CNC TMA AL 550 SolidWorks3D [6]

Figure 3 CNC TMA AL 550 Printed Block Mounted [6]

The model of electrical resistance is cartridge type electric resistance cartridge Ø6,5x40 mm, 160W, 230V, thermocouple type J, S5435/3, which allows us to supply power it through a semiconductor relay from AC voltage.

The semiconductor relay (figure 5) is a relay without mechanical contact, but performs on / off functions in the same way as an electromechanical relay. The relay used is a Fotek brand SSR, with the possibility of power supply from 34-380V alternating current 10A. The supply voltage for the input control is between 3-32V direct current.

The supply voltage for the input control is between 3-32V direct current. Therefore, for a very high frequency temperature reading, a mechanical relay was not chosen, because SSR (figure 5) are capable of extremely fast operations, and these models are capable of generating the desired frequencies, for the greatest possible possibility of maintaining the temperature. of the extruder block in a well-defined range which helps the flow flow of the filament for a desired deposition.

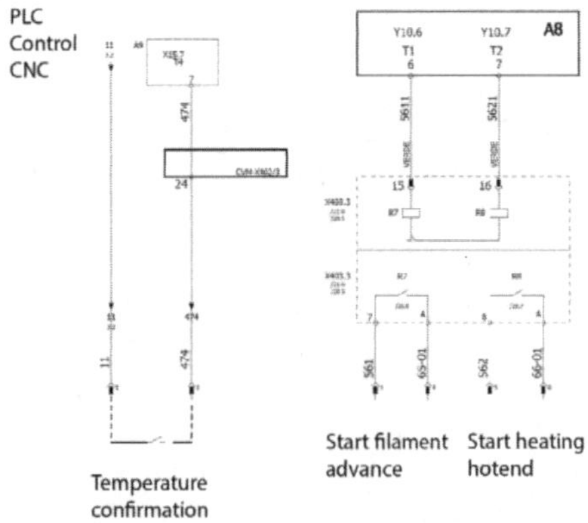

Figure 4 Wiring diagram - connecting the printhead to the CNC

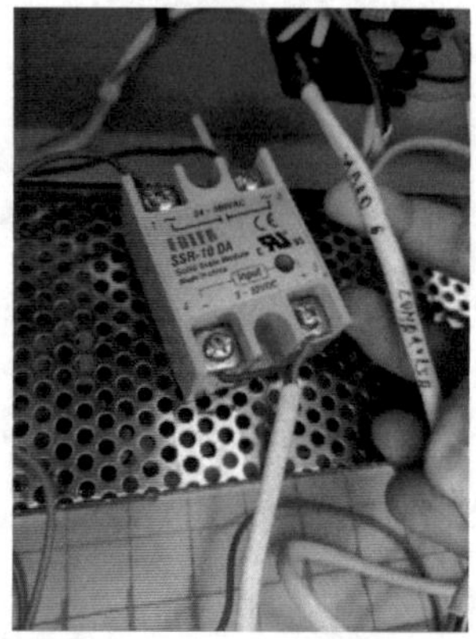

Figure 5 SSR mounted in the control panel

To start and stop the ssr, use the thermocouple (figure 6) located in the extrusion block, which checks the temperature at a certain time set in the controller.

Temperature control can be done at different time intervals, but the PID function of the DELTA DTK (figure 7) controller can also be used, for a precise verification of it. Temperature control via PID mode in the controller by selecting the AT program, which means switching to automatic control, with proportional setting function. Printing is done in the range 210-230℃, depending on the material used, PLA type filaments are printed up to a temperature of 220℃, being an optimal printing temperature up to speeds of 100 mm / s, at higher speeds the temperature must be raised. Printing can be done with PLA starting at 25mm / s at a temperature equal to 190℃, for every 10 mm / s in addition the temperature must be increased by 2.5℃-2.8℃, depending on the ambient temperature or directing the air stream to cool the embossed block, so that the filament passing through the heated block becomes crystallized and can be deposited.

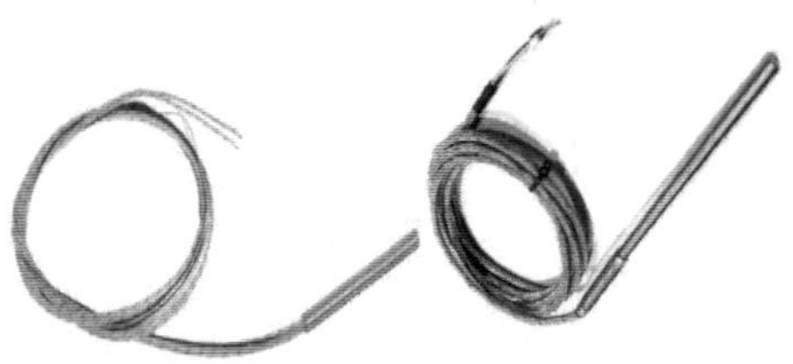

Figure 6 Top-Electric resistant 150W, Bottom-K-type thermocouple probe

Figure 7 Controller temperature DTK

When using the PID function during use, it works on the order of a tenth of a degree, it allows us to use the resistance to heat the extrusion block with great accuracy, because the filament to be extruded is very demanding for fluctuating values.

The high speed sampling of the controller (figure 7) is below 100 ms, which allows us to read at least 10 values per second. The reading of the temperature sensor is very useful to be done as quickly as possible, which allows the output signal regulator to be done automatically to the heating element. To reach the set value, the controller uses the set temperature as the signal point, so each temperature drop is done automatically with a correction for the next output signal for resistance, if set to 210℃, the read temperature is 209.08℃ PID activates the relay for 10ms to reheat the resistance again.

These models of controllers with a hysteresis of the order of a tenth of a degree, help us to keep a lot of the temperature around the set one, these being typical for the temperature adjustment in the case of this model, with a discontinuous output signal. If the unit is left in OUT1Heated mode, it receives

IOP Conf. Series: Materials Science and Engineering 1169 (2021) 012029 doi:10.1088/1757-899X/1169/1/012029

values at a longer time interval of 20 seconds, this reading bringing a lot of problems with the heating of the extruder block, because if the set temperature is 210℃, it before reaching this temperature makes a reading, heats up for another 20 seconds and reaches a temperature of 220℃, and in the next cycle the temperature drops below the set one, which causes a continuous flow of filament, blocking it in the nozzle in the extrusion of the filament during cooling or blockage before the corrugated block, or it may even cause the filament to break, and during the heated heating may increase the flow rate of the filament faster. The difference between the operation of a filament extrusion block with a DELTA DTK4848RI controller using PID and mode without using PID.

3. Connect the FDA system to the CNC

The control part of the extruder head was made based on inputs and outputs (sheets 1) from the CNC PLC, connections (figure 8) that make possible different operating characteristics of the machine.

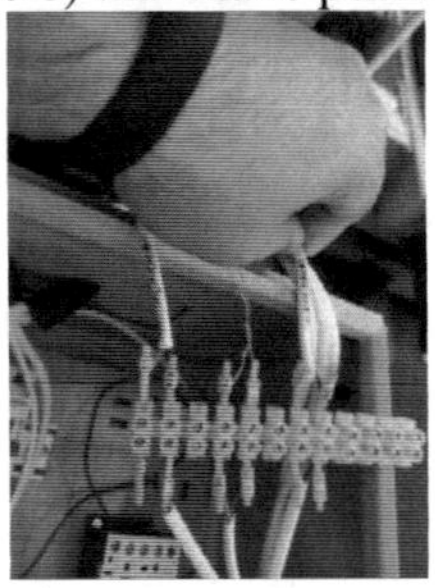

Figure 8 Schneider Contactor LC1D18B5

Figure 9 Driver TB6600

The cnc has as heating output and start filament motor advance on OUTPUT = START HEATING, controll with is made by the DRIVER TB6600 (figure 9)

Alarm 11,474 - FILAMENT ADVANCE START (NEMA stepper motor start-figure 11) Motor advance start 561.65 01;

Input = CONFIRM TEMPERATURE PRINTER HEAD 562, 66 01

In order to be able to control the temperature confirmation input, a Schneider LC1D18B5 (figure 10) contactor was used between the machine and the temperature controller, it keeps the engine feed start control closed.

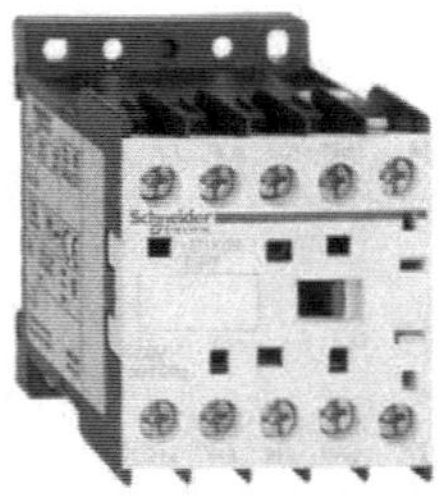

Figure 10 Schneider Contactor LC1D18B5

Figure 11 Nema 23 stepper engine

Filament extrusion is done with a Nema 23 motor (figure 11), it is controlled by a TB6600 driver, which has an Arduino Nano controller connected to tell it the speed and direction of rotation on the pins D2-Dir +, D3-Pul +, D5-En +, and the pins EN-, Dir-, Pul + are grounded on arduino nano. The Arduino controller has the D12 pin connected to the 561 CNC output, which controls starting / stopping the Nema 23 motor.

The contactor and the temperature controller are supplied directly to AC 230V, and the Arduino Nano microcontroller is supplied with a 5V, 2A SMPS source, as well as the filament feed motor driver is supplied with a 24V SMPS source. Their power supply was made separately, because on the 230V AC power outputs of the Cnc we did not have the necessary current for their power supply (figure 12,13).

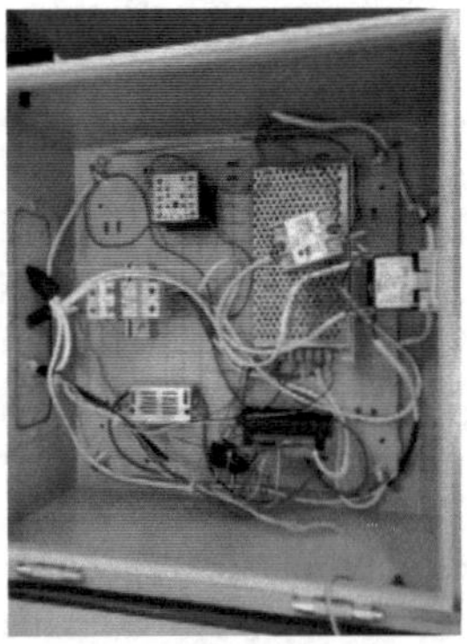

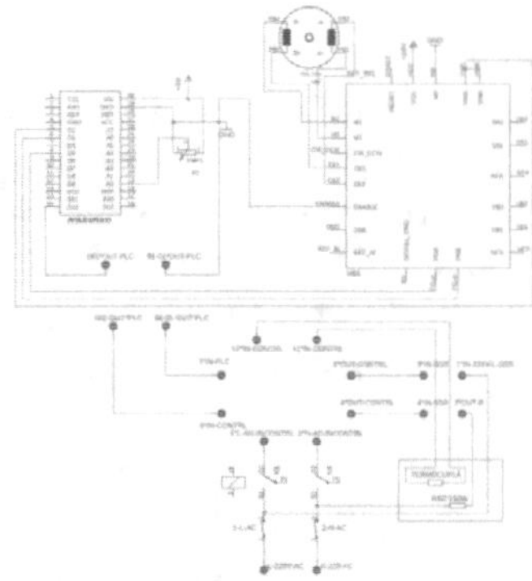

Figure 12 Control panel mounting **Figure 13** Wiring diagram of the 3D printing board

The extruder block assembly (figure 14) was designed to be easy to heat, with the threaded bore for the extrusion head or nozzle. The nozzle was made of CW617N brass (figure 15) with a lathe, its inner hole was ground on the inside to reach the inside diameter of 1mm + -0.01. It helps us to extrude the filament without blocking the carbon fiber particles that are in the composition of PLA filaments, which can have up to 30% insertion of carbon fibers inside them. For the use of these filaments it is very useful to grind the extruder on the inside so as not to create the blockage of carbon fiber particles inside the extruder. When observing a brass nozzle, which printed about 1 kg of PLA filament with carbon fiber insert, after milling it, fine internal scratches are observed under the microscope. These internal scratches occur after the brass is continuously attacked by the carbon fiber insertion flow, leading to internal irregularities in this nozzle.

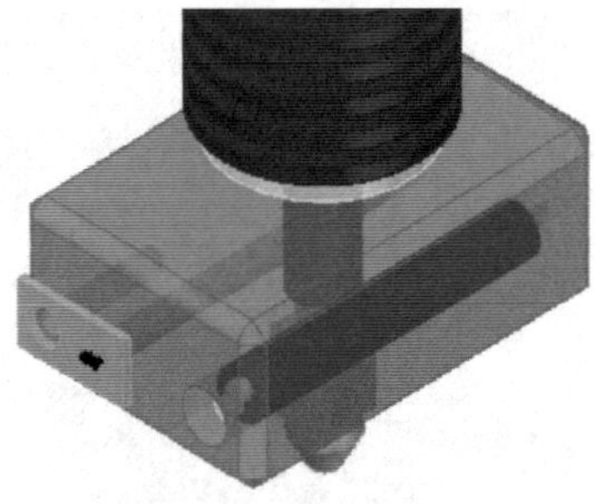

Figure 14 Hot End **Figure 15** Nozzle brass

Following the electrical design of this printhead (figure 16), I generated the G [12] codes for printing, codes involving movements, printing advances, heating of the filaments, we switched to the actual printing (figure 17).

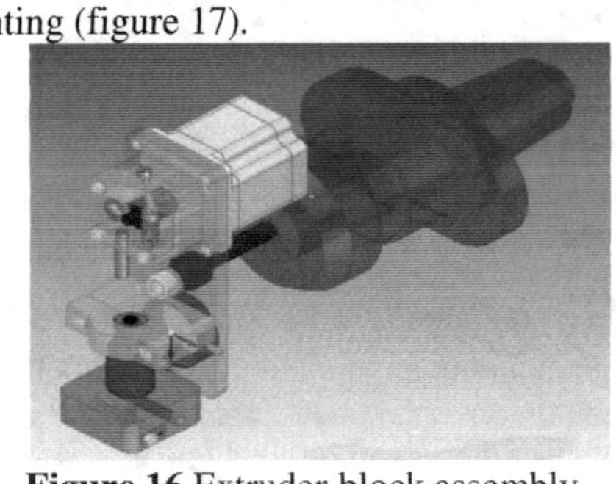

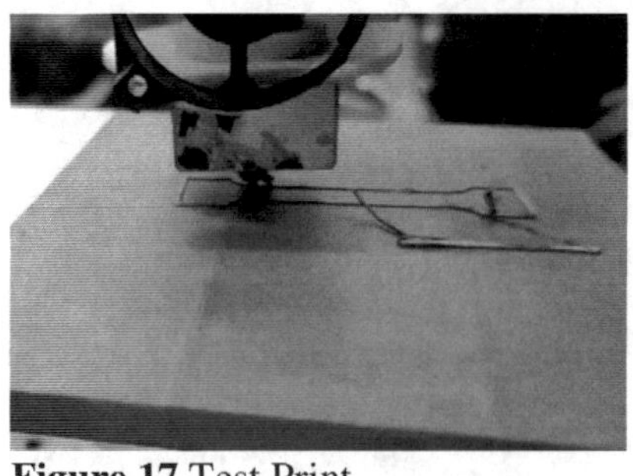

Figure 16 Extruder block assembly **Figure 17** Test Print

4. Conclusion

In conclusion, by combining the knowledge of the mechanical engineer with that of the electronic engineer, such systems as the one presented in this paper can be constructed.

After assembling the mechanical elements with the electronic ones, the CNC machine TMA-AL-550 can print 3D.

In the future we intend to create a post processor for the creation of printing programs.

5. References

[1] I. C. Gherghea, C. Bungau, and D. C. Negrau, "Lead time reduction and increasing productivity by implementing lean manufacturing methods in cnc processing center," *IOP Conf. Ser. Mater. Sci. Eng.*, vol. 568, no. 1, 2019, doi: 10.1088/1757-899X/568/1/012014.

[2] M. Baban, C. F. Baban, T. Buidos, and I. Stanasel, "A reverse engineering approach for the products development," pp. 5–10, 2015.

[3] V. Shanmugam *et al.*, "Fatigue behaviour of FDM-3D printed polymers, polymeric composites and architected cellular materials," *Int. J. Fatigue*, vol. 143, no. October 2020, p. 106007, 2021, doi: 10.1016/j.ijfatigue.2020.106007.

[4] D. Santoro, "Solid Edge 2021 | Siemens Digital Industries Software - Ingenuity for Life," Accessed: Apr. 26, 2021. [Online]. Available: https://solidedge.siemens.com/en/solutions/products/complete-product-development-portfolio/whats-new-in-solid-edge-2021/.

[5] "OxMet Technologies | Home." https://oxmet-technologies.com/ (accessed Apr. 26, 2021).

[6] D. CRACIUN *et al.*, "Design Solution of Rotary Tilting Table for Five Axes Machining," *Ann. ORADEA Univ. Fascicle Manag. Technol. Eng.*, vol. XXII (XII), no. 1, pp. 67–70, 2013, doi: 10.15660/auofmte.2013-1.2783.

[7] "Adding and Subtracting to Make the Next Generation of Hybrid 3DCNC Machines - ASME." https://www.asme.org/topics-resources/content/adding-subtracting-make-next-generation-hybrid (accessed May 04, 2021).

[8] C. I. Indre, F. S. Blaga, C. Micula, and D. C. Negrau, "Research in order to decrease the time manufacturing unit using the combined tools," *IOP Conf. Ser. Mater. Sci. Eng.*, vol. 564, no. 1, 2019, doi: 10.1088/1757-899X/564/1/012007.

[9] I. Ebert-Uphoff, C. M., D. W., and T. Laliberte, *Rapid Prototyping for Robotics*, no. July. 2005.

[10] M. Di Prima, J. Coburn, D. Hwang, J. Kelly, A. Khairuzzaman, and L. Ricles, "Additively manufactured medical products – the FDA perspective," *3D Print. Med.*, vol. 2, no. 1, pp. 4–9, 2016, doi: 10.1186/s41205-016-0005-9.

[11] A. Christensen and F. J. Rybicki, "Maintaining safety and efficacy for 3D printing in medicine," *3D Print. Med.*, vol. 3, no. 1, pp. 1–10, 2017, doi: 10.1186/s41205-016-0009-5.

[12] "G-code - Wikipedia." https://en.wikipedia.org/wiki/G-code (accessed May 08, 2021).

Acknowledgments

Published with the support of the University of Oradea because it provided me with scientific resources and with the support of the ID 123008 - POCU/380/6/13 project, - SMARTDOCT – Programe de înaltă calitate pentru doctoranzii și cercetătorii postdoctorat ai Universității din Oradea pentru creșterea relevanței cercetării și inovarii în contextul economiei regionale.

IOP Conf. Series: Materials Science and Engineering 1169 (2021) 012030 doi:10.1088/1757-899X/1169/1/012030

Researching Diesel car CO_2 emissions in cold and warm-up transient test cycle

L Andrei[1], V Chindea[2] and D L Băldean[2]
[1] Infectious Disease Hospital of Cluj-Napoca, Cluj, Iuliu Moldovan 23, Romania
[2] Automotive and Transportation Dep., Technical University of Cluj-N., Cluj, Muncii 103, Ro.

doru.baldean@auto.utcluj.ro

Abstract. Research of CO_2 emissions is not only important, but it is standard practice for homologation, technical inspections, registration of vehicles and for environmental greenhouse effect. Different agencies are taking into consideration the fact that greenhouse emissions are seriously impacting the climate change and events on the planet. Road traffic and transportation are, beside animal agriculture, the most significant polluters in dense areas, with high values of emissions. For diminishing the impact of automotive sector in the emissions field, the present work approaches the practical side of this topic. During the testing, some results will allow the definition of those conditions in which the pollutant emissions fall into control and are drastically diminished. Main objective of the paper is to research the CO2 emissions generated by a Diesel engine in a transient test cycle, both at cold operation and after warm-up phase. Second objective is to analyse the graphics from low speed and medium speed operation regimes. The CO_2 emission are influenced by fuel consumption, engine speed and coolant temperature. Highway testing of CO_2 emissions are to be made in further and advanced research.

1. Introduction

Internal combustion engines are those machines which transform the mixture of atmospheric O_2 and HC (hydrocarbons) fuel molecules through rapid oxidation reactions at high temperatures inside the combustion chamber in useful work and performance [1, 2].

Digital management system of the engine takes the data from sensors and determines the fuel quantity to be injected [3]. Combustion and exhaust emissions influence the air quality and health risk [4, 5, 6]. Electronic control for fuel supply in dynamic conditions increases the operational efficiency [7]. Health management and environmental risk are closely related to carbon (CO2, CO, and HC) emissions [8]. Injection systems, fuel quantity and air intake mass are important factors to be considered within the electronic control unit [9, 10, 11]. Friction in the bearings and between the wheels and road surface makes it difficult for engine to overpass the resistance torque, thus it takes more fuel to create higher gas pressures in the cylinders and affects the carbon footprint [12, 13]. Advances in software interfaces are supporting the improvement of control and management even through the end-user adaptive interactions and information exchange [14,15]. Electronic Diesel Control and management system provides the data regarding injected fuel quantity on dash-bord screen in real time to facilitate some adjustments in operation and command [16,17]. Development of automated systems with remote control capability is making a step further to optimize the fuel-efficient use and to decrease CO_2 level [18, 19]. Significant advancements and contributions have been made in development of the fuel supply, even through injectors in the combustion chamber, to diminish the losses and increase performance [20].

In the present work is determined CO2 emissions level corelated with engine and vehicle speed.

IOP Conf. Series: Materials Science and Engineering 1169 (2021) 012030 doi:10.1088/1757-899X/1169/1/012030

2. Applied research

Applied method in the present work is supported by the experimental tests which were structured within 3 cycles: Cycle 1 – transient testing in urban traffic using first three gear ratios at 10-12°C; Cycle 2 – transient testing in mixt urban and metropolitan traffic using first four gear ratios at 90°C; Cycle 3 – transient testing using all five gear ratios at 90°C (engine temperature) and 15-17°C.

Figure 1 shows the main steps of the applied research methodology in determining CO_2 emissions.

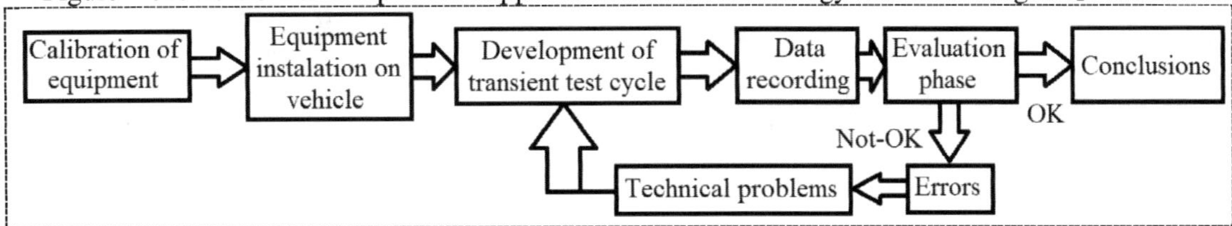

Figure 1. Basic structure of method applied in the present research paper for CO_2 determination

Applied methodology in research is using a Diesel car with technical data given in table 1.

Table 1. Technical data of the test vehicle

Platform	Actual value
Engine type	Diesel 1600 cm^3/ 4 cylinders in line
Fuel supply	High pressure injection
Transmission	5-gear ratio
Powertrain	Front drive, 4x2/VW Golf 6

Figure 2 presents the transient cycles for 3 tests versus the gear ratio, showing the CO2 and speeds.

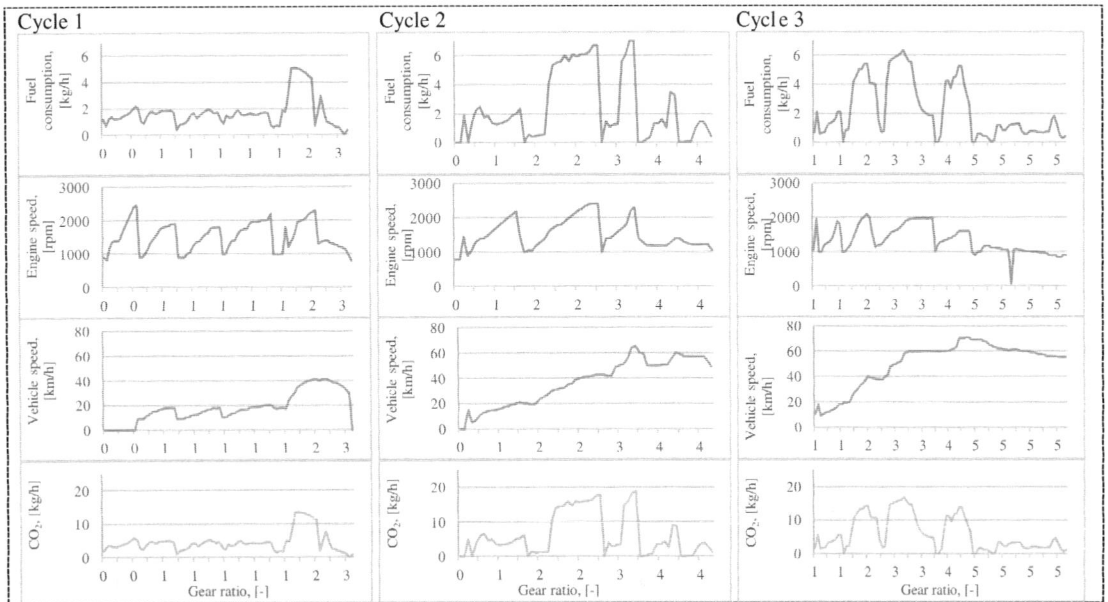

Figure 2. Applied measurement results in three cycles showing the most significant parameters.

In figures 3 to 8 are shown the graph variations for average values of vehicle speed, fuel consumption, engine speed, oil temperature, coolant temperature and CO2 emissions versus gear ratio determine and compared for the 3 cycles. The figures 9 to 14 present the CO2 and engine speed vs. the vehicle speed.

IOP Conf. Series: Materials Science and Engineering 1169 (2021) 012030 doi:10.1088/1757-899X/1169/1/012030

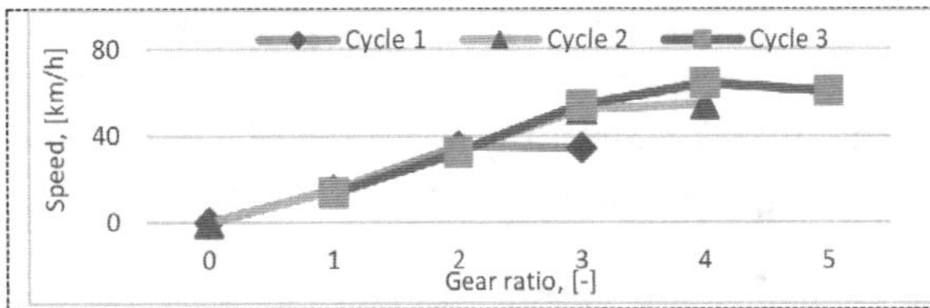

Figure 3. Vehicle speed versus gear ratio values.

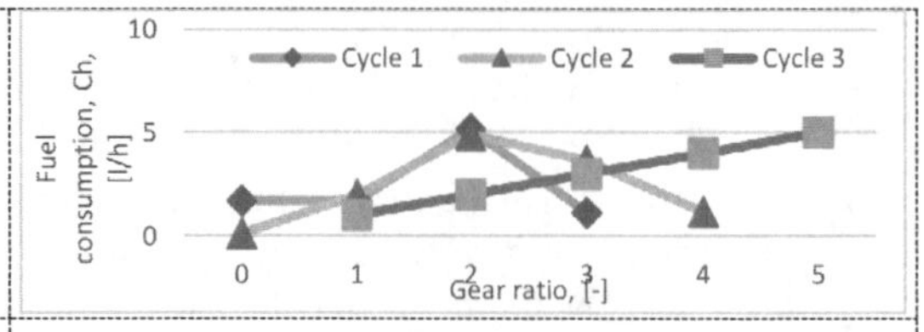

Figure 4. Fuel consumption versus gear ratio.

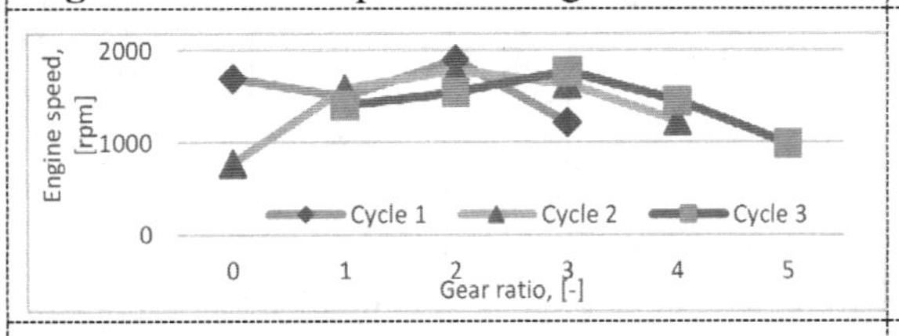

Figure 5. Average engine speed versus gear ratio.

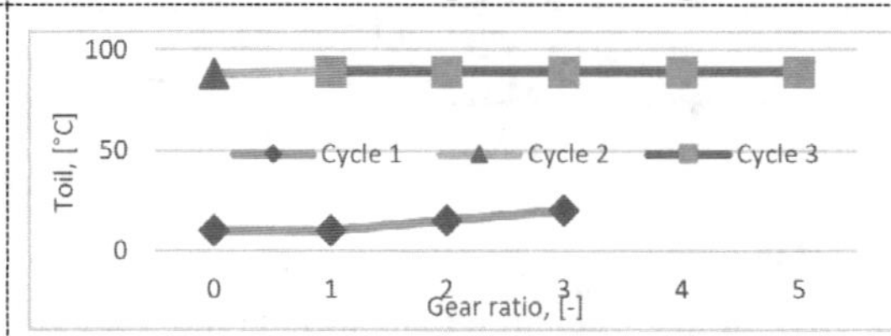

Figure 6. Oil temperature versus gear ratio.

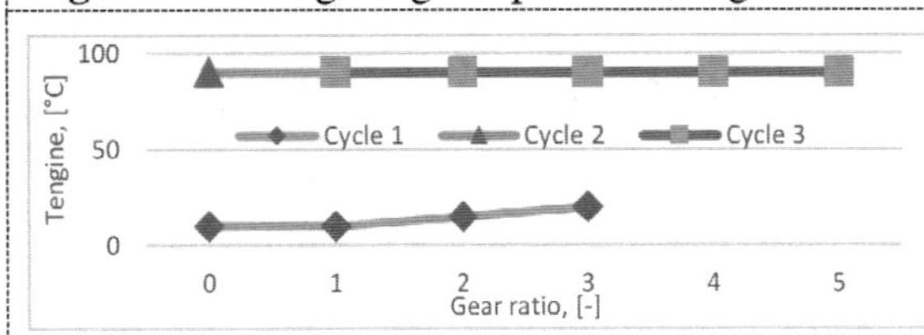

Figure 7. Coolant temperature versus gear ratio.

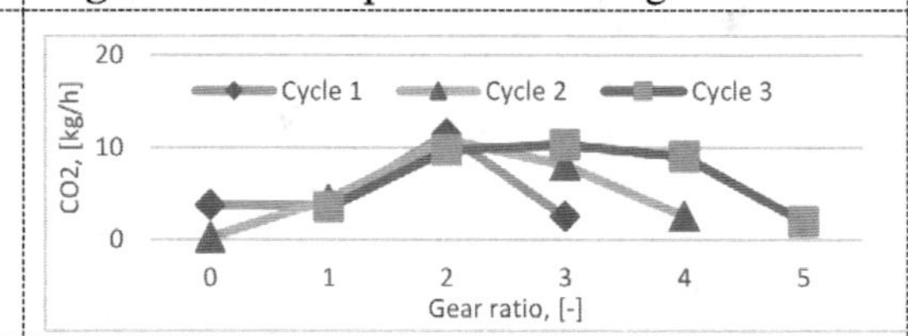

Figure 8. CO2 emissions versus gear ratio.

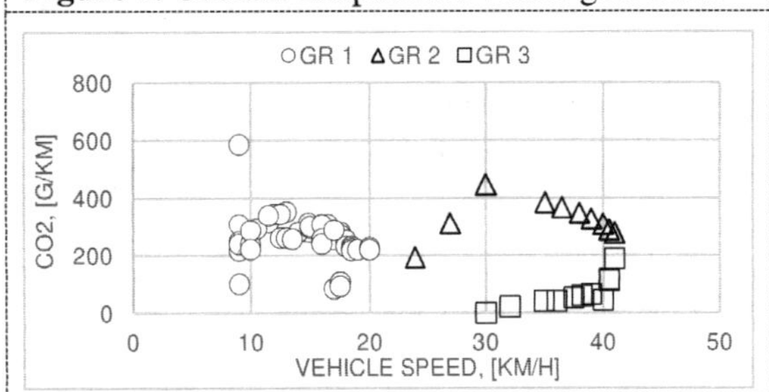

Figure 9. CO2 value vs. vehicle speed-Cycle 1.

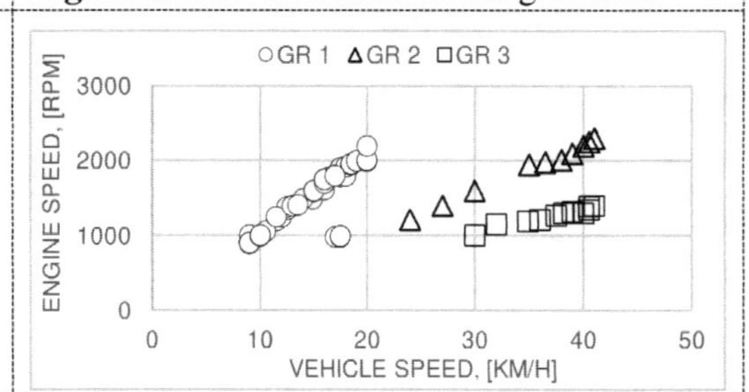

Figure 10. CO2 vs. vehicle speed-Cycle 1.

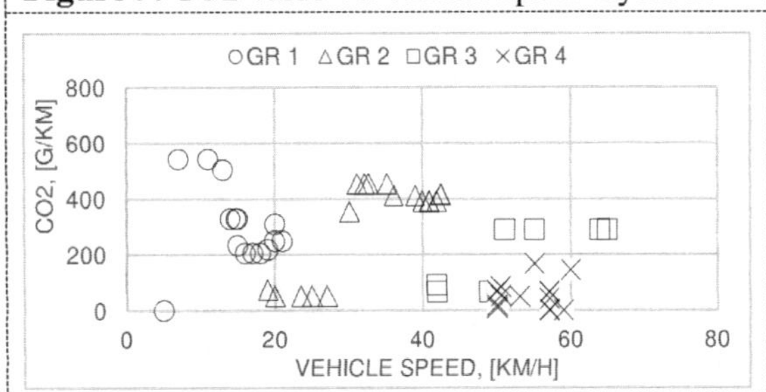

Figure 11. CO2 value vs. vehicle speed-Cycle 2.

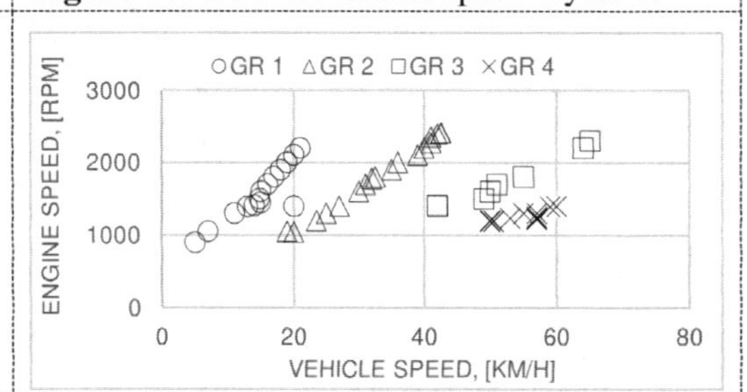

Figure 12. CO2 vs. vehicle speed-Cycle 2.

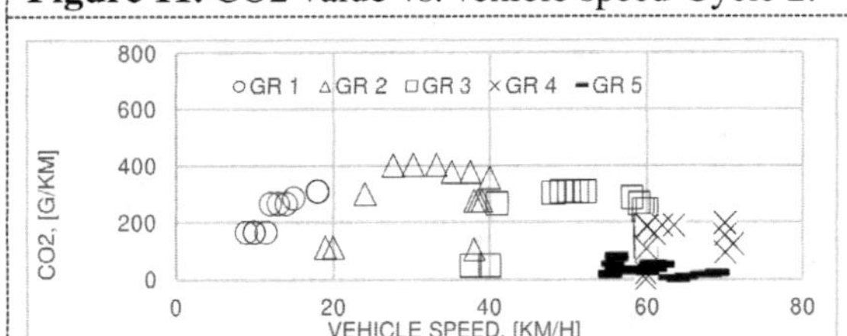

Figure 13. CO2 value vs. vehicle speed-Cycle 3.

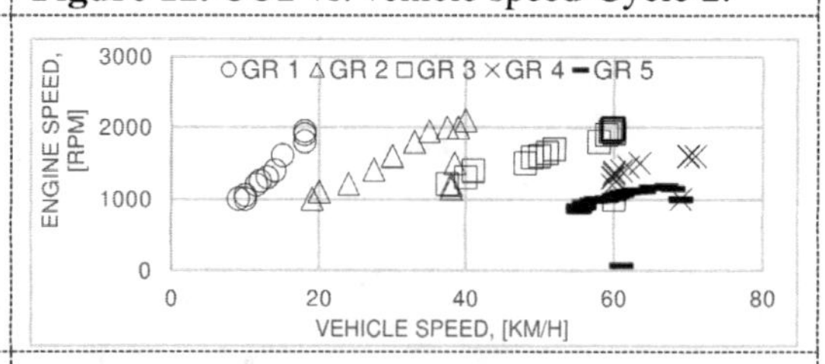

Figure 14. CO2 vs. vehicle speed-Cycle 3.

The CO2 emissions were calculated based on fuel consumption and partial combustion model. All the rest of parameters were measured using the electronic Diesel control system sensors and capabilities.

The average hourly CO2 emissions versus the vehicle speed is given in figure 15. In the rest of graphics from 16 to 20 are presented the engine speed, gear ratio, fuel consumption, oil and coolant temperatures average values versus vehicle speed.

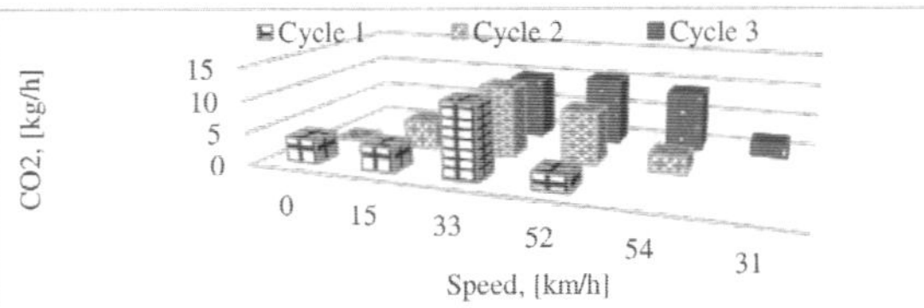

Figure 15. Average CO2 values vs. vehicle speed.

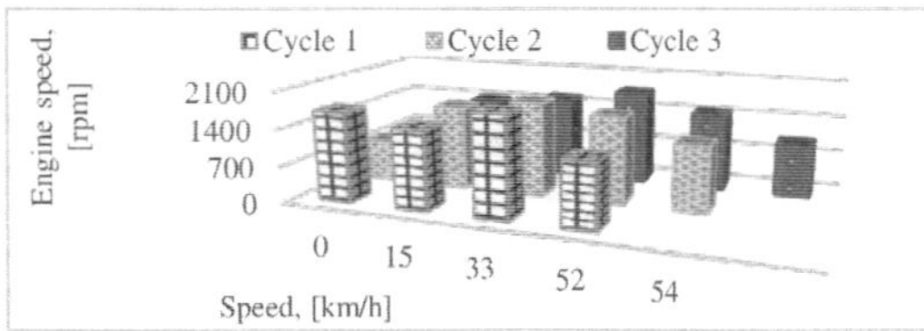

Figure 16. Engine speed vs. vehicle speed.

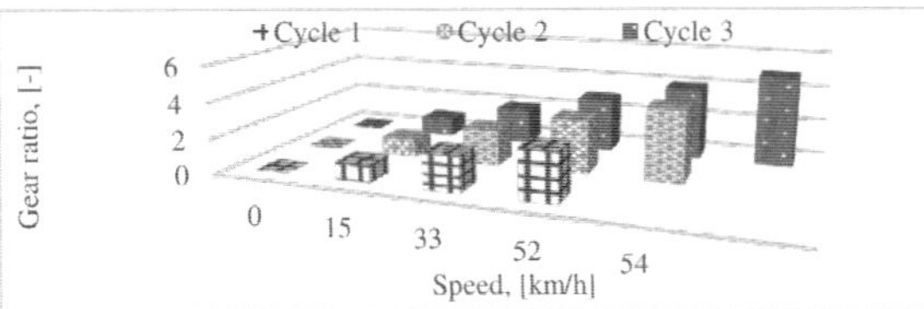

Figure 17. Gear ratios vs. vehicle speed values.

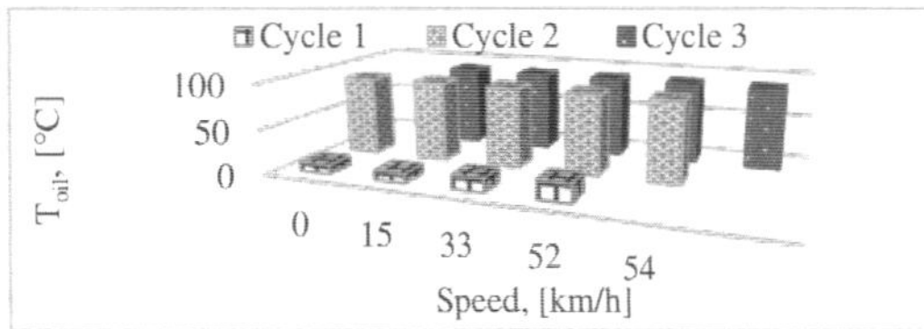

Figure 18. Oil temperature vs. vehicle speed.

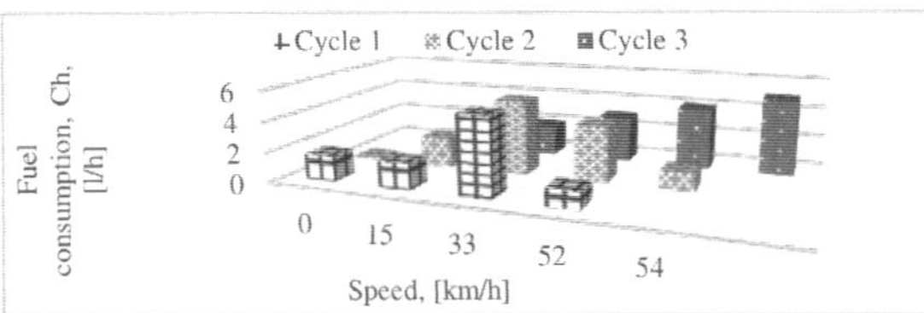

Figure 19. Fuel consumption vs. vehicle speed.

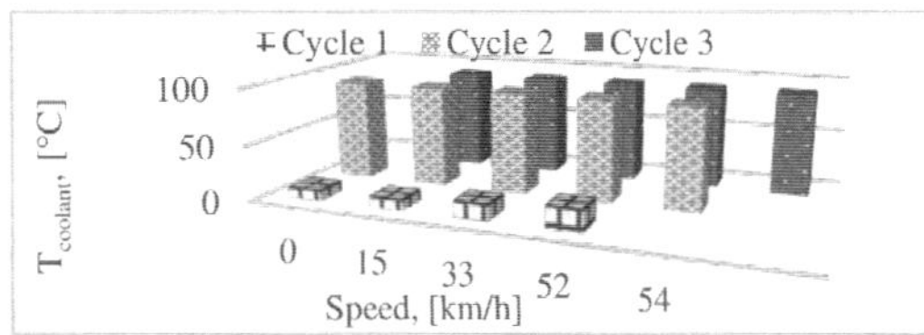

Figure 20. Engine temperature vs. car speed.

CO_2 calculation is supported by air and fuel consumption measured values with the following equation:

$$CO_2 = 0.2306 * 10^{-2} * v^2 - 37.013 * v + 1074.1; \quad CO_2 = C_h * E_c, \text{ [g/km]} \tag{1}$$

where v is the vehicle speed, in [km/h]; CO_2 – carbon dioxide, in [kg/h] or [g/km]; C_h – fuel consumption, in [l/100]; E_c – emission concentration, in [g/l/100km].

3. Observations and conclusions

Density of supplied diesel fuel is 838 g/litre. This type of fuel has 86% Carbon. There are almost 720 g of C chemical element in each litre of diesel fuel. To combine the available carbon molecules to CO_2 are required around 1920 grams of O_2. Addition of the two numbers results in 2640 of CO_2 g/l. Taking into consideration the average consumption of tested car being 6 l/100 km, it leads to 6x2640 g/l/100 (per km) = 158.4 CO2 g/km. Temperature in the cycle 1 was 10 ± 1˚C. Temperatures in the cycle 2 and 3 were 90°C for engine coolant, and 16±1°C for the intake air. CO_2 research indicated that level of emissions is affected by the engine temperature, engine, and vehicle speed.

In cycle 1, at cold start and operation, the average CO_2 value was 5 kg/h, average speed was 21 km/h, average engine speed was 1578 rpm, fuel consumption (Ch) was 2 l/h, average temperature of coolant and oil was 14°C. In cycle 2, the average CO_2 was 5.216 kg/h, average speed 30.8 km/h, average engine speed was 1402 rpm, Ch was 2.35 l/h, 90°C coolant temperature and 88.8°C temperature oil. In cycle 3, the average CO_2 was 6.88 kg/h, v=44.97 km/h, n=1430 rpm, Ch=3 l/h, T_{engine}=90°C, and T_{oil}=88.8°C. Measurements allow the researcher to determine trendlines of CO_2 emissions in different operational regimes. Those trendlines to be defined need more tests and research.

References

[1] Apostolescu N et al 1967 *Internal Combustion Engines / Motoare cu ardere internă* Bucharest/București, Pedagogic and Didactic Publishing House / Edit. Did. si Ped., pp. 7-428.

[2] Barabas I et al 2010 *Performance and emission characteristics of an CI engine fueled with diesel–biodiesel–bioethanol blends* (Fuel 89(12), pp. 3827-3832) DOI: 10.1016/j.fuel.2010.07.011

[3] Băldean DL, Borzan AI 2018 *Contributions on experimental research of fuel consumption and management system in K9K892 Diesel engine from Dacia-Renault*, DOI: https://doi.org/10.1051/matecconf/201818401018

[4] Bec P et al 2020 *Investigation of Failures and Vulnerabilities in Road Traffic Air Quality Management System during 2020 Pandemics*, DOI: 10.1088/1757-899X/898/1/012009

[5] Borza EV et al 2018 *Experimental research of CO2 emission in relation with traveled distance and fuel consumption for Mitsubishi L200 automobile on the Cluj-Pitești road*, DOI: http://stiintasiinginerie.ro/34-61

[6] Borza EV et al 2019 *Research Concerning Fuel Economy Coefficient and Carbon Foot-Print in Various Conditions for a City Compact Size Vehicle with Digital Control for a Green Solution and Method at Tech. Univ. from Cluj-N*, DOI: https://doi.org/10.1007/978-3-319-94409-8_22

[7] Borzan AI et al 2020 *The Development of a New Interface for Intelligent Control of Energy Supply in Dynamic Environment with Process Digitization*, DOI: 10.1016/j.promfg.2020.05.008

[8] Chereches IA et al 2018 *Contributions to experimental research and development of some advanced methods for analyze and report of the carbon oxide for an optimized management of health and life*, DOI: http://stiintasiinginerie.ro/34-69

[9] Chereches IA et al 2019 *Research of Intelligent Control of Injection Systems for Subaru Competition Car*, https://doi.org/10.1007/978-3-030-32564-0_10

[10] Ferenti I et al 2015 *Experimental research regarding performances of engine supply systems from automobiles made for motorsport competitions* (In I.A.T.A., vol. 45, CZU: 629.33.063) DOI: https://ibn.idsi.md/sites/default/files/imag_file/162-165_5.pdf

[11] Ferenti I et al 2019 *Experimental research in mechanic's field of control for further innovation in the hydraulic and electronic operated injection system*, DOI: 10.1051/e3sconf/20198508006

[12] Gabrian C et al 2018 *Experimental study of the possibility of increasing energy efficiency through the control of friction in roller bearings of the wheel train from Dacia Solenza*, DOI: http://stiintasiinginerie.ro/33-57

[13] Gaspar F et al 2017 *Contribution to the experimental research of environmental carbon footprint in the case of 1400 cmc engine from Skoda Fabia vehicle*, DOI: http://stiintasiinginerie.ro/31-74

[14] Jovrea S et al 2017 *Researching on-board display of essential information concerning technical conditions in operation and fuel-economy of a motor-vehicle in operation*, DOI: http://stiintasiinginerie.ro/31-67

[15] Jovrea LD et al 2017 *Study of the semiotic aspects concerning the hydraulics of the injection system in diagnosing 1.8 TDCI Ford Focus engine*, DOI: http://stiintasiinginerie.ro/31-68

[16] Marincaș C et al 2017 *Contributions to the experimental research of EDC module operation in relation with supply of the N47 engine from BMW 320D (E90) automobile*, DOI: http://stiintasiinginerie.ro/31-82

[17] Moldovan A et al 2017 *Experimental research of the management system from the Peugeot 4007 Sport Utility Vehicle*, DOI: http://stiintasiinginerie.ro/31-71

[18] Odenie S et al 2018 *Contributions to research of CO2 emissions in environment pollution management and life*, DOI: http://stiintasiinginerie.ro/34-65

[19] Popescu D et al 2020 *Development of an Automated System for Fuel Tank Level Checking and Machinery Location Management to Optimize Remote Accessibility and Mobile Tracking* (In MDPI, Vol. 63, 1) DOI: https://doi.org/10.3390/proceedings2020063017

[20] Toader GV et al 2017 *Contributions to the experimental research of injection fuel supply system from 3.0i engine of BMW X 5 road vehicles*, DOI: http://stiintasiinginerie.ro/31-70

IOP Conf. Series: Materials Science and Engineering 1169 (2021) 012031 doi:10.1088/1757-899X/1169/1/012031

Researching particulate matter characteristics for mitigating health risks generated by road vehicles

L Andrei[1], **V Chindea**[2] **and D L Băldean**[2]
[1] Infectious Disease Hospital of Cluj-Napoca, Cluj, Iuliu Moldovan 23, Romania
[2] Automotive and Transportation Dep., Technical University of Cluj-N., Cluj, Muncii 103, Ro.

doru.baldean@auto.utcluj.ro

Abstract. Research of PM (particulate matter) emissions is very important in crowded areas with high values of road traffic and industrial sites because may affect the health significantly. To mitigate the health risks generated by motor vehicles, the present paper shows the applied research on this topic with the practical results. The first important objective of the paper is to study the characteristics related to particulate matter in emissions generated by a car in actual road traffic test. The second objective is to compare the results from cold and warm operation regimes. It is important to know the trendline of PM emissions during operation due to their influence upon health. The PM amount is influenced by engine speed, fuel consumption and other operational factors. High levels of PM were recorded at 1500 rpm because it was the upper point in which the gear ratio has been changed.

1. Introduction

Particles in suspension are entering the body and they are interacting with different physical, chemical, and biological structures generating various phenomena. Some of the particles are toxic, meanwhile some of them may be neutral in relation to the organism. There are studies which undertake the problem of multiple micronized and non-micronized particles into human tissue showing flow characteristics [1].

The problematic contact of particulate matter and other pollutants is with the soft tissues, such as eyes and nose [2]. Operation and its inherent problematic aspects may lead to particulate emissions [3]. These particles are retained by specific filters, in standard systems and solutions, for diminishing the public health risks [4]. There are some store limits and regeneration procedures for such systems [5]. Development of electronic interfaces to improve control and management for supplying the systems is making the operation more efficient [6]. Road cars and conventional transport vehicles are generating emissions of particulate matter in suspension which are impacting the air quality. The latter one is under the control of specialized Management System. Particulate characteristics are generating problems not only for living beings but, in some cases, also for the air quality monitoring equipment [7]. "Suspension" in this case is used to define some fluids which consist of suspended particulate matter (gas bubbles, liquid droplets, and rigid particles) in a liquid mass [8]. Multiple solutions and experimental testing methodologies have been proposed during the last 3 decades to properly manage different types of particles [10, 11, 12]. Experimental research of particulate matter in relation to fuel and engine has led to idea that operation regime and energy source are somewhat influential in determining emissions [13]. From modelling the particle dynamics, through production of particle suspensions, the characteristics and behaviour are studied with nonlinear equations [14, 15, 16]. Flow and displacement of liquid and solid particles, studied by rheology is applicable in food engineering and for characteristics' definition of multiple nanofluids [17, 18, 19]. Air quality and environment status are influenced by emissions [20].

2. Applied research

Method of research is based on the experiment and recordings with electronic Diesel control system which has the technical data given in table 1.

Main technical specs concerning the tested car are useful to properly identify the powertrain.

Table 1. Practical specifications of the car used during testing

Parameters	Actual value
Propulsion	Diesel 350 CDI / 3.0 L turbo V6 OM642
Energy supply	Common-Rail Injection system with six injectors
Gearbox	7-ratios / 7G-TRONIC
Axles	Classical solution: 4x2 / Front steering / Rear axle drive
Platform	Mercedes-Benz Bluetec E350 CDI/Model W/A207 (2013)

Figure 1 presents the basic steps of the research methodology in approaching the topic of PM risk.

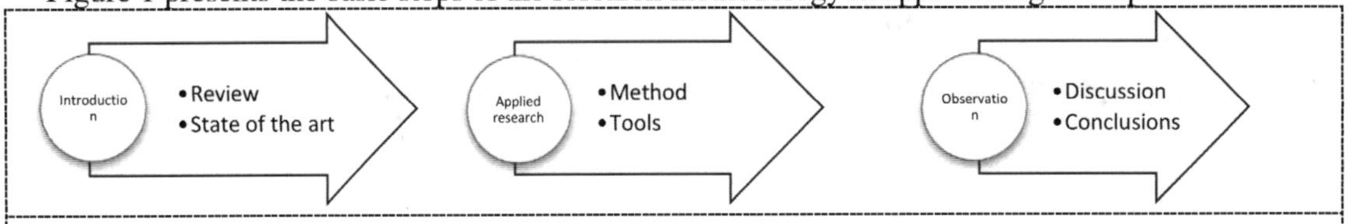

Figure 1. Schematic of the PM emission research, into the environment, when generated by road car

PM assessment starting from air and fuel consumption is based on the instant mixture as shown below:

$$PM = (1\ldots6) * 10^{-7}n^2 - (0.3\ldots8) * 10^{-4}\,n + (0.15\ldots0.3),\ [kg/h] \quad (1)$$

where n is the engine speed, in revolutions per minute [rpm]; PM – particulate matter, in [kg/h].

Figure 2 shows the PM simplified structure in the exhaust gases of a road car. The intake air passes through the front grille 1 to the engine 2. The front axle 3 is used for steering, braking and support, meanwhile the on-board electronic unit 4 allows the control of fuel flow and consumption from tank 5. Soluble organic fraction of hydrocarbons 6 are surrounded by other particles 7 (SO_4) and liquid droplets 8 (HC). Solid carbon 9, combined with soluble fractions 6 and 8, in the presence of vaporised phase 13 of HC, poses a significant carcinogenic health risk due to its specific characteristics. The PM may be produced both by tire 12 friction with the road surface and by exhaust system 11 in the emissions 10.

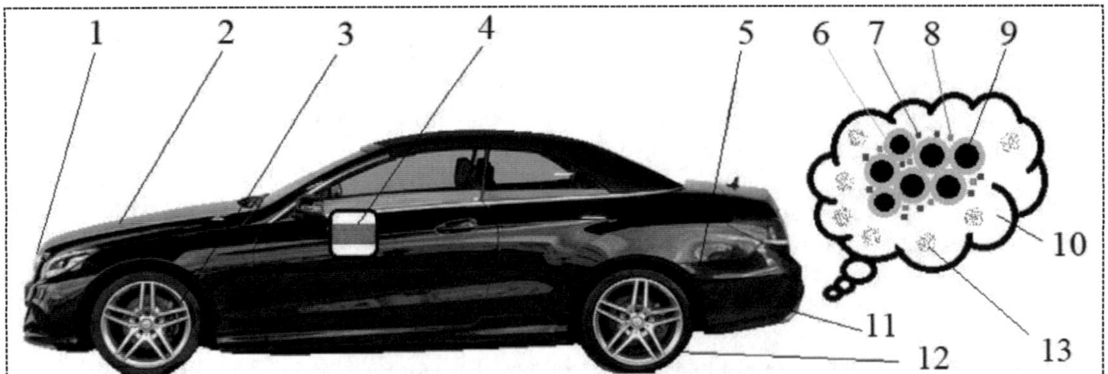

Figure 2. Particulate matter in suspension generated by a road vehicle during the engine operation.

IOP Conf. Series: Materials Science and Engineering 1169 (2021) 012031 doi:10.1088/1757-899X/1169/1/012031

In figures 3 to 14 are shown the graphs for PM variations during the entire transient test cycle of 1800 s duration in conjunction with engine and vehicle speed, gear ratio, engine temperature and autonomy.

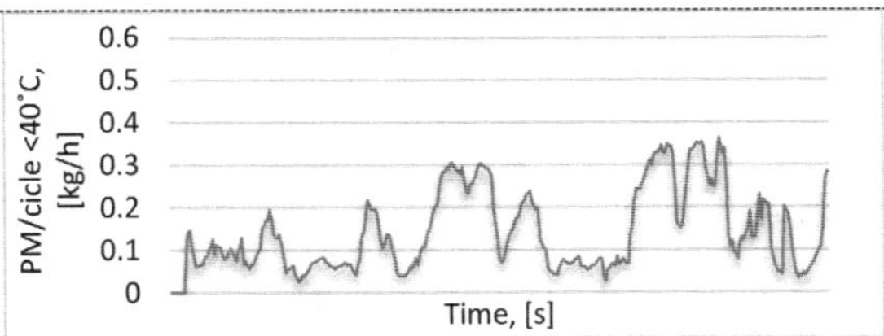

Figure 3. PM per stroke below 40°C during test.

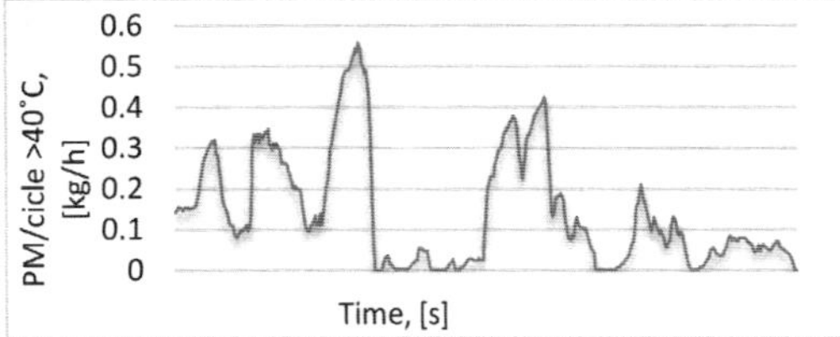

Figure 4. PM per stroke above 40°C on test.

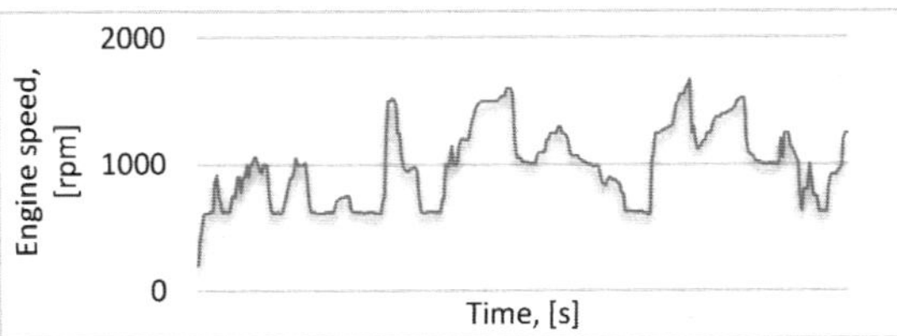

Figure 5. Engine speed below 40°C during testing.

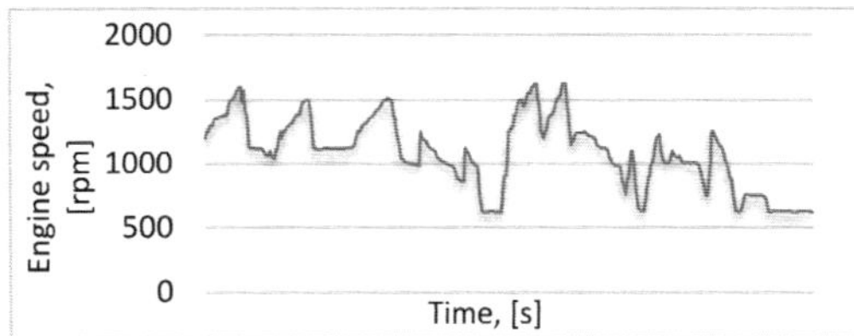

Figure 6. Engine speed above 40°C during test.

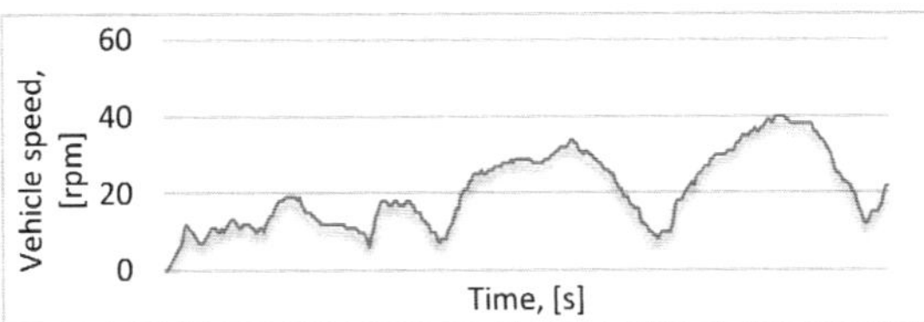

Figure 7. Vehicle speed below 40°C during test.

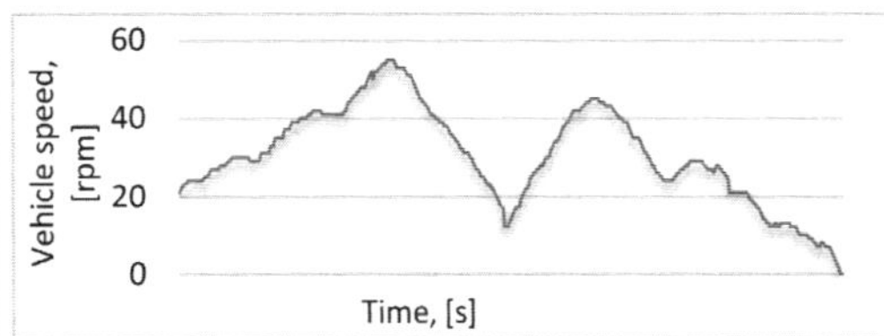

Figure 8. Vehicle speed above 40°C on test.

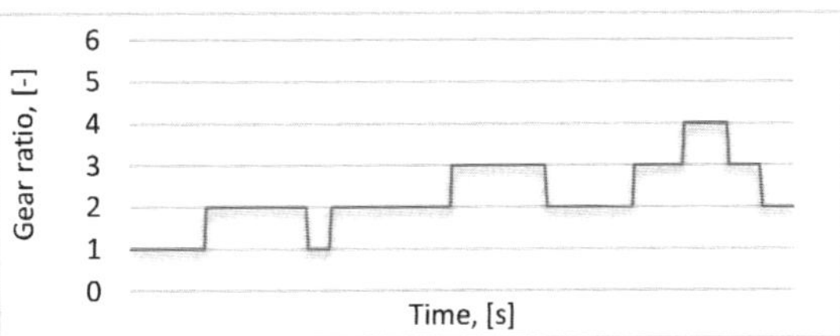

Figure 9. Gear ratio below 40°C during testing.

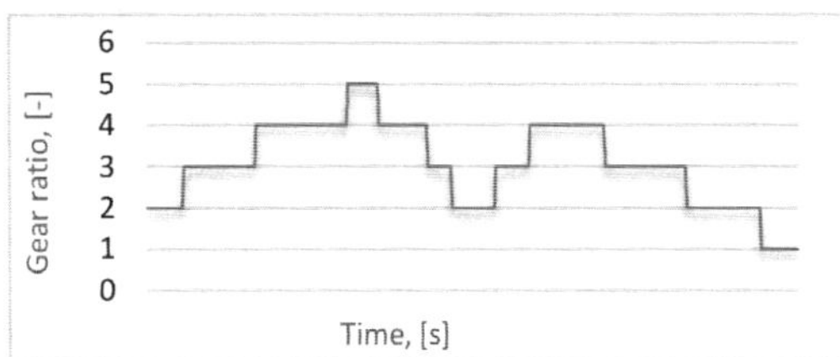

Figure 10. Gear ratio above 40°C on testing.

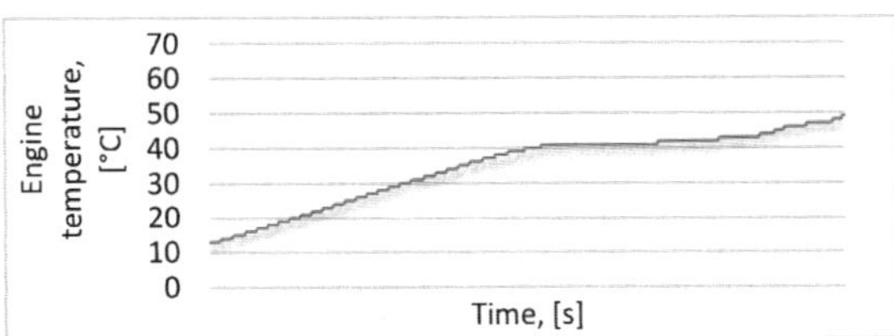

Figure 11. Engine temperature below 40°C.

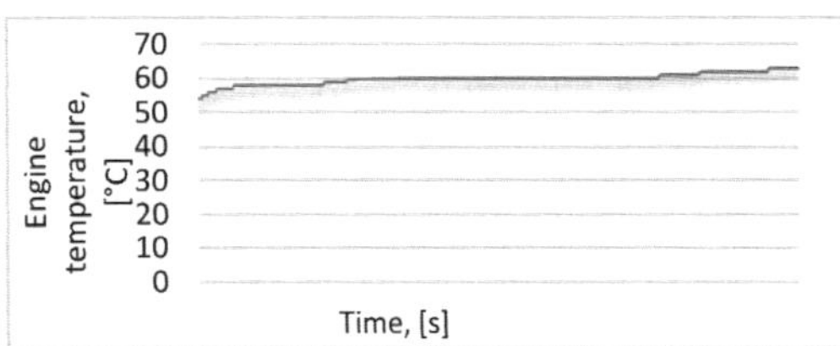

Figure 12. Engine temperature above 40°C.

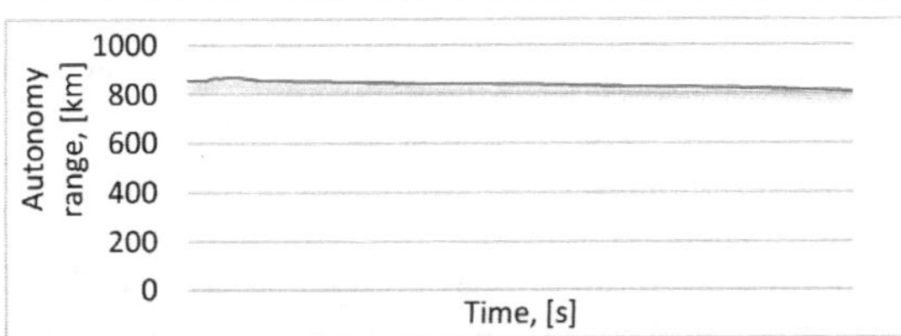

Figure 13. Autonomy range recorded below 40°C.

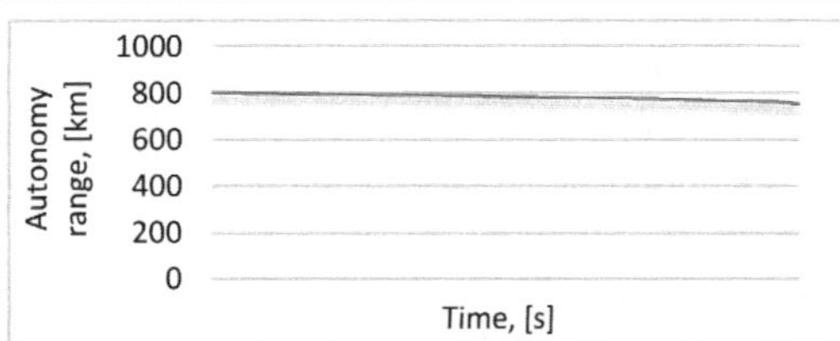

Figure 14. Autonomy range above 40°C.

The PM emissions were determined through calculus. The latter one is based both on the air intake and fuel consumption. It was divided in two sequences: cold start below 40°C and warm-up phase above it.

The hourly PM emission versus the engine/vehicle speed and gear ratio are given in figures 15 to 20.

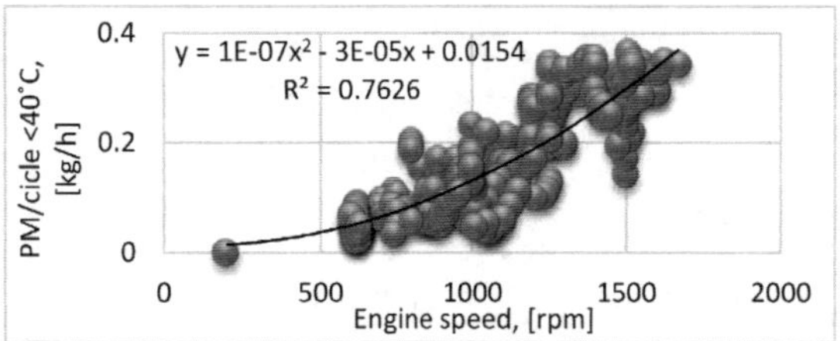

Figure 15. PM values vs. engine speed at cold.

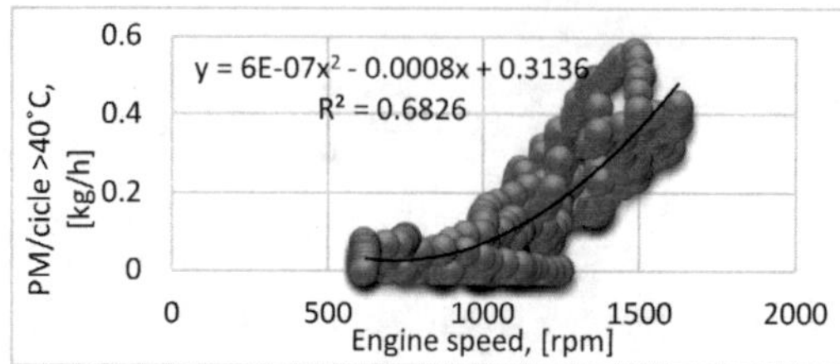

Figure 16. PM vs. engine speed above 40°C

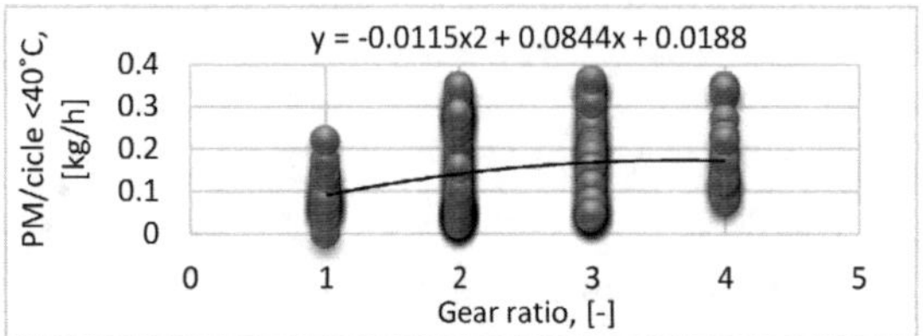

Figure 17. PM values vs. gear ratio at cold start.

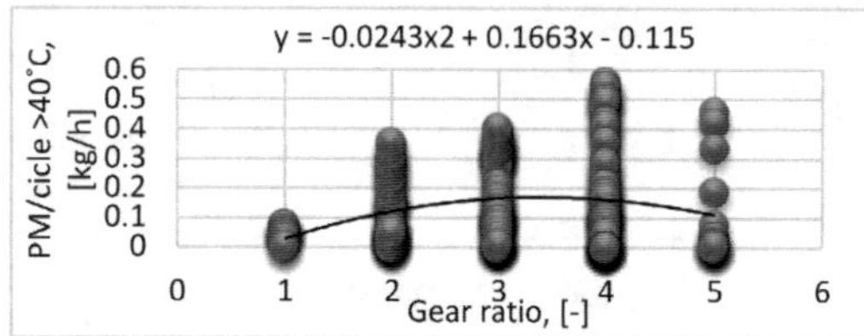

Figure 18. PM vs. gear ratio above 40°C

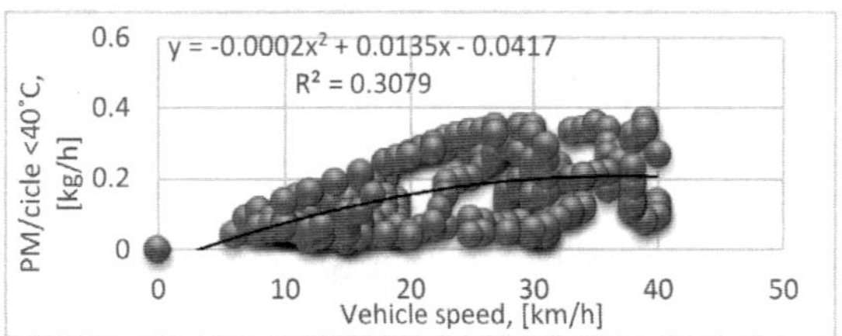

Figure 19. PM vs. vehicle speed below 40°C.

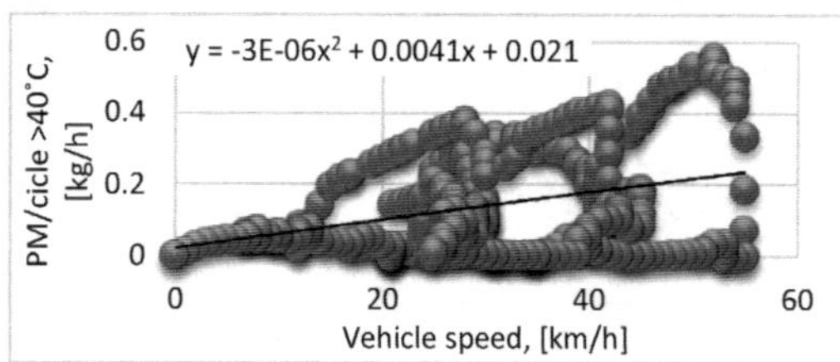

Figure 20. PM vs. vehicle speed above 40°C

PM assessment in the relation with consumption, speeds and gear ratios facilitates some observations.

3. Observations and conclusions

Air contamination with particulate matter is nowadays of high concern due to the negative impact on health and climate. Researching and measuring the non-organic solids and soluble organic fractions, and their variation with different parameters are crucial in comprehending the mechanical, thermodynamic, and chemical transformations in environment and influences in health issues. Air temperature during the tests has been 13°C. The vehicle used in tests for the present research has 7+1 transmission ratios, and 3 changing strategies (E-automatic eco-mode; S-sporty; M-manual). In the current investigation the M strategy for gear change was applied. In the present work, cold start and low temperature operation has been investigated to gain insight about the PM variation with the possibility of partial level control of emissions. Elemental and total particulate matter analyses confirmed that level of analysed pollutant reference samples to be influenced by the engine temperature.

In the first step of the experimental research, when the engine temperature ranged between 13 - 40°C, the average PM value was 2.40 g/min. Between 40°C and 63°C, the second phase of the test, PM average emission value was 2.38 g/min. Polluting activity of PM was the highest when the engine speed, vehicle speed, and the number of injections were at peak values. The average number of injections below 40°C was 495 per minute. In the range 40-63, second stage of testing, there were 544 inj/min. Probability theory allows the calculation of the PM levels increase when coolant temperature changes and the vehicle operate in different engine regimes. These aspects, to be highly defined will require further studies.

References

[1] Allahham A et al 2005 *Factors Affecting Shear Thickening Behavior of a Concentrated Injectable Suspension of Levodopa* (Journal of Pharmaceutical Sciences, Vol. 94, 11, pp. 2393-402) DOI https://doi.org/10.1002/jps.20374

[2] Andrei L et al 2017 *Contributions to the applied research of effects from surrounding environment pollution due to road traffic upon eyes' anatomy and pathology development* (In Science and Engineering AGIR Press) DOI: http://stiintasiinginerie.ro/32-85

[3] Băldean DL 2017 *Researching the nonfunctional aspects of the Diesel Particle Filter (DPF) in the case of Dacia Logan E5 1.5dCI engine* (DOI: http://stiintasiinginerie.ro/31-79)

[4] Băldean DL et al 2020 *Research of NOx and PM10 pollutants in Cluj-Napoca with the mobile system for mitigating public health risks efficiency* (In IOP Conf. Series: Materials Science and Engineering, Vol. 898, Issue 1) DOI: 10.1088/1757-899X/898/1/012003

[5] Borza EV et al 2019 *Research of Diesel Particle Filter Soot Impact and Regeneration Time Trend-Line* (In IOP Springer, Cham) DOI: https://doi.org/10.1007/978-3-030-32564-0_24

[6] Borzan AI et al 2020 *The Development of a New Interface for Intelligent Control of Energy Supply in Dynamic Environment with Process Digitization* (In Procedia Manufacturing 46, 914-921) DOI: 10.1016/j.promfg.2020.05.008

[7] Bec P et al 2020 *Investigation of Failures and Vulnerabilities in Road Traffic Air Quality Management System during 2020 Pandemics* (In IOP Conf. Series: Materials Science and Engineering, Vol. 898, Issue 1) DOI: 10.1088/1757-899X/898/1/012009

[8] Chinesta F, Ausias G 2016 *Rheology of Non-spherical Particle Suspensions Systems* (In Science Direct, ISTE Press - Elsevier) https://doi.org/10.1016/C2015-0-01208-4.

[9] Denn MM et al 2014 *Rheology of Non-Brownian Suspensions* (In Annual Review of Chemical and Biomol. Eng., Vol. 5, pp. 203-228) DOI: 10.1146/annurev-chembioeng-060713-040221

[10] Dong XM et al 2009 *A new variable stiffness absorber based on magneto-rheological elastomer* (In Transactions of Nonferrous Metals Society of China, Vol. 19, 3, pp. s611-s615) https://doi.org/10.1016/S1003-6326(10)60118-5

[11] Ferenti I et al 2019 *Experimental research in mechanic's field of control for further innovation in the hydraulic and electronic operated injection system* (In E3S Web of Conferences 85, 08006) DOI: https://doi.org/10.1051/e3sconf/20198508006

[12] Gast AP, Zukoski CF 1989 *Electrorheological fluids as colloidal suspensions* (In Advances in Colloid and Interface Science, Vol. 30, pp. 153-202) DOI: 10.1016/0001-8686(89)80006-5

[13] Jurchiș BM et al 2019 *Experimental research on Diesel Particle Filter (DPF) in relation to the fuel type,* DOI: 10.1088/1757-899X/568/1/012027

[14] Polfer P et al 2016 *Suspension modeling using smoothed particle hydrodynamics: Accuracy of the viscosity formulation and the suspended body dynamics (*In Applied Mathematical Modelling, Vol. 40, 4, pp. 2606-2618) DOI: https://doi.org/10.1016/j.apm.2015.10.013

[15] Radcliffe AJ et al 2019 *Dropwise Additive Manufacturing of Pharmaceutical Products Using Particle Suspensions* (In: Journal of Pharmaceutical Sciences, Vol. 108, 2, pp. 914-928) https://doi.org/10.1016/j.xphs.2018.09.030

[16] Saintillan D et al 2013 *Active suspensions and their nonlinear models* (In Comptes Rendus Physique, Vol. 14, 6, pp. 497-517) https://doi.org/10.1016/j.crhy.2013.04.001

[17] Shewan HM 2014 *Rheology of Soft Particle Suspensions* (In Food Engineering, The University of Queensland, PhD Thesis) https://espace.library.uq.edu.au/data/UQ_356401/s42402350

[18] Udawattha DS et al 2019 *Predicting the effective viscosity of nanofluids based on the rheology of suspensions of solid particles* (In: Journal of King Saud University – Science, Vol. 31, 3, pp. 412-426) DOI: https://doi.org/10.1016/j.jksus.2017.09.016

[19] Vecera M and Pospíšil J 2012 *Stability and rheology of aqueous suspensions* (In: 20th Internat. Congress of Chemical and Process Engineering) DOI: 10.1016/j.proeng.2012.07.564

[20] *** 2021 *Air quality. Particulate Matter* (In: National Air Quality Monitoring Network) http://www.calitateaer.ro/public/assessment-page/assessment-page/?locale=en

IOP Conf. Series: Materials Science and Engineering 1169 (2021) 012032 doi:10.1088/1757-899X/1169/1/012032

Autonomous vehicles: classification, technology and evolution

F B Scurt[1], T Vesselenyi[1], R C Ţarcă[1], H Beleş[1],G Dragomir[1]

[1]University of Oradea, Faculty of Management and Technological Engineering, Universitatii street, no. 1, Oradea 410087, Romania

E-mail: scurt.florin@gmail.com

Abstract. In the context in which most industries focus on digitalization and automation of production processes and on developing new products, the car manufacturing industry is advancing significantly. The automotive industry together with the IT field, are at an advanced stage in terms of the structure of autonomous vehicles, developing new technologies. The objective of this paper is to present the current state of technology regarding autonomous vehicles. Aspects regarding the devices and technologies used in the construction of these vehicles are approached, but there is also information regarding the way of classifying them according to the driver's assistance level. In addition, some information about the evolution of automated vehicle development is presented.

1. Introduction

Nowadays, when the world is in constant motion, the number of own means of transport (especially cars) or public transport is growing very fast, both due to the needs and due to the current trend. "According to statistics, there are 1.1 billion vehicles, but it is expected to increase to 2 billion in the next 10-15 years" [3]. With the increase in the number of land transport, one thing that has been observed for several years, is the overcrowding of cities. Overcrowding is characterized by the lack of parking spaces and by the traffic jams present throughout the day in the most interesting areas of a metropolis or even small towns. All this leads to the loss of significant amount of time during the day, spent behind the wheel of the vehicle or in means of transport (busses, trolleybuses, etc.), which has direct effects on the state of fatigue and implicitly of the driver's attention. In addition, traffic jams have an adverse effect on the environment, because significant amounts of harmful substances are generated during operation, even when waiting at a traffic light. All these factors mentioned above, often lead to the generation of accidents, which can be material or in the worst case can result in injury or even death of some people. As an alternative for the future, regarding the elimination of the disadvantages of conventional traffic, the autonomous (or automated) vehicle brings a series of facilities in order to facilitate urban and interurban mobility. These facilities range from driver assistance systems to complex system (hardware and software) capable of activities such as: perception, decision and reaction, with the role of streamlining road traffic through better of the driving process a vehicle.

Vehicles equipped with various driving assistance systems are no longer a novelty, they have been on market for several years. These systems have an important contribution to reducing the number of road accidents. According to recent data, approximately 1,2 million people die worldwide due to road accidents, [3].

IOP Conf. Series: Materials Science and Engineering 1169 (2021) 012032 doi:10.1088/1757-899X/1169/1/012032

The autonomous vehicle can be defined as an ordinary vehicle that can move from one place to another without the intervention or assistance of a driver, in other words the vehicle is able to analyze the environment and make favorable decision, to travel alone, due to equipment from endowment.

2. Evolution

Around the 1920s, in the USA, due to the large number of accidents caused by the carelessness of drivers, the development of driverless vehicles was discussed. Thus, in August 1921, the first vehicle without driver on board was presented, which was radio-controlled from an army vehicle located 30m behind it. It can be said that this is the pioneer of cars without a driver. The development of this prototype being closely linked to the technology owned by the army, [3].

Of course, after this presentation, other vehicles appeared either based on the same technology or on something more innovative:

- in 1925, the electronic engineer Francis P. Houdina, equipped an ordinary vehicle with the equipment necessary to perform driving maneuvers, being controlled remotely by radio waves, [1], [2].
- in 1939 during the exhibition "Futurama", an innovative concept propelled by magnetic field generated by the electric circuit incorporated in the running track was presented. It was made by industrial designer Norman Bel Geddes with the support of General Motors, [2].
- in 1958 – the first "automatically guided automobile" completed a one-mile test at the GM Technical Centre (Michigan).It was a Chevrolet that had two electronic sensors mounted in the front that followed a cable stretched along the track, thus, controlling the position of the steering wheel and implicitly the steering wheels, [1].
- in the 1980s, research on autonomous vehicles took off in many countries, both academically and industrially, [1].
- in 1994 in the PROMETHEUS-Project, Ernst Dickman's team developed with the help of Mercedes Benz W140 – a robotic vehicle, which was able to travel on congested highways around Paris at speeds of up to 130 Km/h, [1].

Thus, manufacturers companies such as: Mercedes-Benz, General Motors, Continental Automotive System, Autoliv, Bosch, Nissan, Toyota, Audi, Hyundai Motor Company, Volvo, Tesla Motors, Peugeot, Navya, Google, BMW, Local Motors, Easy Mile, etc., have developed and are still developing new prototypes of autonomous vehicles. Due to this technological boom, the legislative framework on the testing and circulation of autonomous/automated/robotic vehicles in many countries needs to be rethought. In this sense, in 2013 the UK government allowed their testing on public roads, after that, in 2014, the French Government adopted the same legislative measures, followed by other European countries and not only(in 2015, 5 US states allow AV testing on public roads: Nevada, Florida, California, Virginia, Michigan), [4].

3. Classification

According to the J3016 standard published by SAE International, regarding the degree of automation of the driving function, there are 6 levels, from SAE Level 0 (without automation) to SAE Level 5(fully autonomous vehicle). It can be said that this ranking was made on the one hand based on the volume of activities in which the driver is involved, and on the other hand based on the performance and equipment of the vehicle. In addition, this standard (SAE Standard J3016, 2016) corresponds to the levels defined by the Federal Highway Research Institute of Germany (BASt - Bundesanstalt für Strassenwesen), but also to some extent with those formulated by the National Administration for Traffic and Safety (NHTSA), [4].

SAE Level 0 (No Automation): this is where most current vehicles fall. In this case, as it is known, the driver is responsible for all actions,(acceleration/braking, maintaining direction, changing direction, using lights and signals, avoidance of obstacles, etc.), but in some situations, the vehicle may be equipped with warning systems (object warning in the blind spot of the mirror, collision warning).

SAE Level 1 (Driver assistance): systems implemented on the vehicle, at this level, can take control of the acceleration/brake (adaptive cruise control or emergency braking), or the direction of travel, using the information taken with the help of sensors in the environment in which they move (can keep the vehicle on the lane – Lane keeping assist system-LKAS). The driver must be able to take control if the situation so requires.

SAE Level 2 (Partial automation/assistance): the vehicle can control the steering, acceleration and braking, but the driver must be able to intervene when needed, constantly monitoring the surrounding conditions: traffic, weather and road conditions.

Therefore, for the first three levels, the driver is primarily responsible for monitoring the environment (traffic, weather, road condition), he must be prepared at all times to take control of vehicle, when the auxiliary equipment warns him. Starting with level 3, the autonomous driving systems are clearly superior to the assistance equipment present at the previous levels, because they have the capacity to monitor the driving environment.

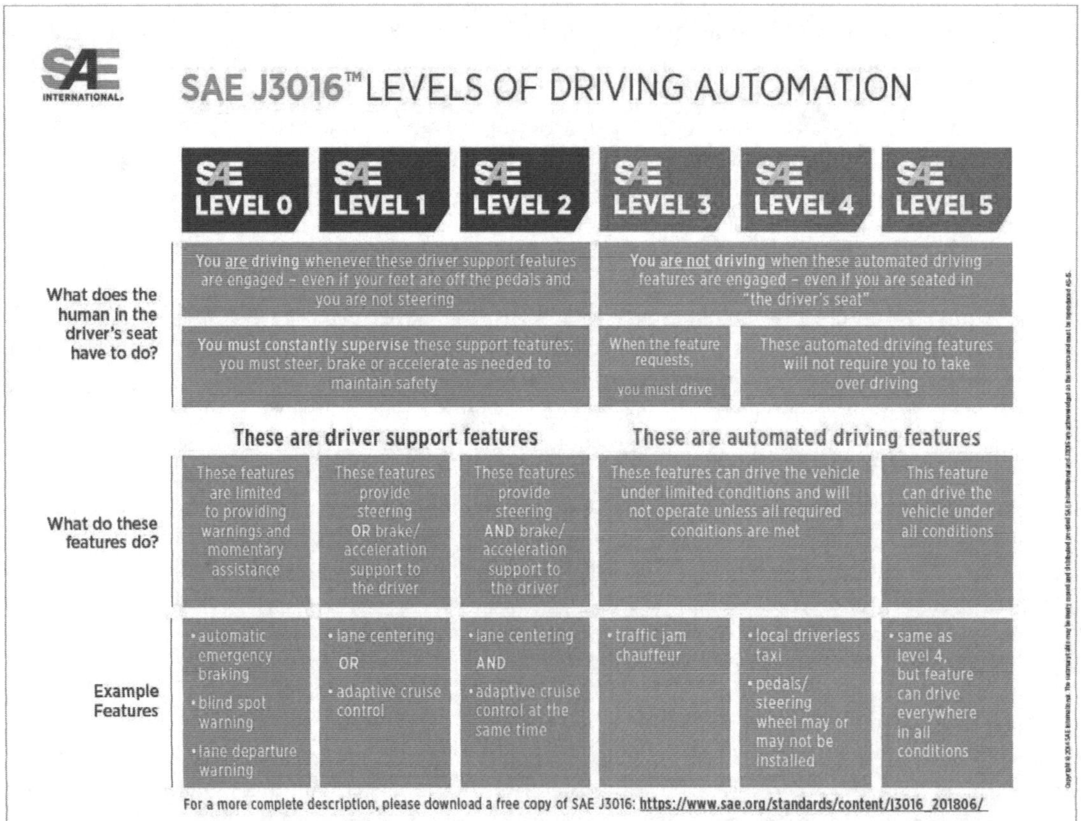

Figure 1. Levels of driving automation according SAE J3016 [12]

SAE Level 3 (Conditional Automation): in this case, the vehicle is able to monitor the surroundings, and control the direction, acceleration and braking in case of highway travel. But, even at this level, the driver must be prepared to intervene if the vehicle asks for this.

SAE Level 4 (High Automation): the activities regarding the controls of the vehicle during the movement can be taken over by an automatic driving system, in some well-defined situations. In some cases, vehicles equipped with such a system no longer have pedals and steering wheel installed.

SAE Level 5 (Full Automation): in this case vehicle is able to manage all the activities necessary for safe driving in any conditions. The only duties of the driver are to start the vehicle and enter the destination.

4. Technology used

In order to be able to travel, autonomous vehicles must be equipped with a variety of sensors and devices that can replace the skills and vigilance of the human factor. The essential devices that are part of an autonomous vehicle, and that help it to move safely are presented in this section. However, it does not mean that all autonomous vehicles have the same type of sensors or technology, these aspects being characteristic of every manufacturer or developer.

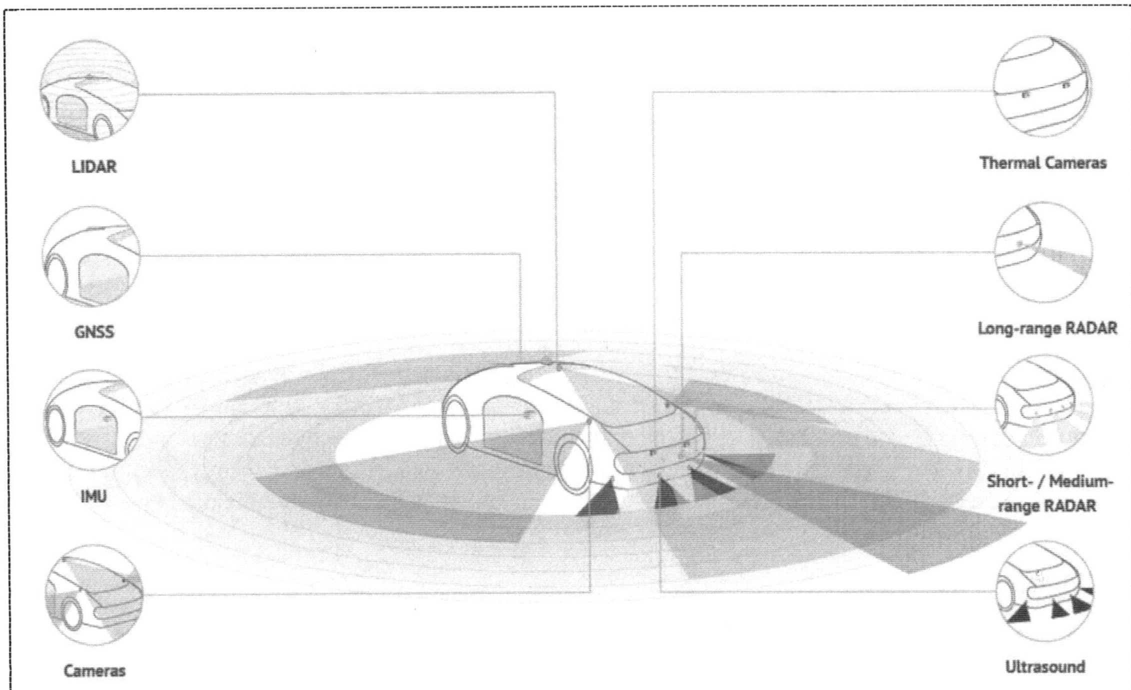

Figure 2. The sensor and device system of an autonomous vehicle [5]

LIDAR (Light detection and ranging) - is a device for measuring distance based on laser beam technology, transmitting light waves (beams) to a target (obstacle), then analyses the reflected light. It can detect static or dynamic objects (vehicles, pedestrian, cyclists, buildings, etc.) in the vicinity of the vehicle. It is capable to create a virtual map (2D or 3D) with these obstacles. It is usually mounted on the roof of the vehicle so that it can have a range of 360 degrees, [2], [4], [6], [9].

RADAR (Radar Detection and Ranging). Radar sensors that equip autonomous vehicles can be grouped in two broad categories: long range radar (LRR) or mid-range radar (MRR). These sensors are mounted on the front, rear and sometimes on the side corners of the vehicles. Radar systems operate on the basis of electromagnetic waves and are used to detect various obstacles or oncoming vehicles. They can also be used to determine the speeds of oncoming vehicles, or for the correct operation of some equipment: adaptive cruise control (ACC), impact warning, for self-parking, etc., [2], [4]. Most often autonomous vehicles have both LRR and MRR radar systems. These systems can have a range of up to about 200m with a dispersion of up to 40° (or even 75°- in this case the range decreases considerably), [2], [4], [7].

GNSS (Global Navigation Satellite System) include all satellite navigation systems, whether we are talking about regional or global systems [8]. GPS (Global Positioning System) the most widely used, is a navigation system that can provide information about time and current position almost anywhere on earth, as long as it is within range of four or more satellites. The system (GPS) helps to keep the vehicle on the desired route with a fairly high accuracy (approximately 30 cm), this being possible also due to the other systems in the endowment of the AV with which it collaborates, [4], [8].

Ultrasonic sensors: these sensors detect objects in the immediate vicinity of the vehicle, being mounted on almost all sides of the vehicle. They are used in parking assistance application, already having a wide spread on the market [4], [10].

Inertial measurement unit (IMU): an inertial measurement unit is a device that can measure the linear and angular motion of the vehicle, as well as the gravitational forces acting on it, using accelerometers and gyroscopes. For high position accuracy, the output data from the IMU are combined with the data obtained from GPS system [2], [4].

Thermal cameras: with their ability to detect heat, they provide a wider range of action than headlights, so they can detect in a timely manner the presence of various obstacles (pedestrians, animals, other vehicles, etc.), on the road or near it. They have the ability to detect even in more difficult conditions of movement: at night, through smoke, on the strong brightness of the sun, etc.

Video Cameras: the video cameras that come with autonomous vehicles are high-resolution stereo cameras, positioned at the front next to the rear-view mirror. Their role is to constantly monitor the road, and to transmit the information (images) for their processing in order to recognize the traffic signs, pedestrian, crossings, traffic lights, cyclists, etc. Some cameras also have the ability to estimate the distance to detect objects. Often autonomous vehicles have video cameras in the back, or mounted on their roof, to be able to see a range of 360°, [1], [2], [3], [4].

ADAS (Advanced driver assistance system) - is one of the most widespread and developed applications in the field of autonomous vehicles, being implemented even in vehicles of the middle and lower class of the market. The advanced ADAS functions used in autonomous driving involve a series of interconnected systems and sensors, thus making it possible to move the autonomous vehicle safely [10]. In other words, ADAS is a piece of equipment that includes both the software solution and hardware part. It is able to carry out the driving process safely without the intervention of the driver, [2].

5. How it works

If we were wondering how we could use an autonomous vehicle to make a trip, compared to a classic vehicle, this is relatively simple: the driver/user/passenger must enter the desired destination, and the vehicle chooses the route or it can be chosen by the driver/ the occupant. This activity is similar to setting a destination on a regular navigation system/device. After setting the destination and the route, the vehicle takes over, [4].

Just as people must have certain abilities to be able to drive a vehicle, so must it be in the case of autonomous vehicles. But in their case, the senses and abilities are taken over by a system of software and hardware devices. The main component of this system is artificial Intelligence through algorithms. Through these algorithms, information is processed and compared, and finally decisions are made regarding the driving commands (acceleration, breaking, steering, etc.) of the vehicle in order to avoid any possible incident.

The system that coordinates an autonomous vehicle can easily be likened to a child's brain and senses, because it is based on approximately the same learning processes (stimulus-perception-processing-classification/comparison-decision making-action). Both, the system and the child need to be initiated in any activity. And with the accumulation of new knowledge, they will be able to better face new challenges in life (in the case of child) or in traffic (in case of the vehicle).

In order to be able to cope in real traffic conditions, the system that ensures the coordination of the vehicle must have a series of data that will represent the database with which it will be able to compare the different situations during the travel. In addition to these data, during the trip the vehicle also records data in order to increase the database, for a better management of the following situations.

Through artificial intelligence, this information is processed so that the vehicle (the algorithms) learns to manage the following situations similar to the previous ones, which were recorded.

Data storage (video/images) and processing is not sufficient for safe driving, so communication with road infrastructure, other vehicles in traffic and even the server are great solutions to improve the safety of manoeuvres.

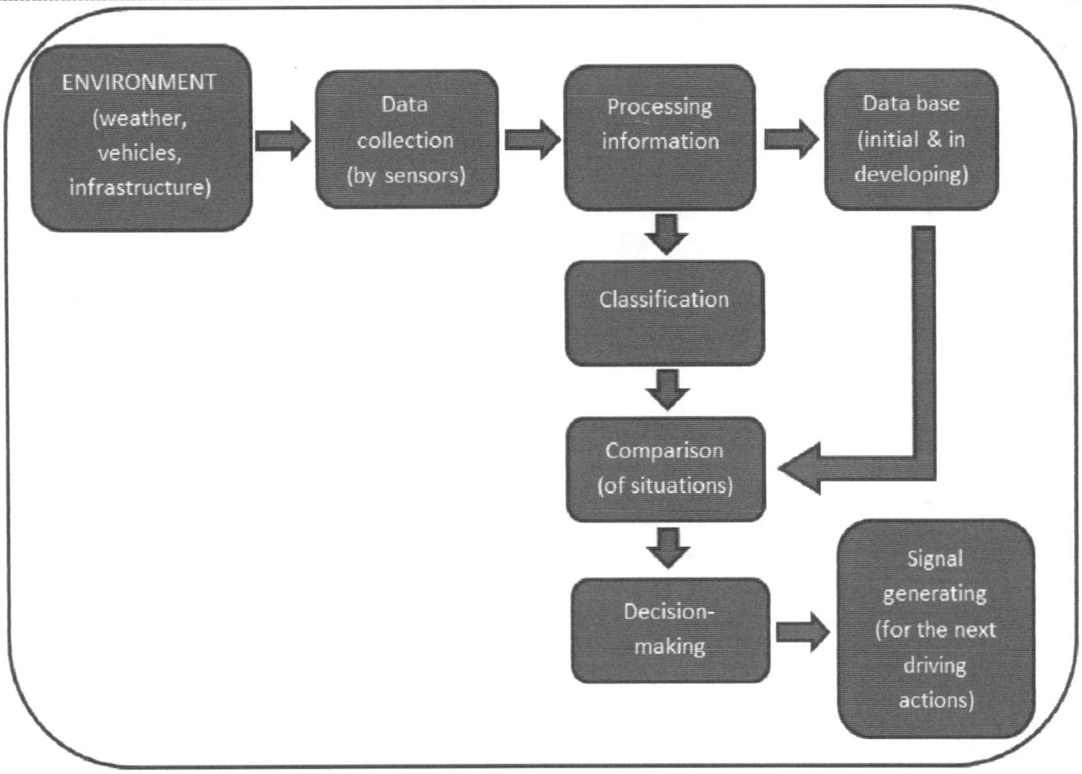

Figure 3. The process behind the every driving action

In other words, artificial intelligence simulates/mimics the complex process specific to human being: perception, processing and decision making, thus realizing the control of the main commands of the vehicle (acceleration, braking, steering, avoiding obstacles, etc.).

Of course, in addition to all the advantages of this technology, there are a number of weaknesses at moment:

- the difficulty of moving in areas under construction that have not been implemented on the digital map;
- the difficulty of moving in off road areas or in chaotic traffic;
- Cyber security, because there is a risk of hacking.

6. The usefulness of autonomous driving technology - possible ways of use

Self-driving technology offers a wide variety of use cases, but only four of them are described below:

Autonomous Valet Parking: refers to the situation in which the vehicle takes control only to move to the first parking lot where there is a free space, or to a parking lot chosen by the driver. Of course, this function can also be used in reverse, for example, the car in the parking lot can be called at the meeting point where it left. Due to the system's ability to manage parking manoeuvres, hard to reach spaces for beginner drivers can be used. Another advantage would be that the size of parking spaces can be reduced, which will provide more space for other purpose in overcrowded cities. Of course, all

IOP Conf. Series: Materials Science and Engineering 1169 (2021) 012032 doi:10.1088/1757-899X/1169/1/012032

these actions (travel to the parking lot or vice versa) must be carried out in strict compliance with the legislation on motor vehicles on public roads, [11], [1].

Interstate pilot: refers to the situation in which the degree of automation of the vehicle allows taking over driving functions (acceleration, braking, steering), through an autopilot, but only in the presence of the driver. This function is intended in particular for travel on the motorway or in stop-and-go traffic jams, thus facilitating the reduction of driver stress in the situations mentioned above. With the help of this function the number of accidents can be reduced, at the same time improving the traffic flow. Activation of the function (transfer of commands) must be performed in a safe manner, [1], [11].

Full automation using driver for extended availability: can be considered a higher version of the "interstate pilot" function, mentioned above. By transferring the driving functions to the automatic system, the driver becomes a passenger, being able to release the steering wheel and pedals, but his/her absence is not allowed. By means of this function the vehicle can move in almost any situation, as long as the movement area is not restricted from a legal point of view, regarding the movement of vehicles equipped with autonomous driving systems, [1], [11].

Vehicle on demand: through this concept and technology, the availability of vehicle in (almost) any location, at the request of an applicant, is possible. This solution is suitable for both passenger and freight transport. By replacing the human operator with an autonomous (automated) driver, the time spent in traffic would be considerably reduced, because the system does not require breaks for rest, being available 24 hours a day. This concept is closest to what an autonomous vehicle (SAE Level 5) means in the future, [1], [11].

7. Conclusion

It is noted that the development of driver assistance systems as well as attempts to make various prototypes of automated/autonomous vehicles has as a starting point the minimization of the number of road accidents (driver and passenger safety, goods safety, and other road users).

In the last 15 years, the technology has developed at a fairly fast pace. The autonomous vehicle and its development bringing together companies from different fields: companies developing navigation system, camera devices, software and hardware solutions, up to the major automotive concerns. But an extremely important factor in the development of this industry is the legislative framework. This is the reason why most often the developers of autonomous vehicles have to meet with the governments of countries of the world, in order to regulate the use of these vehicles in the public space.

The Internet of things (IoT), the giant network of connected devices and users, which collect and share information about the environment where they are, is in continuous development. The evolution of this technology but also of the road telematics (long-range information transfer) made possible the developing of vehicles with a certain level of driving autonomy and increased the safety of using this type of vehicles. Through continuous developing of Internet of Things technology and its integration into automotive filed, the way of interacting between user and vehicle or/and between vehicle and another devices (especially road infrastructure) will be change completely.

In order to increase safety, it will be necessary to intensify the communication protocols between vehicles (V2V) but also the communication with the infrastructure (V2I). Thus, the current (conventional) infrastructure will have to be improved so that it can meet the requirements of autonomous vehicles. And as a considerable improvement in the exchange of information, communication with everything (V2X) – any other device- will further facilitate the movement of autonomous vehicles because they will have enough information to meet the challenges. But all this is possible only with the help of artificial intelligence that requires continuous improvement, so that in the coming years to reach fully autonomous vehicles (SAE-LEVEL 5).

References

[1] Maurer M, Gerdes J C., Lenz B, and Winner H 2015 Autonomous Driving
[2] Iclodean C, Cordos N, and Varga B O 2020 Autonomous Shuttle Bus for Public Transportation:

A Review
[3] Lengyel H , Tettamanti T, AND Szalay Z 2020 Conflicts of Automated Driving With Conventional Traffic Infrastructure
[4] Ondruša J, Kollab E, Vertal P and Šarić Ž 2019 How Do Autonomous Cars Work?
[5] Wevolver. A Review of Autonomous Vehicle Safety and Regulations. Available online: https://www.wevolver.com/article/a.review.of.autonomous.vehicle.safety.and.regulations accessed on (22 January 2021)
[6] Im J H., Im S H, and Jee G I 2018 Extended Line Map-Based Precise Vehicle Localization Using 3D LIDAR.
[7] Bosch Radar Sensor (MRR). Available online: https://www.bosch-mobility-solutions.com/en/products-and-services/passenger-cars-and-light-commercial-vehicles/driver-assistance-systems/automatic-emergency-braking/mid-range-radar-sensor-(mrr)/ (accessed on 21 January 2021)
[8] Min H, Wu X, Cheng C and Zhao X 2019 Kinematic and Dynamic Vehicle Model-Assisted Global Positioning Method for Autonomous Vehicles with Low-Cost GPS/Camera/In-Vehicle Sensors
[9] Chang Y P., Liu Ch N., Pei Z., et al., 2019. New scheme of LiDAR-embedded smart laser headlight for autonomous vehicles.
[10] Advanced Driver Assistance (ADAS) Solutions Guide Ti.com/adas 2015. Available online: https://uk.farnell.com/wcsstore/ExtendedSitesCatalogAssetStore/cms/asset/images/europe/common/applications/automotive/pdf/ti-adas-solution-guide.pdf (accessed on 25 January 2021)
[11] Autonomous Vehicles and Autonomous Driving in Freight Transport Available online: https://tore.tuhh.de/bitstream/11420/1958/1/Fl%C3%A4mig2016_Chapter_AutonomousVehiclesAndAutonomou.pdf (accessed on 23 January 2021)
[12] SAE Standards News: J3016 automated-driving graphic. Available online: https://www.sae.org/news/2019/01/sae-updates-j3016-automated-driving-graphic

IOP Conf. Series: Materials Science and Engineering 1169 (2021) 012033 doi:10.1088/1757-899X/1169/1/012033

The variation of the turnover and its effect on the exploitation risk

Carmen Elena Stoenoiu[1], Florica Mioara Serban[1], Ciprian Cristea[1]

[1]Department Electrical Machines and Drives, Technical University of Cluj-Napoca, Cluj-Napoca, 28 Memorandum Street, Cluj-Napoca, Romania

E-mail: Carmen.Stoenoiu@emd.utcluj.ro

Abstract. Through this study, an analysis was made of the relationship between the result, the volume of activity and the cost structure for a multiproduct enterprise. It was considered that enterprises are constantly facing the need to explore and exploit new production possibilities and for this they need to permanently reconfigure their product portfolio. In order to achieve this, the company must have a control of fixed and variable costs that will allow it to choose the volume of activity correlated with the internal characteristics and information on the market. The statistical calculation was used to calculate the risk associated with turnover through the covariance coefficient. As the size of the risk depends on the company's ability to control the cost structure and to act to reduce possible volume variations, turnover was considered to be the most important indicator that should be given high importance and should be constantly monitored. Thus, from the study it can be seen that the change in turnover determines the change in the risk coefficient.

1. Introduction

The rapid development of new technologies has led companies to find solutions to continuously improve their efficiency in order to cope with potential market threats [1,2]. Thus, ambidexterity emerged among firms, ie a combination of innovations and innovative practices designed to use exploration and exploitation for short-term success and long-term survival [3,4,5]. The problem of ambidexterity has advantages and disadvantages highlighted in the literature [6,7,8], which sometimes affect performance.

Strategic agility is another concept that complements the first through the perspective of organizational renewal without focusing only on strategic orientation [9,10].

Pursuing resource efficiency and cost reduction [11], while exploration seeks greater flexibility and the discovery of new knowledge [12]. Thus, for sustainable growth, exploration is the one that contributes by creating conditions for companies to be receptive to change [13,14], and through exploitation increases the innovative potential of firms and implicitly the degree of risk [15]. Although the risk of exploitation is higher [11,16] the concern for the permanent increase of efficiency, the improvement of products and skills, makes the exploitation activity a performance factor [17]. The risk, encountered in almost all branches of activity, appears as a result of the existence of market factors that cannot be eliminated, such as: demand variability, fluctuations in the acquisition price level of production factors or managers' competence regarding the opportunity of decisions.

The purpose of this study is to perform an analysis of a multi-product enterprise in which products have different weights and which may change due to changes caused by the continuing need to explore new opportunities and exploit them. The analysis between the result, the volume of activity and the structure of the costs highlighted their dependence on any possible variation. Subsequently, the mean square deviation for each product and in total the possible variant was calculated and the increase of the risk coefficient was observed, as a result of the increase of the turnover, through the covariance.

2. Material and method

This study started from an enterprise that makes three products for which the quantity, the selling price, the variable unit cost and the fixed costs are known, according to Table 1.

Table 1. The forecasted situation of the enterprise

No	Elements of study	Unit	Product A	Product B	Product C	Total
1	Quantity	Pcs.	7,520	4,200	2,300	-
2	Unit selling price	RON	700	450	560	-
3	Variable unit cost	RON	620	410	520	-
4	Fix costs	thousand RON				125.00

According to the financial theory at the level of exploitation, the risk can be measured and perceived according to the result indicators. The variability of the result can be manifested by its increase or decrease, the risk can be perceived both as an opportunity to obtain something additional and as a decrease of the initial value until a possible future loss.

The relationship between result, activity volume and cost structure was studied. It is known that fixed expenses affect the flexibility of the enterprise due to the impossibility of adapting them to the volume of activity. Thus, the size of the operating risk depends on the company's ability to control the cost structure and to know the sensitivity to possible fluctuations of the activity level. For this, for the multi-product enterprise, studied, the weighted average price (p_{wa}) of the products was calculated according to their unit price and the value share in the total turnover, the weighted unit margin on variable costs (M_{WV}) according to the individual margin and the value share of each product in the total turnover. Then the equilibrium turnover (Te) was determined according to equation (1) and the safety margin (M_S), according to equation (2).

$$T_e = \frac{CF}{\frac{m_p}{p_p}} \quad (1)$$

$$M_S = \frac{T - T_e}{CA} 100 \quad (2)$$

It was followed that the working hypothesis, to what extent, the variation of variable costs and turnover affects the result. Fixed costs were considered in this case to remain unchanged.

From a statistical point of view, the risk analysis can be obtained by calculating the variance (σ^2) and the mean square deviation (σ) [18]. The analysis used the forecast, as a way for the manager to predict the uncertain revenues, future states and phenomena of the system and possible uncertainties in different degrees and it was admitted from the beginning that factors that could not be identified and on which the manager has no control. For this, a forecast model was made to establish the level of turnover starting from the information from a company related to the previous year and knowing that there may be fluctuations in the economy due to external variables such as inflation, changes in commodity prices and consumables etc. within the limits of $0 \div \pm 10\%$.

The working hypothesis was to determine if the risk increases when turnover increases and if when the share of the product is high it will cause a change in risk.

According to each variant or possible state of evolution of the company's activity, the annual turnover to be forecast was calculated, starting from the individual turnover related to each product, which adjusts with the probability of occurrence, according to equation (3):

$$Tp = X_i \cdot p_i \quad (3)$$

Where: X_i - Projected turnover.

Subsequently, the average related to these possible states was calculated, according to equation (4):

$$X = \sum X_i \cdot p_i \quad (4)$$

The mean deviation was established according to equation (5):

$$\Delta = X_i - X \quad (5)$$

Then we calculated the mean square deviation which multiplied by the probability of occurrence of the state of turnover (0, -5, -10, 5, 10).

The variance was calculated as the mean of the quadratic deviations of a random variable and is the covariance of the random variable (turnover) with itself and measures the dispersion of the result against the mean. When the variant associated with the expected result is higher and the associated risk is high. The risk variation coefficient was determined, which measures the registered risk, the variant being less risky when the covariation coefficient is lower. The covariance coefficient was established according to equation (6):

$$\sigma = \text{cov} = \sqrt{\Delta} \quad (6)$$

3. Results and discussion

Taking into account the calculation elements taken into account after the analysis, the following indicators presented in Table 2 resulted.

Table 2. The situation of cost-volume-result indicators

No	Economic indicators	Unit	Product A	Product B	Product C	Total
1	The company's turnover	thousand RON	5,264.00	1,890.00	1,288.00	8,442.00
2	The share of each product in turnover	%	62.35	22.39	15.26	100.00
3	Operating result	thousand RON	523.66	140.01	72.93	736.60
4	Margin on variable costs (M_{WV})	thousand RON	601.60	168.00	92.00	861.60
5	The share of margin in turnover	%	11.43	8.89	7.14	10.21
6	Unit margin	RON	80.00	40.00	40.00	-
7	Weighted average production price (p_{wa})	RON	436.48	100.75	85.44	622.67
8	Unit weighted margin (M_{WU})	RON	49.88	8.96	6.10	64.94
9	Equilibrium turnover (T_e)	thousand RON	682.01	314.83	267.00	1,198.51
10	Relative safety margin (M_S)	%	85.80			

The analysis of Table 2 shows that product A has the largest share, followed by products B and C. The analysis of the table shows that the company makes a profit when the turnover is over 1,198.51 RON, a value that allows it to bear variable costs determined by the volume of activity and coverage of fixed costs. The contribution of each product to reaching the equilibrium point is given by the equilibrium turnover calculated individually at the level of each product. The relative safety margin shows us that the existing turnover can be diminished by a maximum of 85.8% without the company registering negative results, which tells us that the company has a favorable position on the market, this being less vulnerable to exploitation risk.

Following the calculation of the turnover variation in relation to the percentages of variation taken in the study, the variation was obtained at product level and by total enterprise, according to Table 3.

Table 3. The situation of the turnover variation and the effect on the result

Percentage of variation	The company's turnover			Total	Percentage of variation	Operating result			Total
	Product A	Product B	Product C			Product A	Product B	Product C	
0.00%	5,264.00	1,890.00	1,288.00	8,442.00	0.00%	523.66	140.01	72.93	736.60
-5.00%	5,000.80	1,795.50	1,223.60	8,019.90	-5.85%	493.58	131.61	68.33	693.52

-10.00%	4,737.60	1,701.00	1,159.20	7,597.80	-11.70%	463.50	123.21	63.73	650.44
5.00%	5,527.20	1,984.50	1,352.40	8,864.10	5.85%	553.74	148.41	77.53	779.68
10.00%	5,790.40	2,079.00	1,416.80	9,286.20	11.70%	583.82	156.81	82.13	822.76

The variation of the turnover and of the variable cost with the mentioned percentages determined the variation of the operating result in a much greater proportion.

Table 4 presents a model for calculating the risk coefficient when turnover varies by 0%, according to the result obtained in Table 3.

Table 4. Statistical calculation of the mean square deviation

Product type	p_i	X_i	X_ip_i	X_i-X_{mediu}	$(X_i\text{-}X_{mediu})^2$	$(X_i\text{-}X_{mediu})^2 * p_i$
Product A	0.30	5,264.00	1,579.20	2,760.38	7,619,697.74	2,285,909.32
Product B	0.50	1,890.00	945.00	-613.62	376,529.50	188,264.75
Product C	0.20	1,288.00	257.60	-1,215.62	1,477,731.98	295,546.40
Total	1.00	8,442.00	2,781.80			2,769,720.47

According to the results obtained in Table 4, a value associated with the risk was obtained determined by the risk variation coefficient of 1,664.25 ($\sigma = \text{cov} = \sqrt{\Delta}$).

In order to be able to establish the risk coefficient related to all possible situations for the studied enterprise, the risk coefficient was calculated for each possible variant and for each possibility of variation according to the percentages studied (Table 5).

Table 5. The situation of the risk variation coefficient

Percentage of variation	Option 1 A(0.3), B(0.5), C(0.2)	Option 2 B(0.3), C(0.5), A(0.2),	Option 3 C(0.3), A(0.5), B(0.2)
0.00%	1,664.25	1,541.35	2,080.50
-5.00%	1,564.99	1,488.91	1,926.08
-10.00%	1,476.75	1,447.45	1,779.44
5.00%	1,722.88	1,688.70	2,076.02
10.00%	1,810.27	1,729.33	2,222.05

Following the analysis of Table 5, it is found that option 2 offers for the enterprise the lowest risk in all situations of variation of the turnover studied, compared to options 1 and 3. This tells us that when we assign the lowest probability value the product with the highest share in turnover the risk decreases.

4. Conclusions

Through the turnover forecast, it is possible to schedule the future activity and implicitly of the necessary efforts for its realization. Knowing the possible variants that may exist at the level of the future activity, the management of the enterprise can plan its actions in such a way as to face the uncertainty.

In the study it is observed that any future variation between 0÷ ± 10 at the level of turnover and implicitly at the level of variable costs determines a positive operating result, which varies in a greater proportion than the variable that determined the variation. Because the analyzed enterprise is a multiproduct enterprise, the equilibrium point could not be established quantitatively but only in value through the equilibrium turnover which is observed to be different due to the different weight that each product has in the structure of the enterprise. The analysis of the exploitation result at the level of each product shows us the value level as a result of the variation suffered and implicitly the influence transmitted on the global result.

By analyzing the coefficient of variation of the risk to which the company is subject if a certain event would occur according to the associated probability, it can be seen that it fluctuates, obtaining higher values and an increased risk when turnover is high, the share of the product in the assortment range is high or the probability associated with the main product takes the highest value. This association is justified in terms of the company's dependence on the main product (A) which has the highest share of high risk, the company's dependence on certain customers requesting this product and dependence on

IOP Conf. Series: Materials Science and Engineering 1169 (2021) 012033 doi:10.1088/1757-899X/1169/1/012033

suppliers who deliver raw materials. Any change in product A, determined by technology, customer preferences, market conditions or the price of manufacturers can affect the company to a much greater extent than the changes associated with products B and C.

References

[1] Bican P M and Brem A 2020 *J. Sustainab.* **12** 5239
[2] Li F 2020 *J. Techn.* **92**
[3] O'Cass A, Heirati N and Ngo L 2014 *J. Ind. Mark. Man.* **43** 862
[4] Jurksiene L and Pundziene L 2016 *Europ. Bus. Rev.* **28** 431
[5] Montealegre R, Iyengar K and Sweeney J 2019 *J. of the Assoc. for Inf. Syst.* **20** 1
[6] Lavie D, Stettner U, and Tushman M L 2010 *The Acad. of Man. Annals* **4** 109
[7] Vorhies D, Orr L, and Bush V 2011 *J. of the Acad. of Mark. Sc.* **39** 736
[8] Brem A 2017 *Int. J. of Entrepr. and Innov. Manag.* **21** 261
[9] Hock M, Clauss T and Schulz E 2016 *J. Manag.* **46** 433
[10] Klammer A, Gueldenberg S, Kraus S and O'Dwyer M 2017 *Int. Entrepr. and Manag. J.* **13** 739
[11] O'Cass A, Heirati N and Ngo L 2014 *Ind. Mark. Man.* **43** 862
[12] Raisch S, Birkinshaw J, Probst G and Tushman M L 2009 *J. Org. Science* **20** 685
[13] Stubner S, Blarr H, Brands C and Wulf T 2012 *J. of Small Bus. & Entrepr.* **25** 217
[14] Shin H, Lee J, Kim D and Rhim H 2015 *Int. J. of Prod. Econ.* **168** 181
[15] Ireland R D and Webb J W 2009 *J. Bus. Horiz.* **52** 469
[16] Choi D and Lee K 2015 *J. Comp. in Hum. Behav.* **42** 120
[17] Hughes M, Filser M, Harms R, Kraus S, Chang M and Cheng C 2018 *Brit. J. of Man.* **29** 595
[18] Dallocchio M and Salvi A 2004 *Finanza d'azienda* (Milano: EGEA) p 217

IOP Conf. Series: Materials Science and Engineering 1169 (2021) 012034 doi:10.1088/1757-899X/1169/1/012034

Models for measuring the performance of an organization

G G L Mitrea- Curpanaru

Department of Engineering and Management, FDIMA, Gheorghe Asachi Technical University, 29 Dimitrie Mangeron Bd., Iași, 700050, Romania

gabriela-livia.curpanaru@student.tuiasi.ro

Abstract. The complexity of managing an organization today requires that managers be able to measure performance and analyze the impact of different dimensions of performance on organizational excellence. Many experts in the field have started talking about the importance of measurement as a means of communication. The customer can provide information about the physical quality - the appearance of the product, about the functionality and performance of the product, characteristics that are also related to quality. If the product is used in the performance of a service, then the quality of this service is directly proportional to the quality of the product. A mixed performance management model is concerned with both inputs and outputs. The selection of the best model to improve performance is based on a self-assessment process. Organizational excellence is measured by the organization's ability to achieve and maintain exceptional results simultaneously for all stakeholders.

1. Introduction

Performance management should be understood as a continuous process, reflecting normal management practices, not "special techniques" imposed on managers. Its conceptual framework includes terms such as: "performance management", "performance", "performing organization". Performance management has also been approached as an “adjustment loop”, linking organizational strategies and objectives, performance measurement, performance training, performance guidance and leadership, and rewarding performance. The performance of an organization is determined by: its level of coherence and stability; its ability to procure and use resources; of his reputation and image; the synergy of multiplying determinants such as employee relations, communication, mission, philosophy and leadership style, recognition, quality of organizational processes. Organizations use a wide range of qualitative and quantitative tools that have been accumulated over time to meet specific operational requirements [1]. Increasing interest in performance management has led to an update of accounting systems and an extension of non-cost performance. [2] define the performance management system as a set of indicators used to quantify both the efficiency and effectiveness of actions. [3] argue that the performance measurement system allows for the development of a closed-loop organizational strategy, which provides a structured framework for the flow of information relevant to feedback at close points, which facilitates decision-making and control processes. These emerging performance management systems can be grouped into two distinct categories [4]. The first are those that focus on self-evaluation, for example, the Deming Award in Japan and Asia (Deming 2004), the Baldrige Award in the United States (NIST 2004), and the European Foundation for Quality Award using EFQM Model of Excellence in Europe (EFQM 2004). The second category is those designed to help managers measure and improve business processes, for example, the Performance Pyramid [5] [6], the Balanced Scorecard [7] [8], The Cambridge Performance Measurement Process [9] [10] and the performance prism [11] [12]. The customer can provide information about the physical quality - the appearance of the product, about the functionality and performance of the product, characteristics that are also related to quality. If the product is used in the performance of a service, then the quality of this service is directly proportional to the quality of the product. A mixed performance management model is concerned with both inputs and outputs. The selection of the best model to improve performance is based on a self-assessment process.

Table 1. Evolution of performance measurement systems [13] [14]

Traditional performance measurement systems	Emerging performance measurement systems
Based on the traditional accounting system	Based on company strategy
Based on cost / efficiency	Based on values
Compromise between performance	Performance compatibility
Profit-oriented	Consumer-oriented
Short-term guidance	Long-term orientation
Prevalence of individual instruments	Prevalence of team tools
Prevalence of functional tools	Prevalence of transversal instruments
Comparison with standard	Improvement monitoring
Evaluation as a goal	Evaluation and involvement as a goal
Prevents continuous improvement	Emphasizes continuous improvement

The complexity of managing an organization today requires that managers be able to measure performance and analyze the impact of different dimensions of performance on organizational excellence. Many experts in the field have started talking about the importance of measurement as a means of communication.

Table 2 summarizes the large amount in the literature of performance measurement systems [15] [16], which can be considered to be the main changes and development trends that have been affected by how these systems are viewed.

Table 2. Performance measurement systems

Performance measurement systems and approaches	Performance measurement systems and approaches
Strategic measurement analysis and reporting technique (SMART)	Lynch and Cross 1991; McNair et al. 1990
Performance Measurement Questionnaire (PMQ)	Dixon et al. 1990
Results and determinant matrix (R&DM)	Fitzgerald et al. 1991; Fitzgerald and Moon 1996
The Balanced Scorecard (BSC)	Kaplan and Norton 1992, 1996, 2000
Business Scorecard Comparison (CBS)	Kanji 1998; Kanji and Moura e Sá 2002
Cambridge Performance Measurement Process (CPMP)	Neely et al. 1996, 2000; Bourne et al. 1998, 2000
Consistent performance measurement systems (CPMS)	Flapper and others 1996

IOP Conf. Series: Materials Science and Engineering 1169 (2021) 012034 doi:10.1088/1757-899X/1169/1/012034

Integrated Performance Measurement Systems (IPMS)	Bititci et al. 1997, 1998 a, b
Dynamic Performance Measurement Systems (DPMS)	Bititci and others 2000
Integrated Performance Measurement Framework (IPMF)	Medori 1998a, b; Medori and Steeple 2000

2. The Balanced Scorecard (BSC) Model

Balanced Scorecard is a strategic performance management framework that has been designed to help an organization monitor its performance and manage the execution of its strategy. In a recent global study of the management tools used, the Balanced Scorecard proved to be the sixth most widely used management tool in the world, which also had one of the highest ratings. satisfaction. The concept, having two components BS personal (BSP) and BS organizational (BSO), was introduced in 1992 [17].

The BSO identifies several dimensions of the organization, representing areas where organizations need to achieve results, at the level of departments, teams or individuals. Depending on the type of enterprise, the essential dimensions can be: financial aspect, customers, internal processes, knowledge and learning, quality of service, market share, etc. [17] they stop at four fundamental dimensions, which give four perspectives from which to examine the business of the organization. At the same time, it involves the obligation of managers to find answers to some key questions:

- Financial aspect: its financial health. How do shareholders see the company? What does the company mean to shareholders?
- Customers (external): customer satisfaction. How do customers see the company? What does it mean for our customers?
- Internal processes: process control. How can we control the primary business processes to create value for customers? What processes do we need to excel in to continually satisfy our customers?
- Knowledge and learning: the skills and attitudes of employees and the organization's ability to learn. How can the company remain successful in the future? How should we learn and communicate in order to improve ourselves and thereby achieve our vision?

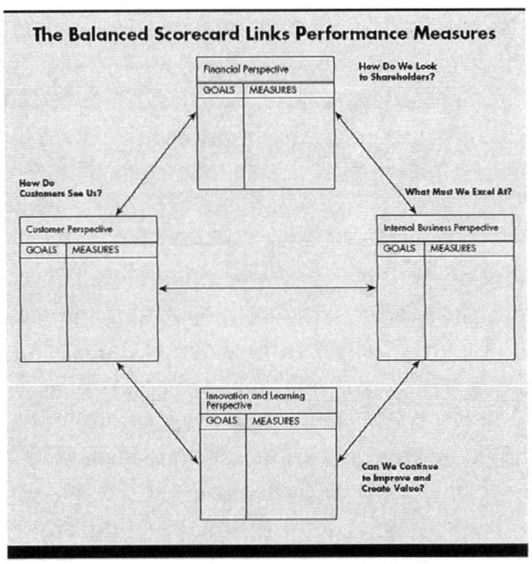

Figure 1. "The Balanced Scorecard Links Performance Measures." [18]

Definition for the four Balanced Scorecard perspectives:

- The financial perspective covers the financial objectives of an organization and allows managers to track the financial success and value of shareholders.
- The customer perspective refers to customer-related objectives, such as customer satisfaction, market share objectives, and product and service characteristics.
- The internal process perspective covers the internal operational objectives and presents the key processes necessary to achieve the objectives related to the client's perspective.
- The Learning and Development perspective covers intangible carriers of success in the future, such as human capital, organizational capital and information capital, including skills, training, organizational culture, leadership, systems and databases.

Research has shown that organizations that use a Balanced Scorecard approach tend to outperform organizations without a formal approach to performance management at the strategic level. The main advantages of using a BSC include:

- Better strategic planning - Balanced Scorecard provides a powerful framework for building and communicating strategy. The business model is visualized in a strategy map that forces managers to think about cause-effect relationships. The process of creating a strategy map ensures that a consensus is reached based on a set of interdependent strategic objectives. This means that performance results, such as key elements or future performance factors (such as intangibles) are identified to create a complete picture of the strategy.
- Improving communication and execution strategy - The fact that the strategy with all interdependent objectives is mapped on a piece of paper allows companies to easily communicate the strategy internally and externally. We've known for a long time that a picture is worth a thousand words. This "one-page plan" facilitates understanding of the strategy and helps engage staff and shareholders in delivering and reviewing the strategy. Finally, it is impossible to execute a strategy that is not understood by everyone.
- Better information management - The Balanced Scorecard approach forces organizations to design key performance indicators for their various strategic objectives. This ensures that companies measure what really matters. Research shows that companies with a BSC approach tend to report high quality management information and gain increasing benefits through the way this information is used to guide management and decision making.
- Improving performance reporting - companies that use a Balanced Scorecard approach tend to produce better performance reports than organizations without such a structured performance management approach. Increasing transparency needs and requirements can be met if companies create meaningful management reports and dashboards to communicate performance both internally and externally.
- Better strategic alignment - organizations with a Balanced Scorecard are able to better align their organization with strategic goals. In order to execute a plan well, organizations need to ensure that all business and support units work toward the same goals. Dividing the Balanced Scorecard into these units will help achieve them and at the same time link the operations strategy.
- Better organizational alignment - A well-implemented Balanced Scorecard also helps align organizational processes such as budgeting, risk management and analysis with strategic priorities. This will help to create a truly strategy focused organization.
- The Balanced Scorecard idea is simple, but extremely powerful if implemented well. As long as we use the basic ideas of the BSC to (a) create a unique and visual strategy on a "cause-and-effect" map, (b) align the organization and its processes with the objectives identified in the strategic map, (c) designing significant key performance indicators; and (d) using them to facilitate learning and improve decision-making; thus the organization has a powerful tool, which should lead to better performance.

To put the balanced scorecard to work, companies should articulate goals for time, quality, and performance and service and then translate these goals into specific measures. Senior managers at ECI, for example, established general goals for customer performance: get standard products to market sooner, improve customers' time to market, become customers' supplier of choice through partnerships with them, and develop innovative products tailored to customer needs. The managers translated these general goals into four specific goals and identified an appropriate measure for each. [18]

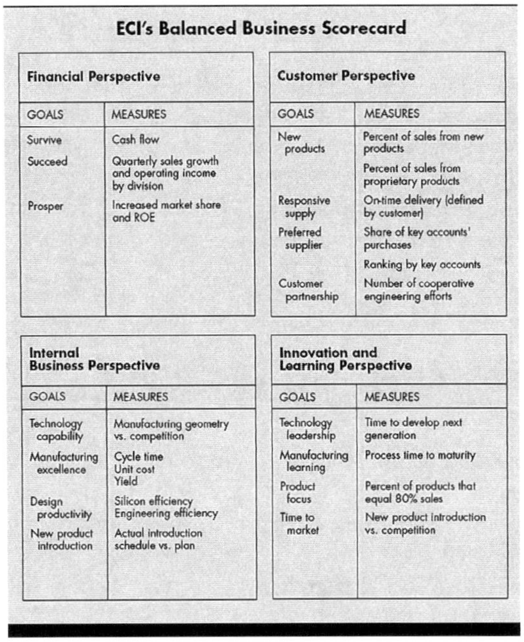
ECI's Balanced Business Scorecard

Financial Perspective

GOALS	MEASURES
Survive	Cash flow
Succeed	Quarterly sales growth and operating income by division
Prosper	Increased market share and ROE

Customer Perspective

GOALS	MEASURES
New products	Percent of sales from new products
	Percent of sales from proprietary products
Responsive supply	On-time delivery (defined by customer)
Preferred supplier	Share of key accounts' purchases
	Ranking by key accounts
Customer partnership	Number of cooperative engineering efforts

Internal Business Perspective

GOALS	MEASURES
Technology capability	Manufacturing geometry vs. competition
Manufacturing excellence	Cycle time Unit cost Yield
Design productivity	Silicon efficiency Engineering efficiency
New product introduction	Actual introduction schedule vs. plan

Innovation and Learning Perspective

GOALS	MEASURES
Technology leadership	Time to develop next generation
Manufacturing learning	Process time to maturity
Product focus	Percent of products that equal 80% sales
Time to market	New product introduction vs. competition

Figure 2. "ECI's Balanced Business Scorecard" [18]

But educational practice does not operate the ideal module with theoretical operational dichotomies, but updates contextually: evaluation involves both normative elements and criterion elements. In addition, both normative and criteria instruments must be designed on the basis of evaluation objectives. Also, the performance can be judged according to the previous results, following the evolution over time. Finally, the interpretation of the results, obtained following the application of an evaluation tool designed according to the specifications of a predominantly normative or criterion type, will be performed accordingly. A normative assessment will distribute the subjects along the entire correction / grading scale, while a criterion assessment will provide a measure of the level of mastery achieved by the group in relation to a clearly specified criterion.

Therefore, suitability for purpose involves choosing the most appropriate way, a set of decisions that are made according to the following criteria: the major purpose of the evaluation / assessment / measurement; logistical constraints and determinations (including the human and material costs involved); the nature and dimensions of the target group investigated; the purposes, nature and characteristics of the curriculum or program underlying the evaluation process; the importance, for all those involved, of the stakes of the expected results; the form of presentation, depending on the audience or the beneficiaries of the final report. Only after making decisions on these issues can we proceed to select the evaluation techniques to be used.

The concept of educational efficiency makes the connection between process evaluation and system evaluation, making visible and explicit the relations between the two. When the one who designs an educational evaluation decides to carry out one of the types of evaluation, a series of questions appear:

- What do we evaluate? The system as a whole or only one of its components, educational performance or efficiency, training / educational process, school results, performance levels, competencies, aptitudes, abilities, attitudes.

- For what purpose do we evaluate? Formative / summative, placement, guidance / counselling.
- Who do we evaluate? Human resources involved.
- How do we evaluate? By establishing a clear and transparent path: goals - objectives - evaluation tools - results - interpretation - communication.
- When do we evaluate? At the beginning of a process, during it, at its end, after a certain time from completion.
- What do we evaluate with? With evaluation tools; Through direct, (semi) structured observation, during the process; Through individual or institutional portfolios; Through self-assessment, peer and group assessment procedures, in order to increase metacognitive reflection and institutional socialization.
- Who benefits from the evaluation results? Human resources involved, decision makers. Once the specific answers to these questions have been established, the situation can be designed and built.

3. Use of Balanced Scorecard - in order to develop the main processes within the school "Alexandru Ioan Cuza" Economic High School Piatra Neamt – Case study

Professors Robert Kaplan and David Norton first introduced the concept of a balanced score card in an article published in Harvard Business Review 1/1992 [18]. The basic model of BSC evaluates the organization from 4 different perspectives and aims to integrate operational control into a long-term strategy.

Performance is assessed from 4-5 different perspectives, which integrate short-term operational control into long-term strategy.

BSC perspectives. According to the study of professors Kaplan and Norton, the 4 perspectives of the basic model are: the financial perspective, the client perspective, the process perspective and the learning and development perspective.

Financial perspective: How should we present ourselves to stakeholders in order to be financially successful?; Stakeholder expectations are also reflected in their profitability and development requirements; Effective use of resources.

Customer perspective: How should we present ourselves to customers in order to achieve the objectives ?; How should we thank our customers ?; Internal processes and development activities must take this perspective into account as the basis of the whole evaluation system.

Process Perspective: What are the processes by which we can create the quality requested by customers and through which we can meet the expectations of stakeholders? This perspective must provide the answer; What are the business processes in which we need to excel to please customers and stakeholders? What areas should we specialize in?

Development and learning perspective: How will we strengthen our capacity to make changes and improvements in order to achieve the objectives ?; What can we do to strengthen and develop our ability to meet the needs of our customers, to ensure the efficiency and productivity of processes that provide quality for customers ?; What are the main areas of expertise that we should develop and focus on?

BSC model applied within the "Alexandru Ioan Cuza" Economic High School in Piatra Neamt. BSC is applied in the Economic High School "Alexandru Ioan Cuza" Piatra Neamt, and the main processes are divided into 4 perspectives that describe the function of an educational organization. The fundamental function is related to the implementation and development of the teaching act. The auxiliary function considers the organization, structures and management of internal services. The development function describes the development of skills and anticipation of students' educational needs. The results of these 3 perspectives are evaluated through results indicators, financial and quality indicators which constitute the fourth perspective, namely, the development perspective.

The strategy, the annual planning and the management system of the quality and of the operations in the chariot of the Economic High School "Alexandru Ioan Cuza" Piatra Neamt were created based on this model. The BSC model is the framework for the whole evaluation of planning and quality. Through this we were able to achieve a practice-oriented model, a model suitable for an educational institution.

An important aspect is that neither the model nor its elements are direct copies, but individual applications. Using this model of quality development, the institution learned and adopted the fundamental idea of the model, its practices and justification:

- Basic function: making a curriculum; pedagogical leadership skills; personalization (individual study plans); student services; correlation with real life.
- Development function: educational institution; active development; quality control. Each student learns the necessary skills in multidimensional work environments, obtains a qualification, finds a job or continues his studies.
- Auxiliary function: systematic planning; inter collegiality.
- Efficiency: result indicators; financial indicators; quality indicators; development indicators.

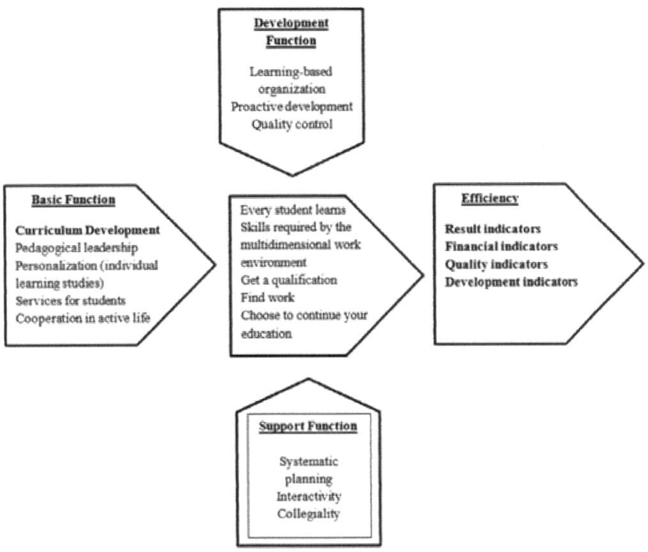

Figure 3. Balanced Scorecard development perspective [19]

Using the BSC Model applied within the "Alexandru Ioan Cuza" Economic High School in Piatra Neamt. Estimated results from the application of the BSC model. The systematic use of BSC over the years has resulted in continuous development and a number of improvements at all levels of activity in the institution. First, it was possible for the institution as a whole to focus development on certain selected sections, to the detriment of unsystematized, individual actions. The model was developed as a basis for the structure and function of work teams. In addition to sector-specific teams, there are also teams that combine leadership competence and pedagogical skills to serve all ancillary and educational functions within the institution. Viable connections with the field of work. The link between the technical sector advisory committees, the sector-specific organizations and the representatives of real life creates an active forum for development depending on the field. The commitment of teams at all levels results in concrete progress at the pedagogical level in favor of students.

Recommendations for implementing the practice. There are 3 key recommendations for success:

- the relevance and systematic activity of the management staff regarding the use of the model;
- continuous feedback on concrete progress;
- achievement → motivation → staff commitment.

The application of BSC in the "Alexandru Ioan Cuza" Economic High School in Piatra Neamt is an adequate tool to be used in quality management and assurance, as well as for strategic planning within a pre-university education institution.

4. Conclusion

In order to create a quality assurance system, a quality management system must first be introduced. It is a complex task, which involves first of all changing the mentality and perception of teachers in pre-university and university education and on the other hand the broad masses, the direct beneficiaries of the reform system. The performance of the educational institution is analyzed on the basis of key internal and external performance indicators, established targets and existing trends, judged constructively, with the establishment of improvement targets. Any performance measurement indicator is also used as a tool to control it, effectively for the implementation of continuous performance improvement actions.

References

[1] Heim, J.A. and Compton W. D., editors, (1992), *Manufacturing systems : foundations of world-class practice*, Washington, D.C. : National Academy Press

[2] Neely, A., Gregory, M. and Platts, K. (1995) "Performance measurement system design: A literature review and research agenda", *International Journal of Operations & Production Management*, Vol. 15 No. 4, pp. 80-116. https://doi.org/10.1108/01443579510083622

[3] Bitici U.S, Carrie A.S and McDevitt, L (1997) *Integrated performance measurement systems: A development guide. International Journal of Operations and Production Management. 17(5): 522-534. DOI 10.1108/09544789710159443*

[4] Wongrassamee, S., Simmons, J.E.L. and Gardiner, P.D. (2003) *Performance measurement tools: the Balanced Scorecard and the EFQM Excellence Model,* Measuring Business Excellence

[5] Lynch R. L., and Cross K. F., *Measure up!: Yardsticks for Continuous Improvement (1st edition, 1991), Blackwell (USA)*

[6] McNair, C.J. Lynch, R. and Cross, K. (1990), *"Do Financial and Non-financial Measures Have to Agree?" Management Accounting*

[7] Kaplan, R.S. and Norton, D.P. (1996), *The Balanced Scorecard, Harvard Business School Press, Boston, MA*

[8] Kaplan, R.S. and Norton, D.P. (2001), *"Transforming the balanced scorecard from performance measurement to strategic management: Part 1", Accounting Horizons*

[9] Bourne, M. C. S., Mills, J. F., Wilcox, M. & Neely, A. D. & Platts, K, W., (2000*),"Designing, implementing and updating performance measurement systems", International Journal of Production and Operations Management*

[10] Neely, A. D., Mills, J. F., Gregory, M. J., Richards, A. H., Platts, K. W., & Bourne, M.C.S., (1996), *Getting the measure of your business, Findlay, London*

[11] Adams, C. and Neely, A. (2000), "The performance prism to boost M&A success", *Measuring Business Excellence*, Vol. 4 No. 3, pp. 19-23. https://doi.org/10.1108/13683040010377818

[12] Neely, A. & Bourne, M. (2000), *"Why measurement initiatives fail", Measuring Business Excellence*

[13] De Toni, A.F. and Tonchia, S. (2001). *Performance Measurement System – Models, Characteristics and Measures. International Journal of Operations and Production Management*

[14] Ghalayini, A.M. and Noble, J.S. (1996), *"The changing basis of performance measurement", International Journal of Operations & Production Management, Vol. 16 No. 8, pp. 63-80. https://doi.org/10.1108/01443579610125787*

[15] De Toni, A. and Tonchia, S. (2001). *Performance measurement systems: models, characteristics and measures. International journal of operations & production management*
[16] Neely, A.D., Gregory, M. and Platts, K. (1995), *"Performance measurement system design: aliterature review and research agenda",International Journal of Operations & ProductionManagement,* Vol. 15 No. 4, pp. 80-116
[17] Kaplan, R. S. and Norton, D. P., *The Balanced Scorecard—Measures that Drive Performance, Magazine (January–February 1992), https://hbr.org/1992/01/the-balanced-scorecard-measures-that-drive-performance-2*
[18] https://hbr.org/1992/01/the-balanced-scorecard-measures-that-drive-performance-2
[19] https://epmromania.wordpress.com/2013/01/20/balanced-scorecard-2/ *Balanced Scorecard*

IOP Conf. Series: Materials Science and Engineering 1169 (2021) 012035 doi:10.1088/1757-899X/1169/1/012035

Enhancing Productivity of CNC Machines by Total Productive Maintenance (TPM) implementation. A Case Study

I C Gherghea[1], C Bungau[1], C I Indre[1], and D C Negrau[1]

[1]Faculty of Management and Technological Engineering, University of Oradea, 1 Universitatii street, Oradea, Romania

cosmin.gherghea@csud.oradea.ro

Abstract. Lean manufacturing is a business system that contributes to the growth of production efficiency and the company profit, by identifying and eliminating waste. TPM is one of the techniques used to eliminate production waste, being focused on the maintenance of production equipment, so as to increase their operation and efficiency, while ensuring the quality of production due to preventing the causes that lead to waste. In the present paper for the continuous improvement of the manufacturing process within the work cells equipped with CNC processing centers, TPM is implemented, which contributes to the optimization of predictive, preventive and corrective maintenance activities of production equipment, thus improving the efficiency of machine tools. TPM ensures maximum equipment efficiency due to the elimination of unforeseen equipment downtime, respectively increasing the speed and availability of the machine. The research is based on a case study from automotive industry. Monitoring the daily activities of the equipment is done through the TPM component, namely Overall Equipment Effectiveness (OEE) which presents the situation in the manufacturing process. The waste related to technological equipment were identified, such as: incorrect operation of the work equipment, unplanned shutdowns, respectively damages that occurred in the manufacturing process.

1. Introduction

The increase of productivity in the work cells equipped with CNC processing centers is directly influenced by the efficiency of the technological equipment and by the flow of raw materials and information.

Lean manufacturing is a concept that contributes both to increasing the efficiency of CNC processing centers and to reducing losses in the manufacturing process, thus optimizing the value flow. [1].

Lean manufacturing is a concept developed by Toyota and its main objective is to identify and eliminate losses. Any activity that does not create value is considered waste (muda in Japanese) and must be eliminated immediately from the operational flow [2].

To exemplify the value flow, a visual management tool is used, called Value Steam Mapping [3].

With the help of this tool the whole manufacturing process is drawn and analyzed, the first time it is drawn the current value stream mapping (which shows the initial state of the operational flow), and then following the proposals for continuous improvement in order to reduce losses it is drawn the future value steam mapping (showing improved operational flow) [4].

In terms of ensuring optimal operation (prevention of interruptions during the production process) of technological equipment, TPM is implemented, which is a world-class production strategy, respectively a Lean manufacturing component [5].

Any unplanned interruption of the production equipment represents a loss in the production process, which must be eliminated immediately, so that the manufacturing process to be a continuous one. [6].

Equipment maintenance is often reactive, intervening to fix it after a failure, but the adoption of specific TPM measures in the context of manufacturing on CNC procssing centers has a different perspective on equipment maintenance, because it applies a proactive approach, performing preventive maintenance activities of equipment [7].

Seiichi Nakajima, the founder of TPM, defines this production strategy as a concept through which a maximum overall efficiency is obtained for machines and installations throughout their lifetime, involving competence in maintenance of all departments within a company, but also of workers in order to ensure the good functionality of the equipment under their management [8].

The first implementation of the TPM concept took place in 1961 [9], by Nippondenso, which is one of the largest companies in Japan, producing components for the automotive industry, but the concept began with the introduction of preventive maintenance in Japan in 1951 [10].

TPM works on the basis of three prevention principles, according to the figure below.

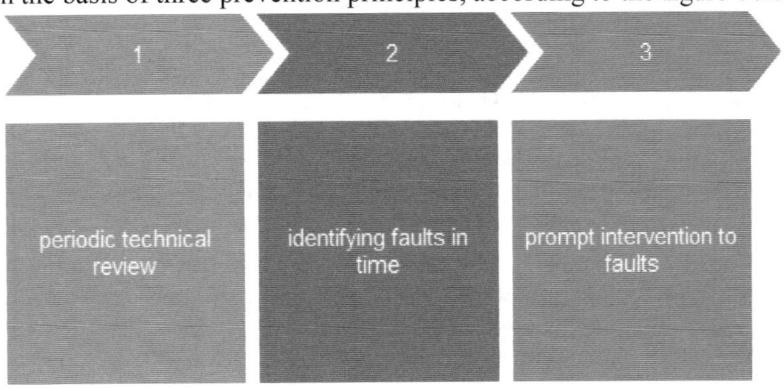

Figure 1. The TPM prevention principles

TPM ensure a stability of equipment including all type of maintenance politics like: preventive maintenance, corrective maintenance and predictive maintenance. All of these drive the equipment to its maximum productive efficiency. [11].

TPM is a technique, which manage the maintenance of all equipment, implemented within Lean Manufacturing, in order to continuous improve (kaizen) degree of usage and equipment lifetime [12].

According to Ahmad M.D. et al, TPM helps maximize equipment performance and prevent outages (due to technical failures) [13].

The purpose of TPM production strategy is to increase OEE by integrating a maintenance management system of the equipment involved in manufacturing process [14], [15].

The implementation of TPM ensures a maximum efficiency of the equipment due to the elimination of unforeseen stops of the equipment in the production process, respectively the increase of the speed and availability of the machine, eliminating the 6 wastes and at the same time obtaining quality products [16].

Some companies have implemented a software to monitor equipment performance (called Total Factory Monitoring), through which they can track in real time the performance of technological equipment and online monitoring. The data provided through this software, contributes to the measurement and evaluation of the operational activities within the manufacturing process, providing an overview of the entire production. Once this information is collected and centralized, from production, wastes are easier to identify, which will implicitly accelerate the process of continuous improvement in production [17]. Also, the implementation of a software that helps to monitor the performance of technological equipment, motivate continuously workers from different departments to ensure that they take the required actions and act accordingly to planned production and set objectives because they can see the impact that its actions has on OEE.

Preparation for the implementation of the TPM consists of the following objectives [16], [18]:

- Direct reporting of the situation to the company's management team;
- Adequate training for workers implied in manufacturing process;
- Implementing the TPM methodology on selected machine.

In conclusion, TPM aims for the following objectives [19]:

- More efficient usage of the equipment;
- Collaboration and Involvement of employees from all departments which own equipment;
- Increasing working hours of the equipment;
- Support from management team;
- Using equipment for continuous improvement;
- Reducing leading time, respectively maintain and increase the quality of the products.

TPM is basing on the following pillars [5], [17], [9]:

- Autonomously equipment maintenance;
- Kaizen (Continuous improvement);
- Planned maintenance;
- Quality of equipment maintenance;
- Training for involved staff in manufacture;
- TPM office
- EHS (Environment, Health, Safety);
- 5S (the basis of TPM).

5S is the base pillar in implementing 5S, which help to create safe and standardized working environment, it ensure obtaining of a perfect quality of the products by eliminating specific maintenance waste (the 6 wastes specific to used equipment).

In conclusion, the implementation of TPM would not be possible without the implementation of the 5S (Sort, Set in Order, Shine, Standardize, Sustain) [20]

2. Equipment monitoring using OEE

The general/global effectiveness of the equipment is a TPM component, through which are monitored daily activities of the equipment, the frequency of stop production caused by technical defections, in order to improve availability and technological equipment performance and implicitly it has the role to ensure the quality of the final product [21], [16].

Also, the most frequent Key Performance Indicator (KPI) in manufacturing cell which influence the implementation of lean practices is OEE. This KPI represent the initial state of the manufacturing process but it is also an instrument for management through which they can monitor and improve the manufacturing process thanks to its capability to measure and evaluate the productivity of the equipment/machine tool [22], [17].

2.1. Types of waste

Because of TPM component, manufacturing companies which own different technological equipment, managed to double(rarely even more) the productivity and performance of their machineries/machine-tool/CNC machining center, reducing, in some cases eliminating completely the OEE specific waste.

In the maintenance process for the equipment are identified 6 wastes, which are integrated in OEE, as following [23]:

- Breakdown losses;
- Setup and adjustment time;
- Speed waste;
- Minor stoppages;
- Defect;
- Rework and quality defects.

2.2. Benefits of OEE in manufacturing process

The OEE value represents the production equipment productivity, respectively, the ratio of maximum productivity capacity of the equipment (planned production) and the real production capacity from the manufacturing process, including any unplanned intervention in production (settings, adjustments, stops, defections, operator's error, changes).

Usually, the optimal OEE value is 85%, but in the most cases, OEE value is between 40% - 60%, which create significant possibilities for improvement and equipment performance increase.

Increasing OEE value has a positive impact on productivity, bringing with it multiple benefits in manufacturing process as following:

- Quick ROI (Return Of Investment) for technological equipment;
- Identification of problems from production process and avoiding wastes caused by those;
- Contribute to reducing production costs, which lead to increasing competitiveness in the market;
- Improvement of the manufacturing process has an impact even on the quality of the manufactured products;
- Increase the company reputation thanks to effectiveness and quality of delivered products;
- Reduce costs for equipment/tool-machine maintenance thanks to preventive maintenance.

2.3. OEE factors

In order to measure the EEO, the following key factors are identified and measured [11]:

a. Availability rate

Availability = (execution time)/(planned production time) (1)

Where:

Execution time is planned production time from which are deducted the times when the equipment was stopped (planned stops – changes in production process or unplanned stops changes, machine/equipment settings).

Planned production time is total time the equipment waits to produce.

Execution time = planned production time-down time (2)

Where:

down time represents the time allocated to the manufacturing process in which the machine / equipment is not in working order.

b. Performance Rate

Performance = (ideal time cycle * total amount of manufactured parts)/(Running time) (3)

c. Quality Rate

Quality = (total amount of quality parts)/(total amount of manufactured parts) (4)

After obtaining the 3 values above, OEE can be obtain using the next formula:

OEE=Availability *Performance * Quality (5)

2.4. OEE implementation

In order to calculate the performance of work equipment, the recording and analysis of production carried out over a well-defined period of time and established together with the production manager / engineer who is responsible for increasing production capacity and also quantifying operational activities, movements, storage, expectations and control.

IOP Conf. Series: Materials Science and Engineering 1169 (2021) 012035 doi:10.1088/1757-899X/1169/1/012035

In some companies, the OEE is measured daily to propose new solutions to improve the manufacturing process, respectively solutions that improve the factors that have an impact on the OEE. Once the values of these activities have been obtained, the main wastes are identified and specific solutions and techniques are proposed to eliminate them.

Given the high cost of investment in production equipment, the aim is to increase the overall effectiveness of the equipment in order to recover the investment in the shortest possible time and also to increase productivity and reduce leading time.

3. Case study

At the basis of this research is a study case, in which are analyzed all the wastes accumulated over 12 months, respectively, the 1st of January 2020- 31th of December 2020, at one of the CNC machining center.

The wastes are closely related to operational activity of the CNC machining center.

In order to implement and evaluate the OEE, the following steps will be performed:

- Establishing the machine - in this case it is a CNC machining center - DMG 55 H;
- The cause of the appearances of the different defects presented in the previous graphs will be identified;
- Relevant data about the production process will be collected;
- The data will be analyzed, respectively graphs will be built in order to have an overview of the current situation;
- Specific TPM actions will be implemented in order to remove and prevent defects registered on the CNC machining center.;

For realizing this case study, all wastes, thanks to lack of preventive maintenance of the CNC machining center DMG 55 H has been centralized, on which are realized different mechanical parts.

On this machining center has been realized machining operation, which involves technological itinerary with all operations specific to manufactured product.

DMG DMC 55 H is an horizontal CNC machining center duoBLOCK with 3 axes (longitudinal travel - x-axis: 560 mm, transversal travel - y- axis: 600 mm, vertical travel - z - axis: 600 mm), the maximum loading dimension is 500kg.

It has been identified and recorded a number of 355 defects because of lack of proper preventive maintenance for CNC machining center DMG DMC 55 H, over the 12 months, structured accordingly to the diagram below.

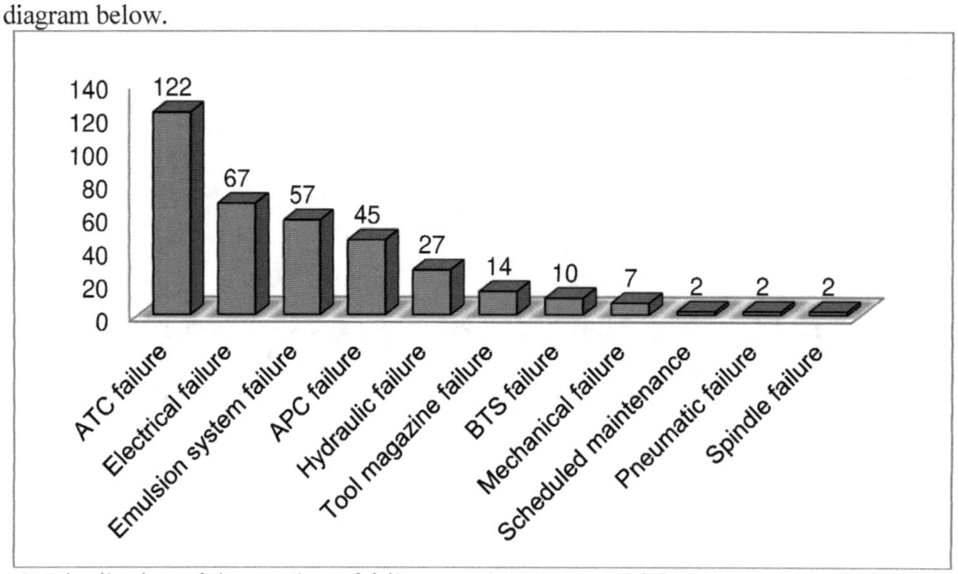

Figure 2. Distribution of the number of failures at the CNC machining center - DMG DMC 55H over a period of 12 months

In the company, the CNC machining center is working over 3 shifts (8hours/shift).

IOP Conf. Series: Materials Science and Engineering 1169 (2021) 012035 doi:10.1088/1757-899X/1169/1/012035

The diagram below, represents the total amount of recorded defects from CNC machining center over 12 month, distributed over the 3 shifts.

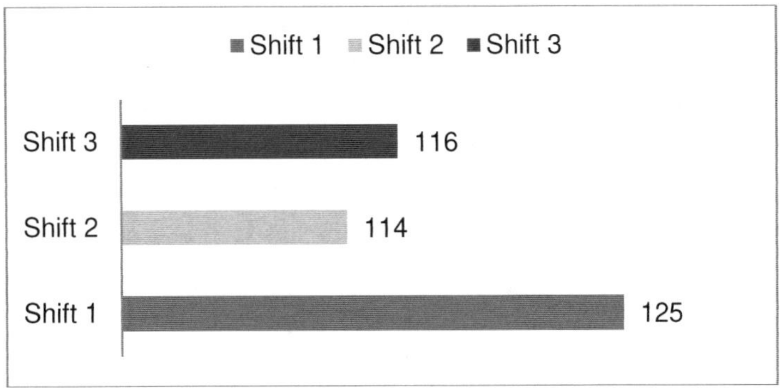

Figure 3. Number of defects distributed over the 3 shifts

First 3 defects (ATC defect, electrical defect and emulsion system defect) which were the most frequently repeated, represents 69.3% from the total amount of registered defects from this CNC machining center.

Once the main causes have been identified, preventive actions for these defects will be taken.

According to the graph in figure 4, the 355 defects occurred due to the lack of a corresponding preventive maintenance at the CNC machining center on the 8 types of products that were processed.

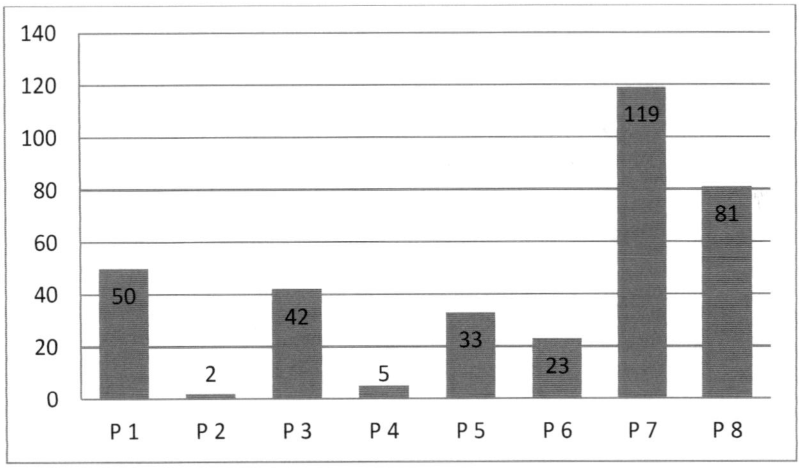

Figure 4. Distribution of defects by product types

According to graph from figure 4, the most defects have been registered at the production of product 7 (P7) and in graph from figure 5, are structured the 119 defects.

IOP Conf. Series: Materials Science and Engineering 1169 (2021) 012035 doi:10.1088/1757-899X/1169/1/012035

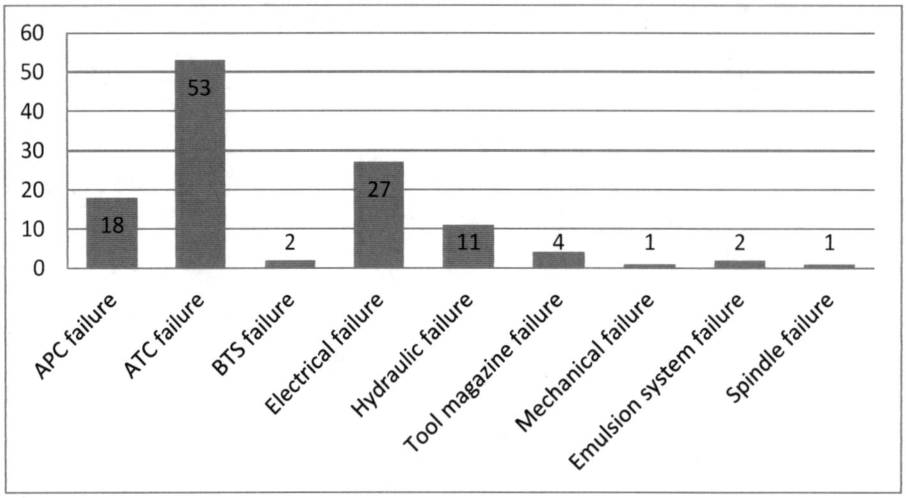

Figure 5. Distribution of defects for P7

3.1 OEE Calculation

It is important to mention the fact that other types of defects took place at the CNC machining center, but those were because of other causes than lack of preventive maintenance.

For determining OEE, obtained data about P7 product have been centralized, from 1st of Jan.2020 to 31th of Dec.2020

Product name P7;
Planned quantity 24940;
Produced quantity 13630;
Unprocessed quantity 11310;
Defective quantity reprocessed 95;
= Down time/year(min) 53640;
Cycle time 17 min/part;
Working days/year 301 (251 working days + 50 Saturdays);
Available minutes over year 301 day*24h/day *60min/h= 433440 min;
Break 0.

In the present study, OEE determination will be performed on product P7. In many cases, for OEE determination it is taken into account a well-defined period of time like: the number of shifts, over working hours or days.

For OEE determination, the 3 factors that determine OEE value will be calculated.

a) Availability

Total amount of time available represent the 301 days (including Saturdays when company was working).

The company works in 3 shifts, representing 24 working hours/day, respectively 433440 minutes available for work (planned production time).

Availability = (execution time)/(planned production time)

Execution time = planned production time-down time=433400-53640=379800 min

Planned production time=433440 min

Availability = (379800/433400)*100= **87.62%**

b) Performance

For determining the second factor, respectively performance, ideal cycle time (how long it takes to produce a part) will be determined, number of manufactured parts and planned running time.

Performance = (ideal time cycle * total amount of manufactured parts)/(Running time)

Performance = (17*13630)/433440= **53.45%**

c) Quality

For determining the last OEE factor, respectively the quality, the total amount of good parts manufactured will be determined, in ratio to the total amount of manufactured parts.

Quality = (total amount of quality parts)/(total amount of manufactured parts)

= (total amount of manufactured parts-scrap)/total amount of manufactured parts

Total amount of quality parts = manufactured quantity- scrap= 13630-95= 13535

Quality=13535/13630 parts=0.9930*100= **99.30%**

OEE = A*P*Q

OEE = 0.8762*0.5345*0.9930= 0.4650*100=**46.50%**

3.2 TPM implementation

After analyzing defects from production process for part P7, for improving the productive efficiency, actions to reduce and eliminate certain defect (accordingly to figure 6) related to machining center will be taken. These defects represent activities that are not add value for the manufacturing process, representing time waste.

By eliminating some of the defects via specific TPM actions, it will significantly influence OEE increase and improvement.

By the measures taken, it is aimed to reduce non-operational times, caused by the different mentioned defects with 50%, which will mean a total of 26820 minutes earned. Therefore, it will remain a total amount of 26820 minutes, non-operational times.

Given this aspect, OEE value would be **55.59%.**

4. Conclusions

This study case presented in this research, presents a tool which helps to measure equipment performance and also identify main wastes which led to a decrease in the efficiency of CNC machining centers.

After defects identification from CNC machining center, it has been implemented specific measures to eliminate these wastes, measure which will contribute to reducing production costs, reducing leading time, increasing the quality of the executed parts,, increasing production capacity and increasing the running time of the technological equipment.

The results of the research, emphasis the impact over increasing productive efficiency of the technological equipment (CNC machining center) because of TPM implementation. The increase of productive efficiency is reflected in measuring OEE, before and after TPM implementation.

According to the study, the OEE has been measured before and after the TPM implementation. Before implementing TPM, the OEE value was 56.50%, and after analyzing and taking some specific lean manufacturing improvement actions, OEE value was 55.59%, which represents an improvement of OEE value by 9.09%.

In graph from figure 6, it can be noticed the differences between the OEE factors before and after TPM implementation.

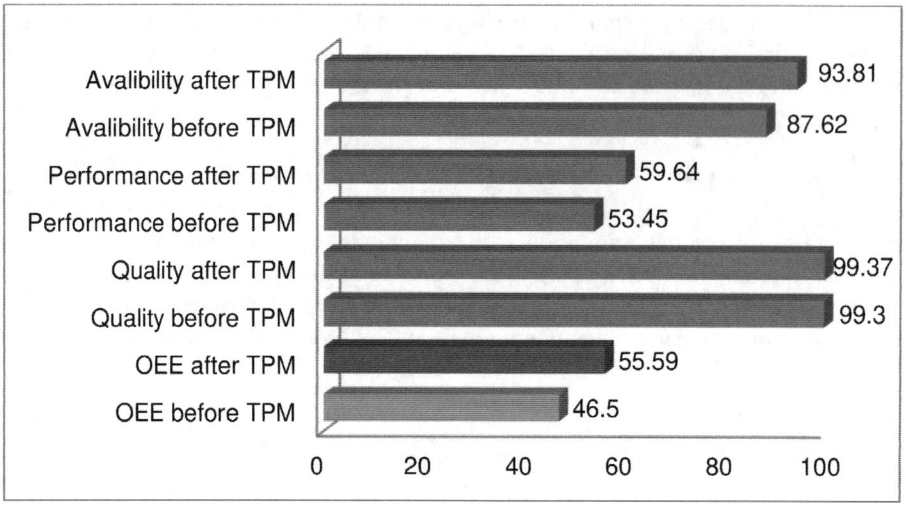

Figure 6. The difference between OEE factors between before and after TPM implementation

Considering the elimination of the non-operation times caused by some defect because of the lack of total preventive maintenance, the obtained improvement is significant.

The main measures taken for OEE grow, were measures for efficient solution to reduce defect like: 5S implementing, employees, respectively operators and CNC programmer trainings to maintain the equipment and also to identify easier the main causes which led to repetitive defects and maintaining the working place clean.

References

[1] Gherghea I C and Bungau C, 2018 Poka yoke application synthesis in manufacturing engineering *Proc. 6th Rev. Manag. Econ. Eng. Int. Manag.* Conf. **6**, pp. 564–571.

[2] Cirjaliu B and Draghici A 2016 Ergonomic Issues in Lean Manufacturing *Procedia - Soc. Behav. Sci.* vol. **221**, pp. 105–110.

[3] Gherghea I C, Bungau C and Negrau D C 2019 Lead time reduction and increasing productivity by implementing lean manufacturing methods in cnc processing center *IOP Conf. Ser. Mater. Sci. Eng.* vol. **568** no. 1.

[4] Gherghea I C, Bungau C and Negrau D C 2019 Best Practices to Increase Manufacturing Productivity - Comparative study *MATEC Web Conf.* vol. **290** pp. 1–9.

[5] Tondato R and Gonçalves M B 2013 Total Productive Maintenance: A casey study *22nd Int. Conf. Prod. Res. ICPR* vol. **12** no. 1.

[6] Arunagiri P Gnanavelbabu A. 2014 Identification of major lean production waste in automobile industries using weighted average method. *Procedia Eng.* **97**, 2167–2175.

[7] Patil J and Raut N 2019 Study Of Total Productive Maintenance And Improving The Production vol. **6**, no. 1, pp. 519–522.

[8] Gosavi A 2006 A risk-sensitive approach to total productive maintenance *Automatica*, vol. **42,** no. **8**, pp. 1321–1330.

[9] Aziz I, Karim S and Hossain 2013 M Effective Implementation of Total Productive Maintenance and Impacts on Breakdown Time and Repair & Maintenance – A Case Study Of A Printing Industry In Bangladesh *Int. J. Eng. Res. Dev.* vol. **8**, no. 1, pp. 1–9.

[10] Sharma K, Gera G, Kumar R, Chaudhary H K and Gupta S K 2012 An Empirical Study Approach on TPM Implementation in Manufacturing Industry *Int. J. Emerg. Technol.,* vol. **3**, no. 1, pp. 18–23.

[11] Dennis P 2015 Lean Production simplified - A plain - Language Guide to the World's Most Powerful Production System, **Third Edit**. CRC Press.

[12] Kulkarni A and Dabade M 2013 Investigation of Human Aspect in Total Productive

Maintenance (TPM): Literature Review *Int. J. Eng. Res. Dev.,* vol. **5**, no. 10, pp. 2278–67.
[13] Ahmad M F, Zakuan N, Jusoh A and Takala J 2012 Relationship of TQM and Business Performance with Mediators of SPC , Lean Production and TPM, *Procedia - Social and Behavioral Sciences* vol. **65**, pp. 186–191.
[14] Riezebos J, Klingenberg W and Hicks C 2009 Lean Production and information technology: Connection or contradiction? *Computers in Industry*, vol. **60**, pp. 237–247.
[15] Laugen B T, Acur N, Boer H, and Frick J 2005 Best manufacturing practices: What do the best-performing companies do? *Int. J. Oper. Prod. Manag.,* vol. **25**, no. 2, pp. 131–150.
[16] Méndez J D M and Rodriguez R S 2017 Total productive maintenance (TPM) as a tool for improving productivity: a case study of application in the bottleneck of an auto-parts machining line, *Int. J. Adv. Manuf. Technol*, vol. **92**, no. 1–4, pp. 1013–1026.
[17] Ng K C, Chong K E and Goh G G G 2014 Total productive maintenance strategy in a semiconductor manufacturer: A case study, *IEEE Int. Conf. Ind. Eng. Eng. Manag.,* pp. 1184–1188.
[18] Arslankaya S and Atay H 2015 Maintenance Management and Lean Manufacturing Practices in a Firm Which Produces Dairy Products *Procedia - Soc. Behav. Sci.*, vol. **207**, pp. 214–224.
[19] Bakri A H, Rahim A R A, Yusof N M and Ahmad R 2012 Boosting Lean Production via TPM *Procedia - Soc. Behav. Sci.*, vol. **65**, no. ICIBSoS, pp. 485–491.
[20] Bungau C, Blaga F and Gherghea C 2011 Method of analysis and audit, used to implement 5s in operational management *2nd Review of Management and Economic Engineering Management, Proceeding: Management of Crisis or Crisis of Management?,* pp. 36–47, vol. **2**.
[21] Wyrwicka M K and Mrugalska B 2017 Mirages of Lean Manufacturing in Practice *Procedia Eng.,* vol. **182**, pp. 780–785.
[22] Kumar A and Rajenthirakumar D 2016 Reducing the Manufacturing Lead Time of Export Pump Components in an Indian Pump, *Industry National J. Technology*, vol. **12**, no. 2 July, pp. 23–31.
[23] Almeanazel O T R 2010 Total Productive Maintenance Review and Overall Equipment, *Jordan J. MMechanical Ind. Eng.*, vol. **4**, no. 4, pp. 517–522.

IOP Conf. Series: Materials Science and Engineering 1169 (2021) 012036 doi:10.1088/1757-899X/1169/1/012036

Finite Element Analysis of the main stresses in Gamma Rod Osteosynthesis

I G Radu[1], E Vultur[1] and C N Drugă[1]

[1]Product Design, Mechatronics and Environment Department, "Transilvania" University of Brasov, Brasov, Romania

ioana.g.radu@student.unitbv.ro

Abstract. The paper presents the results obtained within the Finite Element Analysis of the Gamma rod, used to treat unstable fractures at the intertrohanteric and subtrohanteric level. Trochanteric fractures are a very common type of fracture found worldwide in both young and old people. The paper presents the main stages of the FEA, starting with the creation of the CAD model of the Gamma rod and ending with the Von Mises voltage field, for two of the most used titanium alloys used in the manufacture of rods (Ti-6AI-4V and Ti-6AI-7Nb). Finally, a series of conclusions are presented regarding the advantages of using Gamma rods in treating certain fractures at the trochanteric level.

1. Introduction

Trochanteric fractures are a very common type of fracture found worldwide in both young and old people. These fractures require emergency surgical treatment because patients must be mobilized early, thus reducing the period of postoperative immobilization. Another reason for rapid surgery and careful postoperative care of patients is the avoidance of intraoperative and postoperative complications [1].

The purpose of the Gamma rod is to treat unstable fractures at the intertrohanteric and subtrohanteric levels. This type of rod was developed following studies on corpses in the 80s. Studies show that this type of rod is much more effective than the DHS (Dynamic Hip Screw) method, a method used to treat the same types of fractures, Gamma rod having mechanical strength larger and better weight distribution, being fixed in the middle of the axis of gravity [2].

The Gamma nail involves an intramedullary rod through the end of which enters a nail that is screwed to the level of the femoral head, thus consolidating the head with the rest of the femur. This implant allows the two parts to slide to make an impact [1].

The surgical technique has a number of advantages, with an average duration of 35 minutes, a reduced intraoperative blood loss and reduced tissue damage. The distal screw prevents the rod from rotating and can be inserted without using the imaging method [2].

Over time, however, the Gamma rod has presented many postoperative complications, the most common of which is the femoral shaft fracture caused by the tip of the rod [1].

The main causes of trochanteric fractures: habitual trauma, street trauma, road accidents, industrial traumas [4].

2. Material and Method

Gamma rods are indicated in inverted obliquity trochanteric fractures where they represent the ideal implant. In unstable peritrochanteric fractures, the Gamma system can be used as an alternative to the compression screw [5].

The demands on the femur are: 2460 N on the femoral head, 1700 N on the great trochanter due to the gluteal muscles and 771 N on the small trochanter due to the psoas-iliac muscle [6].

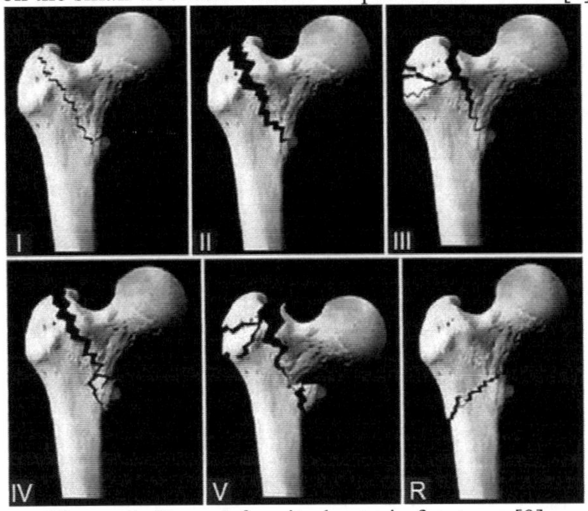

Figure 1. Types of peritrohanteric fractures [9].

2.1. Biomaterials used in the construction of the Gamma rod

The implantable system of the intramedullary short Gamma rod consists of a short rod, also called a nail, having a standard size, one or two nails that are inserted into the head-neck system at the femur. Under the ends of the screws, a plate can be inserted to stabilize the trochanteric mass. On the distal Tuesday of the Gamma rod, another screw is inserted, which has the role of preventing the rotation of the rod [3].

The material from which the short Gamma system for peritrohanteric fractures is made is titanium or titanium alloys [3].

Titanium and its alloys can be processed by advanced powder manufacturing methods, such as additive coating (or 3D printing) or injection molding of metals. This area is receiving increased attention from various production sectors, including the medical devices sector [7].

Titanium is a biomaterial that falls into the category of semi-lightweight materials, with special properties (Table 1). It is completely biocompatible, non-toxic, relatively low density, high specific strength, very good cold and heat resistance, very good fatigue strength, low modulus of elasticity, being compatible with bone structures, not magnetic (Table 1) [4].

Table 1. Properties of titanium

Titanium properties	Stock	Units of measurement
Atomic Mass	49,7	g
Density	4,51	g/cm3
Minimum flow limit	SBO	MPa
Minimum strength of tensile strength	550	MPa

IOP Conf. Series: Materials Science and Engineering 1169 (2021) 012036 doi:10.1088/1757-899X/1169/1/012036

Young module	116	GPa
Poisson coefficient	0,33	
Vicker hardness	970	MPa
Brinell hardness	716	MPa
Melting temperature	1668	° C
Boiling point	3287	°C

By alloying with a number of metals, Al, Mo, V, Mn, Cr, Sn, Fe, Nb, titanium forms various alloys (Table 2). Titanium and titanium alloys are used in a very wide range of medical applications, especially when it comes to dental and orthopedic implants. The conditions of use of titanium and titanium alloys in the medical field are governed by European norms and standard ISO 5832 [4].

Table 2 shows the main categories of titanium and titanium alloys used in medical applications, their characteristics and quality standards. Titanium's corrosion resistance is due to the spontaneous formation of an oxide film on the surface of the material, with a thickness of several tens of nm, which has three main characteristics, extremely useful for medical applications of titanium, namely stability, impermeability, adhesion. [9] Compared to other metallic biomaterials, titanium alloys have significantly higher strength / weight ratios, which is extremely important in the design of orthopedic implants [4].

Table 2. Properties of major titanium alloys

Alloy	Tensile strength (MPa)	Flow limit (MPa)	Elongation (° â)	Modulus of elasticity (GPa)	Standard
Ti-6AI-4V	895-930	825-869	6-10	110-119	IS05832-3
Ti-6AI-7N b	900-1050	880-950	8,1-15	110	IS05832-3
Ti-5AI-2.5Fe	1020	895	15	112	IS05832-3
Ti-6AI-6V-2Sn	965-1103	980	*1k*	110,3	ISO 5832-11
Ti-6AI-6V-2Sn	1280	1210	10	117	ISO 5832-10
Ti-13Nb-13Zr	973-1037	836-908	10-16	77-84	ASTM F1713
Ti-12Mo-6Zr-2Fe	1060-1100	1000	18-22	74-85	ASTM F1813

2.2. *Realization of the CAD model of the Gamma rod*

To create the CAD model, the Catia V5 design software was used, the design process and the final result being presented in Figures 1, 2 and 3.

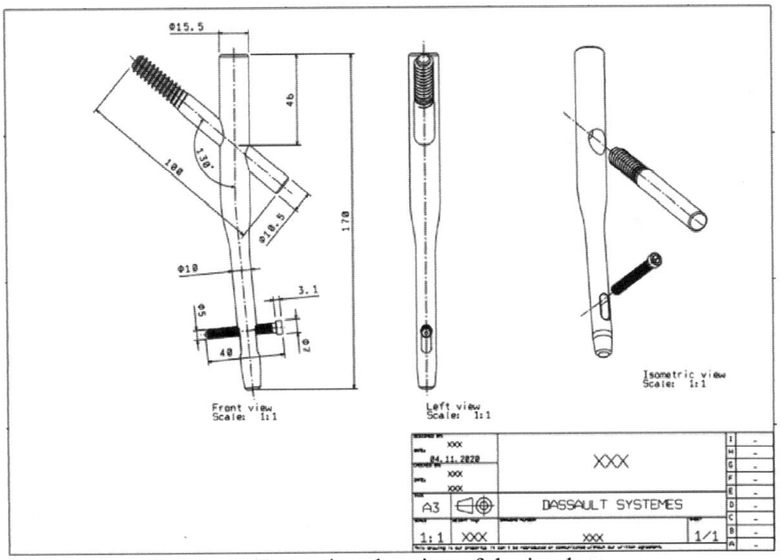

Figure 1. Execution drawing of the implant

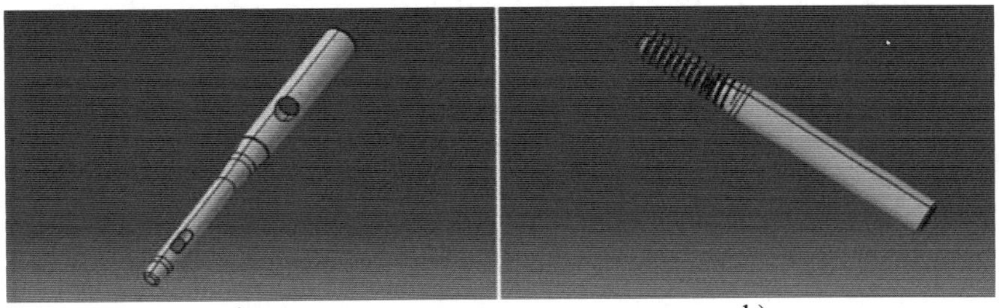

a) b)

Figure 2. CAD model of: a) Gamma rod; b) the proximal screw of the Gamma rod

Figure 2 shows the CAD model of the centromedullary rod inserted at the medullary canal of the femur and the CAD model of the proximal screw of the Gamma rod, inserted at the neck and head of the femur to stabilize peritrohanteric fractures, having the role of antirotation of the rod. The distal screw has the role of fixing the rod to avoid its rotation and sliding.

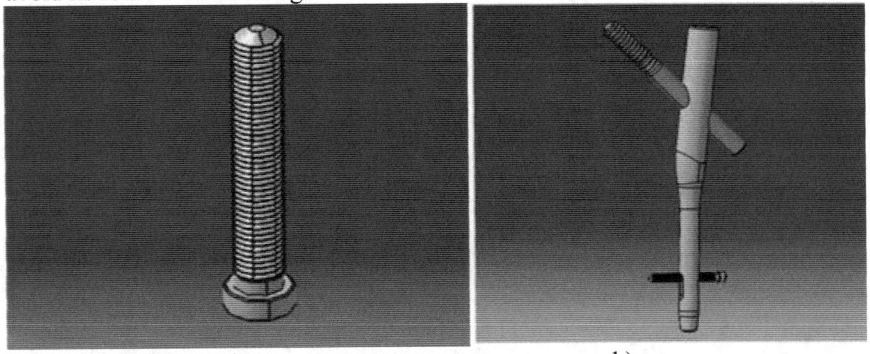

a) b)

Figure 3.a) CAD model of the distal screw of the Gamma rod; b) The assembled CAD model of the Gamma rod

IOP Conf. Series: Materials Science and Engineering 1169 (2021) 012036 doi:10.1088/1757-899X/1169/1/012036

2.3. *Finite element analysis (FEA) of the stresses to which the Gamma rod is subjected*
After making the 3D model of the intramedullary Gamma rod, Finite Element Analysis (FEA) was performed using Catia V5 software, using Titan-based alloys: Ti-6AI-4V and Ti-6AI-7Nb (Table 3) .

Table 3. Properties of alloys used in finite element analysis.

Aloy	Young's module GPa	Poisson's coefficient	Density Kg/m3	Thermal expansion
Ti-6AI-4V	119	0,37	0512	9,1
Ti-6AI-7Nb	110	0,37	0530	9,8

In the Finite Element Analysis, the stress applied to the implant was the compressive force present at the hip joint (which acts on the femoral head and is taken up by the entire femur). The forces exerted by the muscles on the femur were ignored, because they do not act directly on the implant, their influence being insignificant compared to the compression on the femoral head and the implant.

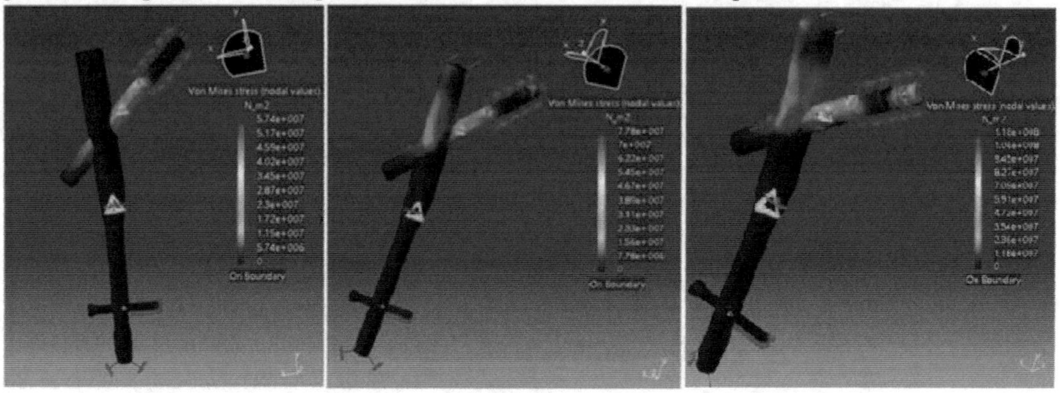

a) b) c)

Figure 4. *Ti-6Al-4V test at: a) 1500 N; b) 2000 N; c) 2500N*

Due to its shape and intramedullary implantation, the Gamma rod takes up this compressive force very well. The force applied to the rod was increased from 1500 N to 2500 N, this being the equivalent of a man of 70 kg and a height of 170 cm (Figures 4 and 5).

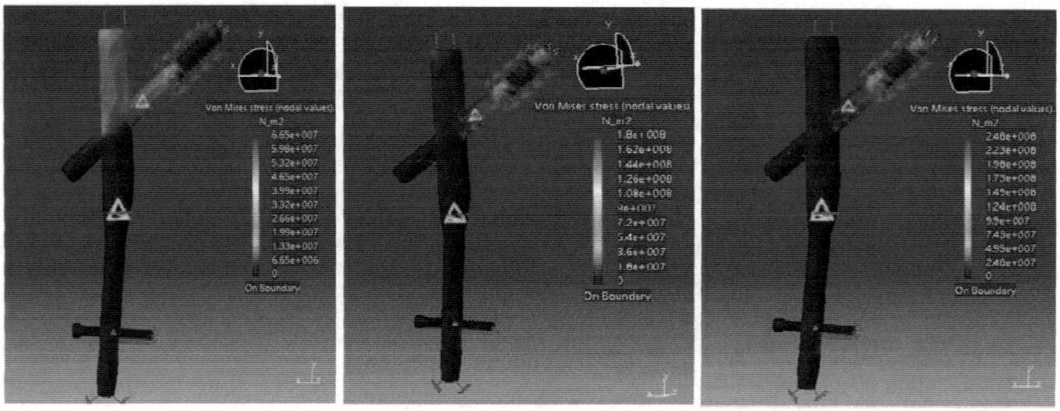

a) b) c)

Figure 5. Ti-6Al-7Nb test at: a) 1500 N; b) 2000 N; c) 2500N

3. Conclusions

Following the Finite Element Analysis to which the Gamma rod was subjected, it can be seen that it passed the tests to which it was subjected, resisting the stresses to which it was tested in the case of both titanium alloys. The results obtained for the two alloys are close, but it can be seen that the Ti-6Al-4V alloy recorded lower figures in the simulation.

The forces applied in the tests are similar to those that occur inside the human body, ie compression on the femoral head in bipodal orthostatism. This compression is transmitted further to the femoral shaft, and in the case of osteosynthesis with Gamma nails, it takes over very well the demands on the femur.

Table 4. Finite element analysis results.

Compressive force applied	Ti-6Al-4V alloy		Ti-6Al-7Nb alloy	
	MAX N/	MIN N/	MAX N/	MIN N/
1500N	5,74*	5,74*	6,65*	6,65*
2000N	7,78*	7,78*	1,8*	1,8*
2500N	1,18*	1,18*	2,48*	2,48*

References

[1] Ghanaat R, *Late results in gamma versus DHS nail osteosynthesis in trochanteric mass fractures,* Carol Davila University of Medicine and Pharmacy Bucharest, Bucharest 2019.

[2] Halder S, *The gamma nail for peritrochanteric fractures,* The Journal of Bone and Joint Surgery, From The Royal Halifax Infirmary.

[3] Medin, *Surgical procedure short reconstruction nail-PFN,* User manual.

[4] Mitu L, *Biomaterials- Titanium and others with titanium base,* Transilvania University of Brașov, Faculty of Product and Environment Design, 2019, course notes.

[5] Huang X, Yu B, Gu Y and Li Z, *Biomechanical comparision of dynamic hip screw and gamma nail for the treatment of unstable trochanteric fractures: a finite element study,* Int J Clin Exp Med, 2017, ISSN 1940-5901/UCEM0047129.

[6] Schipper I, Steyerberg E , Castelein R , *Treatment of unstable trochanteric fractures*, The Bone and Joint Journal, February 2004.

[7] Seral B, Garcia M and Doblare M, *Finite element study of intramedullary osteosynthesis in the treatment of trochanteric fractures of the hip: gamma and PFN*, Injury, March 2004.

[8] Sidambe A, *Biocompatibility of advanced manufacturing titanium Implants- A review*, Materials, June 2014, ISSN 1996-1994.

[9] “The Evan’s classification system” https://www.eoj.eg.net/viewimage.asp?img=EgyptOrthopJ_2014_49_2_96_145303_u1.jpg (accesed June 11, 2021).

IOP Conf. Series: Materials Science and Engineering 1169 (2021) 012037 doi:10.1088/1757-899X/1169/1/012037

Kuka KR5 arc Welding Industrial Manipulator Workspace Modelling Based on Kinematics Study

A Saadah[1], G Husi[1]

[1]Mechatronics Department, Faculty of Engineering, University of Debrecen, Debrecen, Hungary

alaa.saadah@eng.unideb.hu

Abstract In this study Kuka KR5 arc industrial manipulator was study using Denavait Hartenberg (DH) parameters representation and homogeneous matrix in order to obtain the transformation matrix for six degree of freedom articulated robot using MATLAB. The workspace of Kuka robot can be deduce by computing the transformation matrix with each motor workspace range. finally, the study in this paper allows us to control the manipulator in order to achieve any wanted orientation and position.

1. Introduction

In the University of the Debrecen, Department of mechatronics a Cyber Physical laboratory and intelligent robot systems has been established that is to follow up the latest developments in technical field specially in the industrial one. this laboratory contains many industrial and engineering tools and the one of the most important equipment in this laboratory is Kuka KR5 arc industrial welding manipulator controlling this industrial manipulator requires entry into relevant details of industry 4.0 concept also Many industries these days are shifting their focus to adding robots to their assembly line in order to speed up the production process.

It should also be noted that the most challenging issue in robotics systems is the Kinematics study which is considered to be the core of any study in order to start tasks such like control, follow trajectory, handling, painting etc.

Undoubtedly that the kinematics study allows us to create the kinematics equation which represent the robot, these nonlinear equations are used to form the links and joints parameters to the configuration of the industrial manipulator system [1], [2], [3].

2. Forward Kinematics

We can define forward kinematics as a problem which takes the initial position of each of the robot's joints and then gives us the final position of the end effect of the robot, so it's basically considered a relationship between robot's joints and position and orientation of the end effector or the two which located on the end effector.

2.1. Joints and Links

As known the forward kinematics problem is the uses of all given values for the manipulators joints in order to determine the orientation and position of the end effector. The angles between the links called as joint variables in two cases sliding or prismatic [4]. We can specify the joint parameters and computes the chains configuration using the forward kinematics. In case of serial manipulators, we can achieve

IOP Conf. Series: Materials Science and Engineering 1169 (2021) 012037 doi:10.1088/1757-899X/1169/1/012037

that by direct substitution of the joint parameters into direct kinematics equation for serial chain. There is a wide variety kinds of manipulators mechanisms that we can reach by categorizing their joint type and arrangement of joints if we ignore the shape and the size of the link and focus on categorization [5]. There are three types of joints revolute, prismatic, and spherical each one describing the allowed transformations between 2 links to the one it is attached to.
Prismatic: translation among a common axis.
Revolute: rotation about a common axis.
Spherical: rotation about a point.
An industrial manipulator with n joint has n+1 link the two links connected by 1 joint. so we basically have 1 to n joints and 0 to n links, and and we should start numbering from the base. based on this we connect link i-1 to link i by joint i. Therefore, when the joint is actuated the first link does not move.

2.2. Denavait Hartenberg

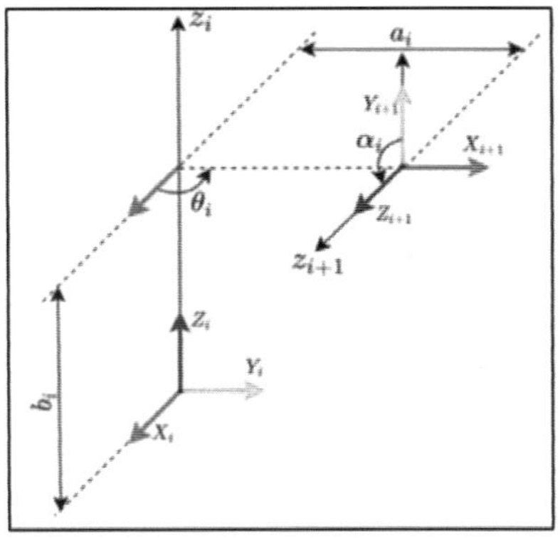

Figure 1. DH Frame Assignment [6].

We can get Ti depend on four primary transformation:

$$T_i = Rot(z, \theta_i)\ Trans(z, d_i) Trans(x, a_i) Rot(x, \alpha_i) \tag{1}$$

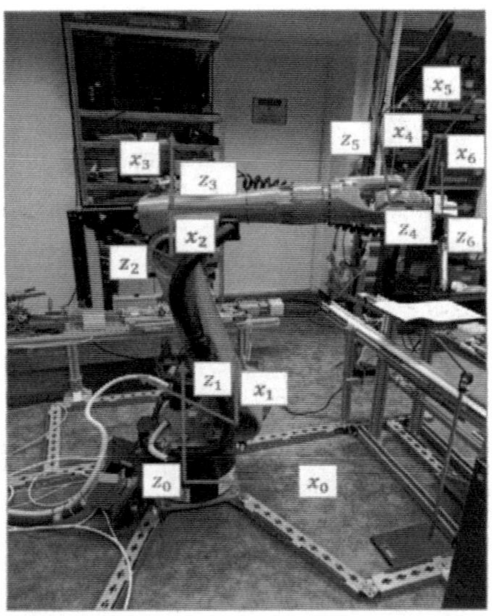

Figure 2. Robotic Arm Frame Assignment
[photographed by authors].

T_i Final matrix will be the result of multiple of those matrixes [7].

$$T_i = \begin{bmatrix} c_{\theta i} & -s_{\theta i} & 0 & 0 \\ s_{\theta i} & c_{\theta i} & 0 & 0 \\ 0 & 0 & 1 & 0 \\ 0 & 0 & 0 & 1 \end{bmatrix} \cdot \begin{bmatrix} 1 & 0 & 0 & 0 \\ 0 & 1 & 0 & 0 \\ 0 & 0 & 1 & d_i \\ 0 & 0 & 0 & 1 \end{bmatrix} \cdot \begin{bmatrix} 1 & 0 & 0 & a_i \\ 0 & 1 & 0 & 0 \\ 0 & 0 & 1 & 0 \\ 0 & 0 & 0 & 1 \end{bmatrix} \cdot \begin{bmatrix} 1 & 0 & 0 & 0 \\ 0 & c_{\alpha i} & -s_{\alpha i} & 0 \\ 0 & s_{\alpha i} & c_{\alpha i} & 0 \\ 0 & 0 & 0 & 1 \end{bmatrix}$$

$$T_i = \begin{bmatrix} c_{\theta i} & -s_{\theta i}c_{\alpha i} & s_{\theta i}s_{\alpha i} & a_i c_{\theta i} \\ s_{\theta i} & c_{\theta i}c_{\alpha i} & -c_{\theta i}s_{\alpha i} & a_i s_{\alpha i} \\ 0 & s_{\alpha i} & c_{\alpha i} & d_i \\ 0 & 0 & 0 & 1 \end{bmatrix} \quad (2)$$

$\alpha_i, \theta_i, d_i, a_i$
a_i: The distance between z_i to z_{i+1} along the x_i axis.
α_i: The angle between z_i to z_{i+1} about x_i axis.
d_i: The distance between x_i to x_{i+1} along the z_i axis.
θ_i: The angle between x_i to x_{i+1} about z_i axis.

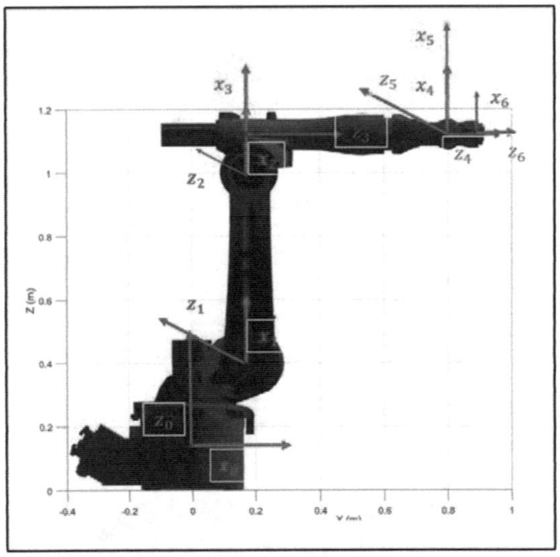

Figure 3. Robotic Arm Frame Assignment

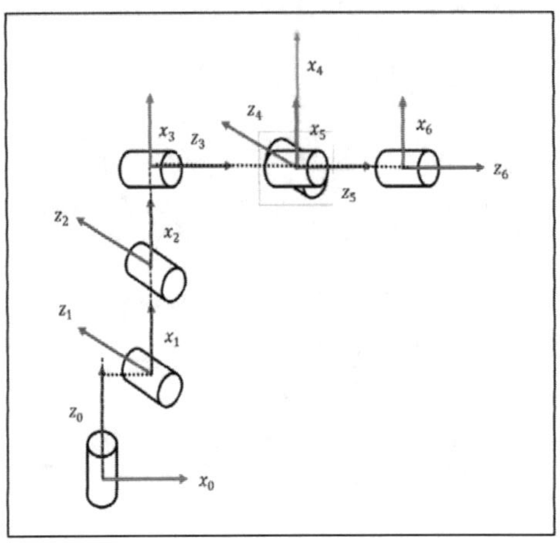

Figure 4. Robotic Arm Frame Assignment

2.3. *Robotic Arm DH Parameters* [8].

Table 1. Denavait Hartenberg Parameters.

i	θ_{i-1}	d_{i-1}	a_i	α_i
1	q_1	d_1	a_1	$-\frac{pi}{2}$
2	$q_2-\frac{pi}{2}$	0	a_2	0
3	q_3	0	a_3	$\frac{pi}{2}$
4	q_4	d_4	0	$-\frac{pi}{2}$
5	q_5	0	0	$\frac{pi}{2}$
6	q_6	d_6	0	pi

$$T_1^0=\begin{bmatrix} c_{q_1} & 0 & -s_{q_1} & c_{q_1}a_1 \\ s_{q_1} & 0 & -c_{q_1} & s_{q_1}a_1 \\ 0 & -1 & 0 & d_1 \\ 0 & 0 & 0 & 1 \end{bmatrix} \tag{3}$$

$$T_2^1=\begin{bmatrix} c_{q_2-\frac{pi}{2}} & -s_{q_2-\frac{pi}{2}} & 0 & c_{q_{2-\frac{pi}{2}}}a_2 \\ s_{q_2-\frac{pi}{2}} & c_{q_{2-\frac{pi}{2}}} & 0 & s_{q_2-\frac{pi}{2}}a_2 \\ 0 & 0 & 1 & 0 \\ 0 & 0 & 0 & 1 \end{bmatrix} \tag{4}$$

$$T_3^2=\begin{bmatrix} c_{q_3} & 0 & s_{q_3} & c_{q_3}a_3 \\ s_{q_3} & 0 & -c_{q_3} & s_{q_3}a_3 \\ 0 & 1 & 0 & 0 \\ 0 & 0 & 0 & 1 \end{bmatrix} \tag{5}$$

$$T_4^3=\begin{bmatrix} c_{q_4} & 0 & -s_{q_4} & 0 \\ s_{q_4} & 0 & c_{q_4} & 0 \\ 0 & -1 & 0 & d_4 \\ 0 & 0 & 0 & 1 \end{bmatrix} \tag{6}$$

$$T_5^4=\begin{bmatrix} c_{q_5} & 0 & -s_{q_5} & 0 \\ s_{q_5} & 0 & c_{q_5} & 0 \\ 0 & 1 & 0 & 0 \\ 0 & 0 & 0 & 1 \end{bmatrix} \tag{7}$$

$$T_6^5=\begin{bmatrix} c_{q_6} & s_{q_6} & 0 & 0 \\ s_{q_6} & -c_{q_6} & 0 & 0 \\ 0 & 0 & -1 & d_6 \\ 0 & 0 & 0 & 1 \end{bmatrix} \tag{8}$$

$$T_7^0=T_1^0.T_2^1.T_3^2.T_4^3.T_5^4.T_6^5.T_7^6=\begin{bmatrix} \mu_x & O_x & \alpha_x & p_x \\ \mu_y & O_y & \alpha_y & p_y \\ \mu_z & O_z & \alpha_z & p_z \\ 0 & 0 & 0 & 1 \end{bmatrix} \tag{9}$$

IOP Conf. Series: Materials Science and Engineering 1169 (2021) 012037 doi:10.1088/1757-899X/1169/1/012037

3. Workspace analysing and modelling.

The industrial manipulators workspace can be determined by set all positions that we can reach using the manipulator depending on dimension of the robot arm taking into consideration the type of movement which based on type of joint. In any robot design the most important consideration is determined any possible points that can be reach. That is also important for workers to set the safety working area.

3.1. *Kuka KR5 arc characteristics.*

Kuka KR5 arc industrial manipulator is designed based on 6 axes its components are base frame, electrical installations, link arm, arm, in-line wrist and rotation column. The base of Kuka KR5arc bolted to the mounting base also there is a tube contain the electrical installations with a cable junction box. on the rotating base there are the motors for axis 1 and axis 2.

The rotational motion of the manipulator around axis 1 cause the column rotation. And with a gear unit screwed to the base the link arm mounted on the column. And between the rotating column and the arm there is the link arm. And the arm located between the link arm and the inline wrist. on the arm we can find the motor of the main axis (A3). and you Motors of the wrist axes (A4, A5, A6). On the inline wrist there are Motors for axis 4 axis 5 and axis 6. The end effector is attached on the In-line Wrist's mounting flange. Axes 4, 5, and 6 are located at the in-line wrist [10].

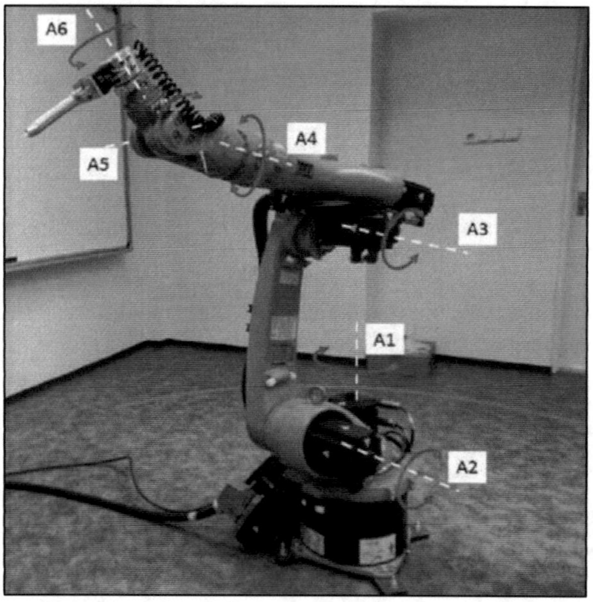

Figure 5. The KUKA KR5 robot mechanism [10].

The following table clarifies the range of each axis depending on manufacturing configurations.

Table 2. Kuka Kr5 arc Axes range

Axis	Range (software)
Axis 1	± 155°
Axis 2	-180° to 65°
Axis 3	-15° to 158°
Axis 4	±350°
Axis 5	±130°
Axis 5	±350°

IOP Conf. Series: Materials Science and Engineering 1169 (2021) 012037 doi:10.1088/1757-899X/1169/1/012037

We can find the diagram in the following figure (fig 6) also this diagram comes with any industrial manipulator's data sheet and from this figure side view of the industrial manipulator we can see the area marked in grey is the reachable area of the industrial manipulator and also we can see the area around the industrial manipulator which is contains unreachable points. and this due mechanical limits on the range of motion of particular joints or because of industrial manipulator control software (to avoid collision between prevent parts of the manipulator with other parts of the manipulator or it's space).

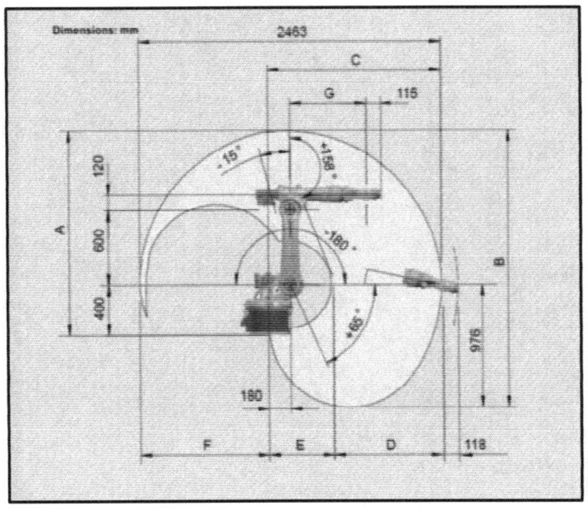

Figure 6. side elevation view of the robot.

3. 2 Workspace analysing and modelling.

We use the following values for modelling the workspace depends on the dimensions previous figure.

Table 3. Kuka Kr5 arc dimensions

	d		a
d_1	0.4	a_1	0.186
d_2	0	a_2	0.6
d_3	0	a_3	0.120
d_4	0.620	a_4	0
d_5	0	a_5	0
d_6	0.115	a_6	0

Finally, depending on previous values axes range, homogeneous transformation matrix and dimensions of Kuka Kr5 arc robot a modelling for manipulator had done using MatLab as seen in the following figure 7, and 8 which clarify the volume coordinate of all points that the end effector can reach.

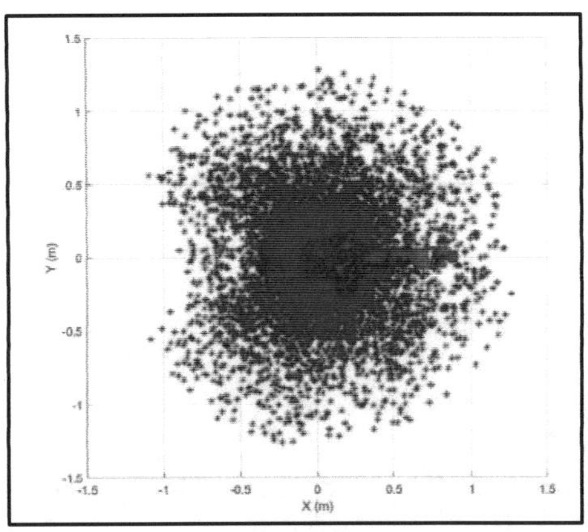

Figure 7. The simulation result of workspace's top view

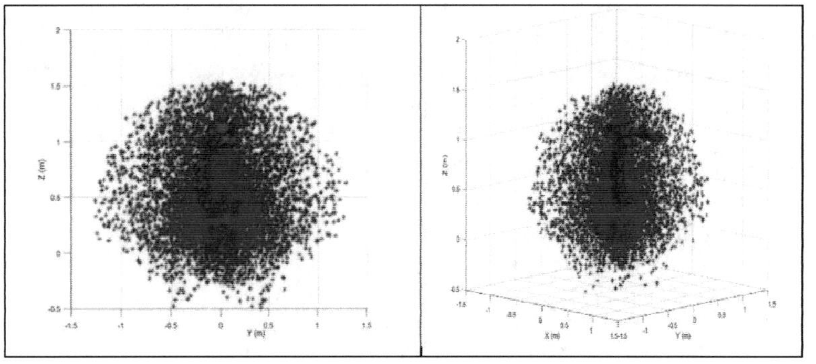

Figure 8. The simulation results of workspaces' side views (x,y) and (x,y,z).

Conclusion

The forward Kinematics model has been validated using MATLAB. The result from the forward kinematics model match with the result from the homogenous transformation matrix script result. The positions obtained from the model have been compared with a practical test of the industrial manipulator to detect the end-effector position and orientation.

References

[1] D. University, Cyber-Physical & Intelligent Robot Systems Laboratory, April 2021, [Online]. Available: https://www.facebook.com/cyberphysicaldebrecen/. [Accessed: 10.05.2021.].

[2] T. Erdei, Z. Molnar, Cyber physical systems in mechatronic research centre, 2017, [Online]. Available:https://www.researchgate.net/publication/320284614_Cyber_physical_systems_in_mechatronic_research_centre. [Accessed: 10.05.2021.].

[3] M. Indri, A. Grau and M. Ruderman, Guest Editorial Special Section on Recent Trends and Developments in Industry 4.0 Motivated Robotic Solutions, April 2018, [Online]. Available: https://ieeexplore.ieee.org/abstract/document/8331177. [Accessed: 10.05.2021.].

[4] J. Iqbal, R. Islam, H. Khan, Modeling and Analysis of a 6 DOF Robotic Arm, July 2012, [Online]. Available: https://www.researchgate.net/profile/Jamshed-Iqbal-2/publication/280643085_Modeling_and_analysis_of_a_6_DOF_robotic_arm_manipulator/links/55c0a56b08aed621de13cf59/Modeling-and-analysis-of-a-6-DOF-robotic-arm-manipulator.pdf. [Accessed: 10.05.2021.].

[5] D. Zhang, B. Wei, A review on model reference adaptive control of robotic manipulators, 2014, [Online]. Available: https://www.sciencedirect.com/science/article/pii/S1367578816301110. [Accessed: 10.05.2021.].

[6] K. Lee, J. Lee, B. Woo, J. Lee, Y. Lee, S. Ra, Modeling and Control of a Articulated Robot Arm with Embedded Joint Actuators, 2018, [Online]. Available: https://ieeexplore.ieee.org/abstract/document/8549903. [Accessed: 10.05.2021.].

[7] Z. Hou, S. Ma, Q.Zeng, A. Li, Kinematics analysis and self - collision detection of Truss type multi-robot cooperative welding platform, 2019, [Online]. Available: https://www.sciencedirect.com/science/article/pii/S2212827119304299. [Accessed: 10.05.2021.].

[8] A. Saadah, G. Husi, Computing The Kinematics Study of a 6 DOF Industrial Manipulator Prototype By Matlab, 2020, [Online]. Available: https://ojs.lib.unideb.hu/rIim/article/view/6190. [Accessed: 10.05.2021.].

[9] N. Obinna, T. Erdei, Z. Molnar, P, Szemes, G. Husi, LabVIEW Motion Planning and Tracking of an Industrial Robotic Manipulator (KUKA KR5 arc): Design, Modelling, and Simulating the Robot's Controller Unit, 2017, [Online]. Available: https://www.researchgate.net/publication/315830009_LabVIEW_Motion_Planning_and_Tracking_of_an_Industrial_Robotic_Manipulator_KUKA_KR5_arc_Design_Modelling_and_Simulating_the_Robot's_Controller_Unit. [Accessed: 10.05.2021.].

IOP Conf. Series: Materials Science and Engineering 1169 (2021) 012038 doi:10.1088/1757-899X/1169/1/012038

Designing and building a remote-controlled 3D printed prototype robot arm implant

M Török[1], T I Erdei[2], Sz Tóth[1], G Husi[1],

[1]Mechatronics Department, Faculty of Engineering, University of Debrecen, Debrecen, Hungary
[2]Doctoral School of Informatics, University of Debrecen, Debrecen, Hungary,

mateka96@gmail.com

Abstract. The basis of this projects are robot arms, and within that topic, implants. This area of science belongs to mechatronics, or more precisely, bionics. The goal of this project was to build and remotely control a 3D printed robot arm implant. The robot arm itself was printed with a custom-built 3D printer using biodegradable materials. The completed implant allows interaction with other physical objects using remote control. This allows to lower direct physical interactions during a pandemic or allow work in an environment where an operator could not stay for a long period of time without health risks.

1. Introduction

Bionics is a field that consists of numerous scientific fields. Its main goal is to implement solutions developed in nature into mechanical creations [1]. Biomechatronics strive to integrate electronics this biological and mechanical design as well. This includes robotics and neurology. Recently, artificial limbs have advanced at a never seen before rate. Nowadays, it is possible for them to precisely reconstruct a real human limb's movements. While these robot arm implants were a great breakthrough, they still have the disadvantage that the movements did not send feedback to the brain, so it was difficult to do precise movements, or to control grip strength [2]. Humanoid robots are a presently active research area. Their goal is to replicate human properties and to reconstruct their movement.
Due to the COVID-19 pandemic, research concerning androids (humanoid robots) has become relevant again. In view of this, a robotic arm was designed and built using additive 3D printing technology, then assembled and made functional using remote control to move it. The robot arm is able to do basic functions, such as grabbing symmetrical objects and push or rotate buttons and knobs, provided that it is fixed to an appropriate mount. In addition, during a pandemic it may be used to lower risk factors, as physical contact with an afflicted person may be avoided, however, cleaning of the surface must still be cleaned periodically with antiviral agents.

2. Robot arm implant design requirements

Planning and implementation were carried out at the Cyber-Physical & Intelligent Robot Systems Laboratory of University of Debrecen, Faculty of Engineering, Department of Mechatronics. Numerous research projects take place in this laboratory because it is possible to test the implementation of these projects in an environment similar to an industrial one [3].
Because designing a robot arm is a complex task, a PhD researcher from the Doctoral School of Pharmaceutical Sciences, University of Debrecen gave us additional support during the project. As multiple Open-Source models were available for research purposes, earlier constructions and designs were re-thought and improved [4]. During the design phase it was vital to pay close attention to the mechanisms responsible for finger movement. The motors responsible for movement were placed in the forearm area, while tendons were replaced with high quality, braided cords. The housing and mounting points of the motors is an original construction. The forearm and the movable wrist are also an original

design. The connection points of the completed fingers and palm also had to be modified. The discs equipped to the motors are likewise an original design.

3. Printing of prototype parts with FDM 3D technology

After designing the robot arm, it had to be 3D printed for prototyping. The model had to be separated into individually printable parts, to allow error-free and efficient additive 3D printing. The work area of the used 3D printer is 200x200x200 [mm]. The extruder head unit is able to move in all three directions of the coordinate system linearly. The model was created in SketchUp [5].
The printing was done using a filament made of PLA, which is a biodegradable, thermoplastic polymer and not soluble in water [6]. The first step of the printing process is separating the model using a specialized program, which process is called splicing. The used program was one called Cura [7]. Using this program, we are able to set numerous printing parameters of a given model. These parameters are layer thickness, fill density 30 [%], printing speed 60 [mm/s], temperature (extruder-210°C, heating bed-60 [°C]. The splicing program and the printed parts are shown on figure 1 and 2.

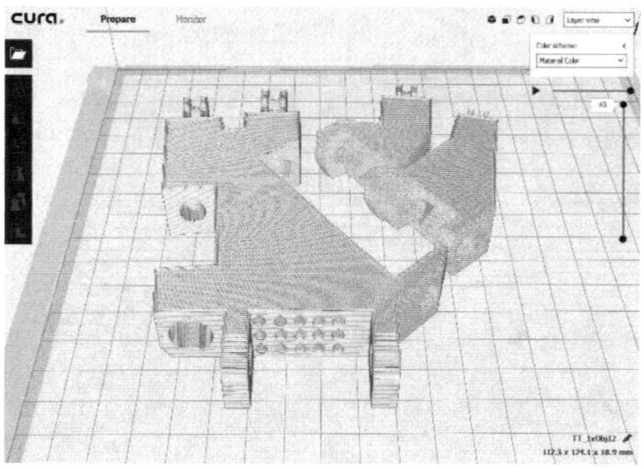

Figure 1. Cura splicer program preview

Figure 2. 3D printed robot arm parts

4. Internal parts and structure of the robot arm

In terms of its construction, the robot arm was modelled after the human arm and hand. The assembly of the printed robot arm began with the fingers. Each finger consists of three parts (phalanges), which were put together, then the assembled fingers were equipped on the hand. Following that, the cords replacing the tendons were lead through the fingers and the palm, responsible for moving the individual fingers. Then the wrist was completed, which is able to rotate in a 90° angle. Later all these cords were connected to the TP MG90S servo motors found in the forearm. The servo motors are responsible for the movement of fingers.

5. ESP NodeMCU-32S & servos

An ESP NodeMCU-32S microcontroller is used for controlling the motors. It should be noted that only the prototype uses this controller for testing purposes, the final design may use an industrial solution. The PWM cycle of the TP MG90S micro servo can be seen on the 3rd figure [8].
The ESP NodeMCU32S module also has the advantage of an on-board Wi-Fi and Bluetooth connection on its PCB. The module itself is shown on figure 4 [9] [10].

IOP Conf. Series: Materials Science and Engineering 1169 (2021) 012038 doi:10.1088/1757-899X/1169/1/012038

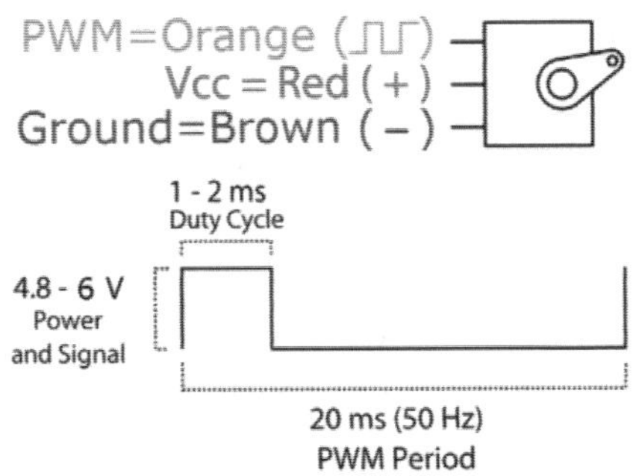

Figure 3. TP MG90S Servo PWM cycle [8]

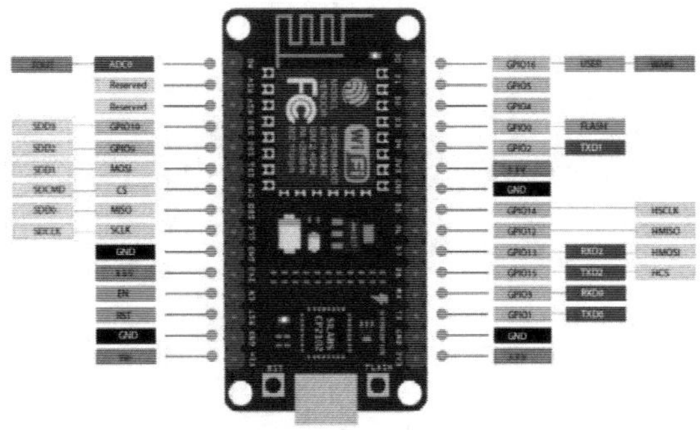

Figure 4. ESP NodeMCU32S pinout diagram [9]

6. ESP NodeMCU32S based remote control

The servos are connected to the ESP32 module, where the finger moving motors are connected to pins 4, 5, 12, 13, 14, while the wrist moving motor to pin 15. The 5V logic power is supplied to the module via the USB connection using an ADATA P20000D Li-Ion power bank with a maximum capacity of 20000mAh [11]. These connections can be observed on figure 5 made by Fritzing [12].
In order to move the fingers individually to a desired position, a GUI (Graphical User Interface) had to be created, which was done in HTML. This interface contains seven sliders: one for each of the five fingers, one for the wrist, and a final slider to move all five fingers at once. Figure 6 shows three states of the hand.

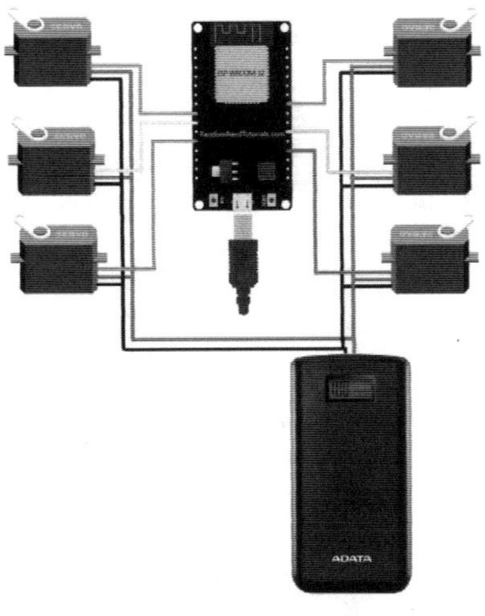

Figure 5. connection diagram

Figure 6. states of the robot arm

This solution allows more precise movement, as the fingers are able to move to and hold any position, not just the end states. The program's flowchart is shown on figure 7. During programming, we first had to load the <Wifi.h> library. Following this, the Wi-Fi network's name and password were configured. Thus, the ESP is able to connect to the existing Wi-Fi network. Following that, the servo motor outputs were declared. Lastly, the HTML page was created. After creating every slider, their respective motors were assigned to them to move the fingers to the desired position at the given angle.
The ESP NodeMCU32S supports the 802.11n Wi-Fi standard, but power-up and reset timing must be considered, which parameters are shown on Fig. 8.

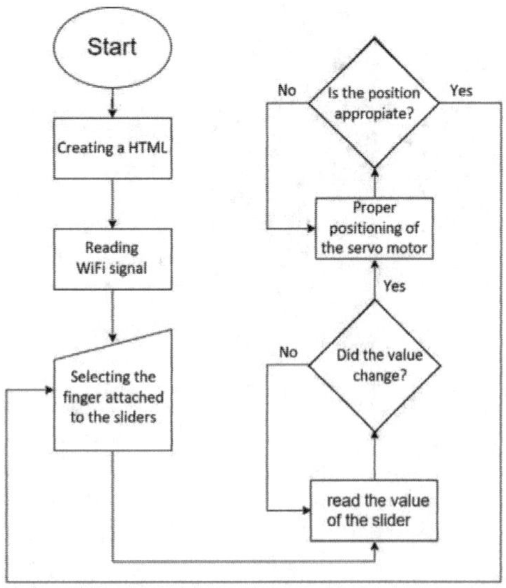

Figure 7. Robot arm program flowchart

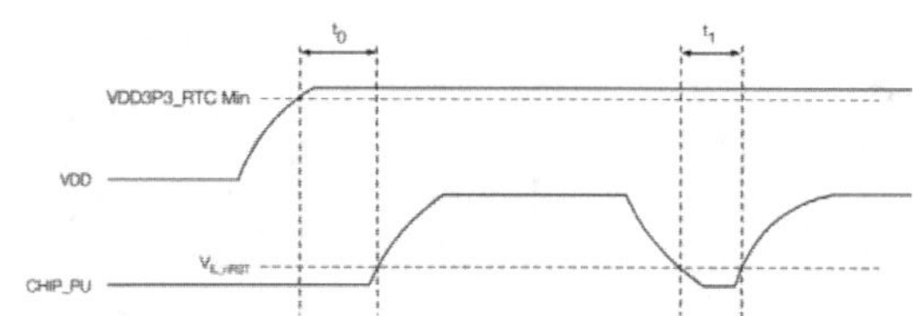

Figure 8. Power-up and Reset Timing [13]

7. Movement of prototype arm phalanges

The method of movement described at the early parts of this article are satisfactory if we only want to control the fingers, and the wrist, however, it was unable to move the phalanges individually. To make this possible, we also used an ESP NodeMCU32S like before. Similarly, an HTML page consisting of sliders was created, to remotely control the finger's phalanges individually. On the 9th figure this new GUI can be seen, where each slider is labelled with the anatomically [14] correct name of the discussed finger parts.

The created prototype is able to grab and hold lightweight symmetrical objects, as seen on figure 10.

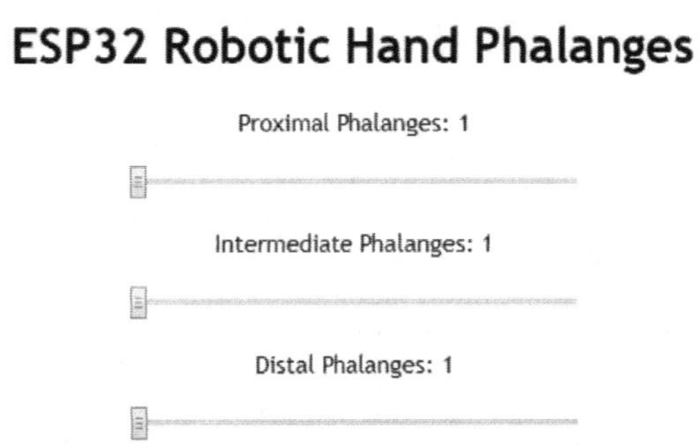

Figure 9. Graphical user interface

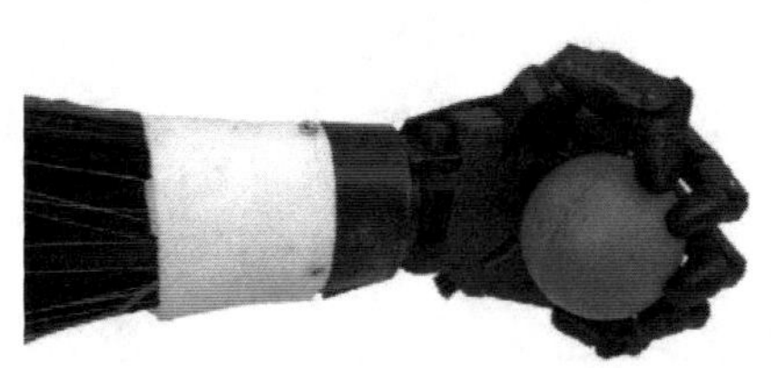

Figure 10. Robot arm prototype holding a light spherical object

Conclusion

The goal of this project was to design, assemble, and remotely control a 3D printed humanoid robot arm. The required parts of the prototype were made locally using a custom-built 3D printer. To control a prototype, the NodeMCU32S developer board was used, which allows the fine control of individual fingers using a GUI. In summary, we achieved our goals further upgrades also could be done.

IOP Conf. Series: Materials Science and Engineering 1169 (2021) 012038 doi:10.1088/1757-899X/1169/1/012038

References

[1] Bionika fogalma [Online]. Available: https://www.wisegeek.com/what-is-biomimicry.htm [Accessed: 09.10.2019.]

[2] C. Quinayas, Doyra Mariela Muñoz-Añasco, Andres Vivas, Carlos Alberto Gaviria Lopez, Design and Construction of the Robotic Hand Prosthesis UC-1 - July 2010Ingenieria y Universidad 14(2):223-237.

[3] T. I. Erdei, G. Husi, „Singularity measurement in the Cyber-physical and intelligent robot systems laboratory," IRASE (International Review of Applied Sciences and Engineering) 2020.

[4] Hand models (COVID-19) pandemic [Online] Available: https://www.thingiverse.com/thing:17773 [Accessed: 12.11.2020]

[5] SketchUp [Online] Available: http://www.sketchup.hu/ [Accessed: 20.10.2019.]

[6] PLA Filament parameters [Online] Available: https://www.simplify3d.com/support/materials-guide/properties-table/ [Accessed: 09.18.2020]

[7] Ultimaker Cura [Online] Available: https://ultimaker.com/software/ultimaker-cura [Accessed: 22.10.2019.]

[8] Szervo motor adalapja [Online] Available: http://www.electronicoscaldas.com/datasheet/MG90S_Tower-Pro.pdf [Accessed: 20.10.2019.]

[9] ESP NodeMCU32S pin kiosztása [Online] Available: https://www.theengineeringprojects.com/2018/10/introduction-to-nodemcu-v3.html [Accessed: 09.16.2020]

[10] Bluetooth Modul HC-05 [Online] Available: https://opencircuit.shop/resources/content/5baa51029ff43/crop/900-600/HC-05-Bluetooth-module.jpg [Accessed: 01.11.2020]

[11] ADATA 20000mAh P20000D [Online] Available: https://www.adata.com/us/feature/425 [Accessed: 12.11.2020]

[12] Frizing [Online] Available: https://fritzing.org/ [Accessed: 20.10.2019.]

[13] VI. ESP NODEMCU32S Datasheet [Online] Available: https://www.espressif.com/sites/default/files/documentation/esp32-wroom-32_datasheet_en.pdf [Accessed: 10.11.2020]

[14] Kézfej izmok [Online] Available: https://www.assh.org/handcare/Anatomy/Tendons?fbclid=IwAR1hXIEttKRtWCkVE1TfC01QC3payzJrFnxBhsCWIyodr9w7bXsypYmqqFM#Finger%20Flexor [Accessed: 20.10.2019.]

Acknowledgements

The project was supported by the cyber-physical & intelligent robot systems laboratory, Special thanks to Petra Arany. Produced in the framework of the EFOP-3.6.1-16-2016-00022 "Debrecen Venture Catapult program".

IOP Conf. Series: Materials Science and Engineering 1169 (2021) 012039 doi:10.1088/1757-899X/1169/1/012039

Performance management – a strategic and integrated approach to ensuring the success of organizations

G G L Mitrea- Curpanaru

Department of Engineering and Management, FDIMA, Gheorghe Asachi Technical University, 29 Dimitrie Mangeron Bd., Iași, 700050, Romania

gabriela-livia.curpanaru@student.tuiasi.ro

Abstract. The performance of an organization is determined by: its level of coherence and stability; its ability to procure and use resources; of his reputation and image; the synergy of multiplying determinants such as employee relations, communication, mission, philosophy and leadership style, recognition, quality of organizational processes. Performance is defined by behaviors, but especially by results. A mixed performance management model is concerned with both inputs and outputs. The selection of the best model to improve performance is based on a self-assessment process. The complexity of managing an organization today requires that managers be able to measure performance and analyze the impact of different dimensions of performance on organizational excellence. Many experts in the field have started talking about the importance of measurement as a means of communication. The institution's performance is analyzed on the basis of key internal and external performance indicators, established targets and existing trends, judged constructively, with the establishment of improvement targets. Any performance measurement indicator is also used as a tool to control it, effectively for the implementation of continuous performance improvement actions.

1. Introduction

Performance management [1] is a system that includes:
- Methodology for setting objectives,
- A performance measurement process,
- A differentiated payment system,
- Career management.

Performance management is a strategic and integrated approach to ensuring lasting success in the work of organizations, by improving the performance of the organization, teams and individuals [2] [3]. Performance management should be understood as a continuous process, reflecting normal management practices, not "special techniques" imposed on managers. Its conceptual framework includes terms such as: "performance management", "performance", "performing organization". The major contribution of performance management is the focus on achieving results, for example products and services for customers inside and outside the organization. The focus is shifted from effort and time to results and effectiveness.

Performance management encompasses the processes of measuring results and evaluating staff [4]. Modern approaches to performance management combine all the aspects mentioned in an integrated framework, unlike past approaches, which limited performance management to the stage of personnel management or to the collection and reporting of organizational performance indicators [5].

The first performance measurement tool was developed by Kaplan and Northon [6] in the 1990's as the Balanced Scorecard, highlighting the relationship between professional performance and the organization's strategy, each evaluation criterion is correlated with each element of the strategy.

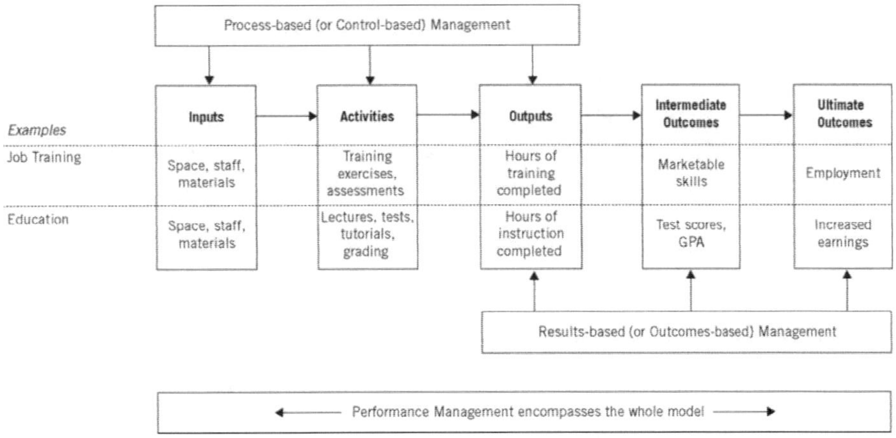

Figure 1. Types of performance management as applied to the elements a logic model. [18]

Performance management was also approached as a "regulation loop," linking organizational strategies and objectives, performance measurement, performance training, performance guidance and leadership, and performance reward. The performance of an organization is determined by:

- its level of continuity and stability;
- its ability to procure and use resources;
- its reputation and image;
- the synergy of multiplying some determinants such as employee relations, communication, mission, philosophy and leadership style, recognition, quality of organizational processes.

Performance management highlights two fundamental aspects [7]:

- How do we measure and evaluate the real level of performance?

- How do we make the level of performance have a weight and how do we ensure the increase of performance?

According to some authors [8] [9], organizational performance refers to:

- Technical performance, which involves the efficient use of all resources in the organization;
- Economic performance, which involves obtaining a superior quality with low costs;
- Social performance, which involves ensuring an optimal climate, favorable to the development of activities in the organization; adaptability and flexibility to customer needs;
- Managerial performance, which requires adaptability and flexibility to the needs of the organization, the needs of the community, the specifics of the organization's activity.

2. The characteristics of the performing organization

Performance is defined by behaviors, but especially by results.

Kane (1996) [10] shows that performance is "something that a person leaves behind".

Bernadin et al. (1995) [11] emphasize that "performance must be defined as the sum of the effects of labor".

According to Kaplan, "without tools, no improvement can be made", it is essential to measure the right things at the right time, so that the action can be decided in a timely manner [12].

Performance measurement tools and indicators do not just measure performance, they are often integrated with policy, strategy and many other behavioral issues.

Campbell (1990) [13] considers that "performance is a behavior and should be distinguished from results or effects". In his view, the results may be affected by contextual factors and, for this reason, do not correctly reflect individual or team performance.

"Performance means both behavior and results. Behavior emanates from the performer and transforms performance from the abstract notion into concrete action. Not only are they tools for achieving results, behaviors are, in and of themselves, results - the product of the physical and cerebral effort made to perform tasks - and can be judged separately from the results. "[14]

Thus, we can evaluate and measure performance both with reference to results (exits from the task performance process) and to behaviors (inputs to the work process).

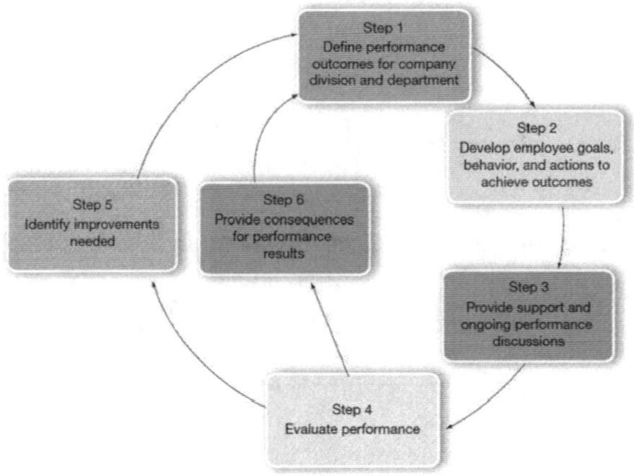

Figure 2. Steps in the Performance Management Process (Noe et al., 2016) [15]

The purpose of measuring organizational performance is to:

- identify the results, respectively the success or failure;
- identify if the client's needs are met;
- help the organization to understand its processes and confirm what is known, or to reveal what is not known;
- identify where there are problems, blockages, losses, etc. and where improvements are needed; ensure that decisions are based on facts, not assumptions, emotion, faith or intuition;
- show whether the planned improvements actually took place [16].

The characteristics of the performing organization, there are also various concepts and approaches:

- *The performance of the organization is expressed by its effectiveness and, in this sense, an effective enterprise is a successful one;*
- *The performance of the organization is determined by four key organizational features, called the "capabilities" of the company*:
    - Adaptive capacity: the ability of the organization to constantly monitor the external environment, responding to its requirements and pressures (for example, by meeting customer requirements) and, at the same time, to constantly change internally in order to respond to these pressures;
    - Leadership capacity: the ability to set directions for action in the future, to provide the necessary resources and to guide people's actions in those directions;
    - Managerial capacity: the ability to use efficiently and effectively the company's resources;
    - Technical capacity: the ability to design and produce goods and services that enhance the services provided to customers and consumers.

To these basic capabilities can be added another, regarding the ability to positively change the external environment through actions of information, education, persuasion of the company's partners.

The company's performance is ensured by improvements in several key areas [17]:

- Organizational stability, reflected in ensuring the continuity of delivery of products and services;
- Financial stability, given by the solvency of the company, which is able to pay its short-term financial obligations;
- The quality of managerial programs;
- Organizational growth, based on attracting resources and providing more goods and services.

3. Conclusion

Performance management involves a systematic approach, at the level of the entire organization, of human resources management and, implicitly, of performance evaluation, based on the objectives and internal needs of the organization, aiming to motivate employees and encourage creative potential.

References

[1] Avasilcai S., (2001), *Organizational Performance Management*, Tehnopress Publishing House, Iasi;
[2] Armstrong, M. & Baron, A. (1998), *Performance Management Handbook*, IPM, London;
[3] Armstrong, M., (2001), *A Handbook of Human Resource Management Practice*, Kogan Page;
[4] Armstrong, M., Murlis, H., (2004), *Reward Management: A Handbook of Remuneration Strategy and Practice*, Kogan Page;
[5] Avasilcai S., (2001), Organizational Performance Management, Tehnopress Publishing House, Iasi;
[6] Kaplan, R.S., Norton, D. P., (2006), *How to Implement a New Strategy Without Disrupting Your Organization,* Harvard Business Press;
[7] Matei, L., (2006), *Public management,* Bucharest, Economic Publishing House;
[8] Zybaczynski, G., Manole V., (2005), *Continuous improvement and management of variations to increase performance,* Publisher: Irecson, Bucharest;
[9] Păunescu, C., (2004), *Merchandise and market globalization*, ASE Publishing House, Bucharest;
[10] Kane, J.S., (1996), The Conceptualization and Representation of Total Performance Effectiveness. Human Resource Management Review, http://dx.doi.org/10.1016/S1053-4822(96)90016-9;
[11] Bernadin, H. K., Kane, J. S., & Ross, S. (1995), *Performance appraisal design, development and implementation.* In Ferris, G. R., Rosen, S. D., & Barnum, D. J. Eds., Handbook of Human Resource Management. Cambridge, MA: Blackwell;
[12] Kaplan, R.S. (Editor), (1990), *Measures for Manufacturing Excellence*, Harvard BusinessSchool Press: Boston, MA;
[13] Campbell, J. P. (1990). *Modeling the performance prediction problem in industrial and organizational psychology.* In M. D.;
[14] Brumbach, G.B. (1988). *Some ideas, issues and predictions about performance management. Public Personnel Management;*
[15] Noe, R. A. et al, (2016), *Human resource management : gaining a competitive advantage, Maidenhead : McGrawHill Education*;
[16] Parker, C., (2000), *Performance measurement.* Work Study;
[17] Blumenthal, B., (2003), *Investing in Capacity Building: A Guide to High Impact Approaches*, Foundation Center, London.
[18] https://ebrary.net/128691/sociology/conceptualiing_performance_management_logic_model_theory_action_production_function,*CONCEPTUALIZING PERFORMANCE MANAGEMENT: THE LOGIC MODEL, THEORY OF ACTION, AND PRODUCTION FUNCTION*

IOP Conf. Series: Materials Science and Engineering 1169 (2021) 012040 doi:10.1088/1757-899X/1169/1/012040

Research of the influence of axial displacement on the wear of hob milling tools

B Sovilj[1], I Sovilj-Nikić[1], S Sovilj-Nikić[2], V Blanuša[3]
[1] University of Novi Sad, Faculty of Technical Sciences, 21 000 Novi Sad, Serbia
[2] Iritel a.d. Beograd, 11 000 Belgrade, Serbia
[3]High Technical School of Professional Studies, 21000 Novi Sad, Serbia

E-mail: sandrasn@eunet.rs

Abstract: The need to increase productivity and cost-effectiveness of gear cutting of cylindrical gears with oblique teeth creates favorable conditions for the development of research in the field of tribology and the increasing application of tribological knowledge in industrial practice. Improving the gear cutting process by hob milling is important and useful for both gear manufacturers and hob milling tools manufacturers. Increasing productivity and increasing of the quality of serration are the basic directions of improving the process of gear cutting by hob milling. Technological parameters significantly affect the wear of cutters and taking into account the parameters for defining the economy of the process affect the achievement of optimal parameters of the cutting regimes. Based on experimental research, the paper presents an analysis of the influence of axial displacement, as one of the technological parameters, on the wear of hob milling tools in the gear cutting of cylindrical gears with oblique teeth.

1. Introduction

Gear pairs are tribomechanical elements for power transmission in which energy transfer is achieved by engaging the teeth of the drive and driven gear. They are characterized by small size in relation to the transmitted loads, high degree of utilization, durability, the possibility of application for very wide areas of power and transmission relations. The negative effect of these tribomechanical elements are vibrations, noise, gaps in couplings and other places of contact, wear and others [1].

Hob milling is the most widely used in the gear cutting of cylindrical gears with oblique teeth due to the high productivity of the process. Complicated kinematic and geometric connections between the hob milling tool and the workpiece create a number of difficulties and problems that prevent the optimal use of the hob milling tool and the hob milling machine [2,3,4]. One of the problems is the maximum and even use of as many teeth of the hob milling tool as possible in the gear cutting of cylindrical gears with oblique teeth. In this paper, certain attention is paid to the mentioned problems.

2. Methods of hob milling and uniform distribution of wear of cutting elements of tools

During gear cutting, with regard to the direction of auxiliary movement, the following types of machining have been developed:

- axial hob milling

IOP Conf. Series: Materials Science and Engineering 1169 (2021) 012040 doi:10.1088/1757-899X/1169/1/012040

- without additional movement of the hob milling tool in the direction of its own axial axis
- with additional movement of the hob milling tool in the direction of its own axial axis

• diagonal hob milling
• radial hob milling
• tangential milling
• radial-axial diagonal hob milling.

Figure 1[2] shows four possible variants of gear cutting of cylindrical gears with oblique teeth by hob milling.

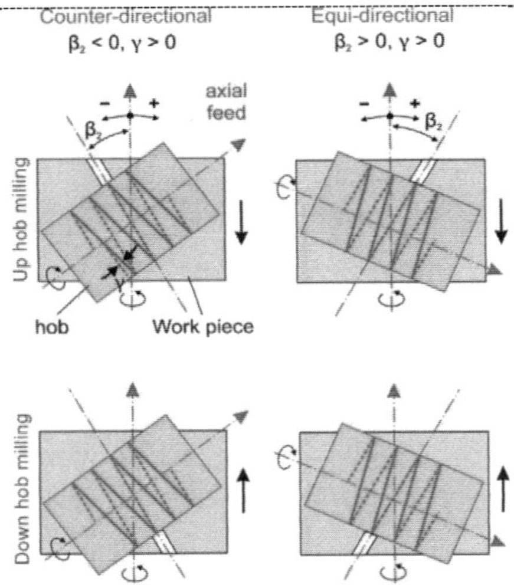

Figure 1. Four possible variants of gear cutting of cylindrical gears with oblique teeth by hob milling

Axial hob milling without moving the hob milling tool along its own axial axis is historically the first developed method of hob milling of gear teeth. The hob milling tool is short and has no possibility of axial movement in the direction of its own axial axis. The length of the active part of the hob milling tool is identical to the width of the grip between the tool and the workpiece, and therefore each tooth of the hob milling tool plays a certain role in removing excess material from the gullet of a certain form and dimensions. This is also the biggest drawback because not all teeth are uniformly loaded and do not wear uniformly. For hob milling tools that work without axial movement along their axial axis, the wear of maximum loaded tooth of the tool is adopted as a wear criterion. Due to that, short hob milling tools were used are uneconomically. The manufacturers of hob milling tools have switched to the production of hob milling tools of larger lengths, thus enabling the hob milling tools to be moved by the appropriate size in the direction of own axial axis when gear cutting in order to economically use one of the most expensive tool for chipping removal. In that way, the maximum load is shifted to the next teeth that were less loaded and lie in the area of the zone that continues to the previous one. The size of the axial displacement of the hob milling tool determines which tooth will be maximally loaded. Figure 2 [3] shows the theoretical distribution of wear on the teeth of the hob milling tool when gear cutting with a certain axial displacement of the tool along the own axial axis.

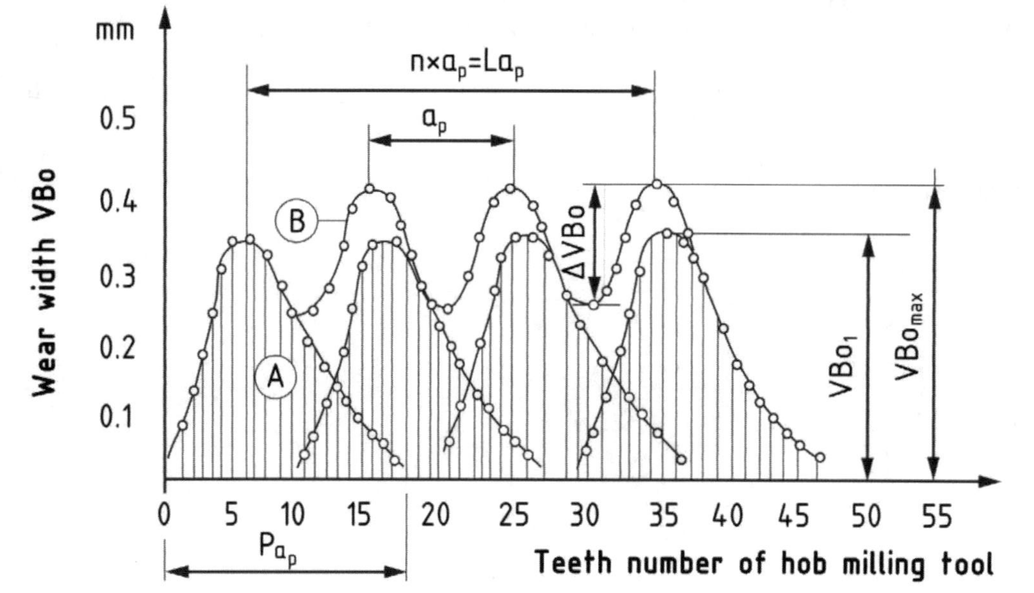

Figure 2. Theoretical distribution of wear on the teeth of hob milling tool during gear cutting with a certain axial displacement along the own axial axis

It is known that modern hob milling machines have the possibility of axial displacement both manually and automatically, but there are still two issues of significant impact on the optimal use of the hob milling tool remain unresolved. The first question is after which amount of chip i.e. the width of the machined serration, the axial displacement of the hob milling tool should be done. Another question is how much that axial displacement amount should be. Figure 3[5] shows the basic forms of flank wear zone of the tooth of the hob milling tool.

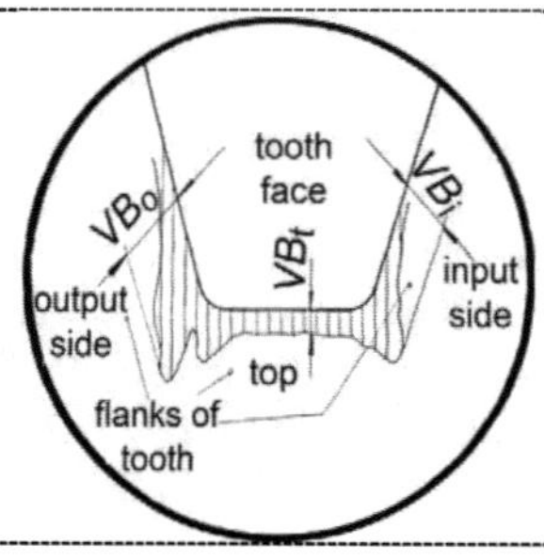

Figure 3. Basic forms of flank wear zone of the tooth of hob milling tool

Based on numerous experimental researches, it has been determined that the parameter VB_o can be adopted with certainty as a criterion for assessing the wear of the teeth of hob milling tool [4]. The wear

IOP Conf. Series: Materials Science and Engineering 1169 (2021) 012040 doi:10.1088/1757-899X/1169/1/012040

curve denoted by A in figure 2[3] refers to the width of the wear zone VB_o on the individual teeth of the hob milling tool when gear cutting without axial displacement of the hob milling tool. After the first axial displacement for length a_p, new wear of the new working area is expected according to the same laws as before the movement. Due to the fact that there is a partial overlap of the working area in the zone marked with P_{ap}, there is a slightly larger width of the wear zone in the area P_{ap} on the individual teeth of the hob milling tool. After n the same axial displacements of the hob milling tool a_p, a greater width of the wear zone occurs, which is shown in figure 2 by curve B. Based on this figure, it can be concluded that by choosing a certain width of the wear zone VB_{o_1} before the first movement and a certain size of axial displacement a_p more uniform wear of all cutting teeth in the area $L_{a_p} = n \cdot a_p$ can be reached. Figure 4 shows the basic values necessary to define the degree of uniformity of the wear distribution of the tooth of the hob milling tool.

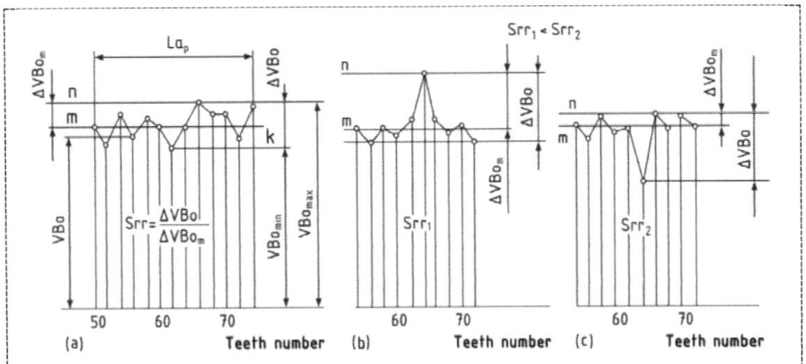

Figure 4. Basic values necessary for defining the degree of uniformity of wear distribution of the tooth of the hob milling tool

Basic values in figure 4 are:

n - tangent line to the maximum width of the wear zone VB_o

k - tangent line to the minimum width of wear zone VB_o

m - midline, parallel to the lines n and k, is the arithmetic mean value of width of the wear zone VB_o at the length L_{a_p}

ΔVB_{o_m} - mean dispersion of the width of the wear zone VB_o measured from line n to line m

Based on these values, an expression for the degree of uniformity of the teeth wear distribution of the hob milling tool can be written as:

$$S_{rr} = \frac{\Delta VB_o}{\Delta VB_{o_m}} \quad (1)$$

Axial displacement of the hob milling tool along its own axis can be performed in two ways:
- displacement by a certain length after machining one or more serrations
- continuous movement of the hob milling tool, i.e. automatically for a certain length during gear cutting using diagonal hob milling

The principle of axial displacement of the hob milling tool is shown in figure 5 [6].

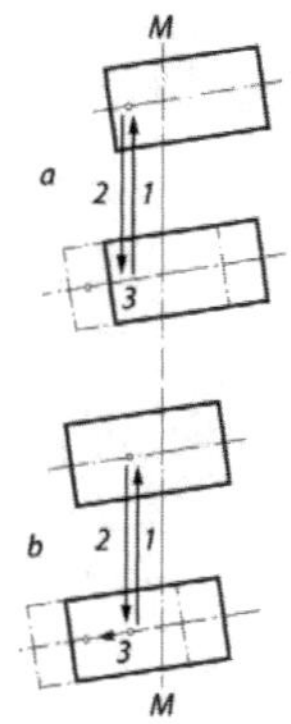

Figure 5. The principle of axial displacement of hob milling tool

One of the basic problems in the application of axial displacement of hob milling tools in the gear cutting is the determination of the optimal size of axial displacement a_p, which depends on a large number of parameters. The influence of the module on the size of the axial displacement can be represented by the expression:

$$a_{p\,opt} = (0.25 \div 0.75)\,m \tag{2}$$

where m is the module. Based on the above expression, a diagram is shown in figure 6 [3].

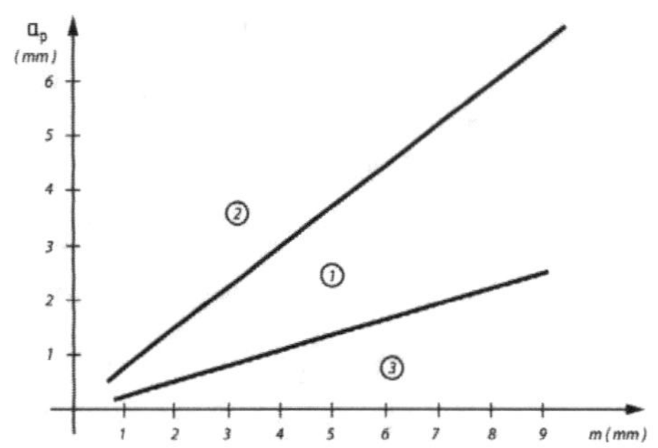

Figure 6. Dependence of axial displacement a_p on modulus m

It can be noticed in the figure that the area of optimal axial displacements also increases with the increase of the modulus. In figure 6, 1 indicates the area of optimal axial displacements, and 2 indicates the area of insufficient utilization of the hob milling tool, and 3 indicates the area of deterioration of the hob milling tools.

The number of axial displacements can be calculated based on the following expression:

$$n_{a_p} = \frac{L_1 - (l_{no} + l_s)}{a_p} = \frac{L_{a_p}}{a_p} \tag{3}$$

- n_{a_p} number of axial displacements
- L_1 length of the working part of the hob milling tool
- l_{no} required length for gear cutting of one serration
- $l_s = 0.4 \cdot t_n$ safety length of the hob milling tool
- L_{a_p} length of the hob milling tool for axial displacement
- a_p axial displacement
- t_n normal pitch

Based on the above, it is possible to define the basic rules regarding the application of axial displacement of hob milling tools in the gear cutting of cylindrical gears with straight and oblique teeth:
- large length of axial displacement will give an ununiform distribution of wear of the tooth of the hob milling tool
- the small length of the axial displacement will result in a large wear of the teeth of the hob milling tools
- when the gears have small serration widths, a small number of teeth and a small module, it is necessary to machine more serrations between two axial displacements
- when the gears are of greater width, a larger number of teeth and a larger module, it is usually done with one workpiece, i.e. one serration between two axial displacements is machined. In extremcases, diagonal hob milling must be applied.

Gear cutting of cylindrical gears with straight and oblique teeth can be performed by the down hob milling and up hob milling.

3. Experimental research and results analysis

Experimental investigations of the influence of axial displacement on the wear of hob milling tools during gear cutting of cylindrical gears with oblique teeth were performed in production conditions. On the original form for monitoring the distribution of wear of the teeth of hob milling tools and the development of tribological processes, data on the workpiece, data on the hob milling tool, data on the hob milling machine, process data, workpiece sketch, sketch of wear of the hob milling tool teeth, data on axial displacement of the hob milling tool, etc. are entered.

Workpiece data are: workpiece outside diameter $d_k = 125.5mm$, number of gear teeth $z_2 = 40$, standard module $m_n = 2.75$, contact angle $\alpha = 20^\circ$, angle of inclination $\beta = 23.33^\circ$, right direction R, gear width $b_2 = 18mm$, serration width $b = 36mm$, material HS-6-5-2-5, coolant REZANOL HP40, hob milling machine MODULE 5WZ-250x5A.

Data on the hob milling tool: integral uncoated hob milling tool number 58, diameter of the hob milling tool $d_y = 110mm$, length of the hob milling tool $L_y = 150mm$, module $m = 2{,}75mm$, basic profile II, normal pitch $t_n = 8.639mm$, direction of thread right R, number of treads $t_1 = 1$, number of grooves

$i = 14$, material of the hob milling tool HS-6-5-2-5, hardness $HRC = 65$, accuracy class A, hole diameter $d = 40 mm$.

Process data: equi-directional milling, $n = 180\ r/\min$, $v = 64 m/\min$, axial feed $s_a = 3mm$, axial displacements $a_{p_1} = 1.1mm$ $a_{p_2} = 4.8mm$ $a_{p_3} = 9.6mm$, displacement after machining of two teeth in a package, etc. Wear criterion: $VB_o = 0.35mm$.

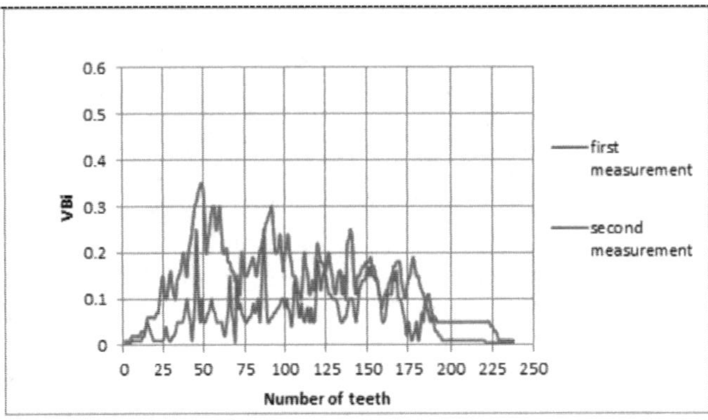

Figure 7. Distribution of teeth wear on the input lateral flank with axial displacement $a_{p_1} = 1.1mm$.

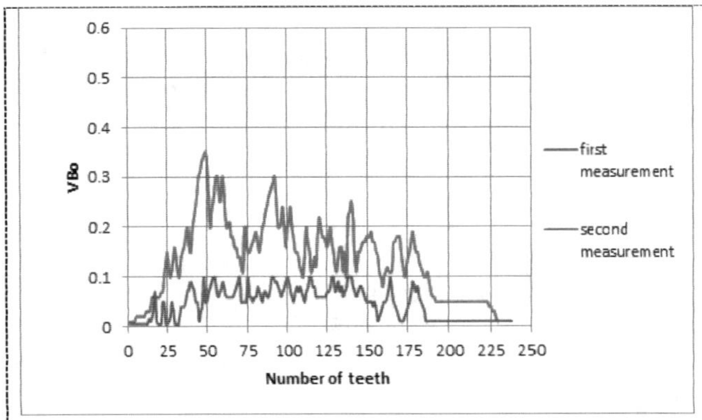

Figure 8. Distribution of teeth wear on the output lateral flank with axial displacement $a_{p_1} = 1.1mm$

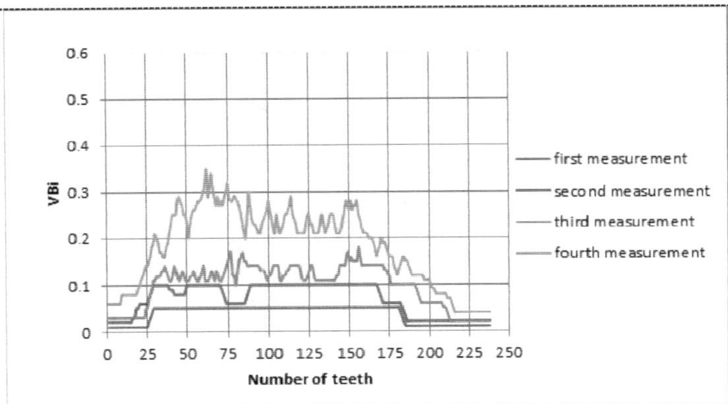

Figure 9. Distribution of teeth wear on the input lateral flank with axial displacement $a_{p_2} = 4.8mm$.

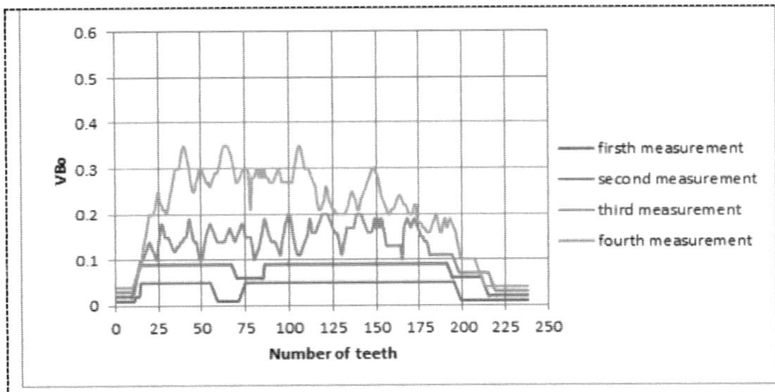

Figure 10. Distribution of teeth wear on the output lateral flank with axial displacement $a_{p_2} = 4.8mm$.

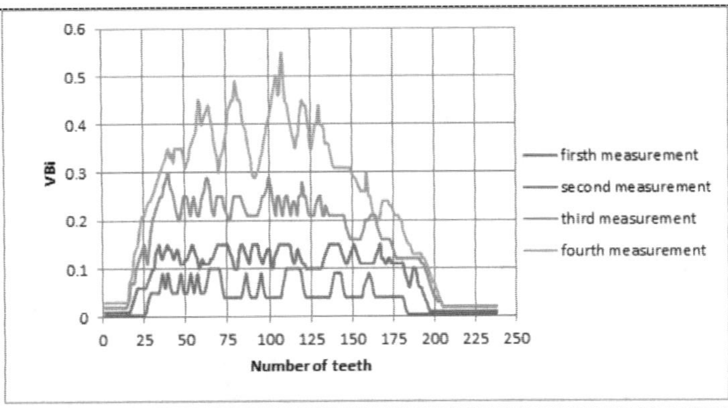

Figure 11. Distribution of teeth wear on the input lateral flank with axial displacement $a_{p_3} = 9.6mm$

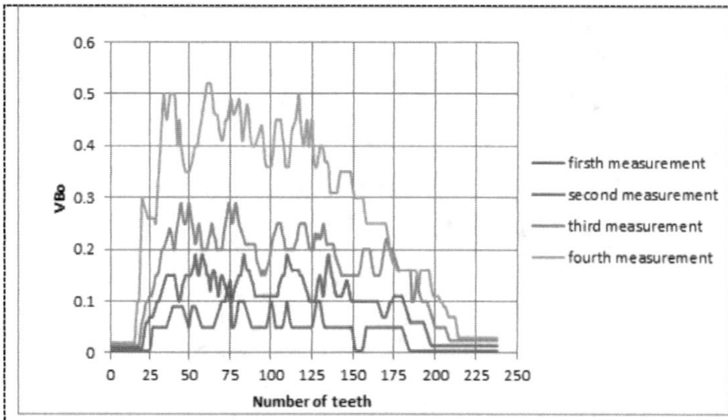

Figure 12. Distribution of teeth wear on the output lateral flank with axial displacement $a_{p_3} = 9.6mm$

Figure 7 and figure 8 show the distribution of teeth wear of uncoated hob milling tool number 58 made of HS 6-5-2-5 material when gear cutting the teeth of a cylindrical gear with oblique teeth on the input and output lateral flank with axial displacement $a_{p_1} = 1.1mm$. Figure 9 and figure 10 show the distribution of wear on the flanks of the same tool but now sharpened for the first time when working with axial displacement $a_{p_2} = 4.8mm$. Figure 11 and figure 12 show wear distribution of the input lateral flank VB_0 and the output lateral flank VB_i when applying axial displacement $a_{p_3} = 9.6mm$ in the process of gear cutting of cylindrical gears with oblique teeth with uncoated cutter number 58 made of material HS-6-5-2-5 which was sharpened for the second time.

When gear cutting of cylindrical gears with oblique teeth with axial displacement $a_{p_1} = 1.1mm$ by hob milling tool there are 140 gear teeth and it was worn-out $VB_0 = 0.3mm$. Applying axial displacement $a_{p_2} = 4.8mm$, the hob milling tool machined 112 gear teeth, and it was worn-out $VB_0 = 0.35mm$. Using the third axial displacement $a_{p_3} = 9.6mm$ the hob milling tool machined 80 gear teeth and was worn-out $VB_0 = 0.5mm$. Based on the above, it can be concluded that better results were achieved with a smaller axial displacement $a_{p_1} = 1.1mm$ compared to larger axial displacements $a_{p_2} = 4.8mm$ and $a_{p_3} = 9.6mm$. The experiment was performed with the same tool, i.e. uncoated hob milling tool made of material HS-6-5-2-5. All technological parameters were constant except of different axial displacement. The hob milling tool number 58 was new when working with $a_{p_1} = 1.1mm$, and it was sharpened for the other two experiments. The axial displacement of the hob milling tool $a_{p_1} = 1.1mm$ depending on the axial division ε is $a_{p_1} = 1{,}78 \cdot \varepsilon$, and depending on the module m is $a_{p_1} = 0.44 \cdot m$ The axial displacement of the hob milling tool $a_{p_2} = 4.8mm$ depending on the axial division ε is $a_{p_2} = 7.79 \cdot \varepsilon$, and depending on the module m is $a_{p_2} = 1.42 \cdot mm$. The axial displacement of the hob milling tool when gear cutting of cylindrical gear with oblique teeth $a_{p_3} = 9.6mm$ depending on the axial division ε is $a_{p_3} = 15.58 \cdot \varepsilon$, and depending on the module is $a_{p_3} = 3.49 \cdot m$.

Considering what is stated in the theoretical part that refers to the magnitude of the axial displacement, it can be concluded that the first axial displacement $a_{p_1} = 1.1mm$ is in the range of optimal magnitudes of axial displacements depending on the module m. The axial displacements $a_{p_2} = 4.8mm$ and $a_{p_3} = 9.6\,mm$ are in the area of insufficient utilization of the hob milling tool when gear cutting of cylindrical gears with oblique teeth.

4. Conclusion

The process of gear cutting of cylindrical gears with oblique teeth is influenced by a very large number of controlled and uncontrolled factors that make this type of machining extremely complex. This paper presents a part of the research of an interesting area of gear cutting of cylindrical gears with oblique teeth by relative rolling, which refers to the analysis of the influence of axial displacement of hob milling tool along its own axial axis after machining a certain number of serration of cylindrical gear with oblique teeth.

The analysis of the obtained research results showed that in this particular case, less axial displacement $a_{p_1} = 1.1mm$ provides less wear and more uniform distribution of wear on all teeth of the hob milling tool when gear cutting of cylindrical gears with oblique teeth.

5. References

[1] Tanasijević S 1989 *The basis of tribology of machine elements (Osnovi tribologije mašinskih elemenata)*, Naučna knjiga, Beograd, Serbia

[2] Bouzakis K-D, Friderikos O and Tsiafis I 2008 FEM-supported simulation of chip formation and flow in gear hobbing of spur and helical gears, *CIRP J. of Manuf. Sci. and Tech.* 1, 18-26

[3] Sovilj B 1988 *Identification of tribological processes during hob milling (Identifikacija triboloških procesa pri odvalnom glodanju)*, Fakultet tehničkih nauka, Novi Sad, Serbia

[4] Sovilj-Nikić I *Modeling and optimization of hob milling process (Modelovanje i optimizacija procesa odvalnog glodanja)*, Fakultet tehničkih nauka, Novi Sad, Serbia, Doctoral Thesis (unpublished)

[5] Sovilj-Nikić I, Sovilj B, Kandeva M, Gajić V, Sovilj-Nikić S, Legutko S and Kovač P 2012 Tribological characteristics of hob milling tools from economic aspect, *J. Balk. Tribol. Assoc.* **18** (4), 577-585

[6] Prapotnik B 1982 *Experimental determination of optimal operating parameters by hob milling tools (Eksperimentalno utvrđivanje optimalnih parametara rada odvalnim glodalima)*, Fakultet strojarstva i brodogradnje, Zagreb, Croatia, Master Thesis

IOP Conf. Series: Materials Science and Engineering 1169 (2021) 012041 doi:10.1088/1757-899X/1169/1/012041

Applications of nonconventional technologies in the current global pandemic

R Veres[1], S Ilea[1], B C Feier[2], G Veres[3], F Corb[1]

[1]Faculty of Managerial and Technological Engineering, Doctoral School of Industrial Engineering, University of Oradea, Romania
[2]Faculty of Medicine and Pharmacy, University of Oradea, Romania
[3]Faculty of Managerial and Technological Engineering, University of Oradea, Romania

ralph.veres@catedral.ro

Abstract. Starting from the first part of the previous year, the mankind faced a very unusual challenge. Starting with the spread of an animal-based virus found in the exotic food markets in Wuhan Province, China, the SARS-COV-2 virus quickly began to spread on a global basis on the entire planet. With such a high contagiousness it was only a matter of time till the virus made its presence in the countries all over the world, infecting many people and, unfortunately, rising the death toll every day. Based on this situation, a global pandemic was instituted by the World Health Organization resulting many restrictions and rising many problems for the medical systems all around the world. In this paper we would like to present some of the applications of modern technologies that helped, in way or another, in the fight against the pandemic, being sometimes a helping hand in the global fight against this virus.

1. Introduction

The Severe Acute Respiratory Syndrome Coronavirus 2 known shortly as SARS-COV-2, it's a virus that caused the COVID-19 global pandemic. It's a respiratory illness that is believed to have a zoonotic origin, having a close genetic similarity to bat coronaviruses. Being a very contagious virus, studies showed that it has a infection rate of 2.39 up to 3.44, if no members of the community are immune or if there not taken any preventive measures like social distancing, wearing masks, vaccination or other methods. [1], [2].

It is known that the virus is spreading mainly via respiratory droplets produced from coughs, sneezes or by speaking. The COVID-19 spread can be reduced inside a community by some measures and simple precautions like physical distancing, wearing a mask, keeping rooms well ventilated, avoiding crowds, cleaning your hands, and coughing into a bent elbow or tissue.

From a global standpoint, the protective equipment were a big concern from many point of views. The high demand of protective equipment started to create a global blockage in the supply chain for every country. [3], [4].

Personal Protective Equipment, also known as PPE, were mandatory for anybody who could have a direct contact with an infected person. Medical personnel were the first ones in the fight against the pandemic are were the first ones to have huge logistics issues with the mentioned PPE.

IOP Conf. Series: Materials Science and Engineering 1169 (2021) 012041 doi:10.1088/1757-899X/1169/1/012041

PPE include gloves, gowns, overalls, face-shields, googles and protective masks. For some of them as presented in [4], new and modern technological processes came in hand to help in the process of manufacturing, resulting thus an advance in the contact beetwen infected people and healthcare practitioners.

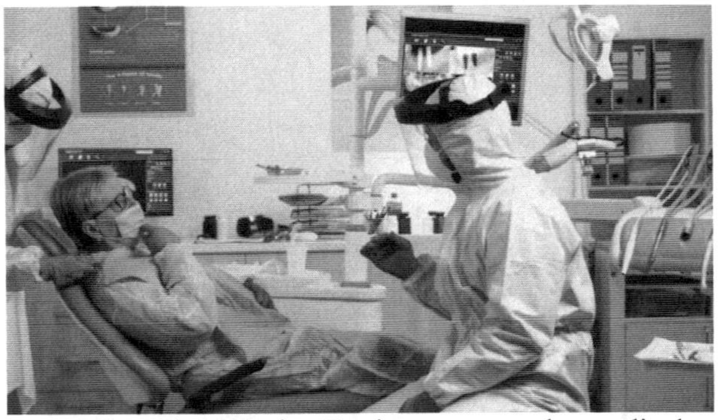

Figure 1. Personal Protective Equipment worn by medical personnel.

2. Methods and cases

Globally, the medical equipment industry went under a huge pressure based on the high demand of them, straining the supply and distribution of them. As a first response on a worldwide scale, engineering branches came in helpful with the manufacturing of protective equipment and many more.

Other nonconventional technologies like Rapid Prototyping or 3D printing came in handy being able to manufacture much needed components for the medical field. It's not unusual for medicine to be in a harmony with engineering, but in this case it was more important and visible than any other time.

3D printers are known for their reliability, efficiency, precision and also cost effectiveness. Medical equipment such as face-shield, protections, head supports were reproduced in a computer aided design software and manufactured in a very short amount of time. In the following figure is presented the equipment used for additive manufacturing in our laboratories. [5],

Figure 2. 3D printing system at University of Oradea – Industrial Engineering Department

Speaking about manufacturing, many manufacturing companies tried to bring their contribution on this fight against the logistic and production stranding. One of the most notable event, even from the start of the pandemic, was related to the production of face-shield on an industrial scale. Caracol-AM company from Northern Italy together with KUKA robots. They developed an automated additive manufacturing process that can process up to 1000 face-shield every day. Also, the drawings, STL files for 3D printers and any other details are open-source, so everybody from anywhere in the world can download and use the models to reproduce their product, solidarity being a keyword in this worldwide collaboration.

Speaking about the idea above, locally, many companies tried to contribute in the same way to support the medical system. Face-shields, accessories, flow regulator and even artificial respirators were reproduce on 3D printers. In Figure 3 is presented a medical practitioner wearing a face shield against respiratory droplets. [5], [6].

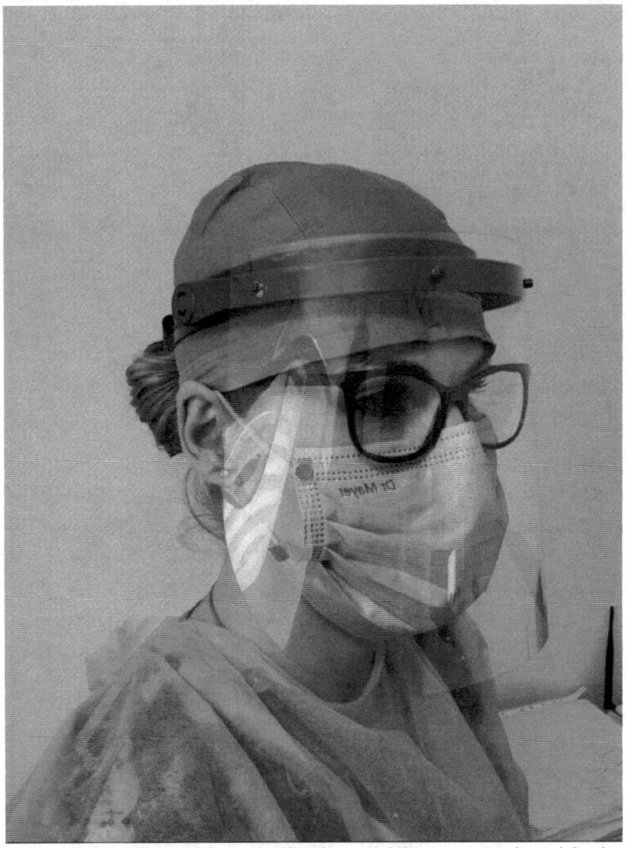

Figure 3. Face-shield used by medical practitioners made with the help of 3D printer – the plastic head support was made with the help of additive manufacturing than the transparent foil was mounted on the front clips.

Due to the hazardous effects of the infection with the SARS-COV-2 virus, patients with consistent medical history were in need of mechanical respirators. Components for this machines were mainly reproduced on 3D printers not only because the cost was more than affordable for those situations, but also for being manufactured in a shorter time.

Open-source projects made by students and teachers have been uploaded on different portals and took part in the efforts to sustain life for patients in critical stages, that were mostly in intensive care units as seen on references [7] and [8].

As for today we are aware of the many issues the COVID-19 pandemic raised on a global scale, increasing the demand for medical equipment and devices, creating thus a new market of products, helping in a way or another solving the supply chain disruptions. One of the reasons of using nonconventional technologies, in this case 3D printing or any procedures of additive manufacturing, represent the advantage of quick manufacturing of complex parts. This require a long period of manufacturing planning and processing. With the help of this new technologies, minimal time is required to deliver small batches of components resulting a quick response in time of crisis. When the pandemic started to gain proportions on a global scale, people started helping as soon they had STL format files. Also another point of view consist of non-complex planning or infrastructure for this additive manufacturing that provides a more efficient crisis response.

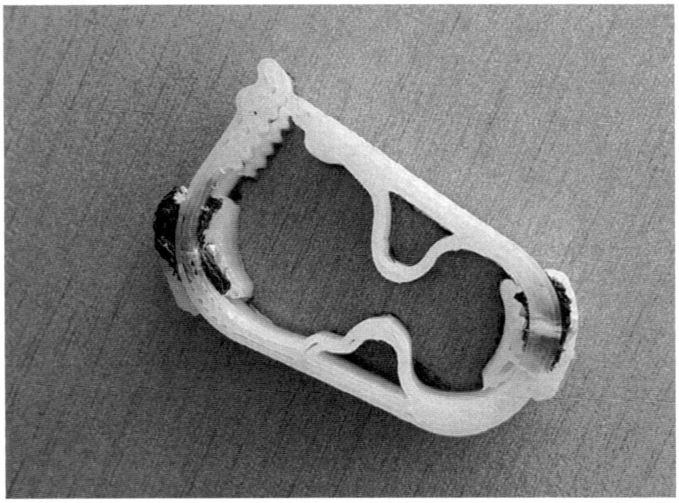

Figure 4. Flow regulator used for artificial respirators and also for intra-venous infusion kits – this one was made on a 3D printer in our laboratories at University of Oradea

3. Conclusion

As for today we are aware of the many issues the COVID-19 pandemic raised on a global scale, increasing the demand for medical equipment and devices, creating thus a new market of products, helping in a way or another solving the supply chain disruptions. One of the reasons of using nonconventional technologies, in this case 3D printing or any procedures of additive manufacturing, represent the advantage of quick manufacturing of complex parts. This require a long period of manufacturing planning and processing. With the help of this new technologies, minimal time is required to deliver small batches of components resulting a quick response in time of crisis. When the pandemic started to gain proportions on a global scale, people started helping as soon they had STL format files. Also another point of view consist of non-complex planning or infrastructure for this additive manufacturing that provides a more efficient crisis response.

Another advantage of this nonconventional procedures is related to its presence at a global scale, 3D printers being very popular machines worldwide. 3D printers do not require high-end knowledge to produce parts. Many printable files can be found on the internet, many open-source or open-design repositories being created in the last year [7], [8].

Additionally, many companies worked and engineered efficient and reliable design solutions for the additive manufacturing workforce worldwide. Also, many scientific papers are a real help, being used in many ways to gather information regarding disciplines that are interconnected in medicine, engineering, researching, etc.

References

[1] A Tino R, Moore R, Antoline S, Ravi P, Wake N, Ionita CN, et all 2020 COVID-19 and the role of 3D printing in medicine. *3D Print Med.* Apr 27;6(1):11. doi: 10.1186/s41205-020-00064-7. PMID: 32337613; PMCID: PMC7183817

[2] A Tsikala Vafea, M., Atalla, E., Georgakas, J., Shehadeh, F., Mylona, E. K., Kalligeros, M., & Mylonakis, E. 2020 Emerging Technologies for Use in the Study, Diagnosis, and Treatment

of Patients with COVID-19. *Cellular and Molecular Bioengineering*, 13(4)

[3] More Advincula, R., Dizon, J., Chen, Q., Niu, I., Chung, J., Kilpatrick, L., & Newman, R. 2020 Additive manufacturing for COVID-19: Devices, materials, prospects, and challenges. *MRS Communications*, 10(3), 413-427. doi:10.1557/mrc.2020.57

[4] Eneko Larrañeta, Juan Dominguez-Robles, and Dimitrios A. Lamprou 2020 Additive Manufacturing Can Assist in the Fight Against COVID-19 and Other Pandemics and Impact on the Global Supply Chain *3D Printing and Additive Manufacturing*.J Volume: 7 Issue 3, Mary Ann Liebert, Inc., publishers, http://doi.org/10.1089/3dp.2020.0106

[5] R Veres[1], B C Feier[2], S Ilea[1] and G Bohm-Revesz Published under licence by IOP Publishing Ltd IOP Conference Series: Materials Science and Engineering, Volume 898, Annual Session of Scientific Papers - IMT Oradea 2020 28-29 May 2020, Oradea, Romania

[6] R. Veres . – 2018 - Rapid Prototyping as an aid in Medicine - *Nonconventional Technologies Review*, *22*(3). Retrieved from http://www.revtn.ro/index.php/revtn/article/view/38 (Sep 2018): 31-37 - Vol. 22, Iss. 3, (Sep 2018): 31-37

[7] 3D printed face shields for medics and professionals—Prusa3D—3D Printers from Josef Průša (2020) https://www.prusa3d.com/covid19/. Accessed 02 May 2021

[8] Open Source Medical Supplies—Open Source Medical Supplies (2020) https://opensourcemedicalsupplies.org/. Accessed 02 May 2021

IOP Conf. Series: Materials Science and Engineering 1169 (2021) 012042 doi:10.1088/1757-899X/1169/1/012042

Research on the biomechanical behavior of dental implants

A Rohan and M Baritz

Design Product and Environment Faculty, University Transilvania from Brasov

E-mail: andreea.rohan@student.unitbv.ro

Abstract. As shown in a large series of researches, dental implants are generally made of titanium because this material allows a stable and functional connection between the bone and the contact surface of the implant. From the point of view of the mechanical behavior of dental implants, it was found that too large or incorrectly directed forces developed in the mastication process can change these connections at the same time as affecting the osseointegration process, compromising the stability of the implant. That is why modelling the behavior of dental implants represents a pre-analysis solution to understand the effect of the correct development of the mastication process. In the first part of the paper a series of researches are presented and their results regarding the importance and effect of the dimensions of dental implants on their behavior. The second part of the paper presents the analysis scheme on dental implants and establishes the procedures for making the 3D model with convertible properties compared to the real implant. The third part of the paper mentions some aspects related to the design of the experimental installation and the design of dental implant testing procedures. The final part of the paper presents the observations and conclusions of this approach of developing a system for behavioral analysis of dental implants.

1. Introduction

The restauration and recovery process of the bone that makes up the orofacial system (through all its components) presents a permanent technical and technological challenge for specialists, especially when it comes to finding materials and techniques for making dental implants. A very important condition underlying the choice of a material for making dental implants is its biocompatibility with the anatomical and physiological characteristics of the oral cavity.

Due to the fact that a dental implant will be subjected to stress in the process of mastication at high compressive and frictional forces and will also operate in a dynamic regime (both biomechanically and biochemically), these parts must be processed using advanced technologies to obtain and maintain the dimensional and mechanical properties for as long as possible.

In order to fulfill these primary requirements, dental implants must meet the following fundamental characteristics [1]:

- the implant's material must not produce harmful toxicological or allergic effects to the patient or operator;
- the physical and mechanical properties of the material should withstand the functional load and the difficult oral environment (e.g., tensile strength, modulus of elasticity, solubility, thermal conduction); the manufacturing technology should be inexpensive and also be feasible for both the medical and technical specialist;

- the shape, color and quality of the implant's material should increase the aesthetic result of the teeth and passively match the rest of the teeth so as not to cause wear at the prosthesis-implant interface;
- the material must allow the use of appropriate oral hygiene measures and prohibit or eliminate the accumulation of oral plaque; the implant should give the opportunity to be repaired when an adverse reaction occurs due to the used material;

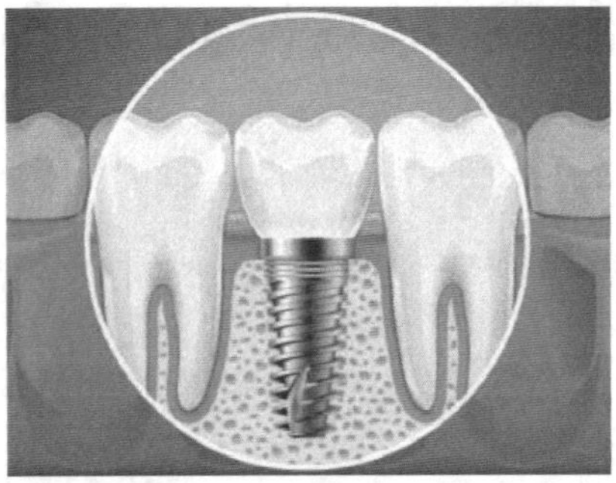

Figure 1. Dental implant [2]

These characteristics take form in the following parameters:

- the modulus of elasticity is important in choosing the material that has a modulus of elasticity comparable to bone (18 GPa), therefore it must be selected to ensure a more even distribution of stress at the implant and to minimize movement relative to the bone interface of the implant.
- the tensile, compressive and shear strength of a material should be high to prevent fractures and improve functional stability.
- the fatigue strength of a material must allow the prevention of brittle fractures under cyclic loading.
- in the case of ductility, this parameter must have a minimum value of 8% for the dental implant, the limit being necessary for the contouring and modeling of an implant.
- hardness and strength - allows the reduction of wear through increasing of hardness of the implant's material and prevents fractures of implants.
- surface tension and surface energy determine the wettability of the implant by blood-fluid wetting and by the cleanliness of the surface of the implant; osteoblasts have improved adhesion to the implant surface, and surface energy also affects protein adsorption.
- the roughness of the surface through its changes at the contact surfaces' level influences the response of the tissues.

To perform a general analysis, the implant's surfaces were classified according to different criteria, such as roughness, texture, and irregular orientation. Thus, from this point of view, the researchers divided the implant surfaces according to the surface roughness as: minimum rough, intermediate rough, rough major.

Furthermore, the surfaces can be analyzed according to their texture and can be classified into concave texture (mainly by additive treatments such as hydroxyapatite-HA coating and titanium plasma spraying), convex texture (mainly by subtractive treatment, such as engraving and sandblasting). In addition, the implant surface can be classified according to the orientation of the surface irregularities into isotropic surfaces (similar topography independent of the measuring direction) and into anisotropic surfaces (clear directionality and varies considerably in roughness) [1]. Another very important aspect in the construction of dental implants is choosing the appropriate material from all points of view: dimensional, biomechanical, surface quality and obviously biocompatibility with the orofacial system, in all its component parts.

The principles of designing an implant must be compatible with the physical properties of the material used to make the body of the dental implant. The materials used to make dental implants can be classified according to their chemical composition or the biological responses when implanted [3].

In terms of materials, dental implants can be made of metals (titanium and titanium alloys, chromium cobalt alloy, gold alloys, tantalum), ceramics (alumina, hydroxyapatite, carbon, bio glass, zirconia, zirconium hardened alumina) or polymers (polymethacrylate, polytetrafluoroethylene).

Each of the categories of materials possible to be used for the construction of the dental implant have been studied both as a material for the structural base and as a coating material leading to the improvement of the osteosynthesis process. Among the most used calcium phosphate coating materials are dense hydroxyapatite and fluorapatite, because these coatings contain partially amorphous regions, but at the same time, preserving the crystalline regions. In addition, the denser coatings are also characterized by higher strength properties and lower solubility [4][5].

2. Theoretical aspects of dental implants stability

One of the most important problems that manifests itself at the level of the dental implant in contact with the facial bone structure of the patient is the process of osseointegration which is in turn supported by the stability of the implant. The primary stability of implants is considered as an essential factor for achieving a proper osseointegration process [6]. Micro-movements higher than the threshold of 50 to 100 μm can lead to the formation of fibrous tissue at the bone-implant interface.

As shown in the research studies, a series of techniques for assessing the level of stability have been initiated and developed, quantifying this level by dental implant stability coefficients (ISQ). Values of this coefficient between 1 and 100 indicate (values at the upper limit of the range) a high level of stability. The variation of the values of the stability coefficient [7][8] is determined by a series of factors such as: the quality and quantity of bone structure in which the implantation is performed, the geometric parameters of the implant (shape, diameters, lengths) and the surgical placement techniques. of the dental implant [9][10].

From a mechanical point of view, any body, including the mandible, obeys Newton's laws, the movements being caused by forces acting on the mandible. For every linear force (F_x, F_y, F_z) there is a moment (angular) or a couple ($M_{azimuth}$, $M_{elevation}$, $M_{rotation}$), the resultant forces and moments generating acceleration according to Newton's second law [11]. The moment of inertia along an axis is defined as the sum of the mass of each particle multiplied by the distance between the particle and the relevant axis squared, for a mandible of about 0.44 kg, the moment of inertia being estimated at about 8,6 kg.cm^2, 2,9 kg.cm^2 and 6,1 kg.cm^2 for $I_{azimuth}$ (axis z), $I_{elevation}$ (axis y) and $I_{rotation}$ (axis x) [11].

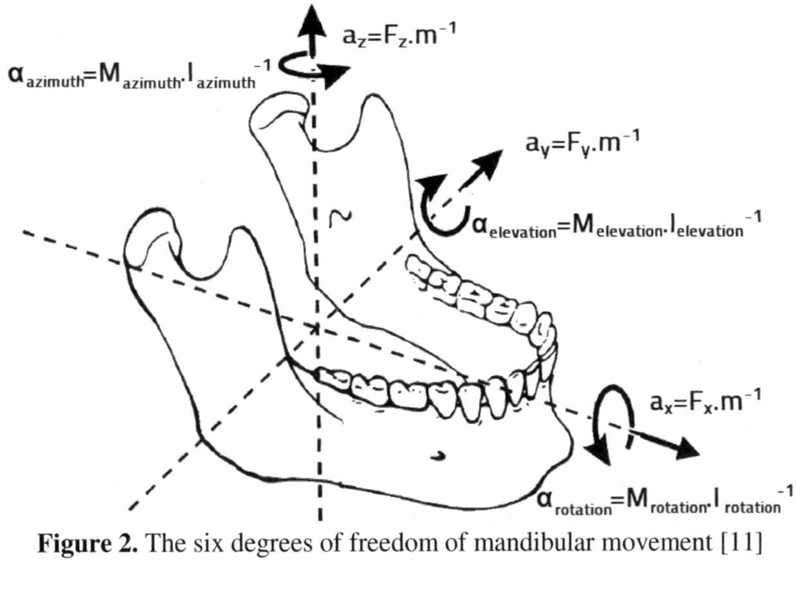

Figure 2. The six degrees of freedom of mandibular movement [11]

About 70% of the population United States population has at least one missing tooth [12]. The first missing teeth usually happen between the ages of 35 and 54 [12]. One of the treatment options for missing teeth is a simple dental implant, which is faster, simpler and cheaper to produce and implant [12]. Because this type of implant has the highest success rate for replacing a single missing tooth, it is the most widely used treatment in this case, especially if endodontics of the adjacent tooth is desired or it needs a crown for support [12]. Thus, the simple dental implant has the highest survival rate for the replacement of a single tooth, has the highest survival rate for adjacent teeth and the lowest complication rate [12].

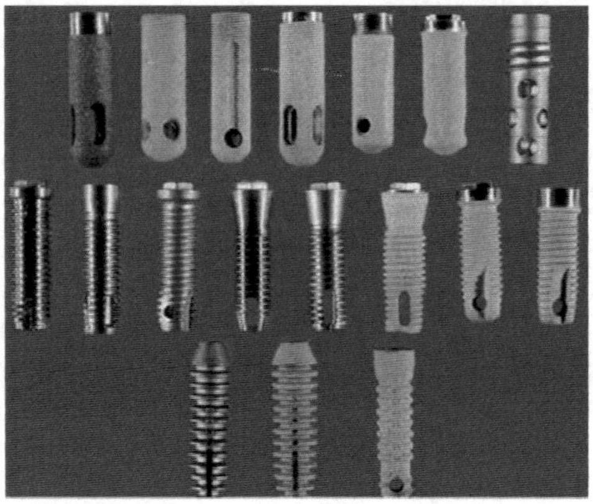

Figure 3. Types of designs for dental implants [12]

From the point of view of the materials used in dental implantology, there are three main basic groups: metals, plastics and silicate materials [14].

Metals and alloys make possible the development of precise and complex shapes of implant structures, in modern implantology there are different types of alloys being used based on metals such as cobalt, chromium, nickel, gold, silver, titanium, palladium, platinum and others [14]. Their disadvantages include the relatively unpleasant aesthetic appearance, as well as the greater possibility of adverse allergic reactions [14].

Dental plastics, based on polyacrylate, are a relatively new option, which has the advantages of being polymerized in the patient's oral cavity, being manufactured in very complex forms and it can be given a relatively simple natural appearance [14]. Polymers are weaker than metals and although they are easier to manufacture, they can only be used for a limited time and for relatively small loads [14].

Silicate materials were developed taking into account the composition of tooth enamel, consisting of 96% inorganic matter (hydroxyapatite (Ca10(OH)2(PO4)6) represents the majority of the inorganic part) and 4% organic matter and water [14]. The first form in which they appeared was the porcelain dental implants formed by casting, but they also developed over time pressed prostheses based on ceramic powder [14].

Titanium and titanium alloys have become the preferred materials for the production of dental implants, pure titanium creating an oxide layer on the surface immediately after exposure to air (9-10 seconds), which can reach 2-10 nm in a second [14]. This stable titanium oxide surface is biocompatible and has high corrosion resistance, high passivity and high chemical resistance, the modulus of elasticity of titanium and its alloys being comparable to that of human bone [14].

To understand the connection between the distribution of occlusal force and clenching strength, we follow the results of mechanical experiments which record a large number of variables that characterize the two elements [15]. Thus, the number of occlusal contacts increases linearly with the clenching strength, the distribution of the occlusal force having the highest values in the molar region [15].

In the process of implanting and determining the stability of the dental implant, osseointegration must also be taken into account, determined by several factors. Osteointegration is a measure of the stability of an implant, which takes place in two stages: primary and secondary [16].

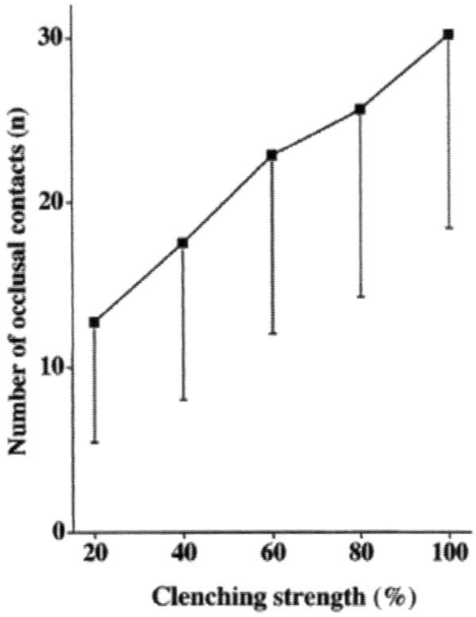

Figure 4. The number of occlusal contacts at different levels of clenching [16]

Mechanical engagement is the source of primary stability, especially the contact with cortical bone [16]. Secondary stability means the process of bone regeneration and remodeling that provides biological stability [16]. Although primary stability is responsible for good secondary stability, the latter is the one that dictates the functional loading time [16].

Among the factors that affect the primary stability are: the quality and quantity of bone tissue, surgical technique, type of implant (geometry, length, surface characteristics) [16]. The elements that affect the secondary stability are: the primary stability, the modeling and remodeling of the bone and the surface conditions of the implant [16].

3. Experimental setup

Because working with in vivo dental implants presents a high difficulty [13] and also subjectivity on the part of participants, in vitro experiments can be used, where different polymer samples with different densities can be used for insertion that simulate the structure of the trabecular bone [13], as well as a computer simulation component, represented by finite element analysis. In the case of digital simulation, the properties of the material in which the fixture is made will be taken into account so that they are as close as possible to those of the trabecular bone, namely: density, Young's modulus (modulus of elasticity), tensile / compressive strength and Poisson's ratio [13].

The determination of maximum masticatory force depends on factors such as occlusion type, age and sex [17]. Stabilization of the maximum voluntary biting force for subjects with full and normal dentition occurs faster in male or older subjects, subjects aged between 15-18 showing a rapid increase in strength [17]. These elements are important due to the change in the characteristics of bone and muscle tissues in the process of aging [17].

For the simulation and finite element analysis the software SolidWorks was chosen (part and assembly modules). When choosing the properties, for the implant we chose a pre-defined material, Titanium Grade 4 UNS R50700 (CP TI75A), for the trabecular bone we used median dates from dedicated research [18]. The differences between individuals with normal dentition and partial dentition is insignificant [18]. The density of mandibular specimens was determined to be in the 0,85 to 1,53 g/cm^3 range, and the mean stiffness value was determined to be 96,2 MPa (approximated in this simulation to 96 MPa).

IOP Conf. Series: Materials Science and Engineering 1169 (2021) 012042 doi:10.1088/1757-899X/1169/1/012042

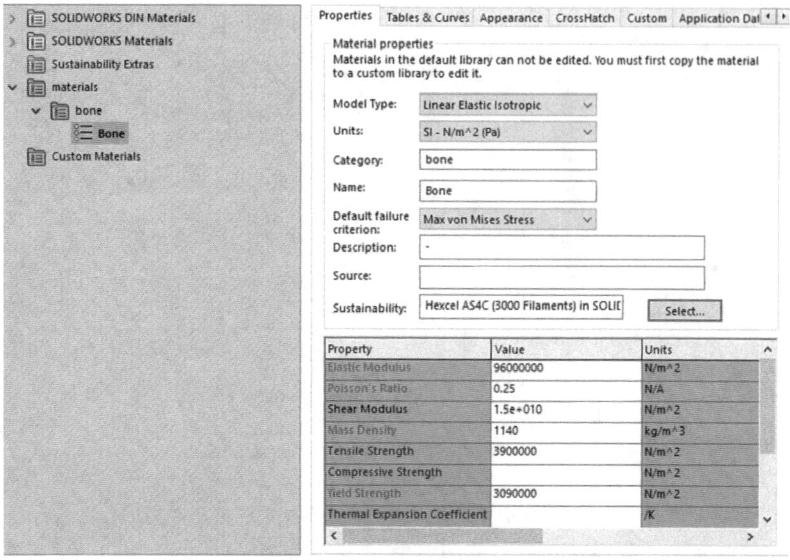

Figure 5. Characteristics used for the simulated trabecular bone

The implant was inserted in the pre-defined material, the connection of the two being made with a soldering rigid material from the list of properties offered by the software in the assembly module. The next step was to choose the fixtures, the contacts and the forces that are applied (the arbitrary value of 400 N was chosen, being considered a maximum force or a force that can even exceed the maximum value depending on the individual). To observe and understand the behavior of the implant, a cross section of the entire assembly was made both before and during the simulation. The analysis was static, the maximum deformation temperature being calculated by the software as 298 ° K (≈ 25 ° C).

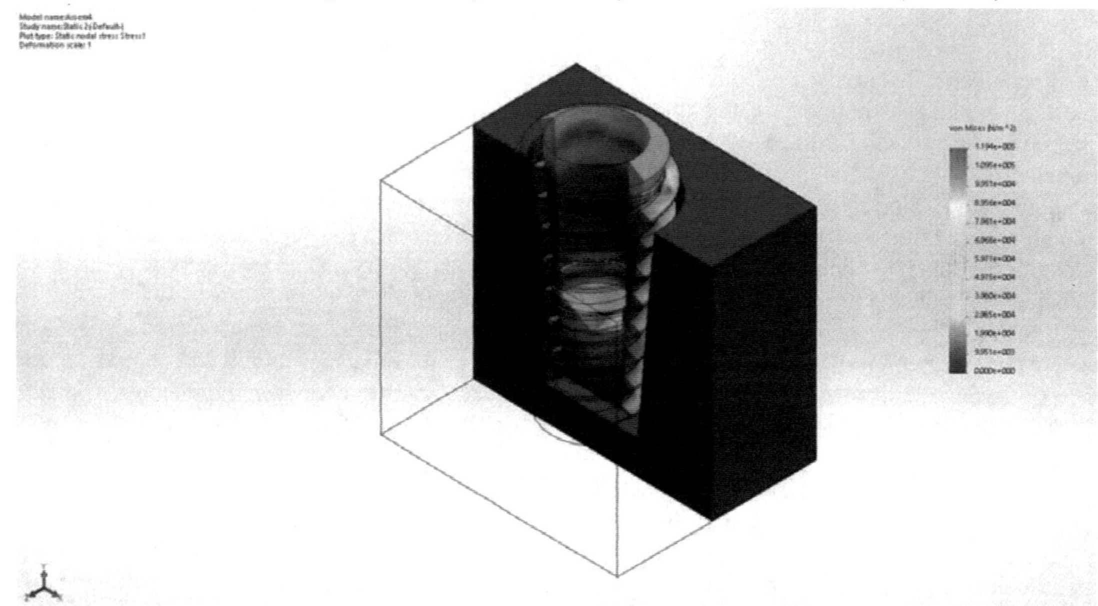

Figure 6. The effect of von Mises stress, cross section

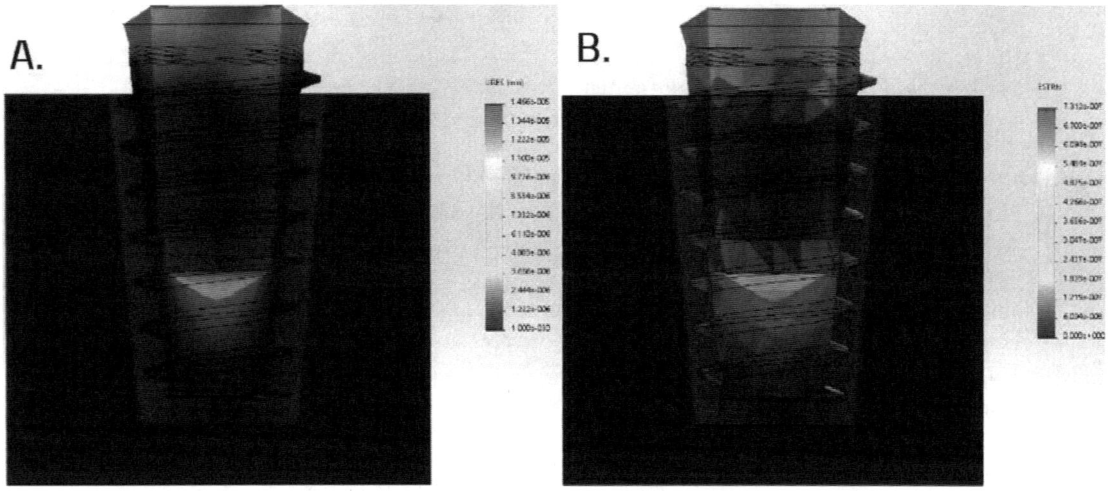

Figure 7. A. Displacement, cross section **B.** Stress, cross section

4. Results and conclusions

The mechanical testing of dental implants involves understanding all stages of the masticatory process, as well as how the morphological characteristics of individuals affect the behavior of implanted bodies. In order to understand this behavior, an ideal case was chosen, simulated in a CAD modelling software on which the finite element analysis was performed.

All the elements subjected to the test show that the movement and stress of the implant appear on the outside in the upper area, and on the inside in the lower area, i.e., the parts used to assemble a composite implant. For this reason, it is important to choose implantable elements that work perfectly with each other, especially for very small implants such as dental ones, but also the assessment of the general health and tissues of patients must be done correctly for in vivo cases.

References

[1] Saini M, Singh Y, Arora P, Arora V and Jain K 2015 Implant biomaterials: A comprehensive review *World J Clin Cases* vol 3(1) pp 52–57
[2] https://gpsdentalsa.com/blog/tag/dental-implants/ [Accessed April 2021]
[3] Osman R B and Swain M V 2015 A critical review of dental implant materials with an emphasis on titanium versus zirconia *Materials* vol 8(3): pp 932–958
[4] Kohal R J, Att W, Bächle M and Butz F 2008 Ceramic abutments and ceramic oral implants. An update *Periodontology 2000* vol 47 pp 224–243
[5] Tinsley D, Watson C and Russell J 2001 A comparison of hydroxyapatite coated implant retained fixed and removable mandibular prostheses over 4 to 6 years *Clin Oral Implants Res* vol 12 pp 159–166
[6] Barikani H, Rashtak S, Akbari S, Badri S, Daneshparvar N and Rokn A 2013 The Effect of Implant Length and Diameter on the Primary Stability in Different Bone Types *Journal of Dentistry, Tehran University of Medical Sciences* vol 10(5)
[7] Sencimen M, Gulses A, Ozen J, Dergin C, Okcu K M, Ayyildiz S and Altug H A 2011 Early detection of alterations in the resonance frequency assess-ment of oral implant stability on various bone types: a clinical study *J Oral Implantol* vol 37(4) pp 411-419
[8] Rokn A R, Ghahroudi A A R, Mesgarzadeh A, Miremadi A and Yaghoobi S 2011 Evaluation of Stability Changes in Tapered and Parallel Wall Implants: A Human Clinical Trial, *J Dent (Tehran)* vol 8(4) pp 186-200
[9] Sabeva E 2018 Comparison between the influence of implant diameter and implant length on the primary stability *Scripta Scientifica Medicinae Dentalis* vol 4(2) p 36

[10] Takahashi K, Shiraishi N, Ishiko-Uzuka R, Anada T, Suzuki O, Masumoto H and Sasaki K 2015 Biomechanical Evaluation of Ti-Nb-Sn Alloy Implants with a Low Young's Modulus *International Journal of Molecular Sciences* vol 16(2) pp 5779-5788
[11] Koolstra J H Dynamics of the Human Masticatory System 2002 *Critical Review in Oral Biology & Medicine* vol 13(4) pp 366-376
[12] Misch C E 2015 *Dental Implant Prosthetics 2nd edition* (Elsevier Mosby)
[13] Bicudo P, Reis J, Deus A M, Reis L and Vaz M F 2016 Performance evaluation of dental implants: An experimental and numerical simulation study *Theoretical and Applied Fracture Mechanics* vol 85 pp 74-83
[14] Shalukho N M, Kuz'menkov M I and Bogdanovich I A 2012 Materials for Dental Prosthetics *Glass and Ceramics (English translation of Stelko I Keramika)* vol 69(7-8) pp 243-245
[15] Kumagai H, Suzuki T, Hamada T, Sondang P, Fujitani M and Nikawa H 1999 Occlusal force distribution on the dental arch during various levels of clenching *Journal of Oral Rehabilitation* vol 26(12) pp 932-935
[16] Atsumi M, Park S and Wang H 2007 Methods Used to Assess Implant Stability: Current Status *The International Journal of Oral & Maxillofacial Implants* vol 22(5) pp 743-754
[17] Varga S, Spalj S, Varga M L, Milosevic S A, Mestrovic S and Slaj M 2010 Maximum voluntary molar bite force in subjects with normal occlusion, *European Journal of Orthodontics* vol 33(4) pp 427-433
[18] Misch C E, Qu Z and Bidez M W 1999 Mechanical Properties of Trabecular Bone in the Human Mandible: Implications for Dental Implant Treatment Planning and Surgical Placement, *Journal of Oral and Maxillofacial Surgery* vol 57(6) pp 700-706

Acknowledgments
In these experiments, we have developed the investigations with equipment from Medical Engineering Laboratory and the research is part of Andreea Rohan's student diploma project work in the Transilvania University from Brasov.

IOP Conf. Series: Materials Science and Engineering 1169 (2021) 012043 doi:10.1088/1757-899X/1169/1/012043

Line follower mobile robots, prototypes of robots and functional of mobile robots

D M Anton[1], S E Szabo[1], C A Mociar[1] and A F Iuhas[1]

[1] Faculty of Managerial and Technological Engineering, University of Oradea, 410087 Oradea, Romania

E-mail: danielanton0227@gmail.com

Abstract. This study presents the working principle of line follower robots. This study starts from the description of the components used in robots of this type to the explanation of the functions used to program robots in the C ++ programming language and the use of multiple cases and exceptions in the programming of robots of this type, because they can be used in different ways and domains and for these, the multiple working characteristics must be determined from the beginning. The study establishes the way these robots work, for a future project to create a family of disinfection robots with UV-C neons.

1. Introduction

Robots have begun to be introduced more and more frequently in most companies in the industry. From storage robots in warehouses, used to load, unload and move goods from the place of arrival, to storage, and then to the work area, to climbing robots on shelves, which can move the goods from heights of 10-20 meters. Robots in various forms become more and more useful in everyday life, such as kitchen robots, vacuum cleaners, lawn mowers, solar panel cleaners, each system becoming more accurate and efficient, using assemblies of systems with sensors, motors, artificial intelligence systems, that are interconnected with each other, such as centralized window closing systems, doors, intelligent control for thermal boilers, rolling panels for windows, curtains, air monitoring systems, temperature, air conditioning, radiant panels, photovoltaic panels, plant watering systems, sprinklers.

2. Components of a assembly set of mobile robots with sensors

These types of robots in the simplest models have sensor systems, control systems, motor control systems, power supply systems and can vary depending on the field of use and the accuracy that is needed for some works.

These mobile line follower robots use sensory systems that can be: distance, proximity, human presence sensors, barcode reading sensors. The type of robots used in this study is based on analog color sensors. They are based on QTR-1A (model sensor) type sensors, mounted on PCBs (printed circuit board) with 8 QTR type A sensors[1]. Types A are listed as some of the most accurate because they read values across a wide spectrum. The signals exported by them are from 0 to 1023. These values are read as signals by their comparators, and the microcontroller receives them in the form of varying voltage. These sensor models are small, have pairs of IR LEDs (ligh-emitting diode) or are also called phototransistors. Each sensor is independent, with its own transmitter and receiver, which can generate a signal for each channel, thus substantially increasing the accuracy.

For line follower robots in most cases they are used as line sensors, but they can also have the function of proximity sensors. These sensors each use a traction resistor to form voltage dividers. They produce an analog voltage at the output, which can be measured between a minimum of 0.2 V and a maximum of 4.8 V. To determine the values, the track of line follower robots are made of white melamine chipboard, and on them the path to be followed by the robot is drawn with a black insulating tape. These conditions may vary, but the operating algorithm will also have to be adapted to the new conditions.

IR sensors are very sensitive to direct sunlight, because it disrupts the sensors and generates incorrect readings, so the robot risks losing the default track (line). When the sensors have a black surface in front of them, the transmitted voltage values are high - they tend to 5V due to the fact that these surfaces reflect a maximum of 10% of the emitted light, this decreases the internal resistance of the receiver. And when the surface is white these values decrease proportionally - they tend to 0V. In figure 1 is presented the schematic wiring diagram of sensors IR (infrared sensor).

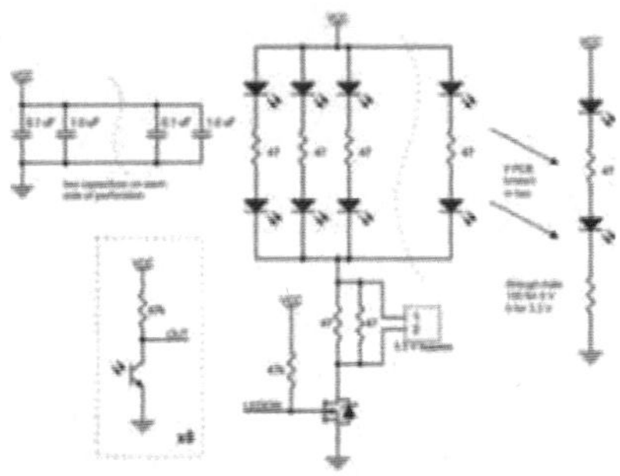

Figure 1. Schematic wiring diagram of sensors IR.

To read the signals from the sensors, the speed is adjusted in 3 steps: 50%, 75%, 100%. For straight line both engines have maximum speed when turning left, the right engine has 100% and the left engine has 50% speed. For turning right it is vice versa. Of course, depending on the radius of the curve, the speed can be adjusted algorithmically.

The robot's microcontroller uses Fuzzy [2] logic to interpret the position on the line. The source code is programmed for sensors 4 and 5 to be fixed on the black line. Those sensors must constantly see the line. If a left curve follows, the line moves to sensors 1 and 2, then the right motor will maintain the speed at 100%, and the left motor will reduce the speed to 50% to regain its position. In figure 2 is presented the reading signals sensor on channel 1 values between 0.2 - 4.8V, real-time reading channel 2 speed.

The robot also has two Pololu micro motors[3] with a 50: 1 reduction, the value of the output speed after the reducer being 600 rpm, for the supply voltage from the 12V driver. This driver is controlled by programming the PWM (pulse-witdh modulation)[4] pins (3, 5 for the left motor and the right motor 6, 9 of the robot) on the Arduino Nano microcontroller[5].

The motors are supplied by time interval pulses (PWM). This allows you to digitally control the speed that can be adapted to the opening of the curve and the accuracy required for the operation to be performed. Figures 3, 4, 5 show the PWM speeds for different duty cycles, at 50%, 75%, 100%. These intervals represent the supply voltage on the vertical axis, and the time in milliseconds for the motor supply on the horizontal axis.

3. Robot construction

In the construction of the robot, the response time of the microcontroller was taken into consideration, in order to determine the accuracy. Its size is 170mm long, the motors are arranged equidistant from the middle of the robot at a distance of 60mm.

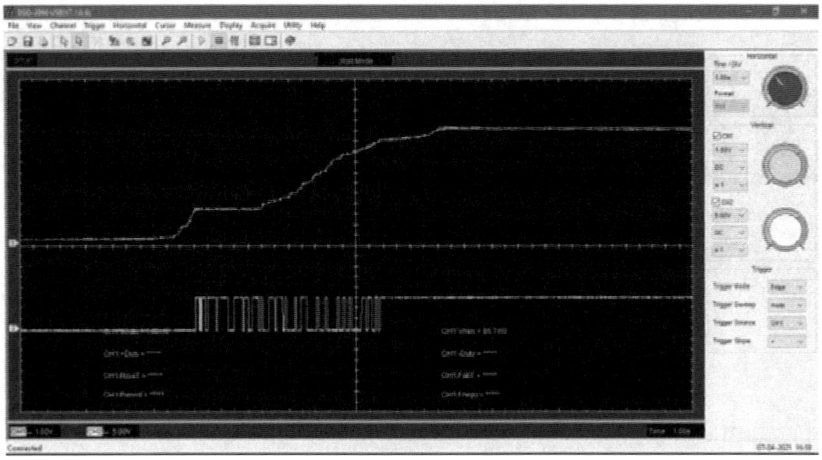

Figure 2. Reading signals sensor.

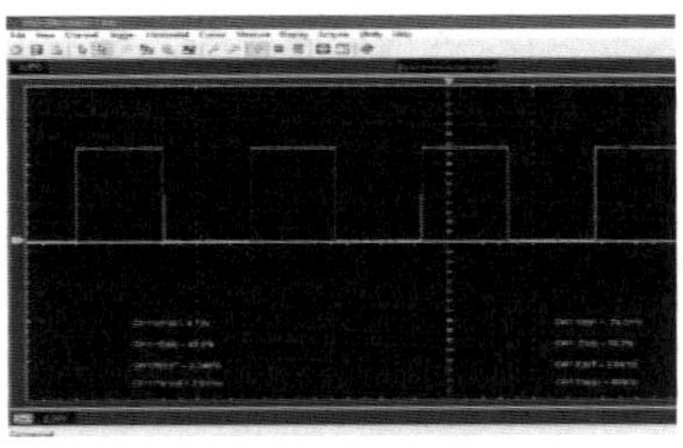

Figure 3. Signal PWM,duty cycle 50%.

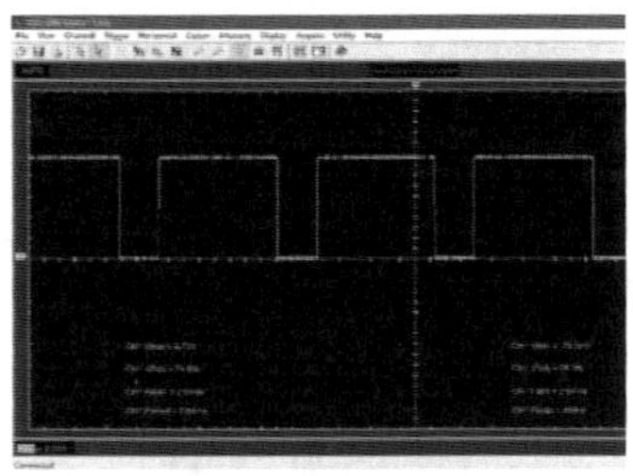

Figure 4. Signal PWM, duty cycle 75%.

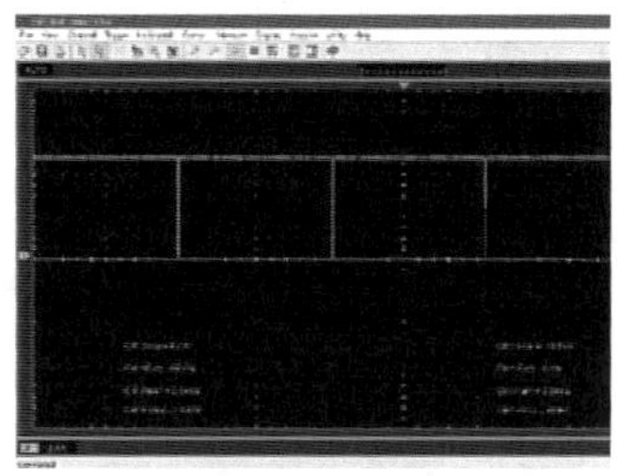

Figure 5. Signal PWM, duty cycle 100%.

The sensors were positioned as far forward as possible and at a distance of 4 mm from the chipboard, in order to receive values as accurate as possible, identical to those of previous measurements.

A set of wheels is mounted on the robot's motors, designed according to the weight and complexity of the route and then made on a 3D printer. These wheels are printed with settings of 0.2 mm per layer and 18% filling, to be as light as possible, with PLA (Polylactic Acid) filament. The printing is done at 210^0C at the extruder, and the bed at 70^0C, the printing speed is 80 mm/s.

The wheels are initially provided with an outer wall, so that after printing a two-component layer of silicone can be poured into these wheels with the syringe. Silicone, when pouring into the wheel must be added to only one part of the wheel in order to eliminate any air bubbles and at the end, the wheel must be either vibrated or lightly beaten by the table so that the air remaining through the silicone rises on top. After the silicone has hardened, the outer wall for the silicone is heated and gently peeled off by the wheel, after which it is left to dry for another two hours. The shorter the outer wall from the wheel, the thinner the silicone layer which is useful if the robot weighs more.

The ADC (Analog to Digital Converter) [6] function of the ATMEGA328 microcontroller converts the analog signal from the sensors into a PWM digital signal so that it can be processed by the microcontroller and sent on to the motor drivers. The working frequency of this ADC converter is 400Hz, being limited by the internal functions of the program.

Figure 7 show the first types of robots built.

Figure 6. The first versions of robot.

Figure 7. Award second place Robotics Championship Oradea.

4. Code source robot

The first part of the program, presented in figure 8, initializes the global variables used in the code, followed by the setup section (void setup). In this section, the input or output modes of the pins on the Arduino Nano board are assigned and pins 3 and 9 are assigned to ground (GND). The next part, shown in figure 9, presents the loop function (void loop), in which the code is repeated endlessly until the power supply is interrupted. In the first part of this function, a vector is built, taking the values read from the sensors, and then with the help of this vector the position of the line is calculated using the Fuzzy algorithm, obtaining a value between 10 and 80 units.

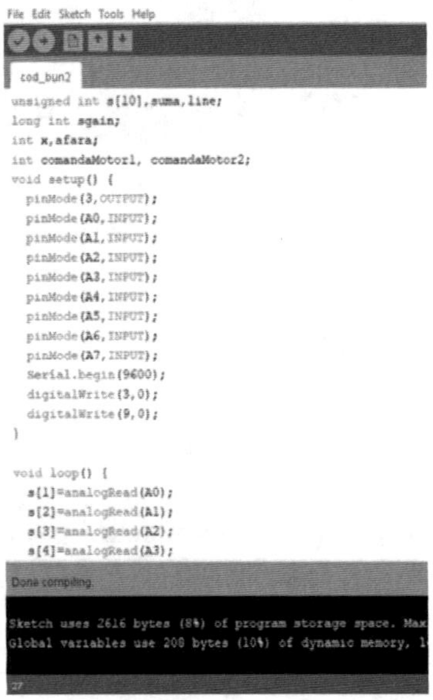

```
unsigned int s[10],suma,line;
long int sgain;
int x,afara;
int comandaMotor1, comandaMotor2;
void setup() {
  pinMode(3,OUTPUT);
  pinMode(A0,INPUT);
  pinMode(A1,INPUT);
  pinMode(A2,INPUT);
  pinMode(A3,INPUT);
  pinMode(A4,INPUT);
  pinMode(A5,INPUT);
  pinMode(A6,INPUT);
  pinMode(A7,INPUT);
  Serial.begin(9600);
  digitalWrite(3,0);
  digitalWrite(9,0);
}

void loop() {
  s[1]=analogRead(A0);
  s[2]=analogRead(A1);
  s[3]=analogRead(A2);
  s[4]=analogRead(A3);
```

Figure 8. Robot source code first part.

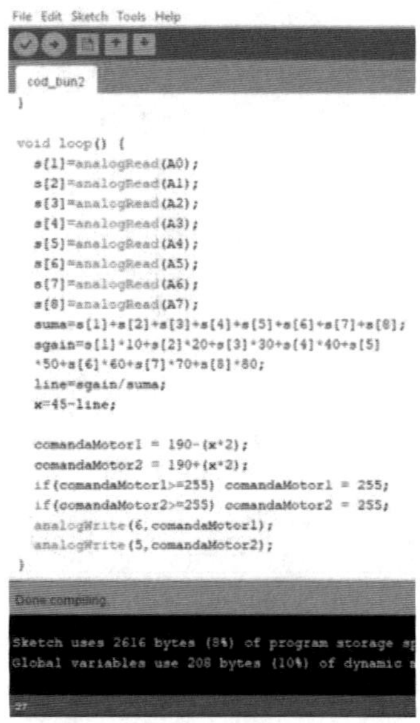

```
}

void loop() {
  s[1]=analogRead(A0);
  s[2]=analogRead(A1);
  s[3]=analogRead(A2);
  s[4]=analogRead(A3);
  s[5]=analogRead(A4);
  s[6]=analogRead(A5);
  s[7]=analogRead(A6);
  s[8]=analogRead(A7);
  suma=s[1]+s[2]+s[3]+s[4]+s[5]+s[6]+s[7]+s[8];
  sgain=s[1]*10+s[2]*20+s[3]*30+s[4]*40+s[5]
  *50+s[6]*60+s[7]*70+s[8]*80;
  line=sgain/suma;
  x=45-line;

  comandaMotor1 = 190-(x*2);
  comandaMotor2 = 190+(x*2);
  if(comandaMotor1>=255) comandaMotor1 = 255;
  if(comandaMotor2>=255) comandaMotor2 = 255;
  analogWrite(6,comandaMotor1);
  analogWrite(5,comandaMotor2);
}
```

Figure 9. Robot source code another part.

Further, in the variable "x" it decreases from the middle value of 45 units, the position of the line calculated above and we obtain a value that varies between -35 and 35. The middle position has the value 0. This variable may or may not influence engine speed. In the variables "comdandaMotor1" and "comandaMotor2" the speed for each motor is calculated, and the PWM signal transmitted to the motor drivers is generated.

5. Results and conclusion

In this form, the robot used the engines at full capacity, the problems that occurred after running the route were overheating of the engines, because for the line follower test the robot must be built to have a low weight and be fast to travel the route in the shortest time. Because a very low weight was desired,

IOP Conf. Series: Materials Science and Engineering 1169 (2021) 012043 doi:10.1088/1757-899X/1169/1/012043

the 3D model was designed to add as little silicone as possible to the wheels, as this greatly increases the weight of the robot. The designed robot uses 8 IR sensors, so reading the line is easy and at the maximum speed of this robot.

These robots participated in the international competitions Robotics Championship Oradea and RoboManiacsTimişoara, and in these competitions the following prizes were obtained:
-Robotics Championship Oradea: Second place in the Line Follower Junior category, as presented in figure (7)
-Robo Maniacs Timişoara: 1st and 2nd place in the Line Follower Baby category

To participate in competitions, it is also necessary to test and simulate the robot in different working functions because some routes are more complicated, for example in the case of 90 degree curves, and the robot loses the line completely, without recovering.

A future study will present the integration of the reading functions of the last sensors that observed the line and will try to implement the PID (Proportional Integrated Derivatives) function, use the bring back function or search for it according to the last reading. It will move to hardware upgrades for robot components, better engine controllers, faster sensor reading and software part optimization with more predefined functions and PID functions.

References

[1] https://www.pololu.com/product/960
[2] https://en.wikipedia.org/wiki/Fuzzy_logic
[3] https://www.pololu.com/product/1093
[4] https://en.wikipedia.org/wiki/Pulse-width_modulation
[5] https://www.arduino.cc/en/pmwiki.php?n=Main/ArduinoBoardNano
[6] https://en.wikipedia.org/wiki/Analog-to-digital_converter

IOP Conf. Series: Materials Science and Engineering 1169 (2021) 012044 doi:10.1088/1757-899X/1169/1/012044

Building a continuous passive mobilization machine (CPM) for the inferior limb recuperation

M G Statache[1], E Vultur[1] and C N Drugă[1]

[1]Product Design, Mechatronics and Environment Department, "Transilvania" University of Brasov, Brasov, Romania

madalina.statache@student.unitbv.ro

Abstract. This paper is presenting the projection and realization of a passive mobilization device (CPM), used for rehabilitate lower knee articulation conditions, such as post-traumatic conditions, degenerative, ligamento-plastic chirurgical interventions, meniscectomy and in special patients who have had been under knee prosthesis chirurgical interventions, this one being the one which causes the biggest issues in the rehabilitation process. Also, in this paper are figured the main steps which are followed throughout the building process of a Low-Cost functional prototype and its way of function.

1. Introduction

Most of the rehabilitation treatments start with the result of the passive mobilization of the affected segments, this one having as purpose to reduce the appearing swelling, reducing pain through special methods, called neuro proprioceptive facilitated methods, keeping the kinesthetic memory and the last but not least growing the movement amplitude at the affected articulation level.

The next steps in the rehabilitation treatment are the muscle force grow at the affected muscle level, coordination recovery, motor system control, balance, muscular resistance growth, skill reset, sensitiveness reset, increase capacity of effort etc. Of course, the rehabilitation objectives and the therapeutical condition are stated depending on many factors like the condition that the patient is suffering, the status of the injury, the severity degree, general patient status, the contraindications to some rehabilitation procedures etc. In this paper we are going to focus on the objectives which are made possible thanks to the articulated passive mobilization, and on its benefits, and especially about the continuous passive mobilization occurred trough medical machines [1],[2],[3].

The Continuous Passive Movement (CPM) devices, are medical devices which are being used to perform a continuous passive movement of the treated articulation, the movement being generated by an external force, with the objective of sustaining and growing the movement amplitude at the affected articulation level, where the mobility is limited. The method is being used on a large scale thanks to the positive clinical experience and its benefits observed by the medics, therapists and patients [4].

2. Material and Method

To build a CPM device was made possible by using mechanical components, electrical and electronical which are available on the market, the final purpose being the completion of a functional prototype, well calibrated for recovering from knee affections, easy to use by the therapist or patient, reliable, qualitative from a construction point of view and with a small production cost. Having in count that the

IOP Conf. Series: Materials Science and Engineering 1169 (2021) 012044 doi:10.1088/1757-899X/1169/1/012044

anthropometry dimensions are different from individual to another (Specially the one from the lower limbs and the bone segment components), the metal support was built with the option of length adjust of the two main segments (thigh and leg) so it could be used for a large number of patients.

2.1. Functioning principle of CPM device

The CPM devise is powered to the 230V network and started through the ON button (figure 1).

The input elements (Amplitude, Speed, Acceleration) will be introduced through the integrated keyboard of the controller, also with the help of the limit switches, this one performing among others the role of regulating of the amplitude of movement realized by the CPM device. The controller will take the input elements, displaying them on the integrated screen and sending them at the same time to the motor driver. The motor driver will command the stepper motor (SM), controlling its direction and the rotation speed. The stepper motor will train the linear platform, which at its time will move the articulated support, making a complete movement happen. The limit switches, besides amplitude regulation, in the moment when they will be executed on the linear platform, will send a feedback to the controller, to command the motor work in the opposite way (figure 1).

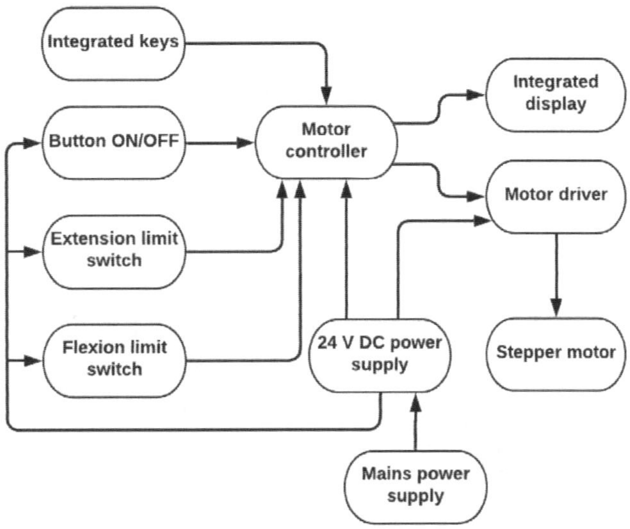

Figure 1. CPM device block diagram.

2.2. Choosing the materials and the components

The base materials and components were chosen to fulfill the tasks planned for constructing and the well performing of the prototype. These were chosen and acquired after an analysis of the different producer and trader offers, the most important criteria was the value for money of the different components.

2.2.1. Motor controller. For the proposed application, the controller has the command unit and control purpose of the continuous passive mobilization device, this one will be the "brain" of the system. This will take the data imputed through the keyboard, will read the data coming from the two limit switches and transmit the order of movement to the motor through the motor driver [4].

The chosen Controller for the accomplishment of this project is one for the stepper motor, from Fuyu (figure 2), with a screw linear platform, this one having the capacity to control systems formed with only one movement axis, as the CPM are as well. The controller has as a component an ARM Cortex-M0 processor, having as a possibility the control of speed and distance covered on the engine platform. The previous mentioned comes as a part with a remote control (figure 2) which can set the motor speed and can stop it at any time. The remote control would be used as a safety system by the patient, having the option to stop the system whenever would appear an above the normal level pain or any other issue related to the therapy or device.

This controller has within its components a display (figure 2) where the system function parameters and the values set by the user will be shown.

Besides the display we will find also 5 buttons through which the user will introduce the data necessary to perform the therapy with the CPM device.

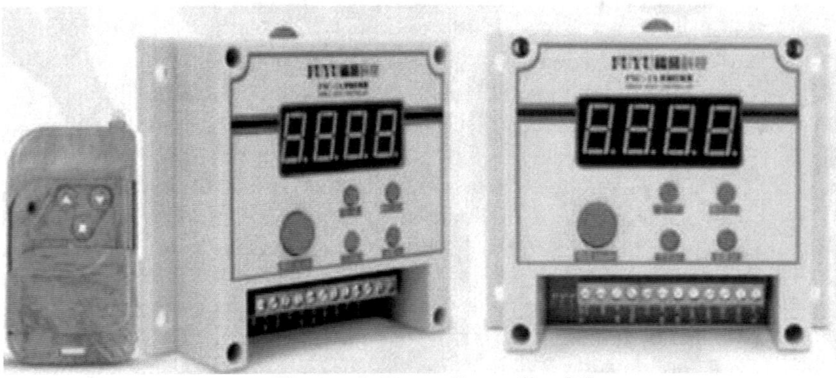

Figure 2. Fuyu Controller [5].

2.2.2 Motor Driver. The motor driver chosen for this purpose comes from the same range of Fuyu products as the Controller and the Engine, having the name Nema 23-Driver for SM, FMDD50D40NOM (figure 3). It is a two phases driver based on DSP control. It comes with twelve contacts with whom the electric circuit necessary for the device functionality works.

Figure 3. Motor Driver Nema 23 [6].

2.2.3 Stepper motor with linear platform. Around CPM, the motor performs the main role, creating the power which makes possible the movement of the mobile elements, achieving on a knee level a Flexion-Extension movement. The movement happens on a maximum amplitude lapse set by the two limit switches [4].

The motor chosen to accomplish the project is also Fuyu built, therefore being a SM, with a linear platform working along with a ball screw system (figure 4). To satisfy the CPM type of devices demanded requirements, regarding the movement amplitude where it will be executing its way, the length of the screw where the platform with the articulated support has to be long enough (In this case is about 900mm).

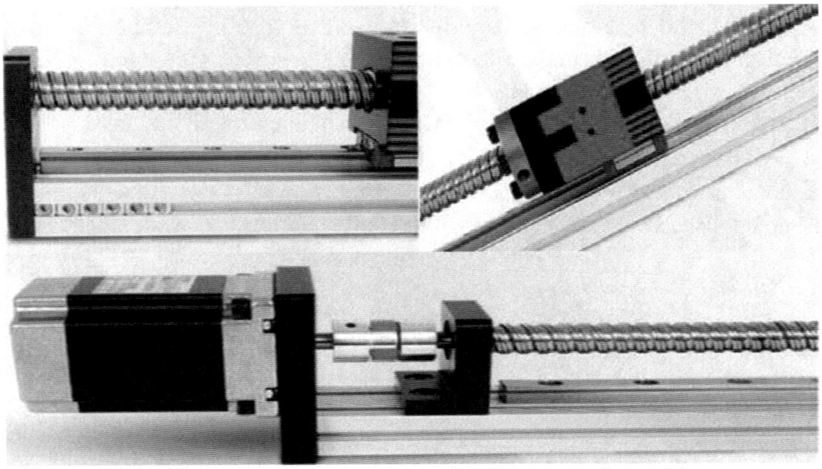

Figure 4. Stepper motor with linear platform [7].

2.2.4 Limit Switch. The device can execute a flexion-extension movement until the maximum set length. This limit switches (figure 5) are frequently used within the linear gliding systems and also as a stop-indicators which stop the engine when engaged. Functionally speaking it is really simple, the switches send a digital feedback, 0 when is not pressed and 1 when it is pressed.

For a correct functioning of the device, the switches must be mounted on the device at the maximum flexion-extension spots. Also for the completed prototype, the switches also accomplish an extremely important role for the correct functioning of the device, which is the movement amplitude regulation. The two switches will be mounted on a sliding support giving the option to move the switch on a scaled band until the desired movement amplitude having in count the patient needs. The movement amplitude regulation is one of the most essential characteristics of the CPM device.

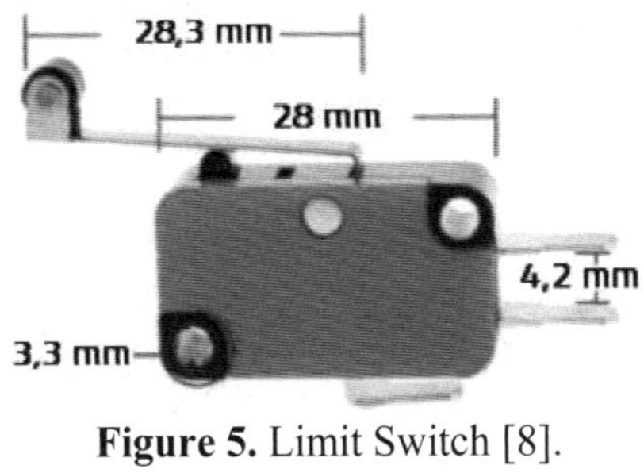

Figure 5. Limit Switch [8].

2.2.5 CC Power Supply. The power supply was chosen so it would satisfy the system requirements, therefore it is a power supply which transforms the network 230V AC into small voltage energy with 24V, having three exits so it can supply all components.

2.2.6 Articulated metal support. To transmit the movement to the patients lower limb, he has to be placed on an articulated metal support. The support articulations at the knee level executes the flexion-extension of the knee articulation, and the articulations at the hip level executes the flexion-extension movement of the hip.

To be functional, the support has to fulfill the following requirements:

- Stability – For a safe and an adequate therapy process.
- Resistance – It must resist to the therapy demands.
- Mobility – The articulations must cover the movement amplitude of the knee's articulation.

- Adjustable size – So it can be used to different patients with a different anthropometric size, the length of the support segments must be adjustable.

Used materials:

- Round pipe made of laminated steel with a 16mm diameter and 1m length;
- Round steel bar with a 16mm diameter and 1m length;
- Round steel bar with a 14mm diameter and a 1m length;
- Steel sheet, polished made of zinc iron 300x1000x0,75mm.

2.3. The device building and assembly

The first step to build the CPM device was to create the device's design, specially the one of the articulated support (figure 6), the element which holds the lower limb and prints the movement to it. The device follows the general design of the actual models of the market.

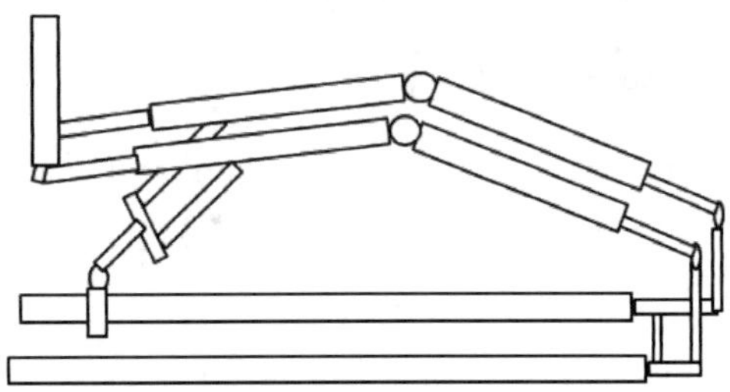

Figure 6. Device's Design.

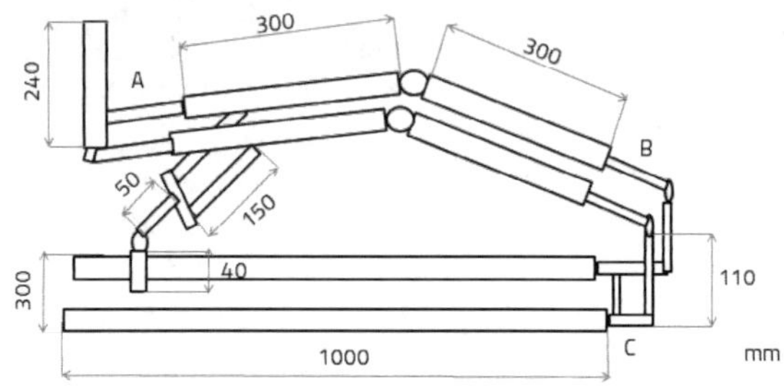

Figure 7. Sketch quotation.

To measure the size of the device's support was used an algorithm to calculate the anthropometric sizes from woman and men. Therefore the sizes of the segments from the support for different size individual were determined.

The calculus formulas for this sizes are the following:

- Leg: (0,246H) for men, (0,244H) for woman;
- Hip: (0,245H) for men, (0,248H) for woman, where H represents the height of the individuals.

After the calculus, the Sketch quotation was determined and the figure 7 support was created.

2.3.1. Articulated support assembly. To assembly the articulated support there used the previous mentioned materials which were manipulated through cutting to reach the required size. The articulations were made out of 16 mm diameter steel bar, being manipulated through turning and milling (figure 8) In the next phase al the articulated segments were assembled, the articulations were introduced and welded at their ends (figure 9).

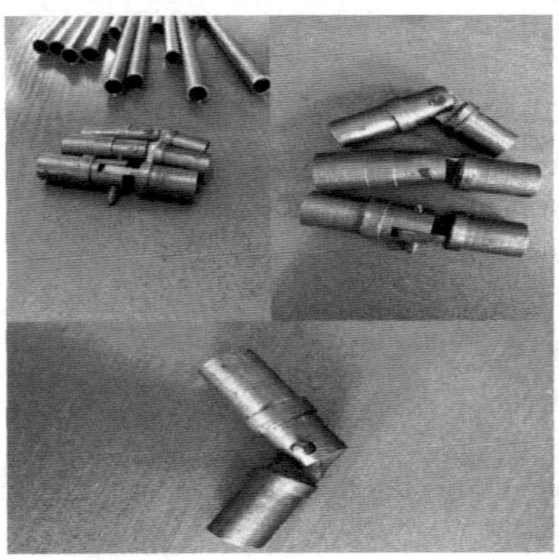

Figure 8. Articulation creation.

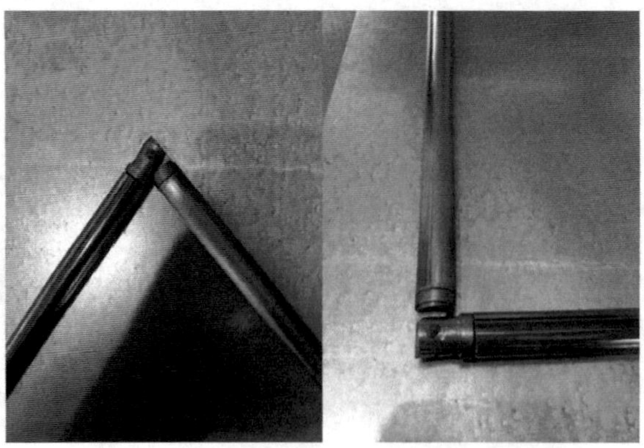

Figure 9. Articulated segment assembly.

The next step was to mount the motor and the linear platform on the steel sheet with the 300x1000mm dimension, this represents the lower part of the device where all other elements will be mounted (figure 10). After fixing the engine in place, the support elements previously manipulated were welded following the chosen design (figure 11).

Figure 10. Mounting the motor and linear platform.

Figure 11. Welding the support.

The next step was building the leg fixing support and welding it on the articulated support (figure 12).

Figure 12. Leg fixing support.

2.3.2 Assembling the electrical components. To mount the control and command components, there was performed an extension of the device's chassis, this extension was mounted on a Plexiglas with 240 x 300 x 15mm dimensions (figure 13).

Figure 13. Assembling the electrical components on a Plexiglas sheet.

Limit switches were mounted on the lateral part of the linear platform, in order to perform the role of setting the minimum and maximum limits of the movement amplitude of the articulated support and implicitly of the knee (figure 14).

Figure 14. Mounting the limit switches.

4. Conclusions

At the moment, the passive continuous movement devices are being used on a large scale in rehabilitation sections, kinesiotherapy clinics, at the patients home and also at the orthopedic sections, surgery, neurosurgery, neurology. Using this passive continuous movement devices is essential to accelerate the rehabilitation process of the patient and reintegrating it in the social life on one hand, and on the other hand eases the work of the medical personal from the rehabilitation sections.

Therefore this low-cost built device is capable to fulfill the necessary functions to recuperate from a variety of articulation conditions at the knee level, to execute the flexion-extension movement that the knee articulation performs on a bigger movement amplitude and fulfill extra functions such as regulating the speed and amplitude. Therefore, it is easy to use by the therapist and patient and it can me easily manipulated fulfilling the size and weight requirements.

The CPM system can by also used for a didactic purpose in the Medical Engineering lab (Environment and Product Design Faculty) and by: Construction and Maintenance of medical devices and prothesis engineering or by the Kinesiotherapy specialized students from the Faculty of Medicine from the Transilvania University from Brașov.

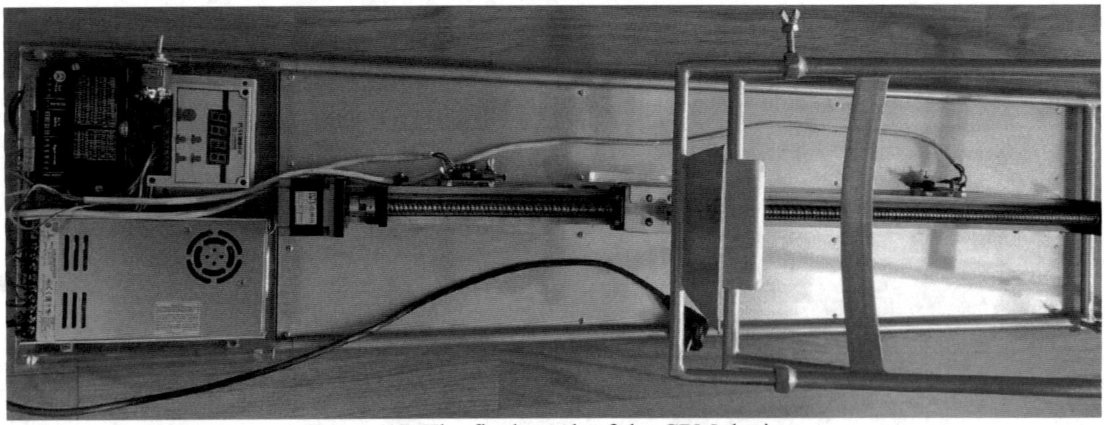

Figure 15. The final result of the CPM device.

Acknowledgements

Practical realization of this project was made possible thanks to funding received in the ***My License Project-2021*** competition held in the Transilvania University of Brasov.

References

[1] V. Papilian and V. Albu, *Human Anatomy (Anatomia omului), Vol I.* ALL, 1998.

[2] C. Costache and D. Costache, *Anatomy course notes (Anatomie note de curs).* Universitatea Transilvania Brasov, 2015

[3] I. A. Tatulea, C. Druga, and I. Serban, "Multicriteria analysis of medical devices for passive recovery of the upper limb" (Analiza multicriterială a dispozitivelor medicale de recuperare pasivă a membrului superior) no. April, 2020, [Online]. Available: https://www.researchgate.net/publication/344171697_Analiza_multicriteriala_a_dispozitivelor_medicale_de_recuperare_pasiva_a_membrului_superior.

[4] P. Erin and S. Robert, "Control system design for a continuous passive motion machine," *ISSN 2502-3632 ISSN 2356-0304 J. Online Int. Nas. Vol. 7 No.1, Januari – Juni 2019 Univ. 17 Agustus 1945 Jakarta*, vol. 53, no. 9, pp. 1689–1699, 2019, [Online]. Available: www.journal.uta45jakarta.ac.id.

[5] "FUYU Ball Screw Linear Guide Actuator Stepper Motor Controller for Industrial Single Axis Pulse Controller[Basic Version]|Motor Controller| - AliExpress." https://www.aliexpress.com/item/4001084967585.html (accessed January 06, 2021).

[6] "Nema23 stepper motor driver for FLS40 FUYU linear motion guide|stepper motor driver|motor drivermotor stepper driver - AliExpress." https://www.aliexpress.com/item/32589549569.html (accessed April 15, 2021).

[7] "Mini Linear Guide Slide Rail CNC Small Stage Actuator Screw Lead Motion Table System Nema 14 Robot Part Motorzied Stepper Motor|Linear Guides| - AliExpress." https://www.aliexpress.com/item/33005212720.html (accessed March 20, 2021).

[8] "The micro switch, Push Button SPDT Momentary Snap Action Limit switch, travel switch|Switches| - AliExpress." https://www.aliexpress.com/item/1005001537363013.html (accessed March 20, 2021).

IOP Conf. Series: Materials Science and Engineering 1169 (2021) 012045 doi:10.1088/1757-899X/1169/1/012045

Area mapping for autonomous disinfection robot

D M Anton[1], R C Milas[1], A M Stiube[1] and L E Stiube[1]

[1] Faculty of Managerial and Technological Engineering, University of Oradea, 410087 Oradea, Romania

E-mail: danielanton0227@gmail.com

Abstract. This article presents an autonomous obstacle avoidance robot with infrared sensors. With the help of these robots, the automotive industry has developed, using various types of sensors, both parking using ultrasonic sensor models, to Lidar sensors that map the immediate proximity of the car in real time. These sensors, through different trail simulations, adjusts the direction of the car to avoid various obstacles, or to avoid the following traffic jams.

1. Introduction

The concept of Artificial Intelligence emerged in 1956, at the Dartmouth Conference [1] organized by Marvin Minsky, Jhon McMarthy and two senior scientist, Claude Shannon and Nathan Rochester of IBM, this dream becoming a reality in recent years, getting more and more in every field. Artificial intelligence by definition states that it resolves new situations or problems based on experience, experience gained on the basis of continuous learning. Artificial Intelligence has begun to take shape in the field of computer science that develops technical systems to solve technical problems related to human intelligence. This artificial intelligence is based on acquired knowledge, and it has no predominant idea to limit it in thinking and developing new ideas based on previous ones [2].

Artificial Intelligence is used also for labyrinth robots, a family of devices that offers the benefits of all PIC18 microcontrollers, namely high performance, cost-effective computers with high-strength addition and flash memory. In addition to these features, the PIC18F2455, 2550, 4455, 4550 series introduces the enhancement design that makes these microcontrollers a logic for many high-performance, application-sensitive users.

2. Robot Maze architecture

This wiring diagram shows the connections of the motors, the sensors, the power supply of the control board, each pin separately.

In figure 1 is presented the electrical scheme of the Pixel Board, designed in Eagle.

IOP Conf. Series: Materials Science and Engineering 1169 (2021) 012045 doi:10.1088/1757-899X/1169/1/012045

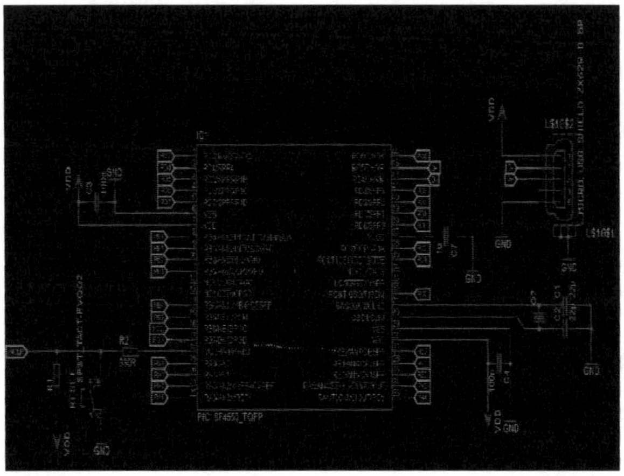

Figure 1. The electrical scheme of the Pixel Board route.

3. Motor driver and motor characteristics

To reduce the size and weight of the robot, it was used a L293D driver, which controls the robot motors at 1000 rpm at the output of the gearbox. This driver supplies two motors because the plate PixelBoard is not capable of delivering a lot of continuous power, so we attach a motor driver that feeds directly to the DC source. This high-frequency bridge is designed to provide bidirectional current up to 600mA at voltages from 4.5V to 36V. Devices are designed to drive inductive loads such as relays, solenoids, DC and bipolar motors, and other high voltage or voltage loads in positive consumption applications. The DC motors [4] of the robot are equipped with reducers, because at the output of the motor we have a speed of 30000 rpm and the torque is very small to move the robot, the transmission ratio is 30: 1, the free running speed is 1000 rotations per minute at 12V. Maximum torque is equivalent to 63.6 mNm. Stall current is reaching up to 800mA in load, no load current settling at 100mA, this being different depending on the weight robot and the frictional force that appear between the wheels and the surface they are running on.

Running the robot is done with the wheels on which the bicomponent silicon care is added and it is very elastic, with a fairly high resistance, if it is stretched, it does not return. Wheel models have an outer mold for the contour of the wheels, designed and printed on the 3D printer. Silicon casting is done with the syringe because it has to take the air out between the mold and the tire and the space required to be filled with silicone does not allow us to introduce silicone quickly because of its sizes.

The analogue distance sensors used are Sharp, with the perception of the distance from 10 cm to 80 cm. These reflective sensors show us the distance with accuracy up to about 0.4 cm in our robot up to a wall, usually these walls are white and the reflection rate is about 90%, but in some cases the walls are gray , this reflection decreases to about 20% because it absorbs part of the light emitted by the diode, and the sensors only receive some of this infrared light emitted by the diode. At the maximum reading threshold, there is a hysteresis of a few millimeters, at minimum the values are almost unreviewable. In figure 2 is presented the actual build of the maze robot, and in figure 3 the electrical scheme.

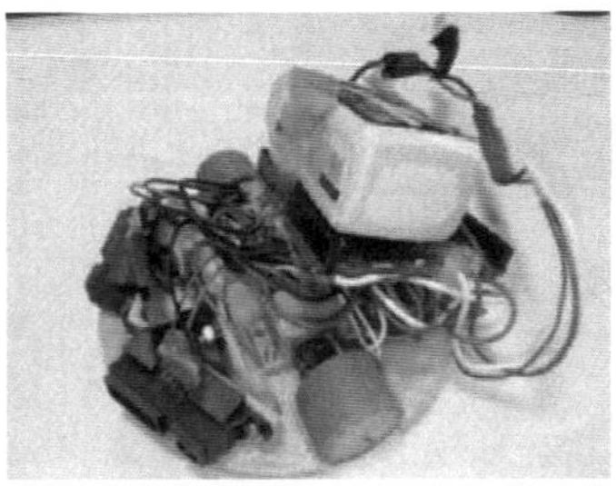

Figure 2. Maze robot.

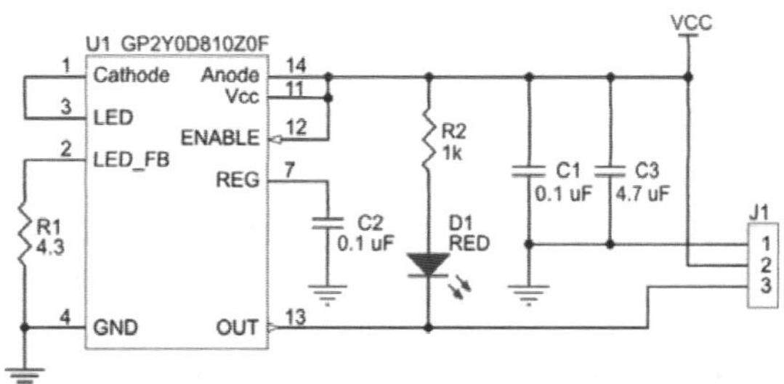

Figure 3. The electric sensor distance diagram. [Pololu]

IOP Conf. Series: Materials Science and Engineering 1169 (2021) 012045 doi:10.1088/1757-899X/1169/1/012045

4. Robot design and process of filament deposition

This chassis has been engineered in SolidWorks, where all the control electronics, sensors, motors, encoders, wheels and battery are mounted, as presented in figure 4.

The robot is made of heated 40 W power filament and a 24 V supply, it heats up to 220^0C, the filament is pushed by a stepper motor, and depending on the engine speeds the ones three axes (x, y, z) automatically adjust the pushing speed of the filament by the motor printer, and on the y-axis, the height of a single layer deposited from a passage of the nozzle extruder, which is a 0, 4 mm through filament care is extruded into layers of at least 0.1 mm high. In figure 5 is shown the scheme, designed in Eagle, with short caption (caption centred).

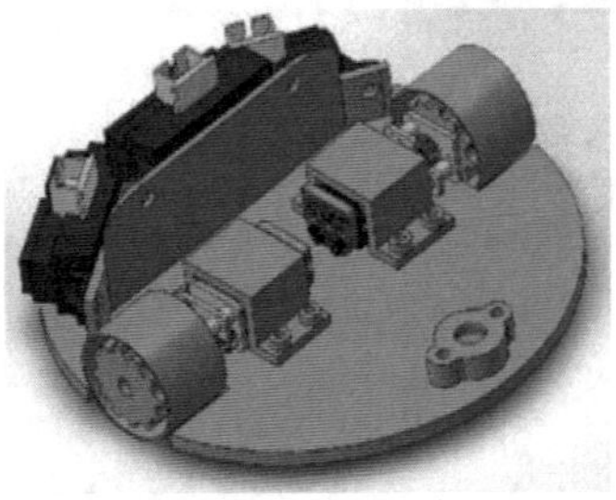

Figure 4. Design robot assembly.

Figure 5. Figure with short caption.

5. Applying the Dijkstra algorithm

Schedule a test area to apply this algorithm based on known distances after completing the route and finding the shortest route by distance and time. Determining the matrix of walls (sqm) this matrix will contain all the walls that the robot will bypass, the starting point and the finish. The declaration of maturity will be done on rows and columns, as in figure 6, all later being the robot working environment.

The other algorithm, also called another engagement algorithm, is sometimes used with the left or right turn function, these being chosen from the beginning for the robot. This is used in order not to get into the problem in which the robot resumes the same route, the other Pledge algorithm is used, which will arbitrarily choose the chosen direction [5].

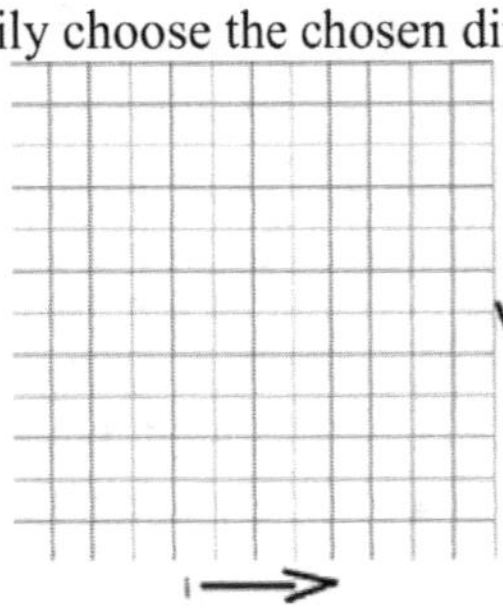

Figure 6. Defining rows and colonies.

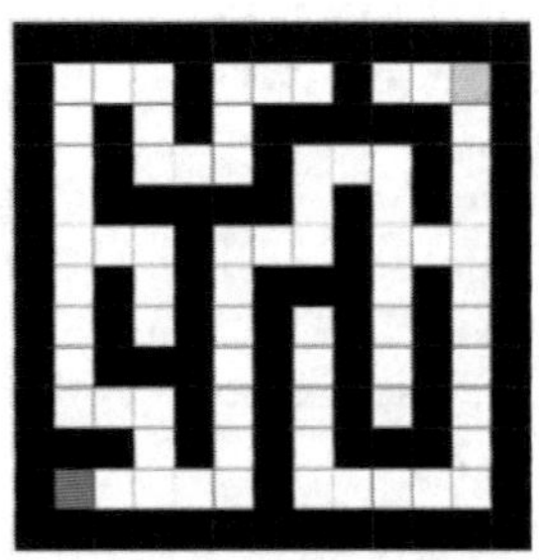

Figure 7. Labyrinth of the robot

The drawing of the interface is based on a matrix in which the walls to be circumnavigated are marked with 1, and with 0 is marked the route where the robot can go willingly to accomplish the task that was imposed at the beginning, namely to choose the optimal route. The starting point and the finish point, the outline of the labyrinth, and the robot's own labyrinth in which to operate all the distances, are set, to find the shortest route [6].

Pentru programarea robotului, pe baza acestui alt algoritm, a fost utilizată generarea de grafică statică, cu implementarea lor bazată pe cod, precum şi coduri open source, care au fost ulterior modificate pentru robot [7,8,9].

Operation of the labyrinth, at each secondary artery blockage, returns when the artery breaks away from the main artery and continues the path until the robot reaches the end point.

IOP Conf. Series: Materials Science and Engineering 1169 (2021) 012045 doi:10.1088/1757-899X/1169/1/012045

In figure 8 and 9, the robot traverses the labyrinth completely to map the maze in order to get a precise map of it, in order to record all possible solution and finding the best suitable one.

At this point, the robot finds a second node from the first second and continues its route until it returns to the main road.

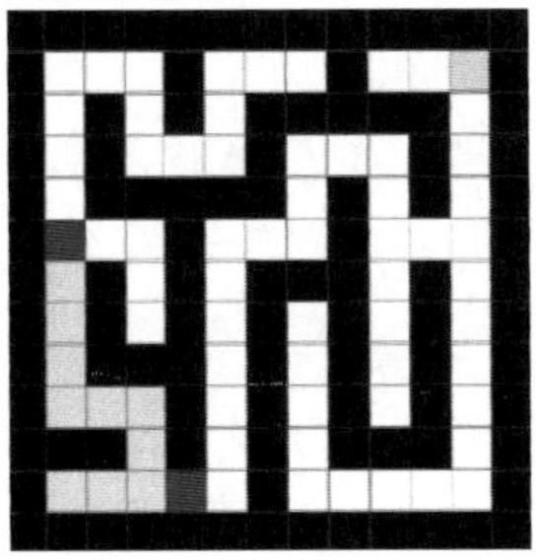

Figure 8. The first nodes traveled, to choose the route

Figure 9. Go through a complete route to a dead end, and return to the last node

In figure 10, the robot has arrived at the desired point, noting the shortest trajectory. The optimal route is due to the two basic rules that make it always choose the shortest and quickest route.

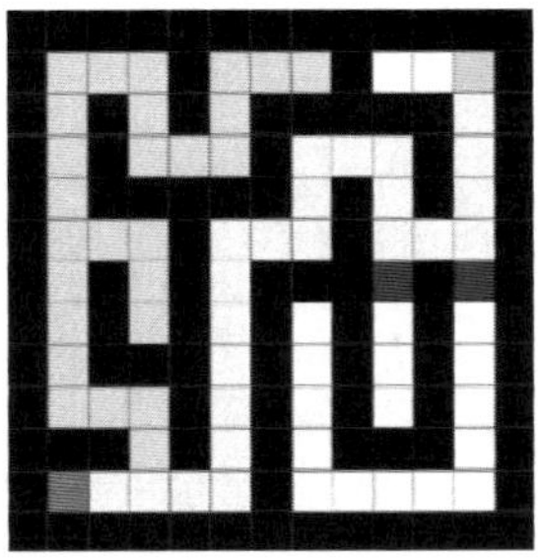

Figure 10. Completion of the route with the shortest route

6. Conclusion

As a conclusion, the maze robot is collecting data from environment where it is placed and utilize these informations to generate an environment map. The algorithm is using the map to generate the optimum path, for the future to traverse all the spaces. This article is the basis for a for a work in progress PhD's work, named Prototipying an autonomous mobile robot family for UV-C neon disinfection. For future work, it is proposed to add fixed state LIDAR and rotative type LIDAR as well, for mapping the environment by generating a cloud dataset.

References

[1] J A Perez, F Deligianni, D Ravi and G Z Yang 2017 *Artificial intelligence and robotics* EPSRC UK-RAS Network

[2] M Singh, R Kumar, V Giradkar, P Bhole and M Kumari 2016 Artificially intelligent maze solver robot *International Research Journal of Engineering and Technology (IRJET)*

[3] E H Kivelevitch and K Cohen 2010 Multi-agent maze exploration *Journal of Aerospace Computing Information and Communication*

[4] https://www.pololu.com/

[5] https://en.wikipedia.org/wiki/Maze-solving_algorithm

[6] https://www.programiz.com/dsa/dijkstra-algorithm

[7] https://www.geeksforgeeks.org/dijkstras-shortest-path-algorithm-greedy-algo-7/

[8] https://cp-algorithms.com/graph/dijkstra.html

[9] https://github.com/Auone31/Dijkstras_Algorithm

IOP Conf. Series: Materials Science and Engineering 1169 (2021) 012046 doi:10.1088/1757-899X/1169/1/012046

Stripping of plated plastic materials

S Ilea[1] . R Veres[1]

[1]Faculty of Managerial and Technological Engineering, Doctoral School of Industrial Engineering, University of Oradea, Romania

sandu.ilea@gmail.com

Abstract. Many of the nowadays materials used in the large sectors of industry are resuming to the plastics. A first hand competitor of the metals, plastics are more and more suitable for many kind of applications due to their improved characteristics, being also more simple to process and also being more cost effective. Starting from automotive sector going through mass industry and finishing in the medical devices sectors, plastic have been proved as efficient materials, taking into account a more supple finishing without sacrificing the utility and endurance of it. In this paper we are presenting the stripping of the plating that can be done to refurbish plastic waste that may occur in the automotive industry.

1. Introduction

More and more plastic workpieces are present nowadays as plastic is widely used in the industry. The can be seen not only for parts that are having a functional role but also for decorative elements used in many assamblies [1],[2].

Plastic materials have a few characteristics that make them a direct competitor for the metals like:

- Low density;
- Low weight – being a very important characteristic in the automotive industry;
- A good combination beetwen rigidity and tenacity;
- High chemical resistance;
- Good isolation characteristics;
- Infinite ways of designing;
- Low processing temperature;
- Low price – high yield.

In the decorative industry, more and more elements made from plastic are used, a good part of them being painted of plated with copper, nickel and chrome, being more resistant and also having a more pleasant touching surface.[3]

As presented in [4], In the automotive industry, furniture and household items are using this kind of plastic materials. As speaking about weight reduction, in the automotive industry, plastics are contributing to a weight loss up to 12-15% from the cars weight and also contributes to a decrease of CO2 emissions.

Plating of plastic materials can be done in two ways:

- Electrolytic process after pretreating in a non-electric process;
- Vacuum metallization, a process that does not use chemicals.

Materials used in the plating process are many but the main two are ABS (Acrylonitrile-Butadiene-Styrene) and ABS/PC (Acrylonitrile-Butadiene-Styrene + Polycarbonate)

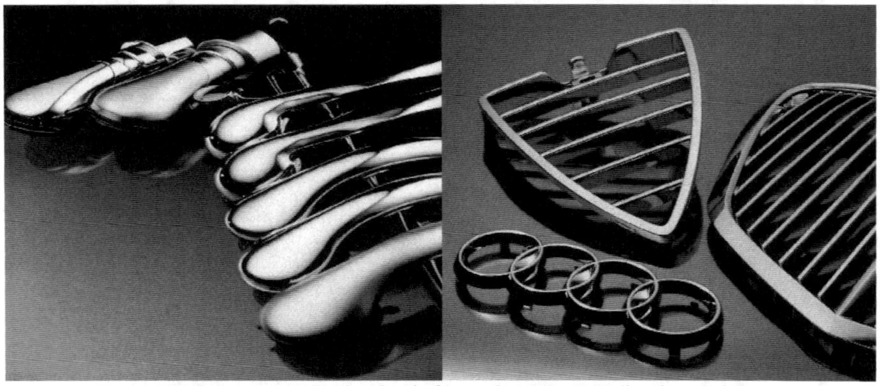

Figure 1. ABS (in the left) and ABS+PC (in the right).

2. Methods for stripping of the metal plating

In the automotive industry, plastic parts are having a very high utility. Having to produce such a large amount of decorative plastic that needs to meet high quality standards results a large quantity of plastic waste that are hard to recycle due to the metal plating done on their surface. For this details the stripping of the metal plating is vital..

The main objective of this paper consists the stripping speed from the plastic parts with the aid of a chemical process.

The stripping of the plating is a process that implies the fast removal of the oxides from the plastic or metal surface that will be go through another process like painting, gluing, welding or plating. All this processes require a clean and adherent surface.[4], [5].

The stripping process can be structured in the following ways, as metioned in [6]:

- Mechanic procedures, made by sanding with inert or compatible materials with steel or plastics. Sanding can be done with siliceous sands, corundum, glass or stainless-steel microspheres
- Chemical procedures, that can apply to metallic or plastic surface and mainly acids are used, like hydrochloric acid HCl, sulfuric acid H2SO4, nitric acid HNO3, flourhidric acid HF;
- Electrochemical procedures, which compared to chemical procedures, spots or haloes are inexistent, does not use aggressive acids listed above and does not contaminate the surface with abrasive particle as in the mechanic procedures.

3. Study case

In our case the samples were collected from the metal plated plastic material, having different plated areas and also different layers.

Figure 2. Sample with a granulation and plated area of 8-11 mm^2, grinded, with layers of Cr 0,5-1 µm , Ni 40-70 µm and Cu 50-120 µm, initially being processed through a mill.

Figure 3. Sample with plated area of 50-60 mm^2, with layers of Cr 0,5-1 μm , Ni 40-70 μm and Cu 50-120 μm.

Figure 4. Sample with plated areas of 60-350 mm^2, with layers of Cr 0,5-1 μm , Ni 40-70 μm and Cu 50-120 μm.

For the determination of the stripping time for the three samples, there have been used equals quantities of plastic materials, chemical substances and also being used the same procedure for all the three samples.

The quantities were the following: 500g of plastic material – ABS/PC T45, 800 ml of chlorhydric acid (HCl) 37% concentration, 400 ml of nitric acid (HNO3) 55% concentration, diluted with 250 ml of water (H2O).[6],[7]

er Removing the Chrome (CR) layers

For removing the first metal plating layer with chrome, the samples were immersed in hydrochloric acid (HCl) till the first layer came off. For the presented samples we had the following results:

- For the samples with surfaces between 8-11 mm2 the plating was removed in 2 minutes
- For the samples with surfaces between 50-60 mm2 the plating was removed in 3.2 minutes
- For the samples with surfaces between 60-350 mm2 the plating was removed in 4.5 minutes

Removing the Nichel(Ni) and Copper (Cu) layers

After the removal of the Chrome layers. The samples are washed and sent to the next phase. The samples were immersed in nitric acid (HNO3) to remove the two layers. Because the reaction produces heat up to 930C it's recommended to cool down the entire process to 70-750C for more efficiency.

- For the presented samples we had the following results:
- For the samples with surfaces between 8-11 mm2 the plating was removed in 26 minutes
- For the samples with surfaces between 50-60 mm2 the plating was removed in 41 minutes
- For the samples with surfaces between 60-350 mm2 the plating was removed in 49 minutes

4. Conclusions

Figure 5. The samples after the stripping process.

In conclusion the results for the metal plating stripping were the following:

- For the samples with surfaces between 8-11 mm2 the total stripping time was 28 minutes
- For the samples with surfaces between 50-60 mm2 the total stripping time was 44.2 minutes
- For the samples with surfaces between 60-350 mm2 the total stripping time was 53.5 minutes

After the metal plate stripping of the three samples we found that the parts with the small area were stripped faster because after cleaning the chrome layer, which is eliminated fast, the second stripping process has also an external and substrate chemical reaction that accelerates the stripping process.

After this we can conclude that higher volumes of parts can be stripped if they are having a much more small area resulting a low consumption of chemical. Also for a more efficient process the temperature must be maintained close to 70-750C.

References

[1] H. Domininghaus, Die Kunststoffe und ihre Eigenschaften, 5. neu bearbeitete und erweiterte Auflage, Springer (1998)

[2] A. Bhowmik, "*Multiple-reflection effects in photoelastic stress analysis*," Appl Opt 40, 2687-2691 (2001).

[3] Eugen G. Leuze Verlag - *Kunststoff-Metallisierung*, , Saulgau Württ. (1991)

[4] M. Brandes, Dr. C. Fels, *MO Nr. 9* (2002)

[5] Ayatollahi, MM. Mirsayar, M. Dehghan, "*Experimental determination of stress field parameters in bi-material notches using photoelasticity*," Opt Lasers Eng 32, 4901-4908 (2011)

[6] Bader, M.G., *Handbook of Materials Selection*, Chapter 35-Composites Fabrication Process, Ed. John Wiley & Soon Inc., New York, 2002.

[7] Balart, R., Lopez, L., Nadal, A., 2001, *Introduction a la ciencia e ingineria de polimeros*, Alfagrafic Editor, Alcoy, Spain.

AUTHOR INDEX